Bill
Strangeway
311

FLUID

MECHANICS for Engineers

FLUID

MECHANICS

ENGLEWOOD CLIFFS, N. J.

MAURICE L. ALBERTSON, Ph. D.
*Director of Colorado State University Research
Foundation and Professor of Civil Engineering,
Colorado State University*

JAMES R. BARTON, Ph. D.
*Professor and Chairman of Civil Engineering,
University of New Mexico*

DARYL B. SIMONS, Ph. D.
*Hydraulic Engineer, U. S. Geological
Survey, Fort Collins, Colorado*

for Engineers

PRENTICE-HALL, INC.

PRENTICE-HALL
Civil Engineering and Engineering Mechanics Series
N. M. Newmark, Editor

Current printing (last digit):
16

FLUID MECHANICS for Engineers

Maurice L. Albertson
James R. Barton
Daryl B. Simons

LIBRARY OF CONGRESS CATALOG CARD NUMBER:
60–14182

PRINTED IN THE UNITED STATES OF AMERICA
25832-C

Preface

This book is an elementary and intermediate treatment of Fluid Mechanics intended primarily to serve the student of fluid mechanics and his teacher. It is intended also to serve the practicing engineer and researcher who need to have a convenient reference to the basic principles and applications of fluid mechanics. We have attempted to organize the book and present the material to provide steady progress from the simplest concepts to the more complex ones—with each step providing basic information for later developments. We have assumed that the student is familiar with the mechanics of solids, including statics and dynamics, and mathematics through integral calculus. In the forepart of each chapter are stated typical examples of the use of the principles to be explained in the chapter. The basic principles are then developed, followed by specific applications—including both example problems and problems to be assigned to the student. At the end of each chapter is a summary to reiterate and emphasize the salient features of the chapter. We hope that the summary will prove to be helpful to both the student and the teacher in determining the most important facts and concepts.

It is our belief that the engineer must be trained to solve the engineering problems of mankind, and that there are three major steps in the complete education of an engineering student. First, the student must be taught the fundamentals of engineering, based upon the important principles of physics. At the same time the student's ability to analyze both the fundamental and applied aspects of a problem must be developed. Finally, the student must develop the ability to apply the basic principles, mathematical concepts, experimental data, and engineering judgment to specific engineering or research problems. Throughout this book a serious attempt is made to accomplish this threefold objective.

In order to explain the fundamental principles and basic concepts as simply as possible, we have used elementary mathematics wherever appropriate. In fact, at times this is done at the expense of what might be considered a more complete, general, and/or elegant form of mathematical treatment. However, we believe it is more important to have the student understand the basic principles, in their most elemental form, than to stress the more complicated mathematical developments which sometimes tend to become an end in themselves.

Fluid mechanics is a combination of an analytical and empirical science. On the one hand, there are certain problems, such as aspects of laminar flow and irrotational flow, which can be analyzed and expressed very accurately, simply from a mathematical approach. On the other hand, most engineering problems involve so many variables it is possible only to approximate a solution by mathematical analysis alone. Consequently, adjustments to mathematical developments must be made on an empirical basis—which results in the often-made accusation that fluid mechanics is a science of coefficients. In order to overcome this criticism as much as possible, we have attempted to explain first the characteristics of fluid properties, based upon the fundamentals of molecular structure, and then to develop the basic theory involved in a particular concept of fluid mechanics. Furthermore, we have attempted to explain and illustrate, on the one hand, the conditions under which the basic theory is directly applicable to an engineering problem and, on the other hand, the conditions under which the mathematical theory must be modified empirically (on the basis of experimental data) in order for it to be applicable to an engineering problem. In short, ideal theory is indispensible for basic understanding, but for a more complete understanding, and for practical use of such understanding, the large number of variables involved frequently requires experimental data to establish a complete relationship—which includes certain coefficients and the range of application of both the theoretical and the empirical aspects. An attempt is made to illustrate these complete relationships through various graphs showing experimental data and transitions from one type of condition to another.

The first six chapters of the book consider the basic principles stemming from the fundamentals of mechanics and the different properties of fluids. Although these first six chapters include many applications of the basic principles, the last six chapters are concerned almost entirely with application of the basic principles to specific problems. The mechanics of solids considers first statics and then dynamics. In like manner, we have considered first fluid statics (which involves the property of fluid weight) and then fluid dynamics. In Chapter 3 on fluid dynamics great emphasis is placed upon the basic principles of continuity, momentum and energy, and the fluid property of density. Chapters 4 and 5 are concerned with the additional fluid properties of compressibility and viscosity. Once these fluid properties and basic principles of mechanics are understood, they can be utilized to establish the fundamentals of fluid resistance in order to complete the foundation material necessary to embark upon the last six chapters of the book.

Obviously, it is not possible to cover this entire textbook in a single course involving from thirty to forty-five one-hour lectures. Therefore, the book is designed to be used on a selective basis for the different branches of engineering as both an elementary and intermediate textbook. If the instructor feels that there is not time to consider fluid compressibility, Chapter 4 can be omitted and the sections on compressibility in later chapters also can be omitted. A mechanical engineering instructor, however, may feel that Chapter 4 is of paramount importance, whereas Chapter 8 on flow in open channels is of secondary importance to his students. In this case Chapter 8 can be omitted without difficulty.

The tables which give the different properties of fluids, and contain other information needed for the solution of problems, are placed at the front and the back of the book for the convenience of the user.

We are very much indebted to the numerous individuals who have made contributions to this textbook. R. T. Shen assisted greatly in the development of the figures, and he made all of the line drawings and sketches. Patsy Bacon and Anji Davar put special effort into typing the manuscript. We are particularly grateful to such local colleagues as J. E. Cermak, A. T. Corey, H. K. Liu, D. F. Peterson, Maxwell Parshall, R. L. Parshall, I. S. Dunn, G. L. Smith, E. V. Richardson, W. W. Sayre, Erich Plate, Niwat Daranandana, and K. B. Murarappa. The manuscript was carefully reviewed by, and excellent suggestions were received from J. S. McNown, W. L. Moore, J. M. Robertson, and N. M. Newmark. Many of their suggestions have been incorporated in the text.

MAURICE L. ALBERTSON
JAMES R. BARTON
DARYL B. SIMONS

Contents

chapter 1

Introduction

Fluid mechanics plays a very important role, not only in the daily lives of everyone, but in fact throughout the entire domain of nature. Air is a fluid which surrounds each and everyone, and the very act of breathing involves many of the principles of fluid mechanics. The circulation of blood in the body and the movement of sap in a plant also involve fluid mechanics. Perhaps more obviously related to the science of fluid mechanics are the fall of rain, the flow of water in a stream, the movement of a ship or a fish through the water, and the flight of an airplane, a bird, or an insect.

The engineer must study the principles of fluid mechanics, so that he can use them to predict and control the behavior of fluids. The science of fluid mechanics is basic to the design of aircraft, ships, submarines, bridges, buildings, rockets, and industrial plants—not only from the structural viewpoint, but also from the viewpoint of design of basic machinery, where such factors as control of fluids, transfer of fluids, power development, lubrication, and fluid pollution control are involved.

Regardless of where one encounters fluids, the laws of fluid mechanics are universally in force, and this text is intended to acquaint the student with these laws in an elementary way. Although in many respects the

1

mechanics of fluids is similar to the mechanics of solids, there are many new concepts and new symbols which must be learned. One of the stumbling blocks to any new science is the task of becoming acquainted with the new symbols which are used to represent basic ideas and principles. Therefore, in the initial stages of learning this new field of science, time can be spent very beneficially studying the symbols and becoming familiar with the new nomenclature (see Table I.)

The subject of fluid mechanics has stemmed in recent years from the subject of hydraulics, which included a study of liquids only. The science of hydraulics was largely an empirical one which did not give adequate consideration to the rational approach to the mechanics of gases and all liquids based upon the classical principles of mathematics and physics. The following history of fluid mechanics traces briefly some of the developments of the subject.

• HISTORY

The history of hydraulics dates back into the ancient past. The Bible is probably the oldest manuscript which includes a record of some of the early hydraulic problems and accomplishments. In addition to the fantastic flood described in the book of Genesis, the second chapter, verse ten, described an early irrigation project: "And a river went out of Eden to water the garden; and from thence it was parted and became into four heads." Although none of the ancient science of hydraulics has been revealed by historical or technical writings, the evidence of its accomplishments is recorded, and some of the actual hydraulic projects and structures still stand today as monuments to the ability of those past civilizations.

In 1712 B.C., Joseph's well was dug in Cairo. This well, which was 295 feet deep, was excavated from solid rock. The upper shaft was 18 by 24 feet in cross section and 165 feet deep, with a helical pathway leading down around the sides of the shaft. The water was raised from the lower shaft 130 feet on an endless chain of buckets operated by mules in a chamber at the bottom of the upper shaft. For a well that is almost 4000 years old, this is a marvelous engineering accomplishment.

From 312 B.C. to 130 A.D., the Romans constructed more than 350 miles of masonry aqueducts to bring water into the city of Rome. There were more than 50 miles of arches on these aqueducts spanning valleys and gullies. Parts of these ancient structures are still serving today. With a water-supply demand of 50,000,000 gallons per day, the city of Rome had a highly-developed water supply system. Although the Roman's structures functioned well, their hydraulic theories were indeed most elementary. They based their calculations for the amount of water flowing on the cross-sectional area, with no consideration being given to the velocity involved. Frontinus, an early water master on the Roman aqueducts, made many changes to measure the water properly; and although some of his methods were sound, his knowledge of the basic principles involved was quite inadequate.

Archimedes, the Greek scientist of about 250 B.C., explained the law of buoyancy of submerged objects, and his contribution is one of the very few that are considered to be as true today as some 2200 years ago, when Archimedes first made his discovery.

The basic principles of fluid motion developed slowly and as the result of work by many investigators. One interesting story concerns the development of the simple equation of flow through an orifice: $V = \sqrt{2gh}$. Extending the work done by Leonardo da Vinci (1452–1519), Galileo (1564–1642) performed extensive experiments on falling bodies. Galileo's time-piece involved a container of water with a hole in the bottom. He observed that the rate at which the height of water decreased in the container was a function of the time. Benedetto Castelli, a student of Galileo, became interested in the hydraulics of the time-piece, and wrongly concluded that the discharge through the orifice was proportional to the head. Torricelli, inventor of the barometer and another student of Galileo, discovered in 1644 that the velocity through the orifice was proportional to the square root of the head so that $V \sim \sqrt{h}$. Huygens (1629–1695) was credited with discovering the magnitude of g, so that by 1700 the information was available to assemble the correct equation—$V = \sqrt{2gh}$. This was done about 1738 by Daniel Bernoulli, who first applied the law of conservation of energy to the flow of water. Thus, in a period of a little over one hundred years, the equation for flow from an orifice, $V = \sqrt{2gh}$, was developed.

After 1700 A.D., many men came forth with substantial contributions to the theory of the mechanics of fluid motion. Fascinating stories are connected with these men and their discoveries, but only a few of the names can be mentioned here.

In the field of experimental hydraulics, the following are well known for their early contributions: Chezy, Venturi, Dubuat, Hagen, Poiseulle, Bazin, Weisbach, Ganguillet, Kutter, Manning, Darcy, Francis, Froude, Russell, and others. Before the beginning of the 20th century, the following scientists, in addition to those named previously, made notable contributions to the theoretical development of the science of fluid mechanics: Euler, Navier, Stokes, La Grange, Rayleigh, Helmhotz, Kirchhoff, Kelvin, Reynolds, and Lamb. Among the contributors to modern fluid mechanics and hydraulics, the following names should be mentioned: Prandtl, Blasius, Bakhmeteff, Freeman, Herschel, Gilbert, Nikuradse, Taylor, Dryden, Durand, Mises, Karman, Woodward, Rouse, Parshall, Lane, and Danel.

Because of the limited space available, these lists of names are very confined; and although these men have all made notable and outstanding contributions to the science, there are numerous others whose work and ideas have also been significant. As stated previously, stories behind these scientists and their discoveries are often extremely interesting, and the student who is interested in personalities as well as scientific facts should read the *History of Hydraulics* by Rouse and Ince.

By the end of the 19th century, the empirical science of hydraulics was being replaced by a new science of fluid mechanics, where both theory and experiment played an important role. The generally recognized founder of present-day fluid mechanics is the German professor, Ludwig Prandtl (1875–1953). His theory of the boundary layer has had a tremendous influence upon the understanding of the problems involving fluid motion. The science, however, is still quite young, and many new and important discoveries are still being made each year in the field of fluid mechanics.

• FLUID CHARACTERISTICS

Matter may exist either in the solid state (which the student has already studied in Mechanics of Solids), or in the fluid state. The behavior of fluids is the subject to be considered in detail in this text. Fluids are divided into liquids and gases, which have many properties that are similar, but certain other properties which are not at all similar. The similarity and lack of similarity among solids, liquids, and gases may be traced generally to the spacing, activity, and structure of their molecules.

In solids and liquids, the *spacing* between the molecules is relatively very small, whereas in gases the spacing is relatively large. In other words, in solids and liquids the molecules are compactly arranged, and in gases they are sparsely arranged. Hence, a gas is much lighter in weight (for a given volume) than either a solid or a liquid.

The *activity* of the molecules becomes progressively greater as a material passes from the solid to the liquid to the gaseous states. This increase in activity is caused by increased heat content and temperature of the molecules. In the solid state, the activity of the molecules is essentially non-existent. In the liquid state, the activity is increased considerably— it being a sort of vibratory movement, without the molecules exchanging places. In the gaseous state, however, the molecular activity is very great— the activity being comparable to so many microscopic billiard balls moving at great velocities but for very short distances, because they are continually bumping into each other and changing their direction of motion.

The molecular *structure* of a solid is rigid, and the bond between the molecules is so great that the molecules do not readily move relative to each other. A solid does have a certain elasticity, however, which means that, for a given stress, there is a corresponding strain or deformation, regardless of whether the stress is compression or tension. Although molecules of a solid do, under certain conditions, change their alignment, they do not migrate or move relative to each other until the elastic limit is reached.

If a liquid is confined, it will exhibit elastic properties with regard to compression, but only under certain special conditions does a liquid withstand stresses of tension. Owing to the sparse spacing of the molecules in a gas, a gas has much less resistance to compression than either a solid or a liquid.

Because the molecular *structure* of a solid is essentially rigid, a force which is applied to the solid is resisted continuously, under static as well as dynamic conditions. In a liquid, the molecules possess a cohesiveness which tends to hold them together. The molecules are not fixed relative to each other, but instead, are free to move and slip past one another, despite the fact that the spacing of the molecules is essentially the same as in the solid state. This means that, if a force is applied to a liquid, the liquid will continue to change its shape until the force is relieved. In other words, unless the liquid is confined on all sides, the liquid resistance to a force is a dynamic rather than a static condition. This resistance is related to the viscosity and the inertia.

Like the molecules of a liquid, the molecules of a gas slip past each other to resist a force against them only under dynamic conditions of motion. Unlike the liquid molecules, however, the gas molecules are sparsely spaced and have much more activity—which results in the gas not only being much lighter in weight, but also having different physical properties of viscosity.

A solid placed in a container will remain rigid and hold its original shape. When a liquid is placed in a container, however, it assumes the shape of the container, and has a free surface at the top like the surface of a lake. A gas which is placed in a container not only assumes the shape of the container, but also completely fills the container, so that the container must be closed to insure that the gas will not escape.

• UNITS OF MEASUREMENT

The dimensions employed in the mechanics of fluids are mass M or force F, length L, and time T—which are the same as the dimensions used in other branches of mechanics. These dimensions, used alone or in combination, are adequate to describe the geometrical aspects of a flow system, the characteristics of the flow, and the properties of the fluid. These fundamental units are related by Newton's second law of motion

$$F = Ma$$

in which a is the acceleration of the mass M caused by the force F.

Two fundamental systems of units, the metric system and the English system, are widely used in physics and engineering. In the metric system, the mass M is expressed in grams, the length L in centimeters, the time T in seconds, and the force F in dynes. The dyne is the force required to accelerate a gram of mass 1 cm/sec².

In the English system, the force F is expressed in pounds, the length L in feet, the time T in seconds, and the mass M in slugs.

One pound of force will accelerate a mass of one slug 1 ft/sec². The force F in Newton's equation is a function of the gravitational system involved. However, the mass of a specific body is the same for every gravitational system. That is, a man weighing 180 pounds on earth would weigh about thirty pounds on the moon, but his mass remains the same. Hence, the mass of a body is the quantity of matter which the body possesses, and its magnitude does not change. Evidently, mass is a more fundamental dimension than force.

On or near the earth's surface, the gravitational force is nearly constant, and g can be assumed equal to 32.2 ft/sec². Therefore, when working on or near the earth's surface, the weight of a given mass is essentially constant, which justifies using the F-L-T system of dimensions throughout this text.

The application of this system of dimensions can be illustrated by a simple example. Assume a ball is thrown through the air. If the ball is spherical, its size can be characterised by one simple dimension of length L, its diameter. In other words, its geometry is described as spherical, and its size is described by the diameter. The ball has a mass M which is acted upon by the gravitational acceleration g to give it a weight of W (a force). As the ball is thrown, it accelerates from zero velocity to some maximum velocity which occurs at the time of release. This acceleration has required a force F (not the same as the force W) on the mass M over a period of time T to attain the velocity at release. The force can be expressed in terms of pounds, the mass in terms of slugs, and the time in terms of seconds. The velocity at any given time is length per unit time L/T, and can be expressed in terms of feet per second, ft/sec, or fps. The acceleration is length per unit time per unit time and is expressed in terms of (ft/sec)/sec or ft/sec².

During its flight through the air, the ball will be a certain distance above the ground at a particular time T. This distance is a length L expressed in feet. Furthermore, its motion is restricted by the drag of the air. This drag is a decelerating force F, and is expressed in terms of pounds.

From the foregoing example, it can be seen that the various aspects of the problem can be expressed in terms of the three basic dimensions—*force*, *length*, and *time.* This is also true for all other problems encountered in this text.

Another important aspect of the three units of measurement is that two of the units, length and time, are scalar quantities; whereas the third, force, is a vector quantity. While magnitudes such as 10 feet or 30 seconds are generally adequate to describe a length and a time respectively, a magnitude of 10 pounds is insufficient information to describe a force completely. Therefore, in addition to a magnitude, a force has a direction and a location, and all three of these properties of a force must be determined in order to define the force clearly and completely.

In this text, all problems for which it is necessary to find a force automatically require determination of both the magnitude and the direction of the force. Although magnitude and direction are adequate for the solution of some problems, there are other problems which require that the location or point of application of the force also be determined.

• DIMENSIONAL HOMOGENEITY

In order for any equation to be physically correct, the combination of variables on each side of the equality sign must have the same dimensions. If this is found to be true for a given equation, the equation is said to be dimensionally homogeneous.

As stated in the foregoing paragraphs, the variables in the equations involved in the mechanics of fluids can be expressed in terms of force, length, and time (F-L-T), as the three fundamental dimensions. As already demonstrated, the dimensions of force and mass are related through use of Newton's equation $F = Ma$. If this equation is dimensionally homogeneous, then the mass M must have dimensions as follows:

$$M = \frac{F}{a} = \frac{F}{L/T^2} = \frac{FT^2}{L}$$

This same procedure can be used to determine the dimensions of any term in an equation, provided the dimensions of all other terms are known.

Another common equation, with which the student has already become acquainted in both physics and the mechanics of solids, is the velocity equation

$$v = v_0 + at$$

This equation may be checked for dimensional homogeneity by inserting the basic dimensions for each term

$$\frac{L}{T} = \frac{L}{T} + \frac{L}{T^2}(T) = \frac{L}{T} + \frac{L}{T}$$

The dimensions of each term are L/T. Therefore, the equation is dimensionally homogeneous.

A differential equation can also be checked. For example, the acceleration equation

$$a = \frac{dv}{dt} = \frac{d^2s}{dt^2}$$

may be expressed as

$$\frac{L}{T^2} = \frac{L/T}{T} = \frac{L}{T^2}$$

In this case, it should be pointed out that, since d^2s is the second differential of s, the dimension remains simply that of length L.

In engineering practice, there are many empirical equations which are not dimensionally homogeneous or which contain coefficients that are not commonly recognized as having dimensions. An example of this is the Chezy equation for flow in open channels (see Chapter 8).

$$V = C\sqrt{RS}$$

in which V is the velocity, R is the hydraulic radius, and S is the slope. This equation can be solved for the coefficient C and checked for dimensional homogeneity to determine whether C is really dimensionless

$$C = \frac{V}{\sqrt{RS}} = \frac{L/T}{\sqrt{(L)(L/L)}} = \frac{L}{T\sqrt{L}} = \frac{L^{1/2}}{T}$$

The coefficient C has been considered by some engineers to be without dimensions, but this check shows it to have the dimensions of $L^{1/2}/T$. Further study of the problem has shown that the equation also involves the acceleration of gravity g and that the Chezy coefficient C should be divided by $\sqrt{g}$,

$$V = \frac{C}{\sqrt{g}}\sqrt{gRS} = C'\sqrt{gRS}$$

to make the coefficient C' dimensionless and the equation dimensionally homogeneous.

In the following study of fluid mechanics, the student will find it extremely helpful to apply frequently the check of dimensional homogeneity. It will save him many lost hours of searching for errors in his computations.

• FLOW PROPERTIES

There are two properties of the flow which are encountered repeatedly in defining the various properties of fluids and in a discussion of dimensional analysis. These are pressure and velocity, which are treated only briefly at this point, but more extensively in Chapters 2 and 3.

Pressure: In general, pressure can be considered as a force distributed over an area (F/L^2). It exists wherever fluids exist either at rest or in motion. That is, a liquid carried in a container exerts a certain pressure on the bottom of the container, the atmosphere blanketing the earth exerts a pressure on its surface, the gas contained inside a balloon or tire exerts an outward pressure on the interior surface, while at the same time the atmosphere exerts an inward pressure on the exterior surface. In this and subsequent chapters, the concept of pressure will be intimately related to hydrostatics, hydrodynamics, and thermodynamics.

Velocity: When an object or body of fluid is moved from one place to another, it must have velocity. Velocity is a vector quantity. That is, it has both direction and magnitude—so that the velocity is not defined completely unless both direction and magnitude are given. Speed has only magnitude and, therefore, is not a vector quantity.

Mathematically, velocity can be expressed as

$$v = \frac{ds}{dt}$$

which means that *velocity can be defined as the rate of displacement with respect to time.* Velocity has the dimensions of length per unit of time L/T, and the units are usually feet per second fps; meters per second mps; or miles per hour mph. If a particle travels along a curved path, the direction (see Fig. 1–1) is changing continuously from point to point, so that the velocity is changing, regardless of whether the magnitude (or speed) is changing. The average path of the particle is a streamline,

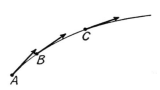

Fig. 1-1. Streamline.

as defined in Chapter 3. As will be seen later, an external force on the object or body is required to change the velocity in either magnitude or direction.

Discharge: The rate of movement or flux of a fluid past a given point is the discharge. For compressible fluids (such as a gas), this is usually represented either by W, the weight discharge which has the dimensions of weight per unit time F/T; or G, the mass discharge which has the dimensions of mass per unit time M/T. For incompressible fluids, the discharge is represented by Q, the volume discharge which has the dimensions of volume per unit time L^3/T. The units for W are usually pounds per second or grams weight per second; the units for G are slugs per second; and the units for Q are cubic feet per second cfs, cubic meters per second cms, gallons per minute gpm, or gallons per day gpd.

• FLUID PROPERTIES

If the basic characteristics of fluids and the system of dimensions and units are known, it is possible to study and evaluate in some detail the specific properties of various fluids. The *fluid properties* which enter problems most frequently in fluid mechanics are:

Density	ρ (rho)	Viscosity	μ (mu)
Specific weight	γ (gamma)	Vapor pressure	p_v
Elasticity	E	Surface energy	σ (sigma)

Each of these properties are now considered in detail.

• DENSITY

The density ρ (rho) of a fluid is the mass which it possesses per unit volume, $M/L^3 = FT^2/L^4$. Since a molecule of a substance has a certain mass regardless of its state (solid, liquid, or gas), it then follows that the density is proportional to the number of molecules in a unit volume of the fluid. The density of a substance in the liquid state is approximately the same as that for the solid state, and is much greater than the density in the gaseous state. It is significant, however, that the density of a substance (under given conditions of temperature and pressure) is a fixed value, regardless of the gravitational system. Density is important in any problem of flow in which acceleration is important.

• SPECIFIC WEIGHT

Specific weight γ (gamma) is weight per unit volume F/L^3 and, in contrast to the density ρ, is dependent upon the gravitational system. Specific weight and density are related by the Newton equation as

$$\gamma = \rho g \qquad (1\text{-}1)$$

which is a fundamental equation used frequently in this text.

Specific gravity is the dimensionless ratio of the specific weight (or density) of the fluid in question to the specific weight, (or density) of pure water at a specified standard temperature—usually 38°F or 4°C. The specific gravity of water at the standard ·temperature, therefore, is 1.0. *Specific volume v* is volume per unit weight L^3/F, which is the reciprocal of the specific weight.

All of the foregoing properties (density, specific weight, specific gravity, and specific volume) are interrelated and vary only slightly, for liquids, regardless of temperature and pressure. For certain studies, however, this slight variation is of importance. Therefore, the variation of density and specific weight with temperature of water is given in Table III, where it will be noted that the variation is less than 5 percent over the range of temperatures from freezing to boiling. The relatively constant density and specific weight of a liquid is attributable to the molecular structure of a liquid, in which the molecules are arranged very compactly.

The specific weight of a gas, in contrast to that of a liquid, changes greatly with a variation of either temperature, or pressure, or both. This variation is attributable, again, to the molecular structure of the gas, in which the molecules have so much kinetic energy that their spacing is relatively sparse. If the temperature is increased and the pressure is held constant, the activity of the molecules increases, which increases the molecular spacing still farther, and the specific weight is decreased. A decrease in temperature, likewise, creates an increase in specific weight. On the other hand, an increase in pressure caused by compression of the gas, so that the total heat content remains the same, causes an increase in specific weight, due to the more compact spacing of the molecules. The basic relationship between specific weight, pressure, and temperature may be stated as a combination of Boyle's law, $p/\gamma = $ constant, and Charles' law, $p/T = $ constant, to yield the ideal-gas equation, which is also known as the equation of state.

$$\gamma = \frac{p}{RT} \tag{1-2}$$

which, since the specific volume is the reciprocal of the specific weight, can be written also as

$$pv = RT \tag{1-3}$$

in which

γ is the specific weight F/L^3;

p is the absolute pressure F/L^2;

v is the specific volume L^3/F;

T is the absolute temperature in degrees;

R is the engineering gas constant L/degrees absolute.

Equation 1-2 is applicable for ordinary temperatures. It should be noted, however, that in the region of liquefaction (where a gas becomes a vapor) Eq. 1-2 does not apply. Avogadro's law, that all gases at the same pressure and temperature have the same number of molecules per unit volume, can be combined with Eq. 1-2 to yield a universal gas constant mR (where m is the molecular weight of the gas), which is essentially a constant at 1544 ft-lb per degree Rankine (Fahrenheit absolute) for all gases. This is particularly true for the simpler monatomic gases—the gases with a more complex molecular structure having a somewhat smaller value of the product mR. Table VI gives approximate values of the gas constant for several common gases.

• COMPRESSIBILITY AND ELASTICITY

Compressibility of a fluid is a measure of the change in volume of the fluid when it is subjected to outside forces. All fluids (both liquids and gases) are compressible, at least to some extent, despite the fact that, for many problems in mechanics of fluids, both liquids and gases are considered (without appreciable error) to be incompressible. Compressibility of a fluid is expressed quantitatively by means of its bulk modulus of elasticity E, which is defined as

$$E = -\tilde{V}\frac{dp}{d\tilde{V}} \tag{1-4}$$

which means that E (having the dimensions of F/L^2) is a measure of the incremental change in pressure dp which takes place when a volume $\tilde{V}$ is changed by the incremental amount $d\tilde{V}$. In other words, if a piston in a cylinder containing a volume $\tilde{V}$ is moved so that the volume is changed by the incremental amount $d\tilde{V}$, then the pressure changes by the incremental amount dp—the magnitude of which depends upon the bulk modulus of elasticity, as expressed in Eq. 1-3.

By multiplying Eq. 1-4 by g/g, the bulk modulus of elasticity is determined in terms of the density and pressure

$$E = \rho\frac{dp}{d\rho} \tag{1-5}$$

Because of their elasticity, fluids are able, when compressed, to store energy which can be recovered when the fluid is allowed to expand to its original volume. In practice, however, this fact is utilized principally in connection with gases. The bulk modulus of elasticity for water and air, for example, is approximately 300,000 psi and 15 psi, respectively, which means that air (a gas) is about 20,000 times more compressible than water (a liquid).

In most problems of fluid mechanics, the compressibility of water is relatively unimportant. However, in studying water-hammer and the transmission of elastic sound waves in bodies of water such as the ocean, the modulus of elasticity is of paramount importance.

Although a gas is very compressible compared with a liquid, the compressibility is of secondary importance in many problems of fluid mechanics. Detailed consideration is given to fluid compressibility in Chapter 4. Despite this fact, however, certain basic aspects of fluid compressibility are considered at this point. Furthermore, the influence of compressibility is becoming increasingly important in engineering problems—particularly for the flow of gases.

On the basis of Charles' law, Boyle's law, and Avogadro's law, the general gas law Eq. 1-2 was established. From this equation, it can be seen that, if the gas in a container is compressed into a smaller volume to increase the specific gravity, the ratio p/T must also increase. An increase in p/T can be accomplished either by increasing the pressure if the temperature is held constant, an *isothermal* process, or by decreasing the temperature if the pressure is to remain constant. In either case, there is a transfer of heat through the walls of the container in order to maintain the conditions prescribed.

If the expansion or compression of a gas is accomplished without heat transfer, the system is said to be *adiabatic* and the pressure-specific weight relation is

$$\frac{p}{\gamma^k} = \text{constant} \tag{1-6}$$

in which

$$k = \frac{C_p}{C_v} \tag{1-7}$$

C_p is the specific heat of the gas at constant pressure, C_v is the specific heat at constant volume. For air, oxygen, hydrogen, and other diatomic gases at usual temperatures.

$$k = 1.4$$

The values of k for certain gases are given in Table VI (see inside the front cover).

This specific heat ratio, which is also known as the adiabatic exponent, is important not only in compressibility and heat transfer problems, but also in problems involving the velocity of sound in a gas.

The modulus of elasticity of a gas is

$$E = p \tag{1-8}$$

for the *isothermal* process, and

$$E = kp \tag{1-9}$$

for the *adiabatic* process, the dimensions of E being F/L^2 in each case.

• VISCOSITY

Viscosity is that property of a fluid which causes resistance to relative motion within a fluid. This resistance to relative motion is discussed in detail in Chapter 6.

Whereas the density and specific weight are properties of a fluid which can be measured under static conditions, viscosity μ (mu) is a property which is exhibited only under dynamic conditions. In other words, a fluid must be in motion for the property of viscosity to be observed.

Viscosity is traceable to the molecular structure of a fluid. In the liquid state, the molecules are packed as closely together as possible, and the viscosity is apparently due to the cohesiveness of the molecules—hence, as the temperature is increased, the cohesion decreases and the viscosity decreases. In the gaseous state, on the other hand, the molecules are spaced very far apart, and the viscosity is due to the activity of the molecules which impinge against each other and bounce about in a random fashion, like so many microscopic billiard balls—hence, as the temperature is increased, the activity increases and the viscosity also increases. The variation of viscosity with temperature is of great importance. Therefore, special tables and graphs are presented to facilitate selection of the viscosity of certain common gases and liquids at a given temperature—see Tables III, IV, V, VI, and Fig. 1-2.

The fundamental term for viscosity is the dynamic viscosity, which has the dimensions of FT/L^2 and is defined more completely in Chapter 5 on Fluid Viscosity and Turbulence. The variations of dynamic viscosity between liquids and gases is very large—being 2.36×10^{-5} lb-sec/ft² for water, and 3.74×10^{-7} lb-sec/ft² for air at 60°F. If the dynamic viscosity is divided by the density, a very useful term, the kinematic viscosity ν is obtained

$$\nu = \frac{\mu}{\rho} \qquad (1\text{-}10)$$

in which the dimensions are L^2/T. The variation of kinematic viscosity between liquids and gases is much less than dynamic viscosity—being 1.22×10^{-5} ft²/sec for water, and 1.58×10^{-4} ft²/sec for air at 60°F. In other words, whereas the dynamic viscosity of water is nearly 100 times that of air, the kinematic viscosity of water is about one-tenth that of air—a fact which is not usually realized.

Under ordinary conditions of pressure, viscosity has been found to vary only with temperature—being independent of pressure. For certain oils, however, the viscosity has been found to vary irrationally for extremely great pressures.

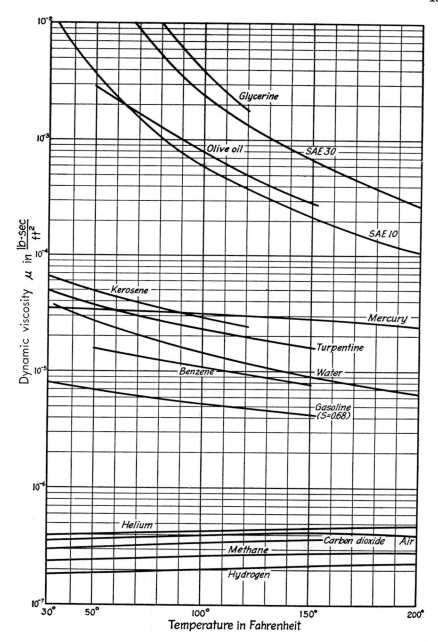

Fig. 1-2. Viscosity graph.

• VAPOR PRESSURE

At the free surface of a liquid, there is a continual interchange of molecules leaving the liquid and molecules entering the liquid to and from the gas atmosphere above. If more molecules leave the surface than enter it, evaporation is occurring; whereas if more molecules are entering the surface than are leaving it, condensation is occurring.

The molecules which impinge on the surface of the liquid create a pressure known as the partial pressure of the liquid vapor. This partial pressure combined with the partial pressures of the other gases in the atmosphere make up the total atmospheric pressure.

The molecules at the liquid surface create a vapor pressure p_v, which determines the rate at which the molecules leave the surface. When the vapor pressure of the liquid is equal to the partial pressure of the liquid vapor in the atmosphere, the number of molecules leaving is equal to the number entering, and the atmosphere is said to be saturated with the vapor of the liquid.

The vapor pressure of a liquid depends upon the temperature, as shown in Table III, on Properties of Water. As the temperature increases, the vapor pressure increases until the boiling point is reached for the particular ambient atmospheric pressure. For water at 212°F and at sea level, the vapor pressure is equal to the atmospheric pressure, so that the water boils and the molecules of vapor are given off at an extremely rapid rate.

The boiling point of a liquid decreases with decreasing atmospheric pressure. Therefore, it is possible for a liquid to boil at room temperature if the ambient pressure is decreased to the magnitude of the vapor pressure of the liquid at that temperature. This fact is very important in connection with the phenomenon of cavitation discussed in Chapter 3.

Mercury has a very small vapor pressure and, hence, is excellent as a fluid for use in a barometer (see Chapter 2). Benzine and others of the more volatile liquids have large vapor pressures. Vapor pressure p_v has the dimensions of F/L^2.

• SURFACE ENERGY AND CAPILLARITY

Surface energy (erroneously known as surface tension) is responsible for many common and interesting phenomena. It is responsible for the facts that: a needle coated with oil or wax will float on the surface of water, certain water insects and lily pads are able to float on water, a small drop of water in air assumes a spherical shape, a drop of melted fat in a bowl of soup floats as a circular disk, water in the soil is able to rise a considerable distance above the elevation of the ground water table, and liquids move up or along a wick or through other porous media great distances from the source of supply. Furthermore, certain insects utilize surface energy to transport themselves along the water surface. This is done by modifying the balance of surface energy with a fluid which they excrete on the water surface to cause a movement of the surface, which in turn gives motion to the insect.

The property of surface energy is applicable to liquids. It is a liquid surface phenomenon which takes place under either static or dynamic conditions. Surface energy is caused by the relative forces of **cohesion** (the attraction of liquid molecules for each other) and **adhesion** (the attraction of the liquid molecules for the molecules of another liquid or a solid).

A liquid molecule in the interior of the liquid body has other molecules on all sides of it, so that the forces of attraction are in equilibrium and the molecule is not attracted more strongly one way than another. A liquid molecule at the surface of the liquid, however, does not have the vast number of other molecules above it to balance the attraction of the molecules below it. Consequently, there is a net inward force on the molecule which is normal to the liquid surface.

The effect of surface energy is illustrated very clearly in the case of a droplet of liquid in which the net force is exerted inwardly over the entire surface of the droplet. The result is to increase the internal pressure within

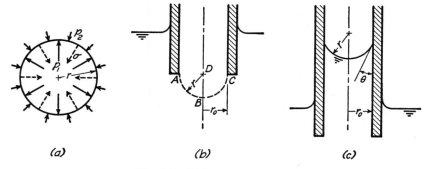

Fig. 1-3. Surface energy diagram.

the droplet, see Fig. 1-3(a). When the droplet of liquid is separated initially from the surface of the main body of fluid, the surface of the droplet contracts, due to the surface energy, to form a sphere—the shape for minimum surface area and for which the inward forces of attraction are in balance with the internal pressure. It is necessary to do work on the drop to change it from this shape. If the shape is changed, the molecules must be separated, and hence, work is done against the inward forces of attraction.

If one liquid is in contact with another liquid, instead of a gas as just discussed, an interface is formed at the junction of these liquids. Furthermore, if these liquids are immiscible (that is, do not mix molecularly), the interface is a sharp plane of demarcation, because the mutual attraction of like molecules of at least one of the liquids is greater than the mutual attraction of unlike molecules. Consequently, an interfacial energy exists which reflects the relative molecular attractions. The attraction of unlike molecules of the two liquids, however, is such that the resulting interfacial energy is less than the surface energy of the liquid having the greater surface energy, when in contact with its own vapor. Furthermore, interfacial energy is always equal to or greater than the difference between the surface energies of each individual liquid in its own vapor.

Within a liquid, there is a thermal agitation which is quite violent for the more volatile liquids such as water. Therefore, at the surface of the volatile liquids, the molecules leave the surface in extremely large numbers. Most of the molecules, however, impinge upon the molecules already in the surrounding atmosphere and bounce back into the liquid. Despite this rather violent movement of the molecules, there is a clear and distinct surface maintained within a thickness of one or two molecules—a truly remarkable fact. A similar condition exists at the interface of two quiet immiscible liquids, which have an unchanging thermal, pressure, and volume relationship. As the attraction of unlike molecules increases and approaches in magnitude the attraction of like molecules, however, the interface becomes progressively less distinct. A gradual blending from one liquid to the other takes place over an appreciable distance. Under these conditions, the liquids are miscible and the interfacial energy is reduced to zero.

The surface energy of a liquid is represented by sigma σ which has the dimensions of FL/L^2—that is, energy per unit area which is commonly simplified to F/L. Since F/L is force per unit length, and since the shapes of the surface and interface are curved in a manner **resembling** that of a sort of flexible skin containing the liquid, the concept of surface "tension" has developed and is generally used in the literature. This concept is fallacious in its basic nature, because the tension in a membrane involves a stress-strain relationship which does not exist in the surface-energy concept. Therefore, the term "surface tension" and the membrane concept are not used in this text.

On the basis of the foregoing concepts of surface energy, it is possible to derive an expression for the pressure difference across an interface such as that of the liquid droplet illustrated in Fig. 1-3(a). The forces shown in this figure are in balance, and consist of the internal pressure p_1 which acts outwardly over the entire surface, the external pressure p_2 which acts inwardly over the entire surface, and the surface energy σ which creates a resultant force acting inwardly, as represented by the broken-line vectors. It is the surface energy which causes the increased internal pressure within the droplet.

If the size of the droplet is expanded an incremental amount dr by a reversible (equilibrium) process, there is an incremental increase in surface area dA and an incremental increase in volume $dA \, dr = d\tilde{V}$. In the process of expansion, the work done by the pressure difference $\Delta p = p_1 - p_2$ is the product of the total force over the surface $\Delta p \, dA$ and the distance dr over which this force is applied

$$\Delta p \, dA \, dr = \Delta p \, d\tilde{V}$$

This work increases the total surface energy by $\sigma \, dA$, and since the work done and the energy increase must be equal

$$\Delta p \, d\tilde{V} = \sigma \, dA \tag{1-11}$$

For a sphere the surface area A is

$$A = 4\pi r^2$$

and
$$dA = 8\pi r \, dr$$

Likewise, the volume $\tilde{V}$ is

$$\tilde{V} = \tfrac{4}{3}\pi r^3$$

and
$$d\tilde{V} = 4\pi r^2 \, dr$$

Consequently, Eq. 1-11 becomes

$$\Delta p \, 4\pi r^2 \, dr = \sigma 8\pi r \, dr$$

or
$$\Delta p = \frac{2\sigma}{r} \tag{1-12}$$

which is the basic equation of surface energy for a droplet or bubble.

The foregoing derivation suggests a simple method of measuring the surface energy σ of a liquid-gas interface, as illustrated in Fig. 1-3(b). The gas is forced under pressure down a capillary tube to form a partial bubble ABC. When the bubble is sufficiently large so that the distance DB is equal to the radius of the tube, the radius of the bubble is a minimum— and the pressure difference Δp across the interface and the radius $r = r_0$ can be substituted into Eq. 1-12 to solve for the surface energy σ.

If the balance of forces of adhesion and cohesion is such that the. meniscus in a tube meets the wall at an angle θ, as shown in Fig. 1-3(c), the radius of curvature of the interface is $r = r_0/\cos\theta$, so that

$$\Delta p = \frac{2\sigma}{r_0}\cos\theta \tag{1-13}$$

in which r_0 is the radius of the tube. This equation can now be applied to small passages between solid boundaries.

When a liquid surface is in contact with a solid, the molecules have a certain attraction for each other. If a solid vertical wall, for example, has less attraction for a liquid molecule (say mercury) than the surrounding liquid molecules, then a contact angle greater than 90 degrees results (130 degrees for mercury and glass, and 105 degrees for water and paraffin wax), and the liquid surface is concave downward, due to the inward attraction of the liquid molecules. This creates a pressure increase across the surface, as it did in the drop of liquid already discussed. Because of the

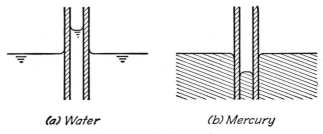

(a) Water   *(b) Mercury*

Fig. 1-4. Capillary tubes.

increased internal pressure, the elevation of the meniscus in a tube is lowered, see Fig. 1-4(b), to the level where the pressure is the same in the surrounding fluid. On the other hand, if the solid wall has a greater attraction for a liquid molecule than the surrounding liquid molecules (water in a glass tube for example), then a contact angle less than 90 degrees results (0 degrees for water and glass), and the liquid surface is concave upward, due to the greater attraction of the solid molecules. The liquid is then said to wet the boundary. This condition creates a pressure decrease within the liquid and the meniscus rises, see Figs. 1-4(a) and 1-3(c), so that the pressure within the column at the elevation of the surrounding liquid surface is the same as the pressure at this elevation outside the column. This rise is known as the capillary rise.

For a circular capillary tube, the height of the capillary rise (or depression) h, based upon the principles of fluid statics and Eq. 1-13 for $\theta = 0$, is

$$h = \frac{2\sigma}{\gamma r_0} \tag{1-14}$$

in which σ is the surface energy per unit area;
  r_0 is the radius of the tube.

As the size of tube becomes larger, the gravitation forces become more appreciable, the meniscus becomes less spherical, and Eqs. 1-13 and 1-14 are less accurate. With distilled water at 70°F in a glass tube, for example, the error in h is approximately 0.02 in. for a 0.2-in. tube and 0.08 in. for a 0.8-in. tube.

The foregoing derivations apply to a spherical interface only. A more general case is represented by a surface which has a radius of curvature r_1 in one plane and a radius of r_2 in a plane at right angles. Solution of this general case results in

$$\Delta p = \sigma \left(\frac{1}{r_1} + \frac{1}{r_2} \right) \tag{1-15}$$

which reduces to Eq. 1-12 when $r_1 = r_2$ for a spherical interface, and to

$$\Delta p = \frac{\sigma}{r_1} \tag{1-16}$$

when $r_2 = \infty$ for a cylindrical surface.

● DIMENSIONAL ANALYSIS

The principal use of dimensional analysis is the organization of a physical investigation. It provides a means of obtaining a functional relationship between the dependent variable and the independent variables. It is not a substitute for theoretical analysis, and it should be noted that skill of application increases with experience.

Earlier in this chapter, it was shown that all terms or variables used in fluid mechanics can be expressed in terms of three fundamental dimensions*—the length dimension L, the time dimension T, and either the force dimension F or the mass dimension M. Furthermore, it was demonstrated that an equation must be dimensionally homogeneous—that is, the dimensions must be the same on each side of an equation. All the foregoing facts play important parts in the development of dimensional analysis.

In most problems confronting an engineer, it is possible for him to assume that certain variables or quantities play an important role in the physical process being studied. Mathematically, this can be expressed as

$$a_1 = \phi(a_2, a_3, a_4, \cdots, a_n) \tag{1-17}$$

in which a_1 is the **dependent** variable whose magnitude depends upon the other variables $a_2, a_3, a_4, \cdots, a_n$. Each of the variables in Eq. 1-17 has dimensions which can be expressed in terms of either Force-Length-Time (F-L-T) or Mass-Length-Time (M-L-T). The F-L-T system will be used because it is somewhat simpler. On account of the need for dimensional homogeneity, a special procedure is used to combine these variables in such a way that a series of dimensionless parameters is determined. This procedure is known as the **pi-theorem,** which was introduced by Buckingham

* Temperature and electrical current can be added as basic dimensions for problems of heat transfer and electricity respectively.

in 1915. The pi-theorem not only arranges the variables into dimension-less groups, but also reduces the number of variables by the number of fundamental dimensions involved more than once. In other words, the number of variables is reduced from n in Eq. 1-17 to $n - m$. If each of the basic dimensions F, L, and T are involved more than once in Eq. 1-17, then $m = 3$ and the net result will be an equation with $n - 3$ dimension-less parameters.

In the application of the pi-theorem, m repeating variables are selected from the n variables listed, and then these are combined repeatedly with each of the other $n - m$ variables. Each combination yields a dimension-less parameter and, hence, there results $n - m$ dimensionless parameters. There is considerable latitude possible in the selection of the repeating variables. However, they must contain, in combination, each of the funda-mental dimensions $(F\text{-}L\text{-}T)$ involved in the process being studied.

Generally speaking, the variables listed in Eq. 1-17 can be divided into three groups:

1. Those variables of length which describe the geometry of the process being considered.

2. Those variables describing the flow, such as velocity V, pressure p, and shear τ.

3. Those variables describing the fluid (or fluids) involved, such as density ρ, viscosity μ, specific weight γ, elasticity E, surface energy σ, and vapor pressure p_v.

In the selection of the repeating variables, the resulting dimensionless parameters are usually most significant if not more than one repeating variable is chosen from each of these groups.

As an example, assume the following specific variables to be applicable to a problem or process under consideration:

$$\phi(a, b, c, V, p, \tau, \rho, \mu, \Delta\gamma, E, \sigma, p_v) = 0 \qquad (1\text{-}18)$$

in which a, b, and c are length variables which describe the geometry. It should be emphasized that in Eq. 1-18 (and also in Eq. 1-17) there must be not more than one dependent variable—all of the others being independent variables.

In Eq. 1-18, there are 12 variables, and the three fundamental dimen-sions $F\text{-}L\text{-}T$ are each found at least twice among these variables. There-fore, there are three repeating variables and, by means of the pi-theorem, these can be combined into $n - m = 12 - 3 = 9$ dimensionless param-eters. For the sake of convenience, and on the basis of previous experience (a most important aspect of the successful use of dimensional analysis), the three repeating variables are chosen as a, V, and ρ—one from each of the three groups of variables. These three repeating variables are now combined one at a time with each of the remaining variables to obtain nine dimensionless parameters, as follows:

1. a, V, ρ, b 6. $a, V, \rho, \Delta\gamma$
2. a, V, ρ, c 7. a, V, ρ, E
3. a, V, ρ, p 8. a, V, ρ, σ
4. a, V, ρ, τ 9. a, V, ρ, p_v
5. a, V, ρ, μ

Each variable in each group is assigned an unknown exponent, except as explained later. The actual magnitude of these exponents is determined so that the resulting parameter for each group is dimensionless. The first group is selected as an example:

$$a^{-1}V^x\rho^y b^z$$

The dimensions of a are L, of V are L/T, of ρ are FT^2/L^4, and of b are L. As a base reference, a is assumed to have an exponent of -1, so that it appears in the denominator to the first power. Although actually any exponent could be used, experience generally indicates what is the most convenient, and it now remains to determine each of the exponents so that none of the dimensions remains. Therefore, the exponents for the length terms are added and set equal to zero

$$L: \quad -1 + x - 4y + z = 0$$

Likewise, for the time terms

$$T: \quad -x + 2y \quad = 0$$

and for the force term

$$F: \quad y \quad = 0$$

These three equations can be solved simultaneously to give $y = 0$, $x = 0$, and $z = 1$, which yields b/a as the dimensionless parameter. In a similar manner, group 2 yields c/a.

A brief study of the variables and their corresponding dimensions in both group 1 and group 2 reveals that the exponent y must be zero, since there is no force dimension in any of the other variables to cancel the force dimension in the density ρ. Furthermore, once ρ is eliminated there are no T-dimensions to cancel the T in the velocity, which makes $x = 0$, thereby leaving only the length terms of b and c to combine with the length term a. ***In a similar manner, many dimensionless pi-terms can be written directly by inspection.*** One's ability to do this develops with practice.

In the third group, experience has indicated that it is desirable to have p appear in the numerator to the first power. Hence, it is assigned an exponent of 1.0, and the pi-theorem is applied as follows:

$$a^x V^y \rho^z p^1$$

$$L^x \left(\frac{L}{T}\right)^y \left(\frac{FT^2}{L^4}\right)^z \left(\frac{F}{L^2}\right)^1$$

which yields

$$L: \quad x + y - 4z - 2 = 0$$
$$T: \quad\quad -y + 2z \quad\quad = 0$$
$$F: \quad\quad\quad\quad z + 1 = 0$$

from which: $z = -1$, $y = -2$, and $x = 0$. These give $p/\rho V^2$ as the dimensionless parameter.

If the same procedure is applied to each of the remaining variables, the dimensionless parameters which result are:

1. $\dfrac{b}{a}$

2. $\dfrac{c}{a}$

3. $\dfrac{p}{\rho V^2}$

4. $\dfrac{\sqrt{\tau/\rho}}{V}$

5. $\dfrac{Va\rho}{\mu} = \text{Re}$

6. $\dfrac{V}{\sqrt{a\Delta\gamma/\rho}} = \text{Fr}$

7. $\dfrac{V}{\sqrt{E/\rho}} = \text{Ma}$

8. $\dfrac{V}{\sqrt{\sigma/\rho a}} = \text{We}$

9. $\dfrac{p_v}{\rho V^2}$

in which Re is the Reynolds number, Fr is the Froude number, Ma is the Mach number, and We is the Weber number, which will be found in later chapters of this text to be highly significant dimensionless parameters. This yields a final equation

$$\phi\left(\frac{b}{a}, \frac{c}{a}, \frac{p}{\rho V^2}, \frac{\sqrt{\tau/\rho}}{V}, \text{Re}, \text{Fr}, \text{Ma}, \text{We}, \frac{p_v}{\rho V^2}\right) = 0 \qquad (1\text{-}19)$$

which contains nine dimensionless parameters—any one of which can be chosen as the dependent variable, and all of the remaining eight variables are then independent variables.

The first two variables express length terms b and c in terms of a characteristic length a. The third parameter expresses the pressure at a point in terms of the stagnation pressure which is proportional to ρV^2 (a term which is developed in Chapter 3 as $\rho V^2/2$). The fourth parameter is the shear velocity $\sqrt{\tau/\rho}$ in terms of the velocity of flow V, and the fifth to eighth parameters are based upon fluid properties. The last parameter, which expresses the vapor pressure in terms of the stagnation pressure, can be shown to be related to the phenomenon of cavitation discussed in Chapter 3.

Various examples of the use of dimensional analysis for specific problems are developed in the chapters which follow.

• SUMMARY

1. The *dimensions* used in fluid mechanics are length, time, force, and mass which are interrelated by $F = Ma$.

2. The combined dimensions of the variables on one side of an equation must be the same as those on the other side in order for the equation to be *dimensionally homogeneous.*

3. *Pressure* is a force distributed over an area, *velocity* is a vector quantity expressing the rate of displacement with respect to time, and *discharge* is the flux or rate of movement of a fluid past a point.

4. *Fluid properties* are dependent upon the molecular structure of the fluid, including the molecular weight, spacing, attraction, and activity. The fluid properties include density, specific weight, compressibility, viscosity, vapor pressure, and surface energy.

5. *Dimensional analysis* helps to organize a physical investigation by providing a means for obtaining a physical relationship of a group of variables.

6. By dimensional analysis the *number of variables* can be reduced by the number of fundamental dimensions involved. In fluid mechanics there are usually three fundamental dimensions: force, length, and time; or mass, length, and time.

References

1. Adam, N. K., *The Physics and Chemistry of Surfaces*, 2nd ed. New York: Oxford University Press, 1938.

2. Binder, R. C., *Fluid Mechanics*, 3rd ed. Englewood Cliffs, N. J.: Prentice-Hall, Inc., 1955.

3. Bridgeman, P. W., *Dimensional Analysis*. New Haven: Yale University Press, 1922.

4. Daugherty, R. L. and A. C. Ingersoll, *Fluid Mechanics*. New York: McGraw-Hill Book Co., Inc., 1954.

5. Dorsey, N. E., *Properties of Ordinary Water-Substance*. New York: Reinhold Publishing Corp., 1940.

6. Langhaar, H. L., *Dimensional Analysis and Theory of Models*. New York: John Wiley & Sons, Inc., 1951.

7. Prandtl, Ludwig and O. G. Tietjens, *Fundamentals of Hydro-and Aeromechanics*. New York: McGraw-Hill Book Co., Inc., 1934.

8. Rouse, Hunter and Simon Ince, *History of Hydraulics*. Iowa City, Iowa: Iowa Institute of Hydraulic Research, 1957.

9. Rouse, Hunter, *Fluid Mechanics for Hydraulic Engineers*. New York: McGraw-Hill Book Co., Inc., 1938.

10. Vennard, John K., *Elementary Fluid Mechanics*, 3rd ed. New York: John Wiley & Sons, Inc., 1954.

Problems

1-1. What are the dimensions of: pressure p; velocity V; acceleration a; specific weight γ; and mass density ρ?

1-2. Verify that each of the following terms is dimensionally equivalent to length: velocity head $V^2/2g$, and pressure head p/γ.

1-3. Determine the physical dimensions of C in the Chezy equation $V = C\sqrt{RS}$, where V is velocity, R is hydraulic radius and has the dimension L, and S is slope which is a ratio of lengths.

1-4. Is the equation $V = (1.5/n)R^{2/3}S^{1/2}$ dimensionally homogeneous, assuming that n is a dimensionless coefficient? If it is not, what dimensions must n have to make the equation homogeneous? Knowing that the constant 1.5 contains the square root of the gravitational acceleration, what dimension must be assigned to n?

1-5. Define power. Show that it is dimensionally equal to (a) velocity times a force, (b) the rate of change of kinetic energy.

1-6. What are the dimensions of modulus of elasticity and moment of inertia?

1-7. Give the physical dimensions of dynamic viscosity and kinematic viscosity. What is the conversion factor relating these viscosities?

1-8. Calculate the physical dimensions of the following combinations of terms: the Reynolds number Re $= VD/v$; the Froude number Fr $= V/\sqrt{gy}$; and the Mach number, Ma $= V/\sqrt{E/\rho}$.

1-9. What are the dimensions of f in the equation $h_f = f\dfrac{L}{D}\dfrac{V^2}{2g}$

1-10. State Newton's second law of motion and give the physical dimensions of each of its terms in both the F-L-T and M-L-T system of dimensions.

1-11. Assuming C, f, and S are dimensionless terms, which of the following equations are not dimensionally homogeneous: **(a)** $F = Ma$; **(b)** $V = C\sqrt{RS}$;

(c) $h_f = f \dfrac{L}{D} \dfrac{V^2}{2g}$; and **(d)** $\dfrac{V_1^2}{2g} + \dfrac{p_1}{\gamma} + z_1 = \dfrac{V_2^2}{2g} + \dfrac{p_2}{\gamma} + z_2$.

1-12. Define the terms **(a)** scalar quantity, **(b)** vector quantity.

1-13. Considering the following terms, state which are scalar terms and which are vector terms: **(a)** velocity; **(b)** acceleration; **(c)** force; **(d)** area; **(e)** pressure; **(f)** momentum.

1-14. Show that the physical dimensions of the engineering gas constant R are $L/°R$.

1-15. Referring to Table I, check the physical dimensions of the terms involved. Find the dimensionless combination of the mass density ρ, the specific weight γ, the depth of flow y, and the velocity V.

1-16. Air is compressed into a steel tank at an absolute pressure of 100 psia when the temperature is 20°C. Find the pressure rise that would occur if the temperature in the tank were raised to 1500°C.

1-17. Compute the specific weight of air, assuming atmospheric pressure is 14.7 psia for a temperature ranging from 0°F to 130°F. Present the results graphically.

1-18. An automobile tire has a bursting strength of 80 psia. If it has a constant volume of 1330 cu. in. and is initially pumped up to a pressure of 45 psia at 20°C, find the temperature at which the tire will burst.

1-19. Explain how the bulk modulus of elasticity for a fluid can be obtained experimentally.

1-20. The volume of a liquid is reduced 0.03 percent by the application of a pressure of 200 psi. Calculate the bulk modulus of the liquid.

1-21. Determine the pressure required to reduce a given volume of water 2 percent, if the initial pressure is standard atmospheric pressure ($T = 60$°F).

1-22. If water boils at 170°F, what is the atmospheric pressure?

1-23. What is the ratio of the kinematic viscosity of water to that of air if the pressure is 120 psia and the temperature of both fluids is 60°F?

1-24. The dynamic and kinematic viscosities of a liquid are 3.229×10^{-5} and 1.664×10^{-5} respectively. **(a)** Determine its density; **(b)** if the liquid is water, what is its temperature?

1-25. The unit of dynamic viscosity in the metric system is the dyne-second per square centimeter. This unit is called the poise. What is the magnitude of the factor which converts this dynamic viscosity to dynamic viscosity in the ft-lb-sec system of units?

1-26. Calculate the conversion factor which relates kinematic viscosity (stokes) in the metric system to its equivalent in the ft-lb-sec system.

1-27. In industry, kinematic viscosity ν is sometimes expressed as seconds Saybolt Universal SSU. These two expressions are related by $\nu = 0.000002433$ SSU $- 0.00210/$SSU. Calculate the viscosity of water at 70°F in terms of SSU units.

1-28. Plot a curve of vapor pressure versus temperature for water from 32° to 212°F.

1-29. Carbon tetrachloride is stored in a closed container. What is the minimum absolute pressure possible in the space above the liquid surface if the temperature of the liquid is 70°F?

1-30. Determine the height of capillary rise of water in a glass tube with a diameter of 0.5 mm, (a) at 50°F, (b) at 100°F.

1-31. A vertical glass piezometer tube is attached to the bottom of an open tank to measure the depth of water in the tank. (a) The depth in the tank being 5 ft, what is the height of the column of water in the piezometer tube, if it has an inside diameter of 1 mm? (b) What percent error results from capillarity?

1-32. Determine the minimum size of glass tubing that can be used to measure water level, if the capillary rise in the tube is not to exceed 0.01 in.

1-33. If spheroid grains having a diameter of 0.1 mm are packed in a column at maximum density, with the base of the column in contact with a free water surface, estimate the height of capillary rise in between the grains, assuming the space is triangular.

1-34. A piezometer tube 1 mm in diameter contains mercury at 68°F. Give the magnitude of the effect of capillary action on the piezometer reading.

1-35. The head loss h_L for flow in a pipe line is a function of the velocity V, the pipe diameter D, the roughness of the pipe e, the absolute viscosity μ, and the density of the water ρ. By dimensional analysis, reduce the problem to an equation where h_L is a function of two dimensionless parameters. Of what significance is this functional equation?

1-36. If it is properly assumed that the velocity of a jet of water issuing from an orifice in the side of a tank is dependent upon the height of the water above the center of the orifice, the specific weight of the water, and the density of the water, what is the functional equation for this problem, and how does it compare with the actual equation $V = \sqrt{2gh}$?

chapter 2

Fluid Statics

Fluid statics is part of fluid mechanics involving certain fundamental, and yet simple, principles of physics which were discovered many years ago. Archimedes, one of the brilliant scientists of antiquity, published a treatise on "Floating Bodies" between 200 and 300 years before Christ. However, this contribution to an understanding of fluid buoyancy stood alone for almost 18 centuries before a Flemish engineer named Simon Stevin (1548–1620) correctly explained the basic principles of fluid statics. Blaise Pascal (1623–1663), a French philosopher, performed the commonly recognized classical experiments in fluid statics, which clearly illustrated the fundamental relationships involved. Because of the simplicity of the fundamental principles of fluid statics, they are the first to be considered in this study of fluid mechanics.

When a person dives into water, his body experiences greater and greater pressure upon it as he goes deeper and deeper into the water. Furthermore, almost everyone has experienced the fact that an object, such as a rock, weighs less while it is submerged under water than when it is removed into the air. As one changes elevation rapidly, such as in an elevator or in an unpressurized airplane, his ears are sensitive indicators of the resulting change in pressure. Nearly every student of engineering has used, or at least observed, a pressure gage (similar to that in Fig. 2-7) to measure pressures in such objects as steam boilers or automobile tires. The sensing element of these gages has a certain similarity to that of the ear. Dump trucks and other power machinery and equipment operated by hydraulic cylinders have been observed to operate with amazing ease, despite the large forces which obviously are involved. Some people have stood above or downstream from a large dam and wondered at the tremendous pressures which the dam must resist to remain in position. Not only small boys, but many adults as well, have stood and watched ships in a harbor being loaded or unloaded and wondered how they keep from capsizing when heavy loads are piled on their decks and they appear so topheavy. Carrying a wide pan of liquid such as water without spilling it is not an easy task. Why is it possible to swing a bucket of water over one's head without spilling it? All of these examples are simple and yet common occurrences illustrating the problems to be studied in the remainder of this chapter.

• FLUID PRESSURE

When an object is submerged in a fluid, it experiences pressures which act perpendicularly to the entire surface area of the object. The pressure acting over a submerged area results in a force acting on that area. The force must be perpendicular to the area, or there would be a tangential component of the force parallel to the area, and this parallel component would represent a shearing stress on the surface of the area. However, no fluid can support a shearing stress unless motion occurs.

Since the fluid under study here is at rest, there can be no shear present and, therefore, the static force on any submerged plane surface must be perpendicular to that surface. The magnitude of the pressure depends upon the weight (which in turn depends upon the height and specific weight) of the fluid or fluids above it. Therefore, an equation can be developed expressing the relation between this pressure and the heights and specific weights of the fluids above it.

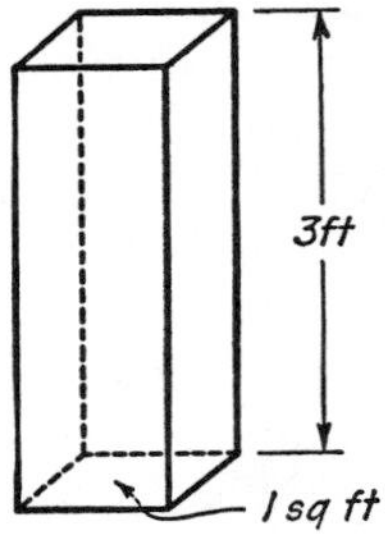

Fig. 2-1.
Prism of liquid.

Consider a container, such as the prism shown in Fig. 2-1, filled with water having a specific weight of 62.4 lb per cu ft. If the height of the container is 3 ft, what total force will be exerted at the bottom? Obviously, the bottom is supporting the entire liquid. Since the volume is a total of 3 cu ft and the specific weight is 62.4 lb per cu ft, the **total force** against the bottom is

$$1 \times 3 \times 62.4 = 187.2 \text{ lb} \qquad (2\text{-}1)$$

But since the area of the bottom is 1 sq ft, the **pressure** against the bottom is 187.2 lb per sq ft. In terms of pounds per square inch (which is the term most frequently used in practice) the pressure is $187.2/144 = 1.3$ psi. Since the pressure is the same at every point on the bottom of the cylinder, this example demonstrates that the **pressure at a point submerged below a liquid surface** exposed to the atmosphere **is equal to the product of the vertical distance below the surface and the specific weight of the liquid in question.**

The foregoing illustration is especially simple, because the horizontal cross-sectional area of the cylinder is the same throughout its length. However, from the experiment of Pascal, we know that even a small tube filled with liquid and projecting upward from the container is sufficient to exert tremendous pressures down below it. In fact, the pressure depends

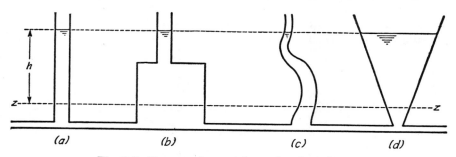

Fig. 2-2. Pressures for containers of various shapes.

only upon the height of column and not at all upon the size of the column. Therefore, in Fig. 2-2(b) the same unit pressure would be exerted against the bottom of the container as in Fig. 2-2(a). Likewise, the same unit pressure is exerted at point z for each of the containers, since each of these points are the same distance h below the free liquid surface.

By means of the foregoing illustration, the following equation can be deduced for incompressible fluids.

$$p = \gamma h \qquad (2\text{-}2)$$

in which

p is the pressure F/L^2

γ is the specific weight F/L^3

h is the vertical distance L of the point of consideration below the free liquid surface.

The general equation for fluid statics can be derived by isolating a free body of fluid and analyzing the forces acting on the free body, Fig. 2-3.

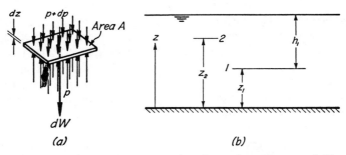

Fig. 2-3. Relationship between elevation and pressure on a fluid element.

The free body in Fig. 2-3(a) is acted upon by the following forces which are in equilibrium—the downward force due to pressure $(p + dp) A$, plus the incremental weight of the body dW is equal to the upward force due to pressure pA. In equation form this is

$$(p + dp)A + dW = pA$$
$$-A \, dp = dW$$

Since dW is the specific weight γ times the volume $A \, dz$ of the body

$$-A \, dp = \gamma A \, dz$$

or
$$-dp = \gamma \, dz \tag{2-3}$$

If the specific weight γ is assumed to be constant with respect to elevation z, Eq. 2-3 can be integrated from 1 to 2, see Fig. 2-3(b), to yield

$$-\int_1^2 dp = \gamma \int_1^2 dz$$
$$p_1 - p_2 = \gamma(z_2 - z_1) \tag{2-4}$$

Equation 2-4 shows that the difference in pressure between any two horizontal planes in an incompressible fluid is equal to the difference in elevation times the specific weight of the fluid. If plane 2 is assumed to be the free liquid surface, then the pressure at the surface is atmospheric pressure, which in terms of gage pressure is zero, Fig. 2-6. The difference in elevation $z_2 - z_1$ then becomes h, and p_2 is zero, so that Eq. 2-4 takes the same form as Eq. 2-2

$$p_1 = \gamma h_1$$

It is sometimes convenient to express the pressure in terms of feet of fluid. For this purpose, the foregoing equation can be written as

$$\frac{p}{\gamma} = h \qquad (2\text{-}5)$$

in which p/γ is known as the pressure head.

Example Problem 2-1: A rectangular tank is filled with oil, sp. gr. = 0.90, to a depth of 10 ft. Calculate (a) the pressure along a vertical element of the wall of the tank, and show (b) the variation of pressure with depth graphically.

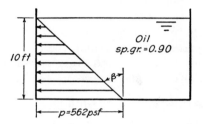

Ex. Prob. 2-1

Solution:

(a) Pressure can be computed by means of Eq. 2-2

$$p = \gamma h$$

At the water surface $h = 0$ and $p = 0$

at the bottom of the tank $h = 10$ ft and

$$p = (0.9)(62.4)(10) = \underline{562 \text{ psf or } 3.9 \text{ psi}}$$

(b) The pressure could also be computed for intermediate depths, but this is not necessary, since

$$\gamma = \text{constant}$$

That is, $p = \gamma h$ is the equation of a straight line making an angle β with the vertical such that

$$\beta = \tan^{-1} \gamma$$

and the pressure diagram is as indicated in the figure.

Pascal's Law: Consider an element of fluid as shown in Fig. 2-4, having a wedge shape and a specific weight the same as that of the fluid in which it is submerged. Further assume that it is an elemental size having the dimensions Δx, Δz, and Δy.

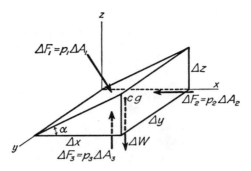

Fig. 2-4. Forces on a fluid element.

The forces on the vertical faces in the $x-z$ planes are not shown, because they are equal and opposite and, therefore, will cancel each other. This fluid element can now be taken as a free body and the sum of the forces in the x and z directions set equal to zero, so that

$$\Sigma F_x = 0, \qquad \Delta F_1 \sin \alpha = \Delta F_2$$
$$\Sigma F_y = 0, \qquad \Delta F_1 \cos \alpha + \Delta W = \Delta F_3$$

As Δx, Δy and Δz approach zero, the incremental mass decreases to a point, and ΔW becomes zero. Then

$$p_1 \Delta A_1 \sin \alpha = p_2 \Delta A_2$$
$$p_1 \Delta A_1 \cos \alpha = p_3 \Delta A_3$$

At the center of gravity cg,

$$\Delta A_1 \sin \alpha = \Delta A_2 \quad \text{and} \quad \Delta A_1 \cos \alpha = \Delta A_3$$

Then

$$p_1 = p_2 \quad \text{and} \quad p_1 = p_3$$

or $$p_1 = p_2 = p_3 \tag{2-6}$$

The foregoing development provides a means of proving **Pascal's law,** which is stated: **At a point in a fluid the pressure is exerted equally in all directions.** Pressure, therefore, is a scalar quantity and not a vector quantity. These characteristics of pressure apply at any point within a fluid and also for moving fluids as well as static fluids.

Example Problem 2-2: The figure shows a hydraulic press. Calculate the magnitude of the force f required to raise a load $F = 20$ tons if the areas a and A are 0.2 sq ft and 10 sq ft respectively.

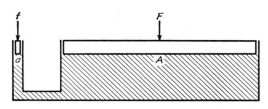

Ex. Prob. 2-2

Solution:

From Pascal's law, an increase of pressure on any part of a confined liquid causes an equal increase throughout the liquid. That is, the pressures on the two pistons must be equal, and

$$\frac{F}{A} = \frac{f}{a}$$

or
$$f = \frac{aF}{A} = \frac{(0.2)(20)}{10} = 0.4 \text{ tons} = \underline{\underline{800 \text{ lb}}}$$

• VARIATION OF ATMOSPHERE PRESSURE

In the foregoing section, it was shown that, when the specific weight γ does not vary with pressure p, the variation of pressure with elevation in a fluid is simply $dp = -\gamma\, dz$, see Eq. 2-3. This same relationship can be applied also to gases, which are compressible fluids, provided the variation in elevation is not great. If the variation in elevations is great, however, the variation of γ with p must be included in Eq. 2-3 before integrating. Such is the case for variation of pressure in the atmosphere.

In Chapter 1, consideration was given to the compressibility of fluids. Eq. 1-6 can be generalized to express a polytropic process as

$$\frac{p}{\gamma^n} = \frac{p_0}{\gamma_0^n} = \text{constant} \tag{2-7}$$

in which n is for air and is determined simply by the specific type of process involved. For an isothermal process $n = 1.0$, and for a dry adiabatic process $n = k = 1.4$.

To determine the variation of pressure with elevation, Eq. 2-7 and Eq. 1-2 can be substituted into Eq. 2-3 and integrated as follows

$$\gamma = \gamma_0 \left(\frac{p}{p_0}\right)^{1/n} = \frac{p_0}{RT_0}\left(\frac{p}{p_0}\right)^{1/n} \tag{2-8}$$

$$\frac{RT_0}{p_0{}^{(n-1)/n}} \int \frac{dp}{p^{1/n}} = - \int dz \tag{2-9}$$

$$\frac{n}{n-1} \frac{RT_0}{p_0{}^{(n-1)/n}} \left(p^{(n-1)/n} - p_0{}^{(n-1)/n} \right) = z_0 - z = -\Delta z \tag{2-10}$$

which can be arranged as

$$\frac{p}{p_0} = \left(1 - \frac{n-1}{n} \frac{\Delta z}{RT_0} \right)^{n/(n-1)} \tag{2-11}$$

in order to relate the relative pressure p/p_0 to the relative change in elevation, which can be written either in terms of the temperature $\Delta z/RT_0$ or in terms of the pressure $\Delta z/(p/\gamma)_0$, since

$$\frac{\Delta z}{RT_0} = \frac{\Delta z}{(p/\gamma)_0} \tag{2-12}$$

Equation 2-11 is applicable for any value of n, except for the particular case of $n = 1.0$, which must be integrated separately to yield

$$\frac{p}{p_0} = e^{-\Delta z/RT_0} = e^{-\Delta z/(p/\gamma)_0} \tag{2-13}$$

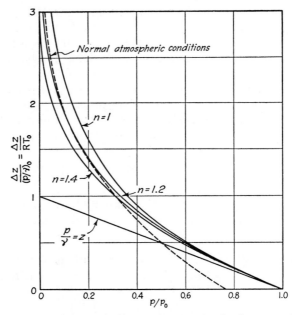

Fig. 2-5. Variations of pressure with elevation in the atmosphere.

In the atmosphere, the actual n-value usually varies between $n = 1.4$ for dry adiabatic conditions and $n = 1.0$ for isothermal conditions. This fact is discussed further, at a later point in this chapter, in connection with stability. Figure 2-5 shows the variation of pressure with elevation for different n-values as compared with an actual variation. Note the actual curve lies between the curves for $n = 1.4$ and $n = 1.2$. In other words, the typical actual variation of pressure with elevation in the atmosphere is adiabatic rather than isothermal and, further, it is between the wet adiabatic ($n = 1.2$) and the dry adiabatic ($n = 1.4$) processes.

Example Problem 2-3: Assuming that atmospheric pressure and temperature at ground level are 14.7 psia and 59°F respectively, calculate the pressure 20,000 ft above the ground assuming (a) no density variation, (b) an isothermal variation $n = 1.0$, (c) a dry adiabatic variation $n = 1.4$, and (d) a wet adiabatic variation, $n = 1.2$. Check the answers obtained by means of Fig. 2-5.

Solution:

(a) For $p = 14.7$ psia and $t = 59°F$,

$$\gamma = 0.0765 \text{ lb/ft}^3 = \text{constant}$$

thus

$$\Delta p = \frac{(0.0765)(20,000)}{144} = 10.6 \text{ psi}$$

and

$$p = 14.7 - 10.6 = \underline{\underline{4.1 \text{ psia}}}$$

(b) For $n = 1$

$$\frac{p}{p_0} = e^{-\Delta z/(p/\gamma)_0}$$

or

$$\frac{p}{14.7} = e^{-\frac{20,000}{\left(\frac{(14.7)(144)}{0.0765}\right)_0}} = e^{-0.723} = 0.478$$

$$p = \underline{\underline{7.04 \text{ psia}}}$$

(c) For $n = 1.4$

$$\frac{p}{p_0} = \left[1 - \frac{(n-1)}{n} \frac{\Delta z}{RT_0}\right]^{n/(n-1)}$$

or

$$\frac{p}{14.7} = \left[1 - \frac{(0.286)(20,000)}{(53.3)\ (519)}\right]^{3.5} = 0.446$$

and

$$p = \underline{\underline{5.91 \text{ psia}}}$$

(d) For $n = 1.2$

$$\frac{p}{14.7} = \left[1 - \frac{(0.167)(20,000)}{(53.3)\ (519)}\right]^{6} = 0.462$$

$$p = \underline{\underline{6.8 \text{ psia}}}$$

• PRESSURE REFERENCE DATA

As indicated earlier in this chapter, pressure, which is force per unit area, is expressed in a number of different ways, such as pounds per square foot (psf) or pounds per square inch (psi). Such a *pressure* however, *must be with respect to some datum or reference pressure.* For example, in the container in Fig. 2-1, the pressure of 187.2 lb per square foot is relative to the pressure surrounding the container and the free water surface. Atmospheric pressure at mean sea level and standard conditions is usually considered to be approximately 14.7 psia (pounds per square inch absolute), or 2117 psfa (pounds per square foot absolute). Therefore, relative to a perfect vacuum, the pressure at the bottom of the container is 14.7 + 1.3 = 16.0 psia. This can be viewed graphically by

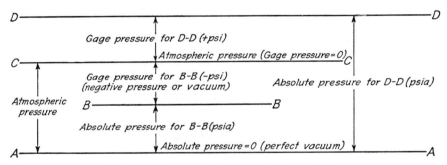

Fig. 2-6. Pressure as related to reference data.

means of Fig. 2-6, in which the datum plane A-A is absolute zero, which is a perfect vacuum. The plane C-C represents the atmospheric or *ambient pressure* which is the zero reference for gage pressure.

Most pressure-measuring devices read in terms of gage pressure— that is, *the pressure relative to the surrounding atmospheric pressure.* For example, if the reading on the gage is represented in Fig. 2-6 by the pressure at plane D-D, the gage pressure will be the difference between the pressure at plane D-D and the pressure at plane C-C (labeled psi or psf) The *absolute pressure* will be the difference between the pressure on plane A-A (which is zero absolute), and the plane D-D, and it will be labeled psia or psfa. When a *vacuum gage* is being observed, however, it may give a negative pressure, which would represent the difference between the plane C-C (atmospheric pressure) and the plane B-B. *Vacuum pressures* in this text represent either negative pressures, using atmospheric pressure as the assumed zero point, or can read in positive absolute pressures, which would be the difference between the pressure at plane A-A and the pressure at plane B-B.

It must be remembered that the **absolute pressure** datum *A-A* is *fixed and never changes*. However, the plane *C-C* simply represents the pressure surrounding the gage. The atmospheric pressure surrounding the gage, may vary from approximately 14.7 psia (greater than absolute zero pressure) when the gage is at sea level, to a value much less, as the gage is taken to higher elevations in the atmosphere. In the following sections, a description is given of several common instruments which are used for the measurement of pressure.

• MEASUREMENT OF PRESSURE

The measurement of pressure is usually accomplished with an instrument which utilizes the effect of a pressure difference on two sides of a sensing element, or on the two ends of a liquid column. In either case, the principle of pressure measurement is based upon the fundamental properties of a known fluid as a reference.

Bourdon Gage: As discussed in the foregoing section, gage pressure is the pressure relative to the surrounding atmospheric pressure. The Bourdon

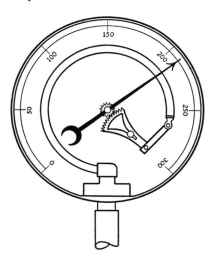

Fig. 2-7. Bourdon gage.

gage, see Fig. 2-7, is an instrument for measuring **gage pressure.** The pressure is introduced into a curved, hollow metal tube which is usually made of brass. One end of the tube is fixed to the frame of the gage, and the remainder of the tube is free to move. As the pressure introduced into the gage is increased, the tube gradually straightens, because of its shape, and the pointer is activated to indicate the pressure on the face of the gage. For a given absolute pressure inside the tube, the gage pressure indicated on the dial could be varied by increasing or decreasing the pressure on the outside of the tube. In other words, it is the **difference in pressure across the wall** of the Bourdon tube that determines the degree of straightening of the tube, and, hence, the indicated pressure. It is for this reason that a Bourdon gage reads in terms of gage pressure, which is referenced to the surrounding atmospheric pressure. Such gages are calibrated to read in terms of pounds per square inch or pounds per square foot, or inches of mercury or feet of water, when dealing with the English system of units.

It is a compact and handy instrument to use, although it is fairly delicate. Usually such a gage is not intended to be used for extremely precise measurements. Greater precision can be obtained by increasing the diameter of the dial or by decreasing the range of the gage. Error due to mechanical friction within the mechanism of the gage can be minimized by tapping the gage with the finger or a pencil just prior to making a reading. If the gage is accidentally submitted to a very high pressure, even for an extremely short period of time, the metal in the tube can be strained beyond the elastic limit and a permanent set take place, so that the calibration of the tube is no longer valid. When such a case occurs, however, it is usually detectable by the indicator hand reading greater than zero when the pressure is the same on both the inside and outside of the tube.

Example Problem 2-4: The Bourdon gage can be calibrated in terms of psi, inches of mercury, and feet of water as well as other units. Calculate the coefficients that will convert (a) psi to ft of water, and (b) psi to inches of mercury.

Solution:

(a) Pressure p can be expressed as the height of a column of *any* fluid by the relation

$$h = \frac{p}{\gamma}$$

Since pressure is generally expressed in pounds per square inch and the value of γ for water is usually given as 62.4 pcf:

$$h = \frac{(144)(\text{psi})}{62.4}$$

or $\text{psi} = \dfrac{62.4}{144} h = \underline{0.433h} = \underline{\underline{0.433 \text{ psi per ft of water}}}$

(b) Since

$$\frac{\gamma_{Hg}}{\gamma} = 13.6$$

then from (a)

$\text{psi} = \dfrac{(0.433)(13.6)}{12} h = \underline{0.49\ h} = \underline{\underline{0.49 \text{ psi per inch of mercury}}}$

Barometers: A barometer is an instrument which is most frequently used for the purpose of accurately measuring the atmospheric pressure. Perhaps the simplest barometer is that illustrated in Fig. 2-8, wherein a glass tube is filled with a liquid (usually mercury), from which all entrained air and other foreign material has been removed, and then inverted into a pool of the same liquid so that the closed end is vertically upward and contains no air. In this case, the atmospheric pressure is being exerted downward on the face of the pool and the liquid in the tube drops to the level h (29.92 in., for mercury at sea level) above this pool. Above the liquid meniscus at the top of the tube is a near vacuum. Actually, the pressure in this upper end is that of the vapor pressure of the particular liquid in the barometer. In the case of mercury, the vapor pressure at ordinary atmospheric temperatures is so small (0.00154 psia at 70°F) that, for most purposes, it can be assumed to be a perfect vacuum.

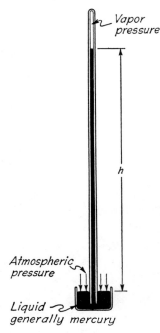

Fig. 2-8. Barometer.

If other liquids were used in the barometer instead of mercury, the vapor pressure could be an appreciable amount greater than absolute zero pressure. Therefore, a correction (which varies with temperature) would be necessary if such a liquid were used. For example, if water were used, the distance h would be 33.08 ft at 70°F instead of 33.92 ft, which it would be if the vapor pressure were zero instead of 0.84 ft of water. Because mercury has a specific weight which is 13.6 times that of water, and because the vapor pressure is so very small, mercury results in a barometer that is much more compact and convenient to use. This type of instrument is used in nearly every meteorological and weather station in the world.

Example Problem 2-5: Referring to the simple barometer of Fig. 2-8, if the liquid is water, the temperature is 100°F, and $h = 25.0$ ft, compute the barometric pressure in psia.

Solution:

Since

$$p = p_v + \gamma h$$

By selecting the correct values for the unit weight γ, and the vapor pressure p_v, from Table III, the equation is

$$p = 0.95 + \frac{(62.0)(25)}{144} = \underline{\underline{11.78 \text{ psia}}}$$

The *barograph* (see Plate 2-1) is a recording barometer. Instead of using mercury and a glass column, however, it utilizes the principle of a pressure difference across a pressure-sensitive element and, hence, has considerable similarity to the Bourdon gage—except that the fluctuating pressure is on

Plate 2-1. Barograph.

the outside of the gage instead of the inside. The inside contains a fixed mass of gas and exerts a constant reference pressure if the temperature is constant. If the temperature varies, a correction must be applied.

The barograph usually consists of a bellows for the sensing element. As the surrounding pressure increases or decreases, the bellows is contracted or expanded, and an indicating arm carrying a pen makes a continuous record of barometric pressure on a chart which is rotated on a drum with a vertical axis. The barograph not only gives a continuous record of barometric pressure, but also is very compact, rugged, and portable.

Manometers: Manometers are very commonly employed for the purpose of measuring the pressure in a tank or in a pipe line, or some other container of fluid. Innumerable designs have been developed to fit the various problems encountered and the particular whim of the designer. Basically, there are two types of manometers, the open manometer and the differential manometer. The simplest open manometer, or piezometer, is the one shown in Fig. 2-9(a).

From simple arrangements in Fig. 2-2 and Fig. 2-9, it can be seen that the pressure in the pipe can be determined by $p = \gamma h$, as shown in Eq. 2-2. If the pressure in the pipe is negative (that is, below atmospheric pressure— a partial vacuum), then a manometer can be used as shown in Fig. 2-9(c).

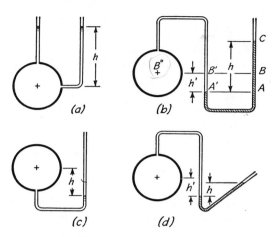

Fig. 2-9. Manometers.

In each of these cases, the piezometer tube is filled with the liquid that is flowing or resting in the pipe or container. Obviously, a piezometer of the types shown in Fig. 2-9(a) and (c) can be used only for the measurement of pressures in liquids, since a gas would be exposed to the atmosphere in the tube, and no visible interface or meniscus could be formed.

Piezometer or manometer tubes which are less than about $\frac{1}{2}$ in. diameter have considerable capillary action (see Chapter 1), which either raises or depresses the meniscus above or below the true value. For water, this rise of the meniscus in inches is approximately equal to $0.0462/D$, see Eq. 1-14. In the case of mercury, the depression in inches due to capillarity is approximately determined by $0.016/D$. In both cases, D is the inside diameter of the tube, in inches.

The connection of the piezometer to the container can be made in any convenient manner, provided there is no flow of fluid past the point of connection. However, when fluid is flowing, as in a pipe, it is very important that the connection on the inside of the pipe be smooth, so that no abnormal reading in pressure is obtained. The reason for such an abnormality is explained in Chapter 10.

When it is necessary to measure relatively large pressures with a liquid column manometer, or to measure pressures of a gas inside a container, a special type of open manometer can be used. Such a manometer is shown in Fig. 2-9(b). In this case, a liquid denser than the fluid in the container is used in the manometer. The container in which the pressure is to be determined is at elevation B. To analyze this problem, it is easiest to begin at the free surface of the heavier liquid in the open tube where the pressure is known to be atmospheric—that is, zero gage pressure. Beginning at the free surface C, determine the pressure changes along the manometer tube to point B. At level A, which is the elevation of the interface of the heavier liquid and the gas or lighter liquid, the pressure is the same on each side of the U-tube. Therefore, it is known immediately that the pressure at this interface is equal to the height h times the specific weight γ of the heavier liquid. Proceeding up the manometer from A' to B', the pressure decreases by $p = \gamma_L h'$. The final step is to consider the change in pressure from B' to B'' in the center of the container. Since the fluid is continuous, the pressure is the same at B' and B''. Therefore, the pressure at elevation B'' in the container is the pressure at the interface A' previously determined, minus the product of h' and the specific weight γ_L of the gas or lighter liquid.

Example Problem 2-6: If the heavier liquid is mercury and the lighter liquid is water, what is the pressure within the container when $h = 2$ ft and $h' = 1$ ft? See Fig. 2-9(b).

Solution:

At point C $p_c = 0$

and at point A $p_A = \gamma_{Hg}$ $h = (847)(2) = 1694$ psf

Since $p_A = p_A'$ then $p_A' - \gamma h' = p_B'$

However $p_B' = p_B''$ so that $p_B'' = p_A' - \gamma h'$

and $p_B'' = 1694 - (62.4)(1.0) = \underline{\underline{1631.6 \text{ psf}}}$

or $p_B'' = \underline{\underline{11.3 \text{ psi.}}}$

Sometimes, very small pressures or changes in pressure must be measured. In this case, it is possible to use a sloping, open tube manometer, as shown in Fig. 2-9(d). In this manometer, the meniscus of the liquid must move over a considerable distance in order to rise in elevation the distance h. For this reason, there is much greater sensitivity, and more refined readings can be made. For example, if the tube is set on a slope of 30 degrees with the horizontal, the meniscus will move a distance of $2h$ as it rises a height of h. Although there are distinct advantages to this type of manometer, there are also certain disadvantages which must be taken into consideration, such as the fact that the meniscus is not symmetrical about the axis of the tube. Therefore, it is sometimes difficult to obtain a precise reading.

• FORCES ON SUBMERGED PLANE SURFACES

On many occasions, engineers are confronted with the problem of determining the forces acting on the walls of containers, such as pipes, tanks, and
concrete forms. In these cases, the forces are due to pressure which is being
exerted outward from the inside of the container. Dams built across streams
to impound water upstream are special cases of a container. On other
occasions, the problem may be to determine the pressure exerted against
submerged objects such as caissons, diving bells, submarines, and balloons.
Another common problem is the determination of forces acting on gates
in the walls of these containers or submerged objects.

Forces acting against containers or submerged objects are due to the
pressure of either a gas, a liquid, or a solid. In the case of gas, pressure
usually does not vary appreciably with elevation in the vertical distances
which are commonly considered. With liquids, however, the pressure will
vary from atmospheric at a free surface to tremendous magnitudes at great
depths, such as in the ocean. The actual absolute magnitude of the pressure
depends on the ambient or atmospheric pressure, the depth of the point
being considered, and the specific weight of the liquid involved. Equation
2-2 expresses the interelationship of these variables in terms of gage pressure, where the atmospheric or ambient pressure is assumed to be zero.

The plane surfaces against which static pressures are acting can range
in their orientation from horizontal to vertical. In the case of a horizontal
surface, as shown earlier in Fig. 2-1, the pressure over the entire bottom of
the container is uniform, because the depth is uniform throughout. In this
case, the total force is simply the product of the average pressure and the
area. Likewise, the pressure could be acting upward against the horizontal
top of a container, in which case the pressure would be uniform and the
total pressure or force again would be
equal to the product of the pressure
at the top and the area of the surface.

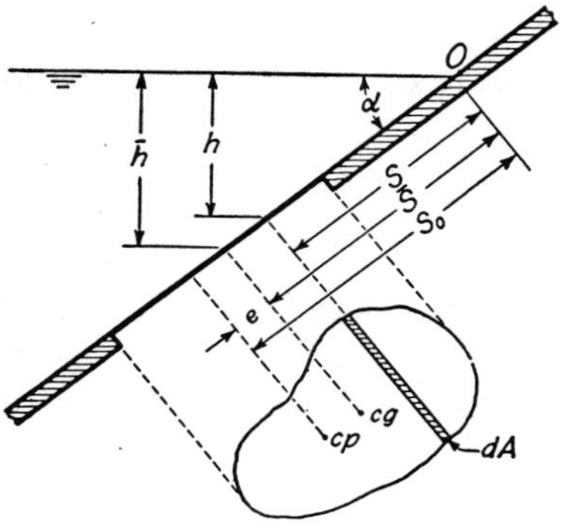

The general solution for forces
on submerged plane surfaces must
include the angle of slope α of the
plane, the depth of the plane, the
shape of the plane, and the specific
weight of the liquid. The derivation
of the general equation is based upon
Fig. 2-10, in which cg is the center of
gravity and cp is the pressure center.
By means of calculus, the product
of an elemental area dA and the
corresponding pressure on it can be

Fig. 2-10. General diagram for
submerged plane surfaces.

integrated over the entire area to yield the total pressure or force F. The differential equations can be expressed as

$$dF = p \, dA = \gamma h \, dA = \gamma S \sin \alpha \, dA \qquad (2\text{-}14)$$

and integration yields

$$F = \gamma \sin \alpha \int S \, dA \qquad (2\text{-}15)$$

The student will recall, from calculus and mechanics of solids, that the integral $\int S \, dA$ is the first moment of the area about the axis O-O, and is equal to the product of the area and the distance from the axis to the center of gravity of the area. In equation form, this is

$$M_0 = \int S \, dA = \overline{S}A$$

Equation 2-15 then becomes

$$F = \gamma \, \overline{S}A \sin \alpha$$

and since

$$\overline{S} \sin \alpha = \overline{h}$$

h – distance from surface to centroid

$$F = \gamma \overline{h} A \qquad (2\text{-}16)$$

It is of interest to note that Eq. 2-16 is independent of direction. In other words, the magnitude of the resultant force on a given area will remain the same, regardless of the orientation of the area, provided only that the centroid remains the same depth below the liquid surface.

From Eq. 2-16, a fundamental law can be stated: The *force F on a submerged plane surface* is the product of the area A of the surface and the pressure $\gamma \overline{h}$ at the centroid of the plane surface.

Not only is it necessary to know the magnitude and direction of the force acting on a submerged surface, but also its location or pressure center. This information is needed in order to compute the moments involved in the analysis of some problems. The differential moments $S \, dF$ about axis O-O in Fig. 2-10 can be summed up by means of calculus for the entire surface and set equal to the product of the resultant force F and the length of the moment arm S_0 from the axis O-O to the pressure center

$$S_0 F = \int S \, dF \qquad (2\text{-}17)$$

However, from Eq. 2-14, $dF = \gamma S \sin \alpha \, dA$, which can be combined with Eq. 2-17 to yield

$$S_0 F = \gamma \sin \alpha \int S^2 \, dA \qquad (2\text{-}18)$$

Solving for S_0 and substituting Eq. 2-15 for F

$$S_0 = \frac{\gamma \sin \alpha \int S^2 \, dA}{\gamma \sin \alpha \int S \, dA} \qquad (2\text{-}19)$$

Again the student will recall, from calculus and mechanics of solids, that the integral $\int S^2\,dA$ is the second moment of the area A about the axis O-O. This second moment is more commonly known as the moment of inertia I_0 of the area A about the axis O-O. Therefore, Eq. 2-19 can be simplified to

$$S_0 = \frac{I_0}{\overline{S}A} = \frac{I_0}{M_0} \qquad (2\text{-}20)$$

This equation can be placed in more usable form by substitution of the transfer equation, which transfers the moment of inertia from the axis O-O to the parallel axis through the center of gravity or centroid.

$$S_0 = \frac{\overline{I} + \overline{S}^2 A}{\overline{S}A} = \overline{S} + \frac{\overline{I}}{\overline{S}A} \qquad (2\text{-}21)$$

The eccentricity e can now be found as

$$e = S_0 - \overline{S} = \frac{\overline{I}}{\overline{S}A} = \frac{k^2}{\overline{S}} \qquad (2\text{-}22)$$

in which k is the radius of gyration, and $\overline{I}$ is the moment of inertia of the area about the centroidal axis parallel to the axis O-O. The equations for $\overline{I}$

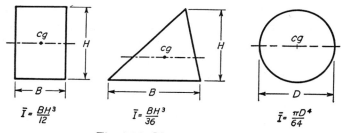

$$\overline{I} = \frac{BH^3}{12} \qquad\qquad \overline{I} = \frac{BH^3}{36} \qquad\qquad \overline{I} = \frac{\pi D^4}{64}$$

Fig. 2-11. Moment of inertia.

for several simple shapes are given in Fig. 2-11, and a more complete listing for other shapes is given inside the back cover.

From Eq. 2-22, another fundamental law can be stated: *The pressure center is always below the centroid of any plane submerged surface that is not horizontal.*

In solving problems based upon the foregoing equation for pressures on submerged plane surfaces, the following procedure may be helpful:

1. Determine *centroid* of area A and its distance $\overline{h}$ below surface of liquid.

2. Multiply *pressure* $\gamma\overline{h}$ by *area* A to obtain *total force* F.

3. Locate *center of pressure* cp by $S_0 = \overline{S} + \dfrac{\overline{I}}{\overline{S}A}$

Summarizing: To obtain the ***total force*** F, first determine $\bar{h}$ and the area A; and to locate the ***center of pressure*** either by S_0 or by the eccentricity e, first determine $\bar{S}, \bar{I}$, and A.

In the special case of a vertical rectangular surface, such as shown in Fig. 2-12, the pressure ranges from atmospheric or zero at the surface of the liquid to γh at the bottom. Because this is a linear variation, it is possible to represent the pressure by a straight line so that the horizontal vector representing p has a length of γh which is equal to the pressure. Furthermore, this method of representation by a pressure diagram can be used to

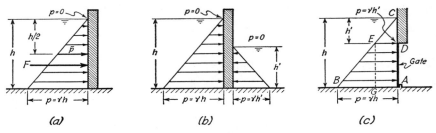

Fig. 2-12. Pressure diagrams.

establish, essentially by observation, certain facts about a problem without the benefit of the general equation, Eq. 2-16. The mean or average pressure is found at half depth, so that $\bar{p} = \gamma h / 2$ and the ***total force F is the product of the average pressure and the area*** hb of the wall against which the force is acting. Thus, for Fig. 2-12(a).

$$F = \bar{p}\, hb = \gamma bh^2 / 2 \qquad (2\text{-}23)$$

in which b is the width of the wall against which the pressure is acting.

In order to design the wall of a tank or container such as shown in Fig. 2-12(a), it is necessary to know where the resultant F of the pressure (the pressure center) acts against the wall. Again, the pressure diagram is useful because, by observation, it may be seen that the greatest pressures are located along the lower half of the wall and the ***center of gravity or centroid of the pressure*** prism is the location of the ***pressure center*** of the resultant force. In the special case of the vertical wall in Fig. 2-12(a), the pressure prism is represented by a triangle. Since the centroid of a triangle is located at a distance $h/3$ above the base, the resultant force F passes through the centroid of the pressure prism, that is, $h/3$ above the bottom. Remember, however, that this is true only when the cross section of the pressure prism is a triangle, and not a trapezoid as in Fig. 2-12(c), where the plane of the gate does not extend up to the surface. Furthermore, if the submerged surface is not rectangular, the general solution Eq. 2-21 must be used.

Sometimes, it is necessary to study the hydrostatic pressures resulting from bodies of liquid on each side of a wall, see Fig. 2-12(b). In this case, the magnitude and location of F is solved independently for each side of the wall, just as for Fig. 2-12(a) and the magnitude of the resultant force is the algebraic sum of the components. The resultant force can be determined also from the resultant pressure diagram which is obtained by adding the two pressure diagrams algebraically.

Example Problem 2-7: The depth of water is $h = 10$ ft on one side of a wall and $h = 5.0$ ft on the other side, as in Fig. 2-12(b). Find the resultant moment about the base of the wall.

Solution:

Analyze a strip of wall one foot wide.

On the left side of the wall $h = 10$ ft

$F = (62.4)(10/2)(10 \times 1) = 3120$ lb acting 10/3 ft from base

On the right side of the wall $h = 5$ ft

$F_R = (62.4)(5/2)(5 \times 1) = 780$ lb acting 5/3 ft from base

$M_{base} = 3120(10/3) - 780(5/3) = \underline{\underline{9180 \text{ lb ft/ft of width}}}$

Another problem frequently encountered is that of determining the *force against a plane* surface located at some depth below the surface of the liquid, for example the gate of Fig. 2-12(c). Again the pressure diagram is useful in visualizing the problem. In this case, however, the triangle ABC includes pressures acting on the wall CD as well as the gate AD. The pressure diagram applicable to the gate alone is $ADEB$. For the sake of convenience, this may be broken down further into $ADEG$ and GEB. Again, the average pressure is that at the centroid of the gate. Furthermore, the pressure center is located at the centroid of the pressure diagram $ADEB$. As will be remembered from mechanics, the centroid of a complex diagram, such as $ADEB$, can be determined by the method of moments. If the gate is not rectangular in shape, however, the general solution Eq. 2-21 must be used.

Example Problem 2-8: Refer to Fig. 2-12(c). If the water depth $h = 15$ ft, the height of gate $h - h' = 10$ ft, and the width of gate is 10 ft, determine: (a) the total force on the gate; (b) the point of application of the total force; and (c) the reaction force of the sill on the gate.

Solution:

(Remember that *finding a force* involves determination of both its *magnitude* and its *direction*.)

(a) The total force is equal to ten times the area of the trapezoidal pressure prism: $ABED \times 10$. For simplicity, the trapezoidal diagram can be reduced to a triangle and a rectangle, giving

$$F = F_1 + F_2$$

in which F_1 is the force due to the triangular diagram and F_2 is the force due to the rectangular diagram.

Then
$$F_1 = \frac{\gamma(h - h')^2(10)}{2} = \frac{(62.4)(10)^3}{2} = 31{,}200 \text{ lb}$$

$$F_2 = \gamma h'(10)(10) = (62.4)(5)(10)^2 = 31{,}200 \text{ lb}$$

and
$$F = 31{,}200 + 31{,}200 = \underline{\underline{62{,}400 \text{ lb}}} \rightarrow$$

(b) To find the location of the force, take moments about A and, noting that F_1 and F_2 act through the center of gravity of the triangular and rectangular prisms respectively,

$$F\,\bar{y} = (F_1)10/3 + (F_2)10/2$$

and
$$\bar{y} = \frac{(31{,}200)(3.33) + (31{,}200)(5.0)}{62{,}400} = \underline{\underline{4.17 \text{ ft}}}$$

in which $\bar{y}$ is the distance above A to the center of pressure.

(c) Taking moments about the hinge at D

$$\Sigma M_0 = 0$$

or
$$(F)(10 - \bar{y}) = 10\,R$$

and
$$R = \frac{(62{,}400)(5.83)}{10} = \underline{\underline{36{,}379 \text{ lb}}} \leftarrow$$

which is the reaction of the sill on the gate.

Example Problem 2-9: The center of a vertical circular gate is located 10 ft below the water surface. If the diameter D of the gate is 6 ft, what is the force on the gate, and where does it act?

Solution:

$$A = \frac{\pi D^2}{4} = \frac{(3.14)(36)}{4} = 28.3 \text{ ft}^2$$

$$F = \gamma \bar{h} A = (62.4)(10)(28.3) = \underline{17{,}630 \text{ lb}} \rightarrow$$

$$S_0 = \bar{S} + \frac{\bar{I}}{\bar{S}A} = 10 - \frac{\pi D^4/64}{10\pi D^2/4} = \underline{\underline{10.23 \text{ ft}}}$$

$$e = S_0 - \bar{S} = \underline{\underline{0.23 \text{ ft}}}$$

Example Problem 2-10: A rectangular gate 4 ft wide and 6 ft high is located as shown. Find the force on the gate and its location.

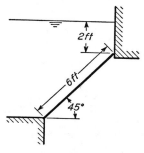

Ex. Prob. 2-10

Solution:

The centroid of the gate is at its center and, therefore

$$\bar{h} = 2 + \frac{6}{2}(\sin 45°) = 4.12 \text{ ft}$$

$$F = \gamma \bar{h} A = (62.4)(4.12)(24) = \underline{6160 \text{ lb}}$$

This force acts perpendicularly to the gate, and its location can be determined as follows:

$$\bar{S} = 2/\sin 45° + 6/2 = 5.83 \text{ ft}$$

$$S_0 = \bar{S} + \frac{\bar{I}_c}{\bar{S}A} = 5.83 + \frac{(1/12)(4)(6)^3}{5.83(4)(6)}$$

$$S_0 = 5.83 + 0.52 = \underline{\underline{6.35 \text{ ft}}}$$

$$e = \frac{\bar{I}}{\bar{S}A} = \underline{\underline{0.52 \text{ ft}}}$$

The force acts 0.52 ft below the center of gravity of the gate, measured along the face of the gate.

• FORCES ON SUBMERGED CURVED SURFACES

The principles set forth in the foregoing sections are applicable only to plane submerged surfaces. Because the analysis of curved surfaces is more complex, another method must be developed for solving problems involving surfaces which are not plane. Since a plane surface is simply a special or limiting case of a curved surface, however, the principles now to be developed are equally applicable to plane surfaces.

Consider the curved surface AB, in Fig. 2-13(a), which is supporting a liquid between a sill at A and a wall at B. Observation reveals that pressures are exerted by the liquid to the right against the wall BC, and down-

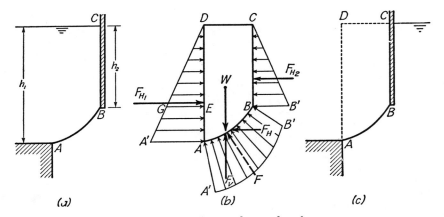

Fig. 2-13. Submerged curved surface.

ward and to the right against the curved surface AB. Since the liquid and the bounding surfaces are in static equilibrium, there must be pressures exerted by the bounding surfaces against the liquid which are equal in magnitude but opposite in direction to those exerted by the liquid against the solid boundaries.

To simplify the analysis of this problem, a free-body diagram Fig. 2-13(b) is taken of the liquid above the curved surface AB and the pressures and resultant forces shown, which are acting on the liquid free body. First, the weight of the liquid is W, which is acting downward. This weight must be supported by the summation F_v of the vertical components of pressure (from the diagram $ABB'A'$) which the curved surface AB exerts upward against the liquid. F_v must be equal in magnitude but opposite in direction and collinear with the weight W. The pressure diagram $ABB'A'$ indicates the distribution of supporting pressures which are acting normal to the curved surface AB which can be resolved into forces in both the horizontal and vertical directions.

Second, along the planes BC and AD, simple hydrostatic pressure diagrams are drawn. These diagrams indicate the pressure which is exerted by the wall against the liquid in the case of plane BC, and the pressure exerted by the liquid outside the free body against the liquid within the free body in the case of the plane AD. The resultants of these pressure diagrams are F_{H_1} to the right, and F_{H_2} to the left. Obviously, the diagram $AA'D$ is larger than the diagram BCB' by the amount $A'AEG$—which is the unbalanced force caused by the pressures on the planes AD and BC. This unbalanced force must be balanced by a force equal in magnitude but opposite in direction, which is caused by the horizontal components of the pressure diagram $ABB'A'$ along the curved surface AB.

From the foregoing discussion, it can be seen that the curved surface AB must support the liquid above it by means of the pressure shown in the diagram $ABB'A'$, which can be resolved into a horizontal force F_H and a vertical force F_v. The vertical force F_v is equal in magnitude but opposite in direction to the weight of the liquid W, and the horizontal force F_H is equal in magnitude but opposite in direction to the difference between F_{H_1} and F_{H_2}—that is $F_H = F_{H_1} - F_{H_2}$. The forces of the liquid against the curved boundary are, therefore, simply W downward and F_H, which is represented by $AA'GE$ to the right.

Once the magnitude and direction of the forces experienced by the free body are determined, the location or line of action of each force must be found, in order to determine whether there will be rotation of the free body. The line of action of the force F_v must be through the centroid of the area $ABCD$, since it must support the weight of this volume of liquid. Likewise, the line of action of the force F_H must be through the centroid of the area $AEGA'$, since it must counteract the difference between the pressure diagrams BCB' and $AA'D$. The resultant F of the forces F_v and F_H can be found in the usual manner (Eq. 2-26).

The forces F_{H_1}, F_{H_2}, F_H, F_v, and F are actually fictitious. There really are no such forces acting at a single point along a curved surface. Instead, there are unit forces or pressures acting as described by the pressure diagrams. Nevertheless, for the sake of convenience in analysis, these forces are used as being equivalent to the summation of the pressure diagrams.

In summary, then,

$$F_v = W = \gamma A b \qquad (2\text{-}24)$$

in which A is the area $ABCD$ and b is the length of the curved surface normal to the plane $ABCD$,

$$F_H = F_{H_1} - F_{H_2} = \left(\frac{\gamma h_1^2}{2} - \frac{\gamma h_2^2}{2} \right) \qquad (2\text{-}25)$$

and

$$F = \sqrt{F_v^2 + F_H^2} \qquad (2\text{-}26)$$

The foregoing analysis is for the forces involved when a curved surface AB supports a liquid above it. It is possible, however, to have such a bounding surface support a liquid on the other side of it, as shown in Fig. 2-13(c). Under such a condition, the curved surface AB is preventing the liquid from occupying the space on the left as in Fig. 2-13(a). Hence, in a sense. it is ***displacing*** the liquid on the left represented by the free-body diagram $ABCD$. Therefore, the pressures which the curved surface AB is exerting against the liquid are just equal and opposite to the pressures exerted along AB in Fig. 2-13(b).

The magnitude and the line of action of the forces involved in Fig. 2-13(c) can be found by assuming an imaginary volume $ABCD$ above the boundary, but the direction of the forces will be opposite. Again, the forces of the liquid against the bounding surface are equal in magnitude but opposite in direction to the forces of the bounding surface against the liquid.

The method of calculating a *force on a curved surface* can be summarized in the following manner.

1. To calculate the **resultant force** against a submerged curved surface, **compute both vertical and horizontal components.**

2. The **vertical component** of force is determined by calculating the weight of liquid (real or imaginary) vertically above the curved surface.

   a. If the curved surface is submerged so that a **real volume** of liquid exists above the surface, then the vertical component of force against the curved surface is downward, and acts through the center of gravity of the liquid volume.

   b. If the curved surface is submerged so that an **imaginary volume** of liquid is used above the surface, then the vertical component of force against the curved surface is upward and, as in (a), acts through the center of gravity of the imaginary volume. This vertical force is also the force of buoyancy, which is equal to the weight of the displaced fluid—see the following section.

3. The **horizontal component** of the resultant force on the curved surface is calculated by horizontally projecting the surface onto a vertical plane and treating the projected area as a submerged vertical plane surface, see Fig. 2-12(c). The **magnitude** of the force F_H is computed by the volume of the pressure diagram, or by Eq. 2-16 rearranged for a vertical surface

$$F_\mathrm{H} = \gamma \bar{h}_p A_p \tag{2-27}$$

in which $\bar{h}_p$ is the vertical distance from the free surface of the liquid to the centroid of the projected area,

and A_p is the projected area of the curved surface.

The line of action of the force F_H is through the center of pressure of the projected area. It is located in the same manner as for a plane submerged surface.

4. Calculate the **magnitude** (from Eq. 2-26), the **direction,** and the **location** of the **resultant force.**

Example Problem 2-11: Assume that the gate AB in Fig. 2-13(a) is the quadrant of a cylinder with a radius of 5 ft and a length of 8 ft. If the depth of water CB is 3 ft, find the total force (including its location) on the gate.

Solution:

Following the outlined method of computing the forces:

1. $F = \sqrt{F_V^2 + F_H^2}$

2. F_V = weight of volume of water above AB.

Volume = Area $ADCB \times 8$ → *quarter circle*

Area $ADCB = (3 \times 5) + \dfrac{\pi 5^2}{4}$ ← = 34.6 sq ft

$F_V = (34.6)(8.0)(62.4) = \underline{\underline{17{,}250 \text{ lb}}} \downarrow$

3. $F_H = \gamma \bar{h}_p A_p$

$\bar{h}_p = 3 + 5/2 = 5.5$ ft $A_p = 5 \times 8 = 40$ sq ft

$F_H = (62.4)(5.5)(40) = \underline{\underline{13{,}700 \text{ lb}}} \rightarrow$

4. $F = \sqrt{17{,}250^2 + 13{,}700^2} = \underline{\underline{22{,}000 \text{ lb}}}$

$\theta = \tan^{-1}\dfrac{F_V}{F_H} = \tan^{-1} 1.26 = \underline{\underline{51.5°}}$ ↘θ

Since the force on each element of the surface area is perpendicular to the surface of the cylinder, each force element, and therefore the resultant, must pass through the center of the circular curve AB.

- # BUOYANCY FORCES

The famous bath tub experiment of Archimedes is well known. From this experiment, he discovered the very important principle of buoyancy and, in his excitement, ran down the street in scanty attire, crying, "Eureka! Eureka!" (which means, I have found it, I have found it—). Although this discovery occurred more than 2000 years ago, the concept of buoyancy as explained by Archimedes has remained unchanged.

Briefly stated, the ***principle of buoyancy*** is: *A submerged or floating object is buoyed up with a force equal to the weight of the fluid displaced by the object.* This statement can be proved mathematically by utilizing the relationship between the surface and volume integrals, as follows:

$$F_b = \int_s p \, ds = \int_{\tilde{V}} \frac{\partial p}{\partial y} \, d\tilde{V}$$

in which $p = \gamma y$ and $\dfrac{\partial p}{\partial y} = \gamma$

Hence
$$F_b = \int_{\tilde{V}} \gamma \, d\tilde{V} = \gamma \tilde{V}$$

since γ is a constant throughout the volume. From Fig. 2-14, it can be observed directly that

$$F_b = F_u - F_d \qquad (2\text{-}28)$$

in which F_b is the force of buoyancy,

$\qquad F_u$ is the summation of the upward components of pressure against the bottom of the body, and

$\qquad F_d$ is the summation of the downward components of pressure against the top of the body.

Fig. 2-14 illustrates four cases of buoyancy of both submerged and floating objects. In each of the four cases, a free-body diagram is drawn for the vertical forces. The summation of the horizontal forces is zero, and is therefore not shown. The resultants of the vertical forces must all be collinear

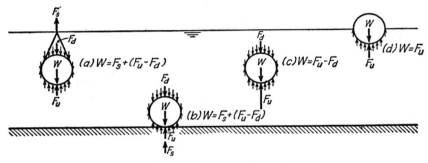

Fig. 2-14. Vertical forces associated with buoyancy.

and pass through the center of gravity of the object, or a torque would result and cause rotation. Furthermore, if the object has neither velocity (which would result in drag forces, see Chapter 9) nor acceleration (which would result in an inertial force $F = Ma$), the resultants must be equal and opposite.

In Fig. 2-14(a) and 2-14(b), the submerged object has a specific weight which is greater than that of the surrounding fluid; while in Fig. 2-14(c) the specific weight of the body is the same as that of the surrounding fluid; and in Fig. 2-14(d) the specific weight of the body is less than that of the fluid. Therefore, the object in Fig. 2-14(a) must be supported from above by a force F_s, which is the weight of the body while submerged in the fluid. Mathematically, this can be expressed as

$$F_s = W - (F_u - F_d) = \gamma_s \tilde{V} - F_b \qquad (2\text{-}29)$$

in which

> W is the weight of the body in air F;
>
> γ_s is the specific weight of the solid F/L^3;
>
> $\tilde{V}$ is the volume of the object L^3

The net buoyant force F_b is the difference between the summation F_u of the upward components of pressure against the bottom of the object and the summation F_d of the downward components of pressure against the top of the object, as shown by Eq. 2-28.

In Fig. 2-14(b), the same equation and analysis apply, but part of the supporting force F_s is supplied by the bottom of the container. The floating object in Fig. 2-14(d) is only partially submerged, because the volume of fluid which must be displaced to support the object is less than the volume of the object, due to the greater specific weight of the fluid. In this case, the external supporting force is zero and the weight of the object is just balanced by the buoyant force of the liquid, that is $F_b = F_u$.

• STABILITY OF FLOATING AND SUBMERGED BODIES

In statics of solids, the student has learned that, for a free body to be in *equilibrium*, the *summation of forces in all directions must equal zero*, and the *summation of all moments must equal zero*. Under these conditions, there is *no acceleration* either linear or angular. Furthermore, if the body in equilibrium has no motion, then it is said to be in

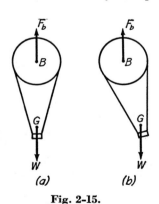

Fig. 2-15.
Stability of a Balloon.

static equilibrium. In this section, consideration is given to floating bodies and submerged bodies which may or may not be in equilibrium.

A body is *stable* if it will remain in a given position and return to this position if rotated through a small angle. The balloon, a *submerged body* in Fig. 2-15(a), is both stable and in equilibrium. It is in equilibrium, because both the summation of the forces in any direction is equal to zero (if the force of buoyancy F_b is just equal to the weight of the balloon W), and the summation of the moments is equal to zero. It is stable because, if rotated through a small angle, as in Fig. 2-15(b), a moment will be created which will return the balloon to its original position of equilibrium in Fig. 2-15(a). This moment is known as a *restoring moment.* The barge, a *floating object,* in Fig. 2-16(a) is also in stable equilibrium, because the forces and moments equal zero and, if rotated through a small angle, as in Fig. 2-16(b), a *restoring moment* is created which will tend to restore the barge to its original position in Fig. 2-16(a).

In the cases of both the balloon and the barge, the force of buoyancy F_b and the weight W are collinear when the body is in equilibrium. However, with the balloon the center of buoyancy for F_b is above the center of gravity for W, and with the barge the center of buoyancy is below the center of gravity. The *center of buoyancy* can be defined as the *center of gravity of the fluid displaced* by the submerged body. When the *center of buoyancy* is *above the center of gravity,* the body can be said without further consideration to be in *stable equilibrium.*

When the center of buoyancy is below the center of gravity, however, the *rotation test* must be applied. For example, if a barge rolls 90 degrees, as in Fig. 2-16(c), it is in unstable equilibrium, because, if rotated in either direction, an *upsetting moment* will be created.

It is possible for a body to be in stable equilibrium with the center of gravity above the center of buoyancy only if it is a floating body. This is due to the fact that when the body is rotated, the shape (but not the volume) of the displaced liquid changes so that the center of buoyancy shifts

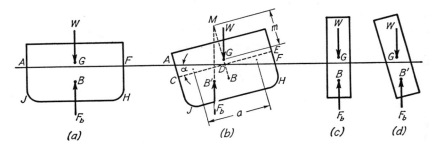

Fig. 2-16. Stability of barges.

and the moment created by F_b is a restoring moment. In Fig. 2-16(b), this change in shape is from $JHEC$ to $JHFA$. The triangle DFE is moved to ADC. For a small angle of heel α, the distance a between the centers of gravity of these triangles is approximately $a = \frac{2}{3}b$, where b is the width CE of the barge. Even at an angle of heel of 20 degrees, the error in using this equation is only 0.2 percent, if the cross section of the barge is approximately rectangular in shape. If the body is *submerged* instead of floating, the location of the *center of buoyancy remains unchanged with rotation* of the body and, therefore, the body is unstable if the center of gravity is above the center of buoyancy.

If the center of gravity and the center of buoyancy coincide, then the object is said to be in neutral equilibrium. In this case, neither a restoring moment nor an upsetting moment will be created, regardless of the orientation of the body.

The point M in Fig. 2-16(b) is known as the **metacenter.** It is located at the intersection of the vertical line of force F_b in the two positions shown in Figs. 2-16(a) and 2-16(b). The **metacentric height** is the distance $m = MG$. It can be computed for small angles of heel as

$$m = \frac{\bar{I}}{\tilde{V}} - GB \tag{2-30}$$

in which $\bar{I}$ is the moment of inertia with respect to the longitudinal centroidal axis of the cross section of the barge, which is defined by the water surface when the barge is in equilibrium, as in Fig. 2-16(a) L^4;

$\tilde{V}$ is the volume of the displaced fluid L^3;

GB is the vertical distance between the center of gravity G and the center of buoyancy B for equilibrium conditions, as in Fig. 2-16(a) L.

If m is positive, the body is stable. If m is negative, the body is unstable. The metacenter is very useful, because it serves as a convenient criterion for stability. If the metacenter is above G, a restoring moment is created, and if the metacenter is below G, an upsetting moment is created. The magnitude of this moment or torque T can be found approximately by

$$T = Wm\alpha \tag{2-31}$$

in which α is a small angle of heel measured in radians;
                W is the weight of the barge or ship F.

The metacenter for the barge in Fig. 2-16 can be used to illustrate the foregoing principles. In this case, the metacenter will remain above G as the barge is rotated counter clockwise until it is in the vertical position of Fig. 2-16(c), where M coincides with G. If the barge is rotated still farther, the metacenter is to the right of G, which is "below" G in the original position of Fig. 2-16(a).

The following list is a summary indicating the relative stability for various conditions involving buoyant bodies.

1. Generally speaking, if the center of gravity of weight is below the center of buoyancy, then the object is stable.

2. If the center of gravity is above the center of buoyancy, the object is top-heavy, but may or may not be unstable—depending upon whether the

couple created by a small angle of rotation causes a restoring moment or an upsetting moment.

3. An outside force may tend to cause tipping and, if the object is top-heavy, the moment may or may not become an upsetting moment. With proper design, such tipping will increase the righting moment.

4. If the metacenter is above the center of gravity of an object, then the couple creates a restoring moment, and the object is stable. However, if the metacenter lies below the center of gravity of the object, an upsetting moment is created and the object is unstable and will rotate toward stability (perhaps a new and undesirable position).

When the foregoing principles are applied to barges, if some external force causes a roll of sufficient magnitude, the angle of heel will become so great that the metacenter will fall below the center of gravity and the barge will capsize. If the barge has liquid in it, the center of gravity of the barge moves when the barge is tipped.

TABLE OF STABILITY CRITERIA

	Submerged body	*Floating body*
Stable:	Center of gravity *below* center of buoyancy.	Center of gravity *below* metacenter.
Unstable:	Center of gravity *above* center of buoyancy.	Center of gravity *above* metacenter.
Neutral:	Center of gravity *at* center of buoyancy.	Center of gravity *at* metacenter.

Example Problem 2-12: Referring to Fig. 2-16(a) and (b) the barge is 8 ft deep, 30 ft wide, and 90 ft long and draws 6 ft of water. If the center of gravity of the barge is 4.5 ft above the bottom at G, determine the righting moment when it has an angle of heel of 4 degrees.

Solution:

The weight of the displaced water is

$$W = (62.4)(6)(30)(90) = 1{,}010{,}000 \text{ lb}$$

The center of gravity of the displaced water is 3 ft above the bottom at B. The distance $BG = 1.5$ ft, and the metacentric height is

$$m = \frac{I}{V} - BG = \frac{(90)(30)^3}{(12)(90)(30)(6)} - 1.5 = 11.0 \text{ ft}$$

and the righting moment is

$$T = (1.01)(10^6)(11)(0.0698) = \underline{\underline{7.75 \times 10^5 \text{ ft-lb}}}$$

• STABILITY OF FLUID MASSES

In the foregoing section, consideration was given to discrete bodies which have a net unit weight either greater than or less than the surrounding fluid. It is possible also to have fluid masses, within the main body of fluid, which differ in unit weight from that of the main body. Such is the case with warmer air that rises in the atmosphere to create a convective thunderstorm, or the warm air which circulates in a room by free convection. Conversely, the cold air which develops near a cold wall or comes from a refrigerator when the door is opened will flow downward and create a draft on one's feet. Such temperature variations (and the corresponding differences in specific weight) are significant not only in air but also in water. Important circulation patterns are established in the ocean and in lakes, as a result of differences in specific weight. These differences are caused by heat exchange through radiation, conduction, and convection.

If an air mass is moved from a position near the ground to a higher position, the reduced surrounding pressure at the higher elevation results in an expansion, which in turn causes a cooling of the air, like the cooling of a tank as a compressed gas is released from it quickly. Conversely, if an air mass is brought to a lower elevation of greater pressure, it is compressed and the temperature increases, like that in a tire pump or compressor. Under these conditions, the air is expanding and contracting as an adiabatic process, and there is no tendency toward buoyancy, despite the gradient of specific weight, because the air mass that is moved has the same specific weight as that of the surrounding air at its new location. This situation is known as a **neutral lapse rate**—the lapse rate being the temperature gradient.

If the temperature of the air at the lower level is increased without increasing the temperature above, an unstable condition develops, because a mass of warmer air moved upward will be warmer and, hence, lighter in weight than the surrounding air. This potential difference in specific weight supplies a buoyant force which causes the warmer air mass to continue to rise. A well-known example of this phenomenon is the creation of a free-convection thunderstorm, which is caused by radiation heating of the ground surface and the air near the ground, and the consequent development of an **unstable lapse rate.** Conversely, radiation cooling of the ground and air causes a **stable lapse** rate known as an **inversion**—in which case there is no instability or tendency to mix. In fact, mixing by turbulence is very much inhibited.

The temperature gradient in the atmosphere can be determined from the basic gas equations. Combining Eq. 1-2, Eq. 2-7, and Eq. 2-3 yields

$$dT = -\frac{1}{R}\left(\frac{n-1}{n}\right)dz \qquad (2\text{-}32)$$

which can be integrated to give

$$\frac{T}{T_0} = -\frac{n-1}{n}\frac{\Delta z}{RT_0} + 1 = -\frac{n-1}{n}\frac{\Delta z}{(p/\gamma)_0} + 1 \qquad (2\text{-}33)$$

in which $T = T_0$ where $z = z_0$, and $\Delta z = z - z_0$. Fig. 2-17 shows the variation of temperature with elevation for various values of the exponent n. Note that the actual or normal temperature variation in the atmosphere

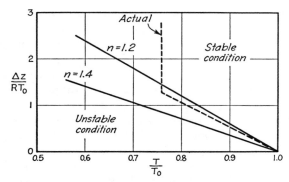

Fig. 2-17. Variation of temperature with elevation in the atmosphere.

lies between the curves for $n = 1.2$ and $n = 1.4$. For wet-adiabatic conditions, the actual curve is closer to the curve for $n = 1.2$, whereas for dry-adiabatic conditions it is the curve for $n = 1.4$.

Since the variation of temperature with altitude (up to about 35,000 ft) has been found to be a straight line, dT/dz is a constant. This is usually assumed to be -6.48 degrees centigrade per 1000 meters, or -3.56 degrees F per 1,000 ft. This value is used for the actual curve in Fig. 2-17. If the temperature gradient is smaller, the air is stable, and if it is greater, the air is unstable. Once the temperature gradient (lapse rate) becomes unstable, the fluid mass usually does not remain static. Instead, the lighter fluid moves upward and the heavier fluid moves downward, which sets up a mixing process known as free convection, and the instability is reduced, as a result of the decreased temperature gradient.

Example Problem 2-13: Referring to Fig. 2-17, determine the value of n which corresponds to the line representing the actual variation of $\Delta z/RT_0$ with T/T_0 when Δz is from sea level to 35,300 ft.

Solution:

From Fig. 2-17, when

$$T/T_0 = 0.85$$

$$\frac{\Delta z}{RT_0} = 0.79$$

which can be inserted into Eq. 2-33 to give

$$0.85 = - \left[\frac{(n-1)}{n} \right] 0.79 + 1$$

from which

$$n = 1.235$$

That is, assuming a linear decrease of temperature with altitude of 3.56°F per 1000 ft, the value of $n = 1.235$ applies throughout the troposphere, elevation $\leq$ 35,300 ft, under standard conditions.

• STATIC FORCES IN PIPES AND CONTAINERS

Static forces are created in pipes and other containers because of the relative internal and external pressures. These pressures can be created by either liquids or gases, or both. Furthermore, the internal or bursting pressure can be more than or less than the external or collapsing pressure.

A pipe line can be subjected to several different static pressures. On the one hand, see Fig. 2-18(a), the internal or bursting pressure in the pipe line can (and generally does) exceed the external collapsing pressure, so that the design of the pipe is usually controlled by the internal pressure—especially in the case of high-pressure lines. On the other hand, see Fig. 2-18(b), an empty pipe or a gas line crossing a river will be subjected to collapsing and also floatation pressure. Likewise, when a pipe line crosses a high point, the internal pressure can become that of the vapor pressure of the liquid flowing and thereby approach zero pressure (vacuum), with atmospheric pressure acting as the collapsing pressure.

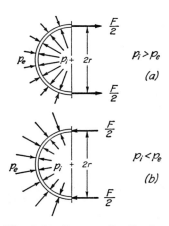

Fig. 2-18. Pressure distribution in pipes.

Most pipes or containers which are designed to withstand large pressures are circular in shape, and the pressures and forces involved are illustrated in the sectional free-body diagrams of Fig. 2-18. In Fig. 2-18(a), the internal pressure p_i exceeds the external pressure p_e, so that the pipe wall is subjected to a tensile stress or a hoop tension.

A general equation expressing the total force F and the unit stress s in the pipe wall can be determined by a summation of forces in the horizontal direction, so that

$$F = 2rL(p_i - p_e) \qquad (2\text{-}34)$$

$$s = \frac{F}{2Lt} = \frac{r(p_i - p_e)}{t} \qquad (2\text{-}35)$$

in which L is the length of pipe;

t is the wall thickness.

When considering the usual pipe line, where p_i is gage pressure and p_e is atmospheric pressure, Eq. 2-35 becomes

$$s = \frac{rp_i}{t} \qquad (2\text{-}36)$$

If the external pressure is greater than the internal pressure, Fig. 2-18(b), the stress s and force F become negative, thereby demonstrating that the pipe wall is subjected to compressive stress instead of tension.

The wall thickness of low-pressure pipe lines and containers is usually determined by factors other than pressure—such as the thickness needed to allow for corrosion and the strength necessary to handle the pipe, or to support external concentrated loads or soil overburden.

Example Problem 2-14: A cast iron pipe 6 in. in diameter is subjected to a pressure head of 200 ft of water. If the allowable stress in the cast iron is 1000 psi, calculate the required thickness of pipe wall.

Solution:

Since

$$p_i > p_e$$

the wall thickness is

$$t = \frac{rp_i}{s} = \frac{(3)(200)(0.433)}{1000} = \underline{\underline{0.25 \text{ in.}}}$$

• INFLUENCE OF ACCELERATION

In Chapter I, it has already been shown that the specific weight of a fluid is due to the mass density and the accelerative force of gravity, which is the result of a mass attraction between the fluid and the earth. When studied from this viewpoint, the influence of other steady or unchanging accelerations can be evaluated in a like manner. See Chapter 3 for a definition and discussion of gravity and other terms such as acceleration.

The specific weight γ is defined as

$$\gamma = \rho g \qquad (2\text{-}37)$$

which shows in equation form that the specific weight is the result of the influence of gravity g upon the density or mass per unit volume ρ. Since g is. a result of the mass attraction of the earth, the acceleration caused by this attraction is vertically downward, and the specific weight γ also acts vertically downward. When an acceleration, in addition to gravity, is also acting on the mass per unit volume ρ, however, the vertical component of the effective specific weight γ_e is either increased if the additional acceleration is acting upward, or decreased if the additional acceleration is acting downward. The specific weight γ is a result of the mass attraction of the earth, and acts in the same direction as the gravitational acceleration. The component of the effective specific weight which is caused by an acceleration other than gravitational is the force resisting the acceleration and, therefore, is acting in the direction opposite to the acceleration superposed on the gravitational acceleration field. The equation for the effective specific weight becomes

$$\gamma_e = \rho(g \rightarrow a) = \rho a_e \tag{2-38}$$

in which γ_e is the effective specific weight acting in the direction of the resultant accelerative field F/L^3;

ρ is the density FT^2/L^4;

g is the acceleration of gravity and assumed as plus L/T^2;

a_e is the effective acceleration L/T^2;

a is the acceleration superposed on the gravitational acceleration field with a component a_y in the vertical direction, and a component a_x in the horizontal direction;

a_y is negative when in the upward direction and positive when in the downward direction;

$\rightarrow$ represents the vector difference.

A simple example of this additional acceleration is found from the study of a container of fluid, say a bucket of water, in an elevator. When the elevator is not accelerating relative to the surface of the earth (that is, when it is either stationary or moving at an unchanging or steady speed), then a is zero and $\gamma_e = \gamma$. However, if the elevator is accelerating upward so that $a = a_y$, the effective specific weight γ_e is greater than the regular specific weight. That is, $\gamma_e > \gamma$, since a is negative, making $a_e > g$. Likewise, if the additional acceleration is downward, the effective specific weight will be less—that is, $\gamma_e < \gamma$. Observation reveals that, if $a = g$ in the downward direction, then $(g - a) = 0$ and $\gamma_e = 0$. Furthermore, if $a > g$ in the downward direction, then $(g - a)$ becomes minus and γ_e is negative, or is upward instead of downward.

Accelerations which have a component in the horizontal direction also may be inserted into Eq. 2-38—provided a is subtracted vectorially from g. The resultant effective weight acts along the same line of action but in the direction opposite to that of the resultant acceleration.

An example of combining a horizontal acceleration with that of gravity is the tank of liquid in Fig. 2-19, where γ_a is the effective specific weight created horizontally by the horizontal acceleration. Since the direction of

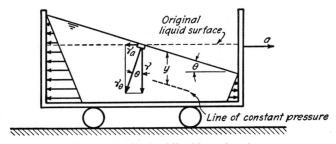

Fig. 2-19. Tank of liquid accelerating.

increasing pressure within the liquid is downward and to the left at an angle θ with the vertical, the liquid surface must adjust to a slope of θ with the horizontal, because the pressure is the same over all parts of the liquid surface. The pressure at any point within the liquid can now be determined by the product $\gamma_e h_e$ in which h_e is measured perpendicularly to the liquid surface. The pressure can also be determined by γh, in which h is the vertical distance to the liquid surface.

Example Problem 2-15: An open tank car, see Fig. 2-19, is 40 ft long, has sides 8 ft high, and is 8 ft wide. If oil weighing 50 lbs/cu ft is 6 ft deep in the car when it is not moving. (a) What is the maximum acceleration that the car can undergo without spilling oil out of the car? (b) What is the force on the back end of the car in (a)?

Solution:

(a) The oil surface goes up two feet at the back and down two feet at the front so that, in the length of the car, the oil surface has a slope of $4/40 = 0.1$.

$$\tan \theta = \frac{\gamma_a}{\gamma} = \frac{\rho a}{\rho g} = \frac{a}{g}$$

$$0.1 = \frac{a}{g}$$

$$a = 0.1g = \underline{3.22 \text{ fps}^2} \rightarrow$$

(b) $$F = \gamma A \bar{h} = 50 \frac{8}{2} (8 \times 8) = 200(64) = \underline{\underline{12,800 \text{ lb}}} \leftarrow$$

Another example involving steady horizontal acceleration is the **rotational vortex** such as a rotating tank of liquid, see Fig. 2-20. From

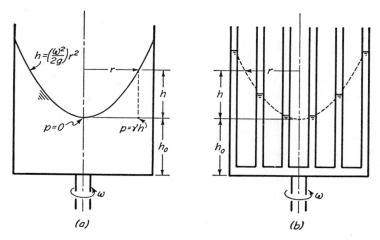

Fig. 2-20. Rotational vortex.

mechanics of solids, at any distance r from the center of rotation there is a normal acceleration a_n acting inward. This normal acceleration is

$$a_n = r\omega^2 \tag{2-39}$$

in which ω is the angular velocity in radians per second. The gravitational acceleration g, together with the normal acceleration, causes the liquid surface to have a slope of a_n/g which in differential form can be set equal to

$$\frac{dh}{dr} = \frac{a_n}{g} \tag{2-40}$$

or

$$dh = \frac{\omega^2}{g} r \, dr$$

and integrated to yield

$$h = \frac{\omega^2}{2g} r^2 \tag{2-41}$$

which is the equation of a parabola with the vertex at the bottom of the concave surface. The distance $h + h_0$ is used to determine the pressure on the bottom of the tank or in the closed container at the point r from the center of rotation.

It is interesting to note that, if the tangential velocity $V = r\omega$ is substituted into Eq. 2-41, the height h becomes

$$h = \frac{V^2}{2g}$$

which should be compared later with the Bernoulli equation, as developed in Chapter 3.

Example Problem 2-16: A circular bucket is 2 ft deep, 2 ft in diameter, and open at the top. When not rotating, the bucket is $\frac{2}{3}$ full of water. If the bucket is rotated about an axis through the center of the bucket, (a) What is the maximum rpm at which the bucket can be rotated without spilling any water out of the bucket, (b) What is the equivalent specific weight and what is its line of action at a point 0.5 ft from the center of the bucket?

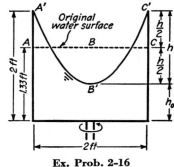

Ex. Prob. 2-16

Solution:

(a) The parabolic curve $A'B'C'$ shown in the sketch is a paraboloid of revolution, and the original water surface ABC is a straight line. When the bucket is rotated, the volume of water above the line ABC is equal to the volume between this straight line and the part of the parabola which is below ABC. The property of a paraboloid of revolution is such that the height h is bisected by the original water surface. Since

$$h = \tfrac{2}{3} \times 2 = 1.33 \text{ ft}$$

$$h = \frac{\omega^2 r^2}{2g}$$

$$1.33 = \frac{\omega^2 (1)^2}{64.4}$$

$$\omega = \sqrt{(1.33)64.4} = 9.25 \text{ radians/sec}$$

$$n = \frac{\omega}{2\pi} = 1.475 \text{ rps} = \underline{\underline{88.5 \text{ rpm}}}$$

(b) At a point 0.5 ft from the center of the bucket, the normal acceleration is

$$a_n = r\omega^2$$

$$a_n = \tfrac{1}{2}(9.25)^2 = 43 \text{ fps}^2$$

Since a_n is horizontal and g is vertical, the equivalent acceleration affecting the equivalent specific weight is

$a_e = \sqrt{43^2 + 32.2^2} = 53.7$ fps² acting down and in toward the center at an angle of 36.9° with the horizontal

$\gamma_e = \rho a_e = (1.94)(53.7) = $ <u>104 lb per cu ft</u>   acting down and away from the center at an angle of 36.9° with the horizontal.

$p = \dfrac{\gamma b}{g}$

• SUMMARY

1. At a point the **pressure** acts in all directions with a magnitude of $p = \gamma h$.

2. **Absolute pressure** is the pressure relative to a complete vacuum which is zero absolute pressure (see Fig 2-6).

3. **Gage pressure** is the pressure relative to the atmospheric or ambient pressure (see Fig 2-6).

4. **Atmospheric pressure variation** with elevation is

$$\frac{p}{p_0} = \left[1 - \frac{n-1}{n} \frac{\Delta z}{(p/\gamma)_0} \right]^{n/(n-1)} \tag{2-11}$$

for which n varies between 1.2 for wet adiabatic and 1.4 for dry adiabatic conditions (see Fig. 2-5).

5. **Pressure measurements** at a point are made:
   a. Relative to surrounding pressure with the **Bourdon gage** (see Fig. 2-7).
   b. Relative to absolute zero with a **barometer** (see Fig. 2-8).
   c. Relative to another point with a **manometer** (see Fig. 2-9).

6. Force on a **submerged plane surface** is

$$F = \gamma \bar{h} A \tag{2-16}$$

and it acts at the pressure center which is a distance along the plane

$$e = S_0 - \bar{S} = \frac{\bar{I}}{\bar{S}A} = \frac{k^2}{\bar{S}} \tag{2-22}$$

below the centroid of the plane surface. To **solve problems** follow the three steps given in the second paragraph following Eq. 2-22.

7. The resultant force on a **submerged curved surface** can be divided into horizontal and vertical components for convenience of calculation. The **method of computing** the resultant force is outlined in the third paragraph following Eq. 2-26.

8. The forces involved in the static equilibrium of a **submerged or floating body** are related as

$$F_s = W - (F_u - F_d) = \gamma_s \tilde{V} - F_b \tag{2-29}$$

9. A **body is stable** in a given position if it will return to this position after being rotated through a small angle. The **metacentric height** for a floating body at small angles of heel is

$$m = \frac{\bar{I}}{\tilde{V}} - GB \tag{2-30}$$

and the **restoring moment** is

$$T = Wm\alpha \tag{2-31}$$

10. **Atmospheric temperature variation** for a neutral lapse rate is

$$\frac{T}{T_0} = -\frac{n-1}{n} \frac{\Delta z}{RT_0} + 1 = -\frac{n-1}{n} \frac{\Delta z}{(p/\gamma)_0} + 1 \tag{2-33}$$

11. The **stress in a pipe wall** is

$$s = \frac{r(p_i - p_e)}{t} \tag{2-35}$$

12. **Accelerations in a fluid** system are combined with gravitational acceleration to determine the pseudo specific weight by

$$\gamma_e \doteq \rho(g \rightarrow a) = \rho a_e \tag{2-38}$$

References

1. King, Horace W. and Ernest F. Brater, *Handbook of Hydraulics*, 4th ed. New York: McGraw-Hill Book Co., Inc., 1954.

2. Powell, Ralph W., *Hydraulics and Fluid Mechanics*. New York: The Macmillan Co., 1951.

3. Rouse, Hunter, *Elementary Mechanics of Fluids*. New York: John Wiley & Sons, Inc., 1946.

4. Shoder, Ernest W. and Francis M. Dawson, *Hydraulics*. New York: McGraw-Hill Book Co., Inc., 1934.

5. Vennard, John K., *Elementary Fluid Mechanics*. 3rd ed. New York: John Wiley & Sons, Inc., 1954.

Problems

2-1. When the water is 400 ft deep in Lake Mead upstream of Hoover Dam, what is the maximum water pressure on the dam? Where is it located?

2-2. The standard atmospheric pressure is 14.7 psi. What is this pressure, expressed in (a) feet of water, (b) inches of mercury?

2-3. If a depth of fluid of 3.1 ft causes a pressure of 1 psi, what is the specific gravity of the fluid?

2-4. If the atmosphere had a constant temperature of 68°F and constant density, how high would the air extend above mean sea level? Would Mt. Everest, 29,141 ft in elevation, be above the air layer?

2-5. A certain sea-going fish does not survive very well if it gets in water where the absolute pressure is greater than 15 standard atmospheres. How deep can this fish go before it experiences trouble? Assume sea water has a specific gravity of 1.03.

2-6. Assume that the atmospheric pressure varies according to the dry adiabatic process with a value of $n = 1.4$. What is the barometric pressure on top of Mt. Everest, elevation 29,141 ft? Assume a mean sea level pressure of 14.7 psi.

2-7. Solve problem 2-6, assuming an n value of 1.2.

2-8. Find the absolute pressure in inches of mercury for the following elevations in a standard atmosphere: (a) 17,000 ft, (b) 38,000 ft, (c) 73,000 ft.

2-9. Find the equivalent pressure in psi for (a) 60 ft of water, (b) 72 in. of mercury, (c) 10 ft of oil (sp. gr. = 0.8).

2-10. A Bourdon gage reads a pressure of 82 psi. What are the equivalent readings in (a) inches of mercury, (b) feet of water, (c) feet of alcohol, and (d) pounds per square foot?

2-11. If a mercury barometer reads 25.7 in. of mercury and a Bourdon gage at point A indicates 175 ft of water, what is the absolute pressure at point A in psia?

2-12. A vacuum gage at point B reads 20 in. of mercury vacuum and an open manometer at point C reads 10 ft of benzene. Find $p_c - p_b$ in psi.

2-13. The absolute pressure at point D has been determined to be 47.0 psia. If a Bourdon gage at point D reads 33.7 psi, what is the barometric pressure at point D in inches of mercury?

2-14. If atmospheric pressure is 14.2 psia, which pressure is the largest, 15 in. of mercury vacuum, or 18 ft of water above absolute zero?

2-15. The pressure on a Bourdon gage at elevation 5,000 ft above sea level reads 75 psi. Convert this to the absolute pressure, assuming that the average air temperature is 68°F and the air conditions are static.

2-16. The Empire State Building in New York City is 1,248 ft high above the street. If the static pressure in a water line at the top of the building must be 25 psi, (**a**) what is the required pressure in psi in the water main in the basement of the building 30 ft below the street? (**b**) In terms of feet of mercury, what is the required pressure?

2-17. The pressure in the air space at the top of a closed oil tank is 10 psi below atmospheric. What is the pressure at a point 15 ft below the oil surface? Specific gravity of the oil is 0.83.

2-18. Find the pressure at point A in psi.

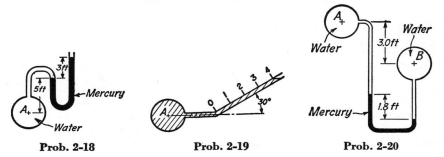

Prob. 2-18   Prob. 2-19   Prob. 2-20

2-19. Kerosene rises 4 ft in a draft gage, as shown. Find the pressure at point A.

2-20. Find the difference in pressure between points A and B.

2-21. To prevent the freezing of the gage in a river measuring station, 6 ft of kerosene is put in the tube, as shown in the figure. A float rides on top of the column of kerosene. Since the float is supposed to record the water surface elevation, what correction C must be made in the records to give the correct water surface?

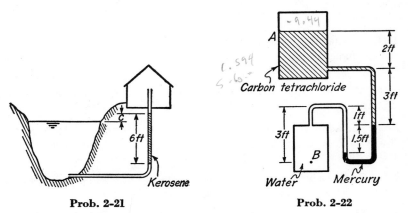

Prob. 2-21   Prob. 2-22

2-22. The pressure at point B is 3 psi. Find the pressure in psi in the closed air space at the top of tank A.

2-23. Two 12-in. diameter pipes are connected by a 3-in. pipe. The upper pipe contains oil, while the lower pipe contains water. Calculate the gage reading in psi on the gage shown on the upper pipe.

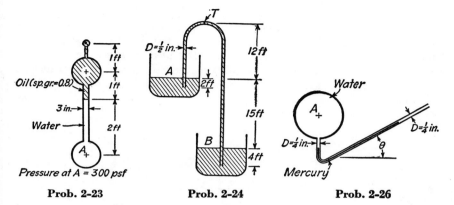

Prob. 2-23   Prob. 2-24   Prob. 2-26

2-24. Gasoline (sp. gr. 0.68) is being siphoned from container A to container B. (a) What is the vacuum pressure in psi at T, if the end of the siphon in A is closed, leaving B open? (b) If B is closed and A is open, what is the vacuum pressure at T in psi?

2-25. A draft gage is connected to a large container, as in Problem 2-19. The fluid in the container and gage is mercury. At what angle should the small tube be inclined, if a differential reading of 8 in. is required for each change of pressure of 1 psi?

2-26. If the pressure at point A is changed from 10 psi to 15 psi and the top of the mercury in the sloping tube moves 8 in. along the tube to the right, what is the angle θ?

2-27. Point A is 6 ft above point B. If the pressure gage at A reads 30 psi, what is the pressure at B in psi?

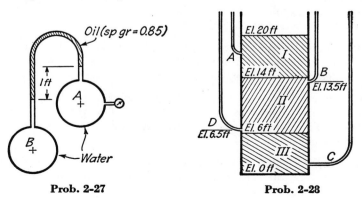

Prob. 2-27   Prob. 2-28

2-28. A container holds 3 liquids, as shown in the figure. Specific gravities are shown after the numbers, I — 0.75, II — 1.0, III — 1.50. What will be the elevation of the liquid surface in each piezometer tube? What is the difference between the liquid elevation of B and D?

2-29. An open cylinder 1 in. in diameter is $\frac{3}{4}$ full of oil whose specific gravity is 0.75. Connected at the very bottom of one side of the cylinder is a straight tube $\frac{1}{4}$ in. in diameter and open at the upper end. If a pressure equivalent to 6 in. of water is put in the open end of the small tube, the free oil surface moves down a distance of 20 in. along the small tube. Find the angle which the small tube makes with a horizontal plane.

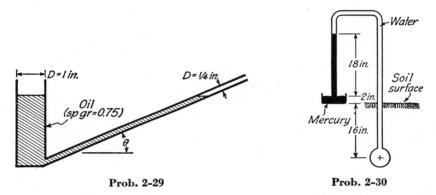

Prob. 2-29   Prob. 2-30

2-30. What is the pressure in the tensiometer cup when the column of mercury is 18 in. above the free mercury surface? The tensiometer is filled with water and the cup is 16 in. below the ground surface. The mercury container is 2 in. deep.

2-31. An open elevated water tank contains water which is 15 ft deep. If the bottom of the circular tank is a plane horizontal surface, what is the total force on the bottom, if the tank is 20 ft in diameter?

2-32. A submerged plane is 10 ft square, and the top of the plane is 6 ft below the water surface. If the plane slopes at an angle of 45° with the horizontal, what is the force on one side of the plane?

2-33. A vertical gate 5 ft high and 8 ft wide is hinged at the bottom. If the water behind the gate is 10 ft deep above the bottom of the gate, what is the force on the gate and what is the moment about the hinge tending to open the gate?

2-34. The container with a rigid bottom is completely full of oil, sp. gr. = 0.8. Find the force on the bottom of the container and also find the weight of the oil in the container. Explain why the force on the inside of the bottom is greater than the force on the outside.

2-35. Length of gate $OA = 10$ ft. Width of gate = 6 ft. Specific gravity of the oil = 0.80. Find the horizontal force F_H applied at A required to hold the gate in equilibrium, as shown.

2-36. The rectangular gate 8 ft by 4 ft weighs 7 tons. Assuming the center of gravity at the center of the gate, what is the maximum depth d that the water can attain without the gate starting to open?

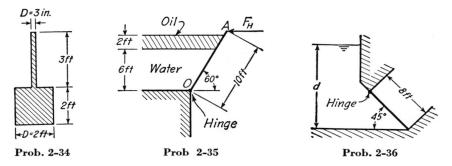

Prob. 2-34   Prob 2-35   Prob. 2-36

2-37. Solve Problem 2-36 if the gate weighs only 4 tons.

2-38. A dam consists of a vertical concrete wall 8 ft high. When the water in front of the wall is 6 ft deep, what is the maximum moment tending to overturn the dam?

2-39. If the centers of gravity of the given plane surfaces are submerged 6 ft below the water surface, find the force exerted on each area.

Prob. 2-39

2-40. A circular gate 3 ft in diameter is pivoted about a horizontal axis through its center. If it is installed in a vertical wall and the water is 10 ft deep above the bottom of the gate, what force must be applied at the bottom of the gate to keep the gate from opening?

2-41. A closed tank 10 ft high and 6 ft in diameter is filled to a depth of 8 ft with oil of specific gravity of 0.80. A 2-ft air space is above the oil. A manometer on the side of the tank is connected 2 ft above the bottom and the oil rises 3 ft in the tube. Find the forces acting on the top of the tank and on the bottom of the tank.

2-42. Assuming that the area of a diver's ear drum is $\frac{1}{10}$ sq. in., what is the force on the outside of the ear drum when a diver goes to a depth of 30 ft in ocean water, specific gravity 1.03?

2-43. A vertical cylinder 6 in. in diameter contains a metal piston also 6 in. in diameter which weighs 20 lb. The piston is supported 1 in. above the floor of the cylinder. A small tube $\frac{1}{4}$ in. in diameter is connected to the bottom of the cylinder under the piston. Water is poured into the small tube. If the cylinder and the tube are filled until the water in the tube is two feet above the bottom of the piston, will the piston start to rise in the cylinder? Neglect friction between the piston and the cylinder wall and assume that the clearance is so small that no water leaks past the piston. Explain the answer.

2-44. A 2-ft diameter circular gate is installed in a vertical wall, as shown. If the depth of water is as shown, what is the torque required on the pivot shaft *AB* to keep the gate from opening?

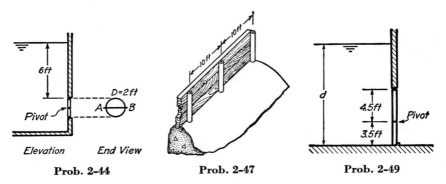

Prob. 2-44   Prob. 2-47   Prob. 2-49

2-45. A common drinking glass has a bottom diameter of 2 in. and a height of 5 in. What is the total force on the bottom of the glass when it is full of milk? Assume sp. gr. of milk = 0.99.

2-46. A closed rectangular tank is 4 ft square and 12 ft high. Water stands in the tank to a depth of 10 ft and the pressure in the air space at the top of the tank is 7 in. of mercury vacuum. The center of a rectangular gate two ft high by 3 ft wide is 2 ft above the bottom of the tank. Find the total force acting on the gate.

2-47. Three feet of flash boards were placed on the top of a small diversion dam, as shown. The pipes holding the boards were placed 10 ft apart across the dam. There were 20 pipes and each one was designed to fail under a bending moment of 6,000 ft-lb. How deep will the water be above the top of the flash board when failure occurs? Assume hydrostatic pressure distribution on the flash boards.

2-48. A vertical wall on a horizontal plane separates two bodies of water. The water is 4 ft deep on one side and 10 ft deep on the other side. What is the moment about the base of the wall?

2-49. The gate pivoted as shown is rectangular in shape. What is the maximum depth *d* the water can attain before the gate will begin to open? Atmospheric pressure is on the right side of the gate.

2-50. Water stands on both sides of a vertical wall which is 10 ft high. The wall acts as a coffer dam, with water 8 ft deep on one side and 4 ft deep on the other side. Find the magnitude, direction, and location of the resultant horizontal force acting on the wall.

2-51. The principle of the hydraulic jack is illustrated by this problem. A small 1-in. diameter cylinder is connected to a large 18-in. diameter cylinder by a small $\frac{1}{4}$ in. copper tube connected to the bottom of both cylinders. Oil of specific gravity 0.80 is in both cylinders and in the connecting tube. Neglecting the weight of the pistons, how much force would have to be applied perpendicular to the 1-in. piston to start raising a load of 2 tons on the large piston?

2-52. A concrete dam has a triangular section, with the base being 0.7 of the height of the dam. Neglecting uplift pressures, if the resultant force passes through the center of the base of the dam, how deep is the water if the dam is 100 ft high? Assume the concrete weighs 150 lb per cu ft. The upstream face of the dam is vertical.

2-53. The cross section of many concrete dams approximates a triangle, as shown in the figure. If the freeboard on this dam is 10 ft, where does the resultant force act on the base of the dam if the following assumption is made for uplift? In designing concrete dams on rock foundations, the following assumption is some-times made. It is assumed that the uplift pressure acts over part of the area of the base, because there is solid contact between many points of the foundation and the concrete, which does not permit the penetration of water to cause uplift pressure at those points. Assume, therefore, that the uplift pressure varies linearly from full hydrostatic head at the upstream toe to zero hydrostatic head at the down-stream toe, but that only 50% of the area is acted upon by the uplift pressures.

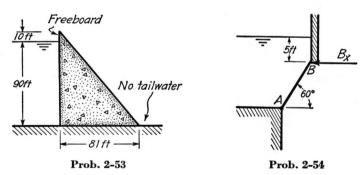

Prob. 2-53   Prob. 2-54

2-54. The length of the gate AB is 10 ft. What horizontal force B_X would the wall have to exert on the gate at B to hold the gate in equilibrium as shown, if the gate weighs 1,000 lb and is 6 ft wide?

2-55. A closed tank is partially filled with water and has air above at a pressure of 288 psf. If the force acting against the vertical face of a 15-in. diameter pressure gate in the side of the tank is 600 lb, find the depth of water in the tank above the center of the gate.

2-56. Find the resultant force and the center of pressure for the submerged areas (a) and (b).

2-57. The gate AB is 10 ft long and 5 ft wide. What is the maximum depth d the water can attain before the gate AB opens, if the weight of the gate is neglected?

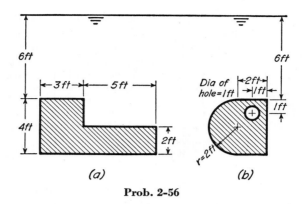

(a)   (b)

Prob. 2-56

2-58. Solve Problem 2-57 assuming that the gate weighs (**a**) 500 lb, (**b**) 1,500 lb.

2-59. Find the three gage readings, h_1, p_2, and h_3 if the pressure at A is 30 psi.

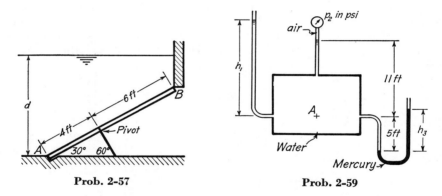

Prob. 2-57   **Prob. 2-59**

2-60. A cylindrical tank is 2 ft in diameter, 6 ft long, and is closed at one end by a flat plate. The tank is lowered into water with its longitudinal axis vertical and the closed end upward, until the closed end is 10 ft below the water surface. Air is forced into the cylinder until the water surface inside the cylinder is only 1 ft above the open end of the cylinder. Find: (**a**) The external force on the flat closed end of the tank; (**b**) The internal force on the closed end; (**c**) What force is required to hold the cylinder submerged if it weighs 500 lb?

2-61. A concrete dam ($\gamma = 150$ lb per cu ft) is triangular in cross section and 100 ft high from the horizontal base. If the water reaches a depth of 90 ft behind the dam, what is the minimum length of the base of the dam such that the resultant force falls within the middle third of the base? Neglect uplift. In order to avoid tensile stresses developing in concrete dams, the resultant force should fall in the middle third of the base of the dam. What is the minimum coefficient of friction required to prevent sliding? What is the stress in the concrete at the toe of the dam?

2-62. Repeat Problem 2-61, assuming that there is an uplift pressure which varies from the full head of 90 ft at the heel to zero at the toe of the dam. What is the minimum coefficient of friction required to prevent sliding?

2-63. Elephant Butte Dam on the Rio Grande in New Mexico was the largest concrete dam of its kind when it was built in 1916. The cross section of this dam approximates a triangle, with a height of 280 ft and upstream face with a slope of 1H:16V and a downstream slope of 2H:3V for the first 200 ft below the top, and 1H:1V for the last 80 ft down to the base. The maximum flood backs up water to a depth of about 270 ft, and the tailwater depth is 80. The masonry in this dam weighs 137 lb per cu ft. Assume the uplift to vary linearly from full hydrostatic pressure at the heel to full hydrostatic pressure at the toe, but assume that these uplift pressures only act over $\frac{2}{3}$ of the area of the base. Using a coefficient of friction of 0.75, determine (**a**) the factor of safety against sliding, (**b**) the point where the resultant force intersects the base of the dam.

2-64. A gravity dam has the cross-sectional dimensions shown in the figure. Assuming that concrete weighs 150 lb per cu ft and the coefficient of friction between the concrete and the base is 0.75, analyze a unit width of the dam and work the following parts to the problem. Determine the following: (1) The resultant reaction

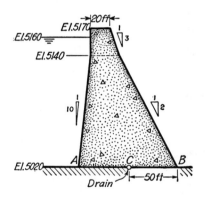

Prob. 2-64

on the base of the dam. (2) The location of the resultant. Check to see if it falls in the middle third of the base of the dam. (3) The factor of safety against sliding. (4) The factor of safety against overturning. Give these answers for each of the following conditions:

(**a**) When there is no seepage under the dam and the uplift is zero.

(**b**) When seepage does exist and the uplift varies linearly from a full hydrostatic head at the heel A to zero at the toe B.

(**c**) When a horizontal ice load of 5,000 lb per foot acts at elevation 5,160 ft in addition to the uplift in Part (b).

(**d**) Repeat Part (c), assuming that the uplift pressure varies linearly from full hydrostatic head at A to zero at the drain at C.

Briefly discuss your findings for each part in connection with the probable safety of the dam.

2-65. A rectangular tank whose bottom is 3 ft square is filled with 2 liquids. In the bottom of the tank is 3 ft of carbon tetrachloride, and on top of that is 2 ft of oil with a specific gravity of 0.80. Find the resultant force on a side of the tank. Since the resultant force has both direction and location, the force is not defined unless these are also given.

2-66. A circular butterfly gate 8 ft in diameter is pivoted about a horizontal axis passing through its center. What force F applied at the bottom is required to hold the gate in position, if water stands 6 ft above the gate on one side and the other side is completely exposed to the air?

2-67. The gate in the figure separates two fluids whose specific gravities are 1.20 on the left and 1.80 on the right. What is the vertical force required to open the 6 ft wide gate?

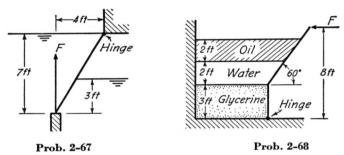

| Prob. 2-67 | Prob. 2-68 |

2-68. Oil, water, and glycerine are standing in a tank, as shown. The tank is 13 ft 5 in. wide. Find the horizontal force F required to hold the gate in equilibrium. The specific gravity of the oil is 0.8.

2-69. The bottom of a vertical cylindrical hot-water tank is a hemisphere. If the tank is 18 in. in diameter, what is the force on the bottom when the water stands 5 ft above the very bottom of the tank, which is open to the atmosphere at the top?

2-70. Repeat Problem 2-69 if the tank is closed at the top and the pressure 5 ft above the bottom is 20 psi.

2-71. The cylindrical gate shown in the figure is 10 ft long.

(a) Find the force on the gate.

(b) What is the minimum weight of the gate required to keep the gate from starting to open? Assume the center of gravity of the gate passes through the center of the cylinder.

2-72. A water tank is filled to the top of the 4 in. pipe. The ends of the tank are half cylinders. Find the total force acting on a curved end of the tank. Give its direction and location.

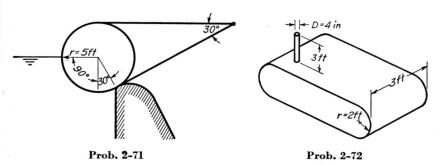

Prob. 2-71   **Prob. 2-72**

2-73. The end of the tank AB is a quadrant of a cylinder and is 8 ft long. Find the force acting on the end of the tank AB.

2-74. A radial gate is used as a control at the crest of a small spillway. If the width of the gate is 10 ft, find the force on the gate. Give its magnitude and direction.

2-75. The equation for the face of a dam is $y = 3x^2$. At A, $y = 3$ and at B, $y = 48$. When the reservoir is full, find the magnitude and direction of the force on the face of the dam per foot width.

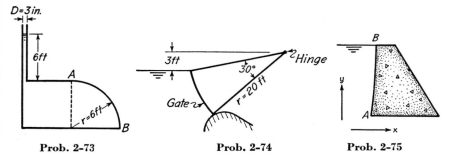

Prob. 2-73   **Prob. 2-74**   **Prob. 2-75**

2-76. A right circular cone made of solid glass (sp. gr. = 2.65) has a height of 18 in. and a base diameter of 18 in. If the cone is suspended on a cord attached to the apex of the cone, what is the tension in the cord when the cone is submerged in carbon tetrachloride (sp. gr. = 1.59)? What is the force acting on the curved surface of the cone, if the base is horizontal and 5 ft below the surface of the carbon tetrachloride?

2-77. A concrete dam similar to the dam in Problem 2-75 has a parabolic face whose equation is $x^2 = y$. When the depth of water upstream is 9 ft, the horizontal distance from A to B is 3 ft. For a unit width of dam, find the magnitude, direction and location of the force due to the water pressure on the front face of the dam.

2-78. What is the submerged weight in water of a cube of metal (sp. gr. = 5.5) which is 3 in. on a side? Give the percent reduction in weight when compared to the weight of the cube in air.

2-79. If a metal sphere 2 ft in diameter weighs 2,500 lb in the air, how much would it weigh (a) submerged in water, (b) submerged in mercury?

2-80. A rectangular barge has the following dimensions: 40 ft long, 15 ft wide, and 7 ft deep. What is the draft in fresh water if it weighs 20 tons empty? What load can it safely carry if it needs an 18-in. freeboard for protection against waves?

2-81. A spherical diving globe 10 ft in diameter is lowered 3 miles into the ocean. Assuming a specific gravity of sea water of 1.05, estimate the magnitude of the total force exerted on the diving globe.

2-82. To save himself in a flood, a small animal jumps on a piece of wood, which sinks uniformly a total of 1.3 in. into the water. The wood is 1 foot square and 2 in. thick. If the effective specific gravity of the wood is 0.52, how heavy is the animal?

2-83. A hollow plastic vessel in the shape of a paraboloid of revolution floats in water at 68°F with its axis vertical and the vertex downward. Neglecting the weight of the vessel, find the height to which it must be filled by a liquid of specific gravity 1.2 for its vertex to be 18 in. below the free surface of the water.

2-84. A circular lid on top of a closed tank is 3 ft in diameter and weighs 250 lb. An open manometer tube is tapped into the side of the tank, and the water rises in the tube to a height of 26 ft, as shown. If a large block of concrete is suspended from the lid to hold it in place, what is the required volume of the block if the specific weight of concrete is 150 lb per cu ft?

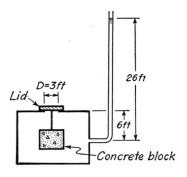

Prob. 2-84

2-85. A barge is 50 ft long, 20 ft wide, and 4 ft deep, outside dimensions. The sides and the bottom of the barge are made of timber with an average thickness of 12 in. The timber weighs 50 lb per cu ft. If the draft of the barge is not to exceed 3 ft, how many cubic yards of sand weighing 100 lb per cu ft can be loaded uniformly into the barge?

2-86. If the specific weight of brass is 535 lb per cu ft, (a) what is the weight of a 1-ft diameter sphere of solid brass submerged in salt water which weighs 64 lb per cu ft? (b) what is its weight submerged in a fluid with a specific gravity of 2.5?

2-87. If the same brass sphere as in Problem 2-86 were submerged in mercury, what force would be required to hold the sphere submerged? If the sphere were released, at what position would it float in the mercury?

2-88. A spherical balloon weighs 700 lb. How many pounds of helium would have to be put in the balloon to cause it to rise. (**a**) at sea level, (**b**) at an elevation of 15,000 ft? Assume standard atmospheric conditions for the air and atmospheric pressure in the balloon.

2-89. A concrete anchor is tied to the end of a long timber which measures 8 in. by 8 in. by 20 ft. The anchor does not touch the bottom, but it submerges 15 ft of the timber, leaving 5 ft extending into the air. The concrete anchor weighs 150 lb per cu ft and its weight in air is 150 lb. What is the specific weight of the timber?

2-90. A hollow metal cylinder 10 ft in diameter and 4 ft long closed at both ends weighs one ton. What volume of concrete hanging from the center of the bottom of the cylindrical tank will cause it to submerge in sea water (sp. gr. = 1.03) so that only 1.5 ft of the cylinder extends above the water surface?

2-91. A lead disk 3 in. in diameter and 1 in. thick will float in a bowl of mercury. How deep will the disk be submerged in the mercury? What size steel cube which weighs 490 lb per cu ft would have to be placed on top of the disk to cause the bottom of the disk to be submerged 1 in. in the mercury?

2-92. A raft is 10 ft square and when a man weighing 220 lb stands in the center of the raft, the water level is 1 in. below the top of the raft. How far can the man move toward the center of one side of the raft before the water surface is level with the top of that edge of the raft?

2-93. The specific gravity of rock used as concrete aggregate is often desirable to know. If a rock weighed 627 grams in the air and 390 grams submerged in water, what would be the specific gravity of the rock?

2-94. Derive the equation for the metacenter height $m = \dfrac{\overline{I}}{\overline{V}} - GB$.

2-95. Prove the equation $a = \frac{2}{3}b$ for floating bodies.

2-96. Prove Eq. 2-31.

2-97. A block made of wood is 6 in. square and 12 in. long. What is the specific gravity of the wood if the metacenter is at the same point as the center of gravity of the wood when the block is floating in water on its side? Would it be stable floating on its end? Explain.

2-98. A wood cylinder is 12 in. in diameter and 3 in. long. If the specific gravity of the wood is 0.6, prove that the cylinder will be more stable, when placed in water, if it floats with the circular cross section parallel to the water surface, than if the circular cross section were normal to the water surface.

2-99. A rectangular barge is 40 ft long, 15 ft wide, and 7 ft deep. The center of gravity is 4 ft from the bottom and the barge draws 5 ft of water at level keel. Find the metacenter height. Find the distance from the metacenter to the bottom of the barge for the following angles of heel: 5°, 10°, and 15°. Find the righting moment for each angle of heel.

2-100. A rectangular barge is 30 ft long, 12 ft wide, and 5 ft deep, and the center of gravity is located at the center of the barge 2 ft above the bottom. When the barge is tipped to one side through an angle of 5°, what is the righting moment if the original draft was 3 ft in fresh water?

2-101. Solve Problem 2-100 if the barge tips endwise through an angle of 5°.

2-102. What concentrated load placed at the middle of one side of the barge would cause water to flow into the barge of Problem 2-100?

2-103. In an air mass with a neutral lapse rate, how does the temperature vary with elevation? Make a plot showing the variation of temperature with elevation, if the air temperature at mean sea level is 0°F. At what elevation does the temperature become essentially constant?

2-104. Assuming a dry adiabatic process, plot temperature against elevation, if the mean sea level temperature is 40°F.

2-105. If the air mass is stable and $n = 1.2$, what is the temperature at an elevation of 20,000 ft, if the temperature at sea level is 60°F?

2-106. A 12-in. diameter pipe line is exposed to an atmospheric pressure of 28 in. of mercury, while the inside of the pipe contains a pressure of 13 in. of maximum allowable internal pressure in the pipe line?

2-107. A steel pipe with a 6-in. inside diameter has a wall thickness of $\frac{3}{8}$ in. If the stress in the steel is not to exceed a design value of 20,000 psi, what is the maximum allowable internal pressure in the pipe line?

2-108. A steamline is operated at a maximum pressure of 1,000 psi. If the pipe is 8 in. inside diameter and is made of steel which will fail at about 60,000 psi, what is the safety factor of the pipe if the wall thickness is $\frac{1}{4}$ in.?

2-109. Find the wall thickness of a steel pipe 18 in. in diameter designed to withstand a head of 1,000 ft of water. The working stress for steel should be 20,000 psi.

2-110. A 24-in. cast-iron main leads from a reservoir whose water surface elevation is 5,215 ft. In the heart of the city, the main is at elevation 4,635 ft. Assuming static conditions, what is the stress in the pipe wall if the wall thickness is $\frac{1}{2}$ in. and the external soil pressure on the pipe at this point is 75 psi. Is the resulting stress excessive for cast iron?

2-111. A car as shown in Fig. 2-19 is rectangular in cross section, 20 ft long, 8 ft wide, and 5 ft high. If the car is $\frac{1}{2}$ full of water, what is the maximum acceleration it can undergo without spilling any water. Neglecting the weight of the car, what force is required to produce the maximum acceleration?

2-112. A cylindrical bucket is accelerated upward with an acceleration of gravity. If the bucket is 2 ft in diameter and 4 ft deep, what is the force on the bottom of the bucket if it contains 3 ft of wet concrete whose specific weight is 140 lb per cu ft? What is the force on the bottom if the bucket is accelerated downward at 32.2 ft/sec²?

2-113. A rectangular car as in Fig. 2-19 is 10 ft long, 5 ft wide, and 5 ft deep. If friction is neglected and the car rolls down a plane which slopes at an angle of 20° with the horizontal, what is the slope of the water surface if the car contained 2 ft of water when the car was horizontal?

2-114. If the car in Problem 2-113 accelerates down the plane at 20 ft/sec², what is the slope of the water surface?

2-115. A bucket like the one in Fig. 2-20 is rotated about the vertical axis until enough liquid spills out to make $h_0 = 0$. At what speed in revolutions per second does this occur if the depth of the bucket equals the diameter of the bucket? If at the instant h_0 becomes 0 the bucket is decelerated to $h = 0$, how full would the bucket be?

2-116. If a bucket is 18 in. in diameter and 24 in. high, at what speed in rpm must the bucket be rotated, as in Fig. 2-20, for h_0 to be 6 in., if the bucket was originally $\frac{1}{2}$ full of water? Find the water surface slope at $r = 6$ in. and $r = 9$ in. What are the maximum and minimum pressures on the bottom of the bucket, and at what points do they occur?

2-117. For Problem 2-116, what is the equivalent specific weight of the fluid and what is its line of action at $r = 6$ in. and $r = 9$ in.?

2-118. A series of open tubes as in Fig. 2-20(b) are mounted as shown. If the tubes are $\frac{1}{3}$ full of water before motion starts, what is the elevation of the water in each tube when the axis of rotation is through the third tube from the left end and the water surface in that tube is at the bottom of the tube, where it connects to the horizontal tube? What speed in rpm is associated with this condition? The tubes are 3 ft long and spaced 3 in. on centers.

2-119. A 1 in. pipe is 3 ft long and is filled with 2 ft of water. If the pipe is capped at both ends and rotated in a horizontal plane about a vertical axis through one end of the pipe, what is the pressure at the other end of the pipe when it is rotating at 200 rpm?

2-120. A hollow sphere 6 in. inside diameter is rotated on a cable at 100 rpm. If the cable is 6 ft long and makes an angle of 30° with the horizontal, what is the pressure in the center of the sphere if it is full of water? What is the equivalent specific weight at that point?

chapter 3

Fluid

Dynamics

Fluid dynamics is a study of fluids in motion, the parts of which move at different velocities and are subjected to various and changing accelerations—both positive and negative. These accelerations occur both in the direction of motion and in the directions normal to the direction of motion. Consequently, the shape of the flow pattern can change with respect to both time and distance. Although static fluids are subjected to accelerations, the accelerations do not change with respect to time or distance along the path of motion. Furthermore, the form or shape of static fluids does not change with respect to time.

Fluid statics is concerned only with the fluid property of weight (which is the effect of gravitational and other accelerations upon the mass of the fluid), whereas fluid dynamics is intimately related to the mass or density of the fluid. From this fluid property arises inertia, which resists acceleration. Therefore, fluid dynamics involves forces of action and reaction—that is, forces which cause acceleration and forces which resist acceleration.

Although an element of fluid must obey the same fundamental laws as a solid, a fluid element can and does change shape readily while in motion—thereby making the analysis of fluid motion more complicated than the analysis of the motion of solids. A large percentage of the problems encountered in fluid motion are solved by the application of three basic equations, used either singly or in combination. These consist of the following:

1. Continuity equation;
2. Momentum equation;
3. Energy equation.

As water in a river flows through a narrowed section such as a bridge, the **continuity equation** shows that the velocity of flow must be increased, so that all of the water can get through the smaller passage. The very act of breathing causes the air to be speeded up and slowed down as it encounters smaller and larger passages.

The **momentum equation** helps to explain the thrust which must be applied in holding a fire hose, or in turning a lawn sprinkler. Furthermore, it explains why blowing will remove dust and debris from a surface such as a table top. Likewise, the momentum equation explains the thrust which a sail boat receives from the wind and the thrust which airplanes and rockets receive from their propellers and jet engines.

The **energy equation** can be used to explain such phenomena as the air entering and leaving the lungs, and the blood circulating through the body with the heart acting as a pump. It explains the use of a pump or turbine to add or subtract mechanical energy to or from a flow system, as well as the Pitot or stagnation tube, which is used to measure the wind speed of aircraft and other fluid velocities.

Both the momentum equation and the energy equation are derived from Newton's second law of motion, $F = Ma$, see Chapter 1. Newton's three laws of motion can be briefly summarized as follows:

1. A body at rest will remain at rest or a body in uniform motion will remain in uniform motion unless acted upon by an external force.

2. A resultant force acting upon a body will cause acceleration in the direction of the force. The acceleration is directly proportional to the resultant force and inversely proportional to the mass of the body. In equation form, this is

$$a = \frac{F}{M} \tag{3-1}$$

3. If a force acting on a body tends to change the state of rest or motion of the body, the body will offer a resistance equal and opposite to the force; or more briefly, for every action there is an equal and opposite reaction.

Definitions

In order to develop the three basic equations, certain terms must first be defined to avoid misunderstandings in the derivation and significance of the equations.

• STREAMLINES AND STREAMTUBES

*A **streamline** is an imaginary line within the flow for which the tangent at any point is the time average of the direction of motion at that point,* see Fig. 3-1(a). In other words, even though there may be fluctuations of direc-

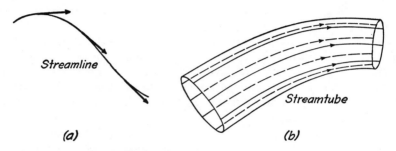

Streamline

Streamtube

(a)   *(b)*

Fig. 3-1. Streamline and streamtube.

tion within the flow, the average direction of velocity at a point is represented by the tangent at that point. Furthermore, if such a definition is true, it necessarily follows that, on the average, there cannot be any **net** movement of fluid across the streamline in any direction.

*The **streamtube*** Fig. 3-1(b), *is an element of fluid bounded by a special group of streamlines which enclose or confine the flow.* Since there can be no net movement of fluid across a streamline, it follows that there can be no net movement of fluid into or out of the streamtube, except at the ends. This fact will be used again in the development of the continuity equation.

• ACCELERATION

Acceleration is the rate of change of velocity with respect to time. Acceleration takes place when the velocity is changed, either in magnitude or direction. If the magnitude changes, then there is acceleration in the direction of motion s and tangential acceleration occurs so that $a_s \neq 0$; and when the direction changes, there is acceleration normal to the direction of motion (which is designated as direction n) and normal acceleration takes place so that $a_n \neq 0$. Mathematically, these accelerations can be expressed as

$$a_s = \frac{dv_s}{dt} \tag{3-2}$$

$$a_n = \frac{dv_n}{dt} \tag{3-3}$$

The total acceleration is the vector sum of a_s and a_n, see Fig. 3-2,

$$a = a_s \nrightarrow a_n = \frac{dv_s}{dt} \nrightarrow \frac{dv_n}{dt} \tag{3-4}$$

and the magnitude of the total acceleration is $\sqrt{a_s^2 + a_n^2}$. The dimensions of acceleration are length per unit of time per unit of time L/T^2, and the units are usually feet per second per second fps², meters per second per second mps², or, in relative terms, the number of gravitational accelerations a/g, known as "g"s.

Equations 3-2, 3-3, and 3-4 express the acceleration in terms of the total derivative for both tangential and normal accelerations. A further breakdown with respect to accelerations is represented by the following partial differential equations for ***tangential*** *acceleration*

$$a_s = \frac{dv_s}{dt} = \frac{1}{2}\frac{\partial(v^2)}{\partial s} + \frac{\partial v_s}{\partial t} \tag{3-5}$$

and for ***normal*** *acceleration*

$$a_n = \frac{dv_n}{dt} = \frac{v^2}{r} + \frac{\partial v_n}{\partial t} \tag{3-6}$$

The first term on the right side of each equation expresses the change in velocity with respect to *distance only*, which is the **convective** *acceleration*. The last term in each equation expresses the variation of the velocity with respect to *time only*, which is the **temporal** *acceleration*.

The convective acceleration term for tangential acceleration in Eq. 3-5 is obtained by combining $v = ds/dt$ with Eq. 3-2 to eliminate time so that the acceleration is in terms of distance only

$$\frac{v\,dv}{ds} = \frac{1}{2}\frac{d(v^2)}{ds} \tag{3-7}$$

The convective acceleration term for normal acceleration in Eq. 3-6 is derived from an analysis of Fig. 3-3. When a fluid is flowing in a curvilinear path, it is subjected to an acceleration normal to the direction of flow, so that the **direction** of the velocity is changing, regardless of whether the magnitude is changing. This acceleration is called normal acceleration a_n, and is illustrated in Fig. 3-2. In Fig. 3-3, the change in direction during the

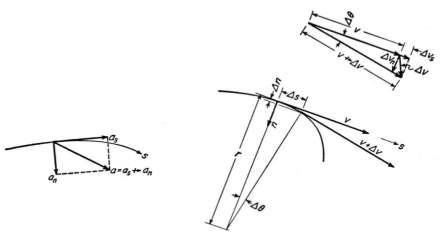

Fig 3-2. General acceleration diagram.   **Fig. 3-3.** Diagram for normal acceleration.

time Δt can be represented by $\Delta\theta$, and the change in the normal component of velocity is $v\Delta\theta$. Hence, as $\Delta\theta$ approaches zero, the convective normal acceleration for steady flow is

$$a_n = \frac{dv_n}{dt} = v\frac{d\theta}{dt} = v\omega = \frac{v^2}{r} \tag{3-8}$$

in which $\dfrac{d\theta}{dt} = \omega = $ angular velocity in radians per second.

Equation 3-5 and 3-6 show that the velocity can change either with respect to time or with respect to distance. If the velocity at a point changes with respect to time, temporal acceleration occurs, and if the velocity at a given time changes with respect to distance, convective acceleration occurs. It is possible for either one or both of these types of acceleration to exist either as tangential acceleration or as normal acceleration. In other words, changes in the magnitude of velocity a_s can occur with respect to time or distance or both, and changes in the direction of velocity a_n can also occur with respect to time or distance or both.

• STEADY FLOW

Steady flow exists if *the velocity at a point remains constant with respect to time.* Conversely, **unsteady flow** exists if *the velocity changes either in* *magnitude or in direction with respect to time.* Mathematically, this may be expressed for a given point of observation as

$$\frac{\partial v}{\partial t} = 0 \quad (for \ steady \ flow)$$

and

$$\frac{\partial v}{\partial t} \neq 0 \quad (for \ unsteady \ flow)$$

in which v is the velocity and t is the time. When $\frac{\partial v}{\partial t} \neq 0$, there is acceleration, as already discussed.

Steady flow and unsteady flow can be illustrated by the common lawn sprinkler. In the most common type, the velocity is continuous and unchanging with respect to time, so that **steady flow** exists. Some lawn sprinklers, however, have a whirling or rotating arm, so that, as one normally observes it in operation, the velocity is continuously changing in direction; hence, **unsteady flow** exists. Another very common example is the operation of a kitchen faucet. As the faucet is being opened or closed, the velocity is changing continuously in magnitude so the flow is unsteady. When the opening of the faucet remains constant, however, the flow is steady.

• RELATIVE MOTION

Unfortunately, engineering problems involving unsteady flow are usually quite difficult to solve mathematically. Therefore, various special techniques must be used to solve them, if indeed they can be solved at all. Most of these techniques are beyond the scope of this text and, therefore, will not be discussed very intensively. However, one method of simplifying the problem is to use the **principle of relative motion,** which simply involves continually changing the point of reference or observation, so that the flow appears steady instead of unsteady.

Consider again the rotating lawn sprinkler. As one observes it while standing on the ground nearby, the flow is unsteady. However, if the observer is above the sprinkler and rotating with it (a practice which is not recommended), the flow appears to be **steady** with the earth rotating beneath. In other words, the flow is steady relative to the observer. The **analysis** of this flow system is complicated, however, by Coriolis forces.

Another more comfortable example of changing unsteady flow to steady flow is that associated with a boat or ship, see Fig. 3-4, which passes an observer standing on a bridge or pier. If a point on the water surface is carefully observed, it is seen to spread outward and forward, form a loop back and inward, and return to approximately its original location as the boat passes, see Fig. 3-4(c). This type of motion is obviously unsteady. To change the flow to steady flow, the observer must ride in the boat and observe the flow past the boat at various points, see Fig. 3-4(b). At the prow of the boat, the water appears to be splitting continuously with no

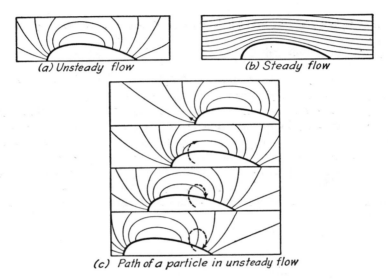

(a) Unsteady flow   (b) Steady flow

(c) Path of a particle in unsteady flow

Fig. 3-4. Steady flow and unsteady flow.

change in the flow pattern—hence steady flow. At the mid-section, the flow again is steady. Even at the stern, the eddies may be steady—the exceptions, where an alternate formation of eddies is sometimes observed, will be discussed later. Again the flow pattern has been changed from unsteady to steady by simply changing the positions of observation.

In short, when a steady motion is involved, unsteady flow can be changed to steady flow if the observer will ride along with the object about which or through which the flow is taking place. The unsteady flow pattern is illustrated by Fig. 3-4.

The principle of relative motion can be utilized quantitatively to change unsteady flow to steady flow by the following rule, provided it is applied from a single point of observation:

The **velocity of object A relative to object B** *is equal to the vector difference between the absolute velocities of these objects*, and hence is equal to the absolute velocity of A minus the absolute velocity of B, see Fig. 3-5.

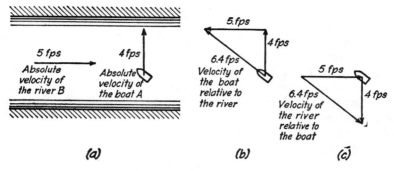

Fig. 3-5. Relative motion.

Example Problem 3-1: A boat is crossing a river, as shown in Fig. 3-5. If, relative to the earth, the river has a velocity of 5 fps, and the boat is to go directly across the river at a speed of 4 fps, then what must be the velocity of the boat A relative to the water B? What is the velocity of the water B relative to the boat A?

Solution:

Fig. 3-5(a) shows the absolute velocities of the boat and the river. Fig. 3-5(b) gives the velocity of the boat A relative to the water B. Fig. 3-5(c) gives the velocity of the river B relative to the boat A. Notice that in Fig. 3-5(c), the boat A corresponds to object B as given in the explanation, and the river B has become object A in the explanation.

• UNIFORM FLOW

If at a given instant the *velocity remains constant with respect to distance* along a streamline, the flow is said to be *uniform*. Expressed mathematically, this is

$$\frac{\partial v}{\partial s} = 0$$

Again, it must be remembered that, if there is a change either in magnitude or in direction along the streamline, then $\partial v/\partial s \neq 0$ and the flow is *non-uniform*. It follows, then, that in uniform flow the streamlines must be straight and the velocity must not change along the streamlines. Examples of uniform flow are the flow in a long straight pipe and the flow in an open channel which is long, straight, and of uniform depth. The flow is non-uniform at bends because the velocity changes direction. At expansions and at contractions, the flow is nonuniform because both the magnitude and direction of the velocity are changed.

• ONE-, TWO-, AND THREE-DIMENSIONAL FLOW

Frequently, the pattern of flow being studied in fluid mechanics involves velocities and perhaps accelerations in all three of the coordinate directions. In other words, the velocity is not restricted to one direction only, but instead the velocity has components in the x direction, the y direction, and the z direction. This type of flow is known as ***three-dimensional flow.***

If three-dimensional flow is to be expressed mathematically, it must include velocities, and perhaps accelerations, in all three coordinate directions, which results in equations that are usually too difficult to solve. Hence, various techniques are applied to simplify the expressions so that a mathematical solution can be obtained. These techniques generally are aimed at reducing the number of coordinate directions from three to two, or even to one. Frequently, the coordinate axes can be shifted or rotated so that the velocity and acceleration in one direction are zero—or nearly so. In this way, one coordinate direction can be eliminated, and ***two-dimensional flow*** is obtained. If the flow system can be arranged so that the velocity and acceleration are symmetrical about one axis, then cylindrical coordinates can be used, which may simplify the analysis, since this too is a form of two-dimensional flow.

If the streamlines are in parallel planes, two-dimensional flow can be expressed in terms of rectangular coordinates. In other words, the flow pattern is identical in each of the parallel planes. Water flowing over the spillway of a dam, wherein the sides of the spillway are parallel, is an example of two-dimensional flow, because all velocities and accelerations are in two directions only—both the velocity and acceleration in the third direction being zero.

If the flow system is restricted to accelerations in one direction only, then it is called ***one-dimensional flow.*** An example of this is flow in a pipe. In other flow systems, however, if the accelerations in the other two coordinate directions are small relative to those in the direction of motion, the assumption of one-dimensional flow is not greatly in error. In other words, the accelerations in the directions normal to the flow are assumed to be negligible, and the ***average*** values of the flow characteristics across any section of the flow are assumed to be representative.

This technique greatly simplifies the analysis and is known as the ***one-dimensional method of analysis.*** It has important limitations, however, because of the simplifying assumptions of negligible velocity and acceleration, except in the general direction of motion. Therefore, for some analyses, certain discrepancies will arise as a result of violation of these assumptions. Nevertheless, much of the science of *fluid mechanics is dependent upon the one-dimensional method of analysis for simplicity.* Otherwise, even approximate answers to many problems would be extremely difficult, if not impossible, to obtain mathematically.

• FLOW NET

In the case of two-dimensional flow, streamlines can be drawn to approximate the flow and give an indication of the flow pattern. However, to locate the streamlines with greater accuracy, and to insure their spacing so that the same discharge flows between any two adjacent streamlines, the flow net is often very useful. The flow net consists of a system of stream-

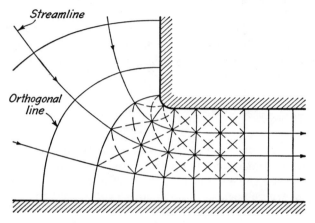

Fig. 3-6. Flow net.

lines and another system of lines which are orthogonal to the streamlines, as shown in Fig. 3-6. This system of intersecting lines describes irrotational flow, which is explained in a later section.

A flow net can be constructed for any two-dimensional boundary system, with the boundaries being used as the outside streamlines. The procedure is to sketch the streamlines approximately and then, beginning in the region of greatest curvature, draw orthogonal lines progressing outward toward the ends of the system. The orthogonal lines must be perpendicular to the streamlines (including the boundaries) at all points, and they should be spaced to make approximate squares throughout the entire flow net. Often, near sudden changes of the boundary shape, it is impossible to sketch good squares. If, in these regions, the quadrangles cannot be made into squares, they should be broken into parts by dividing lines. From the theoretical viewpoint, as the space between the streamlines and orthogonal lines goes to zero, all of the quadrangles will become squares. From the practical standpoint, however, this is very difficult to attain, and the drawing of the flow net always involves considerable erasing and redrawing. Once the flow net, or any part of it, appears to be a system of squares, a check can be made quickly by drawing diagonal lines through all squares in both directions. These diagonal lines should also form squares. Once the student has become accustomed to drawing flow nets, he can do it with surprising ease and speed. However, frequent erasing and reconstruction are necessary for even the most skillful.

Although a flow net can be drawn for any system of boundaries, the boundaries must represent actual streamlines, or the flow net will not represent the true flow pattern. In the case of separation of the flow from the boundary, the general line of separation (if it is known or can be estimated) can be used as a streamline, and the flow net will still define the flow pattern reasonably well.

For special cases, flow nets can be computed mathematically by applying the theory of conformal mapping. An adequate treatment of conformal mapping is given in most texts on complex variables, hydrodynamics, or advanced fluid mechanics.

• SEPARATION

As stated in the foregoing section, a flow net can be constructed for any boundary system, but it does not actually represent the flow pattern if the flow separates from the boundary. As seen in Fig. 3-7, separation occurs at both A and B, where pockets of fluid simply rotate and contribute nothing

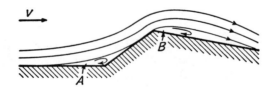

Fig. 3-7. Flow separation.

to the flow of the fluid in the general direction of motion. Separation zones develop between the main flow and the boundary where the boundary geometry would cause the streamlines to diverge or spread apart, if the pattern followed the boundary in the manner that the flow net would predict.

Basically, separation of the flow from the boundary is caused by the inertia of the fluid, as at point B, or by the drag of the boundary reducing the velocity, as at point A. At point B, the inertia of the fluid is so great it cannot travel around the sharp corner and along the diverging boundary. Furthermore, if the flow net were applicable at such a sharp corner, the acceleration would become infinitely great—which is a physical impossibility. The edge of a zone of flow separation is similar to the path of a vehicle which is traveling at considerable speed and attempting to turn a sharp corner. The zone of flow separation, however, is essentially constant in size for all velocities of flow around a sharp corner.

The study of separation zones yields important information about the technique of streamlining a boundary form. If the zones of separation are filled in so that the boundary follows the outside edge of the zone, then the separation can be eliminated, the flow pattern improved, and the drag reduced. Furthermore, as stated in the previous section, the flow net can be drawn with reasonable accuracy, if the boundary is assumed at the edge of the separation zone.

• IRROTATIONAL FLOW

Use of the flow net automatically includes the assumption of irrotational flow and, if the flow is rotational, the flow net is not applicable (except for certain conditions of flow such as through porous media). Therefore, it is important to distinguish between rotational and irrotational flow.

Basically, *irrotational flow exists if each fluid element in the flow system has no net rotation about its own mass center.* In other words, if any part of a continuous fluid element rotates in one direction, another part of the element rotates in the opposite direction, so that the *net* rotation is zero, and only distortion (rather than rotation) of the fluid element occurs.

Figure 3-8 shows rotational and irrotational flow. With rotational

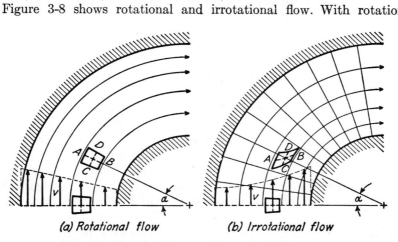

(a) Rotational flow   *(b) Irrotational flow*

Fig. 3-8. Rotational flow and irrotational flow.

flow, the velocity is greatest at the outside, whereas for irrotational flow, the velocity is the greatest on the inside. In Fig. 3-8(a), the fluid element moves through the angle α, and both of its coordinate axes rotate through the same angle. Hence, the flow is rotational. In Fig. 3-8(b), on the other hand, the element moves along a streamline through the angle α, and the longitudinal axis CD also rotates clockwise through this same angle. The other axis AB, however, rotates counterclockwise through an angle greater than α, to compensate for the rotation of the axis CD, and the **net rotation is zero.**

From the foregoing discussion, it is concluded that *for* **irrotational** *flow there is no net rotation* of a fluid element as it moves from point to point along a streamline, and hence *each fluid element has* **zero angular velocity** *about its own mass center.*

An *example of* **rotational motion** is the liquid in a rotating tank where the velocity varies directly with the distance from the center, see Fig. 2-20. Figure 3-8(a) may be considered as a section of this flow system. In this case, an object floating on the surface can be seen to rotate in the same manner as the tank itself.

An *example of* **irrotational motion** is the vortex or whirlpool which develops above a drain in the bottom of a stationary tank. In this case, the velocity varies inversely with the distance from the center, and a small object floating on the surface can be seen to move in a circular path but not to rotate relative to the tank. Figure 3-8(b) is a section of the irrotational vortex.

Within a separation zone, the flow is obviously rotational, but outside the separation zone, the flow frequently is essentially irrotational, in which case the flow net can be applied, provided the outside or bounding streamline of the flow net is taken as the outside edge of the separation zone.

Continuity Equation

The continuity principle stems from the law of conservation of mass, and is one of the three fundamental principles of fluid motion. It is used extensively in all problems involving a moving fluid. The continuity idea is simple and quite easily understood.

At this point, the development of the continuity principle is limited to steady flow, because its application to many problems of unsteady flow becomes too complex for elementary students of fluid mechanics. Applying this limitation of steady flow, however, does not decrease to any significant extent the usefulness of the continuity idea, since most practical problems either are, or can be reduced to, problems of steady flow.

In the definition of a streamtube, it was stated that there is no net flow across the boundaries of the tube, see Fig. 3-1. From the law on conservation of mass, then, which states that matter can be neither created nor destroyed, the mass of fluid entering one end of a streamtube during a given time interval must be equal to the mass leaving the other end of the streamtube during that same time interval. Since in steady flow no mass is stored or lost anywhere along the streamtube, the mass of fluid passing any cross section of the streamtube in a given period of time must be equal to the mass of fluid passing any other cross section of the same streamtube in the same period of time.

• DERIVATION OF CONTINUITY EQUATION

Expressed mathematically, the law of conservation of mass for the streamtube becomes

$$\frac{\Delta M_1}{\Delta T} = \frac{\Delta M_2}{\Delta T} = \frac{\Delta M_3}{\Delta T} = \text{constant} \qquad (3\text{-}9)$$

in which ΔM is the mass of fluid passing sections 1, 2, and 3 of the stream-tube in Fig. 3-9 during the time interval ΔT. As ΔT approaches zero, the equation takes the differential form

$$\frac{dM_1}{dt} = \frac{dM_2}{dt} = \frac{dM_3}{dt} = \text{constant} \qquad (3\text{-}10)$$

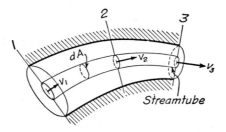

Fig. 3-9. Streamtube for continuity equation.

If Eq. 3-10 is integrated over the cross section of the flow, the result states that the total mass of fluid passing any cross section of the total flow in a given period of time is equal to the total mass passing any other cross section in the same period of time.

A brief analysis of the mass flow will indicate that the amount of matter passing a given section of the streamtube is a function of the cross-sectional area of the tube and the velocity of the flow (see Fig. 3-9). The volume of fluid passing a cross section dA of the streamtube, in a unit period of time dt, is

$$\frac{ds}{dt}\, dA = v\, dA = \text{volume per unit of time}$$

The symbol used for volume flow per unit of time is Q and, since the flow in the streamtube is incremental in magnitude, the discharge in the tube is dQ

The equation for volume flow in the streamtube then becomes

$$dQ = v \, dA \qquad (3\text{-}11)$$

in which v is the average velocity acting over the differential area dA. Integration of Eq. 3-11 across the entire cross section of the flow gives the following equation

$$\int dQ = Q = \int v \, dA \qquad (3\text{-}12)$$

• VOLUME, MASS, AND WEIGHT DISCHARGE

Since the velocity appears to the first power in Eq. 3-12 the integral can be evaluated in terms of the average velocity through the cross section, even though the velocity is not constant across the cross section, and Eq. 3-12 then becomes

$$Q = AV \qquad (3\text{-}13)$$

in which V is the average velocity through the cross section A. Since Q represents the volume of flow per unit time, multiplying by the mass density ρ gives ρQ, which represents the mass of flow per unit time, and the *continuity equation for mass flow* becomes

$$G = \rho_1 Q_1 = \rho_1 A_1 V_1 = \rho_2 Q_2 = \rho_2 A_2 V_2 = \text{Constant} \qquad (3\text{-}14)$$

in which ρ is mass per unit volume M/L^3 or FT^2/L^4;

                 Q is volume discharge or volume flux L^3/T.

The term ρAV applies either to steady flow or to unsteady flow at a particular cross section, and for compressible or incompressible fluids. Equation 3-14, however, is the general continuity equation equating the term ρAV from section to section along the flow. It applies to the *steady flow* of fluids which are either compressible or incompressible.

If Eq. 3-14 is multiplied by the gravitational acceleration g, it becomes the continuity equation of *weight flow*

$$W = \gamma_1 Q = \gamma_1 A_1 V_1 = \gamma_2 A_2 V_2 = \text{constant} \qquad (3\text{-}15)$$

in which W is weight per unit time $- F/T$;

                 γ is weight per unit volume $- F/L^3$.

For incompressible fluids where $\rho_1 = \rho_2 = \rho_3$ and $\gamma_1 = \gamma_2 = \gamma_3$, both Eq. 3-14 and Eq. 3-15 reduce to the general continuity equation for the *volume discharge* for steady incompressible flow.

$$Q = A_1 V_1 = A_2 V_2 = A_3 V_3 \qquad (3\text{-}16)$$

An important observation regarding Eq. 3-16 can be made at this time. Since the product AV must be a constant from one section to another in the flow system, it can be concluded that where the cross-sectional area is large the velocity is small. In other words, a decrease in area from section to section results in an increase in velocity, and an increase in area results in a decrease in velocity from section to section.

Example Problem 3-2: A pipe line carrying 1 cfs changes from a 6-in. diameter to a 9-in. diameter. What are the velocities in each pipe?

Solution:

$$Q = A_6 V_6 = A_9 V_9 \qquad A_6 = 0.196 \text{ sq ft} \quad A_9 = 0.441 \text{ sq ft}$$

$$V_6 = 1/0.196 = \underline{\underline{5.1 \text{ fps}}}$$

$$V_9 = 1/0.441 = \underline{\underline{2.27 \text{ fps}}}$$

Momentum Equation

The momentum equation is based upon Newton's second law of motion applied to the entire force system. It provides a simple means of relating the hydrostatic forces and the boundary forces to the change of momentum flux—that is, the momentum which passes a point during a unit time interval. The forces and resulting accelerations are analyzed from the overall viewpoint of the changes which occur, rather than the viewpoint of the details of the flow processes and internal mechanics. There are a number of problems in fluid mechanics which are solved rather simply by using the bulk-solution approach of the momentum equation, instead of attempting to analyze the details of pressure and velocity distribution. Examples of these problems are: force on a pipe elbow, solution of the hydraulic jump, forces on a propeller blade, the force of a jet of water impinging on a wall or on the ground, and jet propulsion. In this section, the momentum equation is first derived, followed by discussion and application of the equation.

• DERIVATION OF MOMENTUM EQUATION

From Newton's second law of motion

$$F = Ma$$

it is seen that a force is required to change the velocity of a body, either in magnitude or in direction. Since the force F and the acceleration a are vector quantities, they both have magnitude and direction. In mechanics of solids, the impulse-momentum relationship was developed on the basis of this same fundamental law of Newton's.

The impulse-momentum relationship of the mechanics of solids is developed for finite or discrete bodies, whereas in mechanics of fluids a continuous stream of flow is involved, in which it is assumed that the stream is infinite in extent, at least in the direction of motion. For example, on the one hand, if a block impinges against a fixed object and is stopped in a given period of time, there is involved only a single object and all the action to be analyzed is completed within the given time. On the other hand, if a jet of water is striking a fixed object, the action takes place continuously and is not completed within a given period of time. Therefore, in the mechanics of fluids, the Newton equation $F = Ma$ must be applied to an action which occurs continuously.

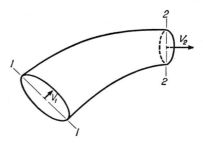

Fig. 3-10. Streamtube for momentum-flux equation.

Consider steady flow of the element of fluid in Fig. 3-10 which passes section 1 in the interval of time Δt. The fore part of the element passes section 1 at the beginning of the time interval and reaches section 2 at the end of the time interval Δt. This element will have a mass M equal to the product of the volume of the element and its density ρ. The volume in turn is the product of the discharge Q and the interval of time Δt. The mass involved in the time Δt can now be stated as

$$M = \rho Q \, \Delta t \tag{3-17}$$

Because the momentum equation considers only the bulk conditions or total force system at sections 1 and 2 and in between, the manner in which the velocity varies between these sections is not of concern—hence, the finite increment of time Δt can be used. Since the acceleration is the change in velocity ΔV_x during an interval of time Δt, the Newton equation $\Sigma F_x = Ma_x$ now becomes

$$\Sigma F_x = \rho Q \, \Delta t \, \frac{\Delta V_x}{\Delta t}$$

which then simplifies to

$$\Sigma F_x = \rho Q \, \Delta V_x = \rho Q (V_{x2} - V_{x1}) = \rho Q V_{x2} - \rho Q V_{x1} \tag{3-18}$$

This equation is the basic *force-momentum flux equation* used in mechanics of fluids. It relates the force system acting on the flow to the change in momentum-flux caused by this force. It corresponds to (but has important differences from) the Impulse-Momentum equation used in mechanics of solids. In brief, the *force-momentum flux* equation states that *the change in momentum flux* between sections 1 and 2 *is equal to the sum of the forces causing this change.*

It is important to understand the distinction between the momentum principle applied to the mechanics of fluids and the momentum principle applied to the mechanics of solids. Solids are discrete and finite, so that an action can take place and be completed in a finite period of time. Steady flow of a fluid, however, involves a motion which is continuous and of infinite extent in the direction of motion—hence, the action is continuous and is not completed in a finite period of time. Therefore, in momentum considerations of a fluid, a continuous force is involved (rather than an impulse, which is force over a finite period of time), and a continuous flux of momentum of a fluid stream is involved (rather than the momentum of a discrete and finite object).

Although Eq. 3-18 was derived for the x direction, it can be applied similarly to the other two coordinate directions in order to determine the total force. Hence,

$$\Sigma F_x = \rho Q \Delta V_x = \rho Q(V_{x2} - V_{x1}) \tag{3-18}$$

$$\Sigma F_y = \rho Q \Delta V_y = \rho Q(V_{y2} - V_{y1}) \tag{3-19}$$

$$\Sigma F_z = \rho Q \Delta V_z = \rho Q(V_{z2} - V_{z1}) \tag{3-20}$$

The forces ΣF_x, ΣF_y, and ΣF_z in Eqs. 3-18, 19 and 20 are those forces acting against the fluid which are required to accelerate the fluid and thereby change the momentum-flux of the flow in the three corresponding coordinate directions.

The foregoing equations express exactly the forces involved in a change in momentum-flux, provided the distribution of velocity at section 1 and at section 2 are both uniform across the flow. When the velocity distribution is not uniform, however, a more general equation must be used.

Strictly speaking, the momentum-flux M in the x direction which is passing a given section of a streamtube, as shown in Fig. 3-10, is represented by the integral

$$M_x = \int \rho v_x \, dQ = \int \rho v_x v \, dA \tag{3-21}$$

As a matter of convenience, it is desirable to express the velocity as the average velocity. If the density ρ does not vary across the area A, and the average velocity V_x in the x direction is used, Eq. 3-21 can be written as

$$M_x = K_m \rho Q V_x = K_m \rho A V V_x \tag{3-22}$$

In Eq. 3-22, the momentum-flux coefficient K_m replaces the integral. It can be defined by combining Eq. 3-21 and Eq. 3-22 as

$$K_m = \frac{1}{Q} \int_A \frac{v_x}{V} \, dQ = \frac{1}{A} \int_A \frac{v_x}{V_x} \frac{v}{V} \, dA \tag{3-23}$$

in which K_m is the momentum-flux correction coefficient;

v_x is the velocity in the x direction at a point L/T;

v is the total velocity at a point in the flow;

V_x is the mean velocity in the x direction across the section;

V is the mean velocity in the general direction of motion;

A is the cross-sectional area across the streamtube L^2;

Q is the discharge in the streamtube L^3/T.

As can be seen in Fig. 3-11, however, when the flow is oriented in the x direction, $v_x/V_x = v/V$. Hence, Eq. 3-23 can be simplified to

$$K_m = \frac{1}{A} \int_A \left(\frac{v}{V}\right)^2 dA \tag{3-24}$$

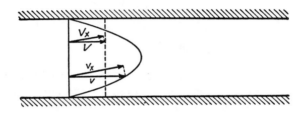

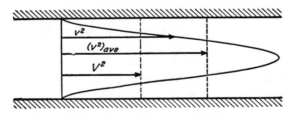

Fig. 3-11. Distribution and components of velocity.

which shows that the momentum-flux correction coefficient K_m is required because of the difference between the square of the mean and the mean of the squares of the velocity distribution. This should be contrasted with the continuity equation, in which the discharge varies with the velocity to the first power, and the energy equation (which is presented later in this chapter), in which the kinetic energy flux varies with the velocity cubed.

In actual practice, the velocity distribution is usually assumed to be uniform as a first approximation, so that $v \approx V$ and the momentum-flux coefficient K_m is usually taken as 1.00. In reality, however, this assumption can be markedly in error, since under ordinary conditions $K_m = 4/3 = 1.33$

for laminar flow in a pipe, see Fig. 3-11 (where, as will be seen in Chapter 5, the velocity distribution is found to be parabolic), and K_m varies from 1.01 to 1.07 for turbulent flow, with a usual value of about 1.03. With corrections of this order of magnitude, it is readily understandable that, when considerable refinement or accuracy is desired, it is necessary to include the correction coefficient in Eq. 3-18, which then becomes

$$\Sigma F_x = (K_m\rho QV_x)_2 - (K_m\rho QV_x)_1 \tag{3-25}$$

for the x direction, with corresponding use of K_m in Eqs. 3-19 and 3-20 for the other directions. If the velocity distributions are different at sections 1 and 2, then the K_m values are also different.

• DISCUSSION OF MOMENTUM EQUATION

Strictly speaking, Eqs. 3-18, 19, and 20 are the force-momentum flux equations. However, for simplicity and because of standard practice in usage, they are hereafter termed the momentum equations.

The forces F in the momentum equation are vector quantities, because they have both magnitude and direction. Likewise, the velocity is a vector, which makes the momentum-flux a vector, despite the fact that the density ρ and discharge Q are scalar quantities. Because the momentum equation consists of vector quantities, it is of utmost importance to carefully observe the signs throughout the equation. It also means that the discharge to be used is the entire discharge which reaches the body supplying the force to the fluid, and not just that part of the discharge moving in the direction under consideration. In other words, the entire discharge experiences the change in momentum-flux, and it is the **change in velocity** in a given direction **of the entire discharge** that determines the force in that direction. If the body is moving, the flow pattern is unsteady. Therefore, the discharge must be corrected to values relative to the body, as though it were fixed.

Although the total force F_B of the boundary acting against the fluid can be determined by

$$F_B = \sqrt{F_{Bx}^2 + F_{By}^2 + F_{Bz}^2}$$

it is possible, usually, to simplify the problem to one involving not more than two coordinates, by breaking the problem into parts or by careful selection of the coordinate system.

In applying the momentum equation, as already stated, it is necessary to keep careful account of the signs. This can best be done by selecting in advance the direction for positive forces and positive velocities. The sign of the momentum-flux term ρQV is determined by V because, as already mentioned, ρ and Q are scalar quantities and always positive.

It is helpful, but not necessary, to note that the direction of the force exerted by the boundary against the fluid can usually be predetermined. This is possible since the boundary force exerted on the fluid consists of a vectoral summation of unit pressure forces which act in a direction normal to the boundary. That is, if the direction of the pressure vectors acting on the fluid are known (either inward or outward normal to the boundary), the direction of the total force exerted by the boundary against the fluid, as well as the direction of its components, are also known.

To keep account of the signs for the velocities and forces in the momentum equation, the student should *always use an unbalanced free-body diagram,* see Fig. 3-12, particularly with the more complicated problems.

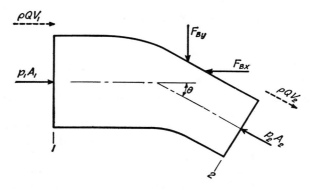

Fig. 3-12. Unbalanced free-body diagram for momentum equation.

The use of this diagram is illustrated in the examples for this section found on the following pages.

As the reader will note, use of the free-body diagram results in the selection of a portion of the flow which can be considered as a streamtube with essentially uniform flow across each end. Furthermore, the problem involved is a study of the change in momentum-flux, which has taken place from one end of this streamtube to the other, together with the forces which must be applied to the streamtube to create this change. In other words, a solution of the problem is not concerned with an analysis of what is taking place between the ends of the streamtube, but rather with an evaluation of the flow at the two ends of the streamtube, where the flow usually is uniform. In many problems, the velocity and pressure distributions may be so complex that a mathematical evaluation of these factors along the streamtube is very difficult, if not impossible (see the section on the energy equation). Therefore, the momentum equation plays a very important role in helping to evaluate the *bulk characteristics* of certain flow systems.

In the ordinary use of the momentum equation, the distribution of velocity across the end sections of the free-body or streamtube is not considered. Instead, an average velocity is used. Despite the fact that the velocity distribution is ignored, however, the agreement between theory and experiment is usually very close, as will be seen later. When greater refinement is required, however, the momentum-flux correction coefficient K_m must be used as shown in Eq. 3-25.

The solution of problems involving the momentum equation may be simplified considerably by utilizing the following steps. Refer to Fig. 3-12.

1. Isolate an unbalanced free-body of fluid so that the free-body includes the region of nonuniform flow where a change of velocity (either magnitude or direction or both) occurs. Select the end sections of the free-body so that they are in regions of uniform flow and are normal to the direction of flow.

2. Determine the direction of the force of the external boundary against the fluid whenever possible and, if this cannot be determined, then assume a direction.

3. Place vectors on the free-body which represent all forces acting upon it. These forces will usually include pressure forces, and external forces transmitted to the fluid from the boundary which confines the flow. Also put broken line vectors at sections 1 and 2 of the free-body to represent the momentum-flux at these sections. These vectors help to visualize the direction of the force causing the change of momentum-flux which occurs in the region of flow represented by the free-body. The forces represented on the free-body do not show the exact location of the force, because the location may be difficult to determine. The free-body is used only to indicate the direction of the forces acting.

4. Select a coordinate system which will simplify the problem as much as possible. Indicate positive and negative directions.

5. Use Eqs. 3-18, 3-19, and 3-20 to solve for the forces causing the change in momentum-flux. If the simplest coordinate system is selected, then usually only one or two of these equations is necessary.

The example, Fig. 3-12, of flow in a closed conduit will illustrate the procedure for setting up the momentum equation in the x direction for a problem which contains most of the factors involved in even complex momentum problems.

The momentum equation in the x direction is:

$$\Sigma F_x = \rho Q (V_{x2} - V_{x1})$$

which, from Fig. 3-12, becomes

$$p_1 A_1 - p_2 A_2 \cos \theta - F_{Bx} = \rho Q (V_{x2} - V_{x1})$$

Explanation of each term:

a. p_1A_1 is given a positive sign because, when placed on the free-body diagram as acting toward the free-body, it is to the right or positive.

b. $p_2A_2 \cos \theta$ is negative because, when placed as acting toward the free-body, the x component is to the left or negative.

c. F_{Bx} is the force exerted by the boundary on the free-body. This force is a result of the normal pressure against the fluid distributed around the periphery of the boundary and, although it is a nonuniformly distributed force, it is the resultant of the distributed unit forces of pressure in the x direction, and is represented with a single vector. It can be observed, by studying the unit forces of pressure exerted against the fluid by the boundary, that F_{Bx} acts to the left or in the negative direction, or its direction can be assumed and then checked by computing both its magnitude and sign.

6. In the case of problems involving relative motion between the fluid flowing and the boundary in contact with the fluid, all calculations for momentum-flux must be based on the relative velocity and the relative discharge between the fluid and the boundary. The relative discharge Q_r is computed by the discharge equation of $Q_r = A V'$ in which V' is the relative velocity acting through the cross section A. The problem of relative motion is explained in connection with Fig. 3-15.

• APPLICATION OF MOMENTUM EQUATION

As already stated, the momentum equation is very useful in the bulk solution of many problems in which the net forces causing a change in momentum-flux are to be determined. Certain of these problems are now discussed in detail, together with illustrative examples. It should be pointed out that the energy equation, which is developed in the next section, is necessary to study in detail the internal velocity and pressure distribution. Therefore, several of the following applications are given further consideration in the section on the energy equation.

Consider a jet impinging upon a flat plate or a curved surface, as shown in Fig. 3-13. The plate in both cases deflects the jet through an angle of 90°. If the assumption is made that there is no energy lost by the jet when it hits the plate, then the magnitude of the velocity V_1 approaching the plate is equal to the magnitude of the velocity V_2 leaving the plate. If no energy is lost as the jet strikes the plate, the above statement of $V_1 = V_2$ is true for jets striking moving plates as well as stationary ones. In the cases of the 90° turns shown in Fig. 3-13, the final velocity in the x direction V_{x2} is zero and Eq. 3-18 becomes

$$\Sigma F_x = -\rho Q V_{x1}$$

Because the momentum-flux $\rho Q V_{x1} > \rho Q V_{x2}$, the force ΣF_x causing the change of momentum flux is negative to the left, as shown in Fig. 3-13. The force F_x is always in the direction in which there is an increase in momentum-flux. In Fig. 3-13(a), the force F_y in the y direction is balanced

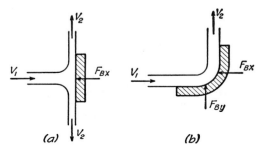

Fig. 3-13. Jet turning through 90 degrees.

and hence equal to zero because the jet is split in radial directions. But for Fig. 3-13(b), the jet is turned entirely in one direction. Therefore,

$$\Sigma F_y = \rho Q(V_{y2} - 0)$$

in which the force of the boundary against the jet in the y direction is positive upward.

The foregoing examples can be complicated by having the angle of deflection either more or less than 90°. The general case can be illustrated by a curved-vane deflector of any angle α, see Fig. 3-14. In this case, $V_{y2} = V_2 \sin \alpha$ and $V_{x2} = V_2 \cos \alpha$ so that

$$\Sigma F_x = \rho Q(V_2 \cos \alpha - V_1) \qquad (3\text{-}26)$$

$$\Sigma F_y = \rho Q(V_2 \sin \alpha - 0) \qquad (3\text{-}27)$$

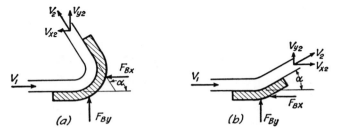

Fig. 3-14. Jet turning through angle.

In Eq. 3-26, α is greater than 90° and $\cos \alpha$ is negative, so that both velocity terms inside the parentheses are negative, and the boundary force causing the change of momentum-flux is negative. In accordance with the same procedure, the momentum-flux term in Eq. 3-27 is positive, and hence the boundary force is positive.

Example Problem 3-3: Find the force exerted on the fixed curved blade in Fig. 3-14 if $\alpha = 135°$, $V_1 = 100$ fps and $Q = 0.5$ cfs of water.

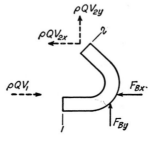

Ex. Prob. 3-3

Solution:

The unbalanced free-body diagram is first constructed for the data given and then the assumption is made that the magnitude of the velocity approaching the blade equals the magnitude of the velocity leaving the blade, so $V_1 = V_2 = 100$ fps. The momentum equation applied in the x direction is

$$\Sigma F_x = \rho Q(V_{x2} - V_{x1})$$

or $$\Sigma F_x = -F_{Bx} = \rho Q(V_{x2} - V_{x1})$$

in which F_{Bx} is the force of the boundary or the blade on the free-body of water. Therefore,

$$-F_{Bx} = (1.94)(0.5)[100(-0.707) - 100] = -165.5 \text{ lb}$$

so that the force of the blade against the water is $F_{Bx} = 165.5$ lb $\leftarrow$ and the force of the water against the blades is equal and opposite 165.5 lb $\rightarrow$

In the y direction

$$\Sigma F_y = \rho Q(V_{y2} - V_{y1})$$

so that the force of the blade against the water is

$$\Sigma F_y = F_{By} = (1.94)(0.5)[100(0.707) - 0] = 68.6 \text{ lb} \uparrow$$

and the force of the water against the blade is equal and opposite 68.6 lb $\downarrow$

The problem can be complicated still further by having a blade which is moving like the one in Fig. 3-15(a). In this case, the problem is greatly simplified if the velocities and the discharge are all taken relative to the blade. To accomplish this, a velocity of V_b must be superposed on the entire system so that Fig. 3-15(b) represents the flow system for steady flow with a stationary blade.

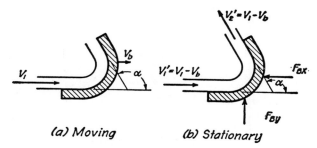

Fig 3-15. Jet and moving boundary.

Equations 3-26 and 3-27 now become

$$\Sigma F_x = \rho Q_r (V'_2 \cos \alpha - V'_{x1}) \tag{3-28}$$

$$\Sigma F_y = \rho Q_r (V'_2 \sin \alpha - V'_{y1}) \tag{3-29}$$

in which V' is the velocity relative to the blade L/T;

$V'_1 = V'_2$ in magnitude;

Q_r is the discharge relative to the blade L^3/T.

It should be noted that, when a single moving blade is considered, the discharge which is reaching the blade Q_r is not the same as that which reaches a stationary blade. The discharge relative to the blade is the velocity relative to the blade times the cross-sectional area of the jet $Q_r = V'_1 A_1$. If the blade is moving in the direction of V_1, then V_b is positive and $Q_r < Q$, whereas if the blade is moving against V_1, then V_b is negative and $Q_r > Q$. The magnitude of Q_r to be used in Eqs. 3-28 and 3-29 can be determined by Eq. 30.

$$Q_r = Q \frac{V_1 - V_b}{V_1} = \frac{QV'_1}{V_1} \tag{3-30}$$

When a series of blades is used, however, such as on an impulse turbine (see Chapter 12), then $Q_r = Q$, because all the discharge is intercepted by the blades. In other words, one blade after another moves into position to intercept its part of the discharge, so that the summation of the discharges intercepted by all the blades is equal to the total discharge reaching the turbine.

Example Problem 3-4: A jet of water with a velocity of 40 fps strikes a single blade moving at 30 fps in the same direction as the jet in Fig. 3-15. Find the x and y components of the force on the jet, if $Q = 4.0$ cfs and $\alpha = 60°$.

Solution:

The momentum equations are

$$\Sigma F_x = \rho Q_r (V'_{x2} \cos \alpha - V'_{x1})$$

and

$$\Sigma F_y = \rho Q_r (V'_{y2} \sin \alpha - V'_{y1})$$

From the free-body diagram, Fig. 3-15

$$\Sigma F_x = F_{Bx} \quad \text{and} \quad \Sigma F_y = F_{By}$$

then

$$V'_1 = V_1 - V_b = 40 - 30 = 10 \text{ fps.}$$

$$V'_1 = V'_2$$

and

$$Q_r = V'_1 A_1 = (10)(0.1) = 1.0 \text{ cfs.}$$

or from Eq. 3-30

$$Q_r = \frac{Q(V_1 - V_b)}{V_1} = \frac{(4.0)(40 - 30)}{40} = 1.0 \text{ cfs.}$$

By substitution in the momentum equations

$$-F_{Bx} = (1.94)(1.0)[10(0.5) - 10]$$

$$F_{Bx} = \underline{\underline{9.7 \text{ lb}}} \leftarrow$$

and

$$F_{By} = (1.94)(1.0)[10(0.866) - 0] = \underline{\underline{16.8 \text{ lb}}} \uparrow$$

Both F_{Bx} and F_{By} are the forces of the boundary exerted against the jet which cause the change of momentum-flux.

Jet propulsion is an important engineering principle which is being used more and more widely to propel aircraft, surface craft, and submarine craft. Basically, the thrust for jet propulsion is accomplished by directing a jet of fluid out of the rear of a body. The jet is created by internal combustion, or by an impeller which pumps the fluid through the body. Two simple cases will be considered as examples, one in which the initial velocity is zero, and the other in which the initial velocity is not zero.

If a jet is issuing from a tank so that the initial velocity of the fluid is zero, then $V_1 = 0$ and

$$\Sigma F = F = \rho Q (V_2 - 0)$$

in which the force F is that force against the fluid required to increase the velocity from zero to V_2. The thrust on the body F, according to Newton's third law of motion and d'Alembert's principle, is the force of inertia and is equal in magnitude but opposite in direction to F. The force developed by the jet is entirely independent of the fluid into which it discharges.

Example Problem 3-5: A 3-in. diameter jet is issuing from a large tank with a velocity of 50 fps. Find the reactive force on the tank.

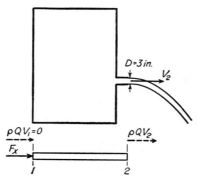

Ex. Prob. 3-5

Solution:

$$\Sigma F_x = \rho Q(V_{x2} - V_{x1})$$

$$\Sigma F_x = \rho Q(V_{x2} - 0)$$

therefore $F_{Bx} = 1.94(50 \times 0.0491)(50 - 0) = 47.6 \text{ lb} \rightarrow$

The force on the tank is equal and opposite $\underline{\underline{47.6 \text{ lb} \leftarrow}}$

(see Prob. 3-110 for further detail).

If a part or all of the fluid in the *jet* is taken into the body through an upstream intake, then $V_1 \neq 0$ and the equation must be modified to the form

$$\Sigma F = (\rho_j Q_j V_j)_2 - (\rho_i Q_i V_i)_1$$

in which the subscripts j and i are for the jet and the intake, respectively.

The **slipstream** downstream from a propeller or a fan can be considered as a jet. In this case, the propeller increases the momentum-flux of the surrounding fluid by an amount equal in magnitude but opposite in direction to the **thrust** of the fluid against the propeller.

Example Problem 3-6: Find the thrust against the air exerted by a propeller or fan which is moving 5000 cfm of air at a temperature of 60°F and at standard atmospheric pressure. The approximate dimensions of the air stream are as shown in the sketch.

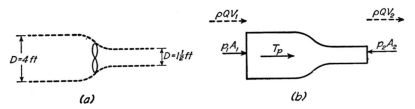

Ex. Prob. 3-6

Solution:

Since this is a free jet of air, $p_1 = p_2 = 0$, and since

$$\Sigma F_x = \rho Q (V_{x2} - V_{x1})$$

$$T_p = \rho Q (V_2 - V_1)$$

$$V_1 = Q/A_1 = \frac{5000}{60\pi(2)^2} = 6.55 \text{ fps}$$

$$V_2 = Q/A_2 = 46.6 \text{ fps.}$$

At 60°F, $\rho = 0.00237$ lb-sec²/ft⁴ for air.

Therefore, $T_p = (0.00237)(83.3)(46.6 - 6.55) = \underline{\underline{79.1 \text{ lb}}} \rightarrow$

The thrust against the propeller is equal in magnitude and opposite in direction to the thrust T_p which the propeller exerts against the air.

The **transition in a closed conduit** also is encountered frequently in engineering design. The transition may be an expansion, a contraction, or a bend which changes the velocity either in magnitude or direction or both. Although the complete analysis of most of these problems involves also the energy equation (which is developed in the next section of this chapter), there are many problems which can be solved from the bulk viewpoint by means of the momentum equation.

The problem of a jet issuing from a **nozzle** at the end of a fire-fighting hose, for example, is very similar to the problem of jet propulsion, and the thrust can be so great that even two men have difficulty holding the nozzle steady. This backward thrust of the jet can be balanced by a forward thrust exerted by the firemen. The following example with the free-body diagrams illustrates the solution of such a problem.

Example Problem 3-7: A nozzle fastened to the 3-in. diameter hose discharges 300 gpm of water in a $1\frac{1}{8}$-in. diameter jet. Analyze the problem and determine the external force required to balance the backward thrust of the jet. The analysis of the problem can best be illustrated by a series of three free-body diagrams.

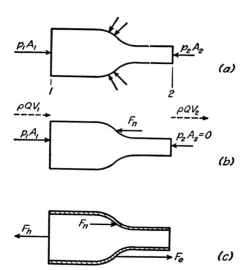

Ex. Prob. 3-7

Solution:

The first sketch shows the pressure forces exerted on a free body of liquid in the nozzle. This free-body is not in equilibrium, because $p_1A_1 > F_n$ and $p_2 =$ atmospheric pressure, making $p_2A_2 = 0$. F_n is the force exerted by the nozzle on the liquid. The difference between p_1A_1 and F_n is $\rho Q(V_2 - V_1)$, which is the change in momentum flux from 1 to 2. All of the forces and the momentum vectors are shown in the free-body (b) of the figure and, from the momentum equation,

$$\Sigma F_z = p_1A_1 - F_n = \rho Q(V_{z2} - V_{z1})$$

The next step is to analyze the free-body of the nozzle. If the external force F_e applied by the firemen to balance the backward thrust of the jet is applied to the nozzle, then from the free-body of the nozzle:

$$F_h = p_1A_1 = F_n + F_e$$

in which F_h is the resultant of the stress acting around the circumference of the nozzle to hold the nozzle on the hose. Solving for F_e

$$F_e = p_1A_1 - F_n$$

and substituting this expression for F_e in the momentum equation yields

$$F_e = \rho Q(V_2 - V_1)$$

Since   1 cu ft = 7.48 gallons,

$$Q = \frac{300}{7.48 \times 60} = 0.668 \text{ cfs}$$

$$V_2 = \frac{0.668}{0.0691} = 96.6 \text{ fps}$$

$$V_1 = \frac{0.668}{90441} = 13.6 \text{ fps}$$

$$F_s = 1.94(0.668)(96.6 - 13.6) = \underline{\underline{107.7 \text{ lb}}} \rightarrow$$

A *sudden expansion* in a closed conduit can be analyzed by the momentum equation to determine the change in pressure. Furthermore, experimental data show that this theoretical analysis is an accurate approximation to the actual conditions.

Example Problem 3-8: With a discharge of 2 cfs flowing in the sudden expansion shown in the figure, what is the change of pressure which occurs as the fluid flows from the small pipe to the large pipe? The free-body is chosen as shown, and the pressure acting across section 1 is assumed to be the pressure which exists in the 6-in. pipe.

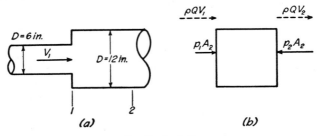

Ex. Prob. 3-8

Solution:

Since the momentum flux decreases from section 1 to section 2, the direction of the decreasing momentum flux is to the right. The force causing the change of momentum flux is the pressure force p_2A_2 minus the pressure force p_1A_2.

For equilibrium, $\Sigma F_x = \rho Q (V_{x2} - V_{x1})$

which from the free-body diagram becomes

$$p_1A_2 - p_2A_2 = \rho Q(V_{x2} - V_{x1})$$

Therefore, $(p_1 - p_2)A_2 = \rho Q(V_{x2} - V_{x1})$

$$p_1 - p_2 = \frac{(1.94)(2.0)}{0.785}\left[\frac{2.0}{0.785} - \frac{2.0}{0.196}\right]$$

$$p_1 - p_2 = \underline{\underline{-37.8 \text{ psf}}}$$

which means that $p_1 < p_2$.

Therefore, there is an increase in pressure from section 1 to section 2, and this pressure rise is accompanied by a loss of energy, although this loss is not reflected directly in the momentum equation.

Rearranging the equation yields an equation of the form

$$p_1 A_2 + \rho Q V_{z1} = p_2 A_2 + \rho Q V_{z2}$$

from which it can be seen that the total pressure plus the momentum flux at section 1 must equal the total pressure plus the momentum flux at section 2.

A problem for flow in open channels which is very similar to the foregoing sudden expansion in closed conduits is the **hydraulic jump,** wherein the total pressure plus momentum flux concept is again helpful. The hydraulic jump is a natural phenomenon which is used extensively by

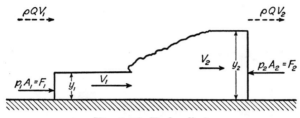

Fig. 3-16. Hydraulic jump.

engineers to dissipate the kinetic energy in the high-velocity water at the bottom of a spillway. An unbalanced free-body of a hydraulic jump is shown in Fig. 3-16.

Since there is a change of momentum flux from section 1 to section 2, the difference in the pressure forces is the resultant force responsible for that change. For equilibrium

$$\Sigma F_x = \rho Q (V_{z2} - V_{z1})$$

which becomes

$$p_1 A_1 - p_2 A_2 = \rho Q (V_{2x} - V_{1x}).$$

Since both V_1 and V_2 are in the same direction, the subscripts for x can be dropped, and

$$p_1 A_1 - p_2 A_2 = \rho Q (V_2 - V_1).$$

By analyzing a section of the jump which has a unit width normal to Fig. 3-16, it can be seen that

$$A_1 = (y_1)(1.0) = y_1$$

and

$$A_2 = (y_2)(1.0) = y_2.$$

The average pressures acting on the areas are equal to $\gamma y_1/2$ and $\gamma y_2/2$, so that the momentum equation becomes

$$\frac{\gamma y_1^2}{2} - \frac{\gamma y_2^2}{2} \doteq \rho q (V_2 - V_1)$$

in which q is the discharge per unit width of the channel. From the continuity equation

$$q = V_1y_1 = V_2y_2$$

By combining these two equations, the following equation for the hydraulic jump can be derived:

$$y_2/y_1 = \frac{1}{2}\left(\sqrt{1 + \frac{8V_1^2}{gy_1}} - 1\right) \tag{3-31}$$

This phenomenon of the hydraulic jump is discussed in greater detail in Chapter 8 on Flow in Open Channels.

The forces involved in the **bend** of a closed conduit such as a pipe, see Fig. 3-17, come from two sources:

1. The internal pressure, as studied in Chapter 2, and

2. The change in momentum flux.

By using the unbalanced free-body diagrams, these forces can be distinguished clearly for analysis.

The thrust due to the internal pressure p is determined as in Chapter 2, and the thrust due to the change in momentum flux is determined by Eqs. 3-18 and 3-19, as for a free jet striking a vane. These two thrusts must be added vectorially to obtain the total thrust involved.

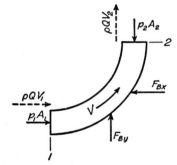

Fig. 3-17. Flow in pipe bend.

Example Problem 3-9: Find the components of the total thrust and the resultant thrust on the horizontal elbow shown in Fig. 3-17 if $Q = 10$ cfs, the pressure in the bend is 30 psi and the pipe is 12 in. in diameter.

Solution:

Using equation 3-18,

$$\Sigma F_x = \rho Q(V_{x2} - V_{x1})$$

$$p_1A_1 - F_{Bx} = \rho Q(V_{x2} - V_{x1})$$

$$-F_{Bx} = \rho Q(V_{2x} - V_{1x}) - p_1A_1$$

Substituting values in the foregoing equation and solving for F_{Bx} gives

$$-F_{Bx} = \rho Q(0 - V_1) - p_1A_1 = -(1.94)(10)\frac{10}{0.785} - 30(144)(0.785)$$

$$F_{Bx} = 3617 \text{ lb} \leftarrow$$

Using the same procedure

$$\Sigma F_y = \rho Q(V_{y2} - V_{y1})$$

$$F_{By} - p_2A_2 = \rho Q(V_{y2} - V_{y1})$$

$$F_{By} = \rho Q(V_{y2} - V_{y1}) + (p_2A_2)$$

$$F_{By} = (1.94)(10)\frac{10}{0.785} + (30)(144)(0.785) = 3617 \text{ lb } \uparrow$$

The magnitude of the resultant force is

$$F_B = \sqrt{F_{Bx}^2 + F_{By}^2} = 5100 \text{ lb. } \nwarrow 135°$$

and the direction of the resultant force is

$$\theta = \tan^{-1}\frac{F_{By}}{F_{Bx}} = \tan^{-1} -1.0 = 135°$$

Therefore, the thrust of the boundary on the free-body of water is "upward" (in a horizontal plane) and to the left at an angle of 45° with the x direction, so the thrust of water on the boundary (both static and dynamic) is equal in magnitude and opposite in direction.

$$F_B = \underline{\underline{5100 \text{ lb}}} \searrow 45°$$

Energy Equation

At the beginning of this chapter, the statement was made that there are three very important equations used for solving problems in fluid mechanics—*the continuity equation, the momentum equation, and the energy equation.* The first two have been developed and applied to various problems in the previous sections of this chapter. The energy equation now will be derived mathematically and applied to problems in which it is used either alone or in combination with the continuity and momentum equations.

Whereas the continuity and momentum equations considered the bulk characteristics or net results of the flow, the *energy equation* is concerned more with the *mechanics of internal energy changes* and the velocity and pressure variations involved in fluid flow. While the momentum equation contains vector quantities, the energy equation contains scalar quantities. Like the momentum equation, the energy equation is developed assuming that the approximation of one-dimensional flow is sufficiently accurate for the purposes of this elementary text. A more exact treatment is necessary, however, when greater refinement is required and more complicated problems are encountered.

The complete treatment of the energy principle involves all of the various forms of energy. If all of these forms are considered, then the physical law of conservation of energy is applicable throughout this text. One form of energy is heat, and in fluid mechanics the terms *energy dissipation* or *head loss* are not the destruction of energy but usually the conversion of kinetic energy into heat through viscous shear, or into mechanical energy through a machine such as a turbine.

In this chapter, the principle forms of energy considered are the kinetic energy E_k per unit weight (called the velocity head $V^2/2g$), the pressure energy E_p per unit weight (called the pressure head p/γ), and the potential energy E_e per unit weight (called the elevation head z). In addition heat energy E_H and mechanical energy E_M, can be added or subtracted from the flow system by equipment such as heat exchangers, pumps, or turbines. Expressed in equation form, this is

$$E_k + E_p + E_e + E_M + E_H = \text{constant} \tag{3-32}$$

because, from the law of conservation of energy (which the student has studied in physics), energy can be neither created nor destroyed. Instead, the sum of the various energy forms is a constant, and energy is converted or transformed from one form to another. Pumps and turbines are treated in greater detail later in this chapter and in Chapter 12. Heat energy is of particular importance in the study of flow of compressible fluids, which is discussed in the next chapter.

Essentially then, this chapter is a study of fluids which are both non-viscous (which is associated with irrotational flow) and incompressible. In the derivation of the energy equation, it is also assumed that neither mechanical energy nor heat energy are added to the system or subtracted from the system.

• DERIVATION OF ENERGY EQUATION

In the mathematical development of the energy equation, Newton's second law of motion

$$F = Ma$$

again will be employed. Perhaps a more elegant derivation of the energy equation could be made by utilizing immediately the basic differential equation. For the purpose of clarity, however, and in order to tie the development more directly into the student's knowledge of solid mechanics, the problem will be studied first from the viewpoint of *work-equals-change-of-energy*, utilizing finite distances over which the work is done.

Assuming steady flow, consider an incremental section of a streamtube having a length Δs which at the limit is ds. This is the distance the fluid moves during the time dt, see Fig. 3-18. The work required to move the fluid from section 1 to section 2 is the product of the force F, which does the work, and the distance ds, over which the force is applied. Therefore, Newton's equation becomes

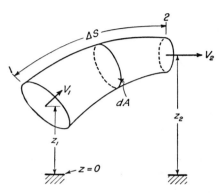

$$F \, ds = Ma \, ds \qquad (3-33)$$

which means that the force F is equal in magnitude and opposite in direction to the forces it must overcome. In moving the fluid element from 1 to 2, there are two forces which must be overcome. One of these forces is the weight $W = \gamma dA \, ds$, which is the force due to gravity,

Fig. 3-18. Streamtube for energy equation.

and the other is the force due to the difference in pressure, $(p_2 - p_1)dA$, between sections 1 and 2. The work done to overcome the weight is the product of the weight and the vertical distance over which it is moved, $\gamma dA \, ds(z_2 - z_1)$; and the work done to overcome the pressure difference is the product of the net force $(p_2 - p_1)dA$, due to the difference in pressure at section 1 and section 2, and the distance ds over which this difference takes place. Therefore, the **work** part of the equation becomes

$$F \, ds = -(p_2 - p_1)dA \, ds - \gamma dA \, ds(z_2 - z_1) \qquad (3-34)$$

The mass of the fluid to be moved is $\rho dA \, ds$, and the average acceleration $(V_2 - V_1)/dt$ is the change in velocity which takes place in the time dt during which the fluid moves from section 1 to section 2. Therefore, the **change in kinetic energy** which takes place, due to the work done, is

$$Ma \, ds = \rho dA \, ds \, \frac{(V_2 - V_1)}{dt} \, ds = \rho dA \, ds \, \frac{(V_2 - V_1)}{dt} \frac{(V_2 + V_1)}{2} \, dt \qquad (3-35)$$

since
$$ds = V \, dt = \frac{(V_2 + V_1)}{2} \, dt$$

in which V is the average velocity over the distance ds.

Equating Eqs. 3-34 and 3-35 yields

$$\frac{\rho dA \, ds(V_2^2 - V_1^2)}{2} + (p_2 - p_1)dA \, ds + \gamma dA \, ds(z_2 - z_1) = 0 \qquad (3-36)$$

Each of the terms in Eq. 3-36 has the dimensions of energy FL which, in the English system of units, are foot-pounds (ft-lb). For use in fluid mechanics, however, it is usually preferable to have the equation in terms of energy per unit weight FL/F (ft-lb per lb), which then simplifies the dimensions to length L (ft) only. Therefore, upon division by the weight $\gamma dA\ ds$, Eq. 3-36 becomes

$$\frac{V_2^2 - V_1^2}{2g} + \frac{(p_2 - p_1)}{\gamma} + (z_2 - z_1) = 0 \tag{3-37}$$

which can be rewritten as

$$\frac{V_1^2}{2g} + \frac{p_1}{\gamma} + z_1 = \frac{V_2^2}{2g} + \frac{p_2}{\gamma} + z_2 \tag{3-38}$$

for convenient use. Equation 3-38 has been derived along the streamtube and, therefore, it is applicable along a streamline because, as dA goes to zero, the streamtube becomes a streamline.

This equation is the **basic work-energy equation for steady flow of incompressible fluids,** where neither heat nor mechanical energy are either added to or subtracted from the system between the sections 1 and 2 under consideration. The equation expresses the relationship between the velocity, the pressure, and the elevation at various points **along a streamline or streamtube.** The case where the sum of the terms in Eq. 3-38 is being reduced and converted into heat energy will be considered later in this chapter.

Equation 3-38 is commonly termed the Bernoulli equation in honor of the Swiss mathematician, Daniel Bernoulli, who published in 1738 one of the first treatises concerning fluid flow, which later lead to the development of this equation. The dimensions of the equation are energy per unit weight, which reduces simply to length or $LF/F = L$, and the different terms are commonly known as the **velocity head** $V^2/2g$, the **pressure head** p/γ, and the **elevation head** z. The sum of the pressure head and the elevation head is termed the **piezometric head** $h = (p/\gamma) + z$, and the sum of all three heads is the **total head** H_T. Hence,

$$H_T = \frac{V^2}{2g} + \frac{p}{\gamma} + z = \frac{V^2}{2g} + h \tag{3-39}$$

If the *fluid flowing is a gas,* then the effect of variation in elevation head is usually insignificant, compared with the other terms in the equation, so that z can be deleted and Eq. 3-38 can be multiplied by γ and written as

$$\rho\frac{V_1^2}{2} + p_1 = \rho\frac{V_2^2}{2} + p_2 = \text{constant} \tag{3-40}$$

Although the energy equation, as just presented, has been derived for use only *along* a streamline, there are certain problems in which it is desirable to use the energy equation *across* the streamlines, Such use of the energy or Bernoulli equation is permissible only if the flow is irrotational. If the flow is not exactly irrotational, then the results will be correspondingly inexact. Problems in which it is permissible to use the Bernoulli equation across the streamlines usually involve rapidly converging flow.

In the remainder of this chapter, Eq. 3-39 and Eq. 3-40 are discussed, followed by a number of examples of their use both along streamlines and across streamlines. First, however, the influence of velocity distribution upon the term kinetic energy per unit volume $\rho V^2/2$ or velocity head $V^2/2g$ is analysed.

The foregoing development expresses exactly the velocity head for a given section and flow, provided the velocity distribution is uniform across the section. When the average velocity is used and yet the velocity distribution is not uniform, however, a correction factor K_E must be employed, as shown in the following presentation.

It has been pointed out that Eq. 3-38 is applicable along a streamline, and that each term represents the energy per unit weight which exists at any given point along the streamline. Therefore, the term $V^2/2g$ is the kinetic energy per unit weight which the flow possesses at a particular point. If the velocity distribution varies across a section of the flow, then the actual kinetic energy per unit time which passes the section must be obtained by integrating (from streamline to streamline across the flow section) the product of the kinetic energy per unit weight and the weight of fluid passing per unit time.

$$\text{Energy flux} = \int_A \frac{v^2}{2g}(\gamma\, dQ) = \gamma \int_A \frac{(v^2)}{2g} v\, dA = \frac{\gamma}{2g}\int_A v^3\, dA \qquad (3\text{-}41)$$

in which γ is the specific weight F/L^3;

v is the velocity at a point L/T;

dQ is the incremental discharge L^3/T passing through the incremental area dA.

For the sake of convenience, the one-dimensional method of analysis is usually employed, which requires that the average velocity for the cross section be used. The energy flux, therefore, can be represented by

$$\text{Energy flux} = K_e \frac{\gamma}{2g} A V^3$$

which may be substituted into Eq. 3-41 to yield

$$K_e = \frac{1}{A}\int_A \left(\frac{v}{V}\right)^3 dA \qquad (3\text{-}42)$$

in which K_e is the correction coefficient required (because the cube of the average velocity is not equal to the average of the cubes of the velocities taken over incremental areas across the flow section) to convert the kinetic energy obtained by use of the average velocity V to the actual kinetic energy.

The coefficient K_e varies from 1.0 for a uniform velocity distribution, to values greater than 1.0 as the distribution of velocity becomes less and less uniform. For turbulent flow, K_e varies from about 1.02 to 1.15, with an average value of about 1.06, and for laminar flow K_e is 2.0. Usually, the kinetic energy coefficient K_e is not used in fluid mechanics computations, because the additional degree of refinement is usually not needed. However, if greater refinement is needed, the coefficient K_e should certainly be used. Occasionally, the engineer is confronted with a discrepancy in his computations or analysis simply because through habit he has neglected K_e.

Equation 3-42 should be compared with the similar equations for the continuity principle, Eq. 3-12, and for the momentum principle, Eq. 3-24. In these three equations, the continuity, the momentum-flux, and the energy equations, the velocity appears at an ever-increasing power to be integrated across the flow section. In the continuity equation, the velocity appears to the first power, in the momentum equation to the second power, and in the energy equation to the third power. Therefore, since discharge is a scalar quantity, momentum-flux is a vector quantity and energy is a scalar quantity. This is based upon the fact that the product of a scalar quantity and vector quantity is a vector quantity and the product of two vector quantities is a scalar quantity.

• DISCUSSION OF ENERGY EQUATION

As discussed previously, the energy equation is derived to evaluate the distribution of head, or energy per unit weight $LF/F = L$, within the flow. This energy is divided into three forms, the velocity head, the pressure head, and the elevation head. In this way, the energy equation permits a detailed study of the internal mechanics of flow, whereas the momentum equation is concerned only with the bulk characteristics of the flow. The energy or Bernoulli equation, with its variations, is perhaps the equation most widely used in fluid mechanics. Its use can easily be abused, however, unless tempered by a clear understanding of the assumptions involved in its derivation and the resulting limitations of its use. *These **assumptions** in the derivation, and the corresponding **limitations**,* will now be discussed in detail.

1. The flow is assumed to be steady, so that, at a given point, there is no variation of velocity with respect to time.

2. The flow is assumed to be either essentially one-dimensional or irrotational. Although many of the problems solved satisfactorily by the Bernoulli equation are both far from one-dimensional and not exactly irrotational, the flow is either sufficiently one-dimensional or sufficiently irrotational to permit use of the equation.

3. The variation of the velocity across a given section of the flow is important in determining the applicability of the energy equation. In the derivation of the equation, the velocity distribution across a single streamtube is assumed to be uniform. If the streamtube is small in cross-sectional area, this assumption is permissible and the energy equation is applicable along the streamtube. However, it must be remembered that, if there is an appreciable variation in the velocity distribution, the correction coefficient K_e must be applied to obtain the exact amount of energy to be found at a given point.

4. The assumption is made that the flow is incompressible and possesses a uniform density ρ or specific weight γ. Although this assumption may seldom be exactly the case for gases, the resulting error for many problems involving gases is sufficiently small to permit use of the equation. Chapters 4, 7, and 9 consider compressible fluids in greater detail. This limitation does not apply to liquids, because they are essentially incompressible.

5. The assumption is made that no energy is added to or subtracted from the flow system along the streamtube in question. If other types of energy are considered, such as mechanical energy E_M or heat energy E_H, additional terms must be added to the left side of Eq. 3-38, so that it becomes

$$\frac{V_1^2}{2g} + \frac{p_1}{\gamma} + z_1 + E_M + E_H = \frac{V_2^2}{2g} + \frac{p_2}{\gamma} + z_2 \qquad (3\text{-}43)$$

If energy is added, the terms will be positive, and if it is subtracted, the terms will be negative.

6. The assumption is made that no shear or drag exists along the flow system. If shear does occur, energy is converted to heat, so that a term for this head loss H_L must be added to the right side of Eq. 3-43

$$\frac{V_1^2}{2g} + \frac{p_1}{\gamma} + z_1 + E_M + E_H = \frac{V_2^2}{2g} + \frac{p_2}{\gamma} + z_2 + H_L \qquad (3\text{-}44)$$

In this chapter, the heat energy and the loss of energy due to friction are not considered, so that heat energy E_H and the head loss H_L are zero, and in most of the examples the mechanical energy E_M is also zero. Subsequent chapters are devoted to detailed consideration of these additional terms.

• APPLICATION OF ENERGY EQUATION

The correct use of the energy or Bernoulli equation depends upon a thorough understanding of the foregoing assumptions and limitations. Therefore, a number of examples are now considered as applications of the Bernoulli equation. In the following applications, any deviation from the assumptions and limitations will be explained in detail.

The Bernoulli equation is derived to give the distribution of the piezometric head $h = (p/\gamma) + z$ and the velocity head $V^2/2g$ along a streamline in the flow (and across the streamlines if the flow is irrotational). The applications are designed to begin with a simple case, in which each of the terms is easily visualized, and then to analyse progressively more and more complicated uses of the Bernoulli equation.

One of the simplest applications of the Bernoulli equation is the analysis of a *free liquid jet* in the atmosphere, as shown in Fig. 3-19, in which the effect of air resistance is neglected. Because the variations of pressure head with elevation in the atmosphere are normally insignificant compared with the elevation and velocity heads of the liquid jet, the jet is considered to be

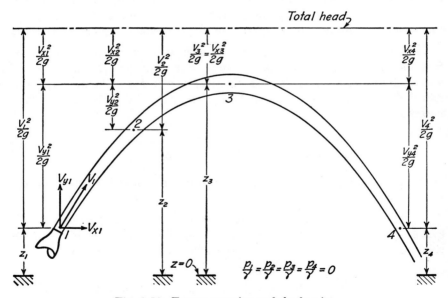

Fig. 3-19. Energy equation and the free jet.

surrounded by air at uniform pressure (zero). Therefore, since atmospheric pressure is zero gage pressure, and $p/\gamma = 0$ throughout, the total head involves only the sum of the elevation and velocity heads.

The Bernoulli equation now can be written as

$$\frac{V_1^2}{2g} + z_1 = \frac{V_2^2}{2g} + z_2 = \frac{V_3^2}{2g} + z_3 = \frac{V_4^2}{2g} + z_4 \qquad (3\text{-}45)$$

in which $V_1^2 = V_{x1}^2 + V_{y1}^2$, $V_2^2 = V_{x2}^2 + V_{y2}^2$, $V_3^2 = V_{x3}^2 + V_{y3}^2$, etc.

The piezometric head h in this equation is only the elevation head z, because the pressure head p/γ is considered to be zero throughout.

Before analysing the trajectory and the velocity distribution of the jet in Fig. 3-19, attention must be called to the assumption which is made that the jet does not disintegrate. The entire flow system is easily visualized and analysed by means of Fig. 3-19, in which are drawn the **total head line**, the **line of constant** $V_x^2/2g$, the **reference datum**, and the **dimensioning** for each of the terms in the Bernoulli equation. The total head line shows the location of the total head above the reference datum at all points along the flow system. Because $V_x^2/2g$ is a constant, the line for $V_x^2/2g$ is drawn horizontally a distance below the total head equal to $V_x^2/2g$.

At point 2, the dimensioning is shown for each of the terms in the Bernoulli equation, including z_2, $V_2^2/2g$, $V_x^2/2g$, $V_y^2/2g$. At points 1 and 4, the elevation $z_4 = z_1$ is a minimum, and the velocity head $V_1^2/2g = V_4^2/2g$ is a maximum (if the jet falls below point 4, the velocity head $V_y^2/2g$ will continue to increase). At point 3, however, the velocity is a minimum and the elevation is a maximum, which occurs when the entire velocity head in the y direction $V_y^2/2g$ has been converted to elevation head, so that $V_y = 0$.

Not only is the Bernoulli equation useful in analysing Fig. 3-19, but the continuity equation $A = Q/V$ also can be employed to determine the area of the cross section of the jet at any point along its course.

Example Problem 3-10: A 3-inch diameter jet leaves a nozzle at an angle of 30° upward from the horizontal. If the discharge in the jet is 2 cfs, how high does the jet go, and what is the diameter of the jet at its maximum elevation?

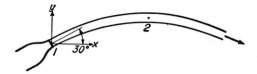

Ex. Prob. 3-10

Solution:

Applying the Bernoulli equation from 1 to 2

$$\frac{V_1^2}{2g} + \frac{p_1}{\gamma} + z_1 = \frac{V_2^2}{2g} + \frac{p_2}{\gamma} + z_2$$

Since $p_1 = p_2 =$ atmospheric pressure $= 0$, and by letting $z_1 = 0$, rearrangement of the equation results in

$$z_2 = \frac{V_1^2}{2g} - \frac{V_2^2}{2g}$$

$$\frac{V_1^2}{2g} = \frac{V_{y1}^2}{2g} + \frac{V_{x1}^2}{2g}$$

The horizontal component of the velocity is constant.

Therefore, $$\frac{V_{x1}^2}{2g} = \frac{V_2^2}{2g}$$

By substitution, the equation for z_2 becomes

$$z_2 = \frac{V_{y1}^2}{2g}$$

$$V_{y1} = V_1 \sin 30°$$

$$V_1 = Q/A_1 = \frac{2.0}{0.0491} = 40.7 \text{ fps}$$

$$z_2 = \frac{20.35^2}{64.4} = 6.43 \text{ ft}$$

$$V_{x1} = V_2 = V_1 \cos 30° = 40.7(0.866) = 35.3 \text{ fps}$$

$$A_2 = \frac{Q}{V_2} = \frac{2}{35.3} = 0.0566 \text{ sq ft}$$

$$d_2 = \left(\frac{0.0566 \times 4}{3.14}\right)^{1/2} = 0.268 = 3.22 \text{ in.}$$

A *jet of gas* issuing from the end of a pipe is another problem which is simple to analyse by the Bernoulli equation, see Fig. 3-20. In this case, the assumption is made that the compressibility effects are insignificant and that the total head line is great enough so that the change in pressure head over the diameter of the pipe at section 1 is negligible by comparison with total head. Hence, the elevation head z can be considered constant. For the sake of convenience, this constant is chosen as zero, $z = 0$. Therefore, the piezometric head h consists only of the pressure head p/γ, and hence the Bernoulli equation, Eq. 3-38, can be written as

$$\frac{V_1^2}{2g} + \frac{p_1}{\gamma} + 0 = \frac{V_j^2}{2g} + \frac{p_j}{\gamma} + 0 = \frac{V^2}{2g} + \frac{p}{\gamma} = \text{constant} \qquad (3\text{-}46)$$

which can be rearranged as

$$\frac{V_j^2 - V_1^2}{2g} = \frac{p_1 - p_j}{\gamma} = \frac{\Delta p}{\gamma} \qquad (3\text{-}47)$$

This equation shows that the increase in velocity head through the orifice is equal to the corresponding decrease in pressure head through the orifice.

The limiting streamline in Fig. 3-20 is along the wall of the pipe and the inside face of the end plate. Therefore, the pressure distribution over the end plate, from the corner at B to the orifice at C, may be determined by Eq. 3-46. Since the sum of the velocity head $V^2/2g$ and the pressure head must be a constant, a straight line can be drawn from A to A' to represent the distribution of total head over the end plate. Two limiting cases can be stated immediately. On the one hand, at the corner B the velocity is zero

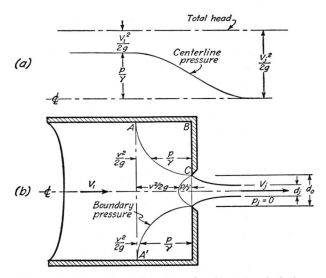

Fig. 3-20. Irrotational flow through orifice at end of pipe.

and, hence, on the basis of Eq. 3-46, the entire velocity head has been converted to pressure head. This is known as stagnation pressure, since there is no velocity. On the other hand, at the edge C of the orifice, the pressure must be atmospheric, which is zero gage pressure. Hence, as the fluid moves along the end plate from B to C, the total head is gradually converted from entirely pressure head to entirely velocity head.

It is well to note that an unbalanced free-body of the flow from section 1 to the **vena contracta** can be set up using the actual distribution of the pressure over the end plate, instead of the total force which was used in the previous section, while considering only the total external forces causing a change in momentum flux. The **vena contracta** is the point along the jet where the streamlines become parallel and the internal pressure is atmospheric.

All along the outside of the jet issuing from the orifice, the pressure is atmospheric and, hence, the velocity remains a constant—having the same magnitude as at the edge of the orifice. Inside the jet, however, say along the centerline, the pressure gradually decreases from p_1 upstream to $p_j = 0$ downstream at the **vena contracta**—where the streamlines are no longer converging but have become parallel. Again, the Bernoulli equation applies, with the sum of the velocity head and the pressure head remaining a constant along the centerline (see Fig. 3-20.)

Since the streamlines are converging as the flow approaches and passes through the orifice, the flow can be assumed to be essentially irrotational in the vicinity of the orifice, and the Bernoulli equation can be applied across the streamlines with reasonable accuracy. By this means, the pressure can be seen in Fig. 3-20 to increase from zero at C to a maximum at the centerline. Likewise, the velocity is a maximum at C and a minimum at the centerline. At the **vena contracta,** however, the pressure within the jet is zero, and the velocity across the jet is essentially uniform and equal to the total head.

The foregoing discussion illustrates the use of the Bernoulli equation to analyse the internal distribution of the velocity head and the pressure head within the flow and along the boundary. The flow net can be used to determine the distribution of velocity, from which the Bernoulli equation can be used to determine the distribution of pressure at any point in the fluid. The results of this analysis can also be used to determine the discharge through an orifice. Hence, an orifice is used frequently as a flow measuring device, as described in Chapter 10 on Flow and Fluid Measurements.

Equation 3-38 can be used to determine the velocity of the jet V_j if the other terms V_1, p_1, and p_j are known. To determine the discharge Q, however, the continuity equation also must be employed, in the following form

$$Q = A_1 V_1 = A_j V_j \tag{3-48}$$

Since the area of the jet A_j usually is not known, a coefficient of contraction C_c is used, which relates the area of the jet at the point where the streamlines first become parallel and the internal pressure becomes atmospheric (the **vena contracta**) to the area of the orifice A_o. Thus

$$C_c = \frac{A_j}{A_o} \tag{3-49}$$

Figure 3-24 shows the variation of C_c with the diameter ratio for an orifice at the end of a pipe.

Equations 3-47, 3-48, and 3-49 can be combined to obtain an equation which will give the discharge from an orifice, based upon the continuity and energy equations. This yields an equation involving a discharge coefficient which is dependent only upon the geometry of the pipe and the orifice. The derivation of this equation, however, is left as an exercise for the student.

The simplest form of Eq. 3-47 is used for analyzing efflux from a large tank (A_1/A_j in Fig. 3-20 becomes very large) wherein $V_1 = 0$, and Eq. 3-47 can be simplified to

$$\frac{V_j^2}{2g} = \frac{\Delta p}{\gamma} \tag{3-50}$$

or

$$V_j = \sqrt{\frac{2\Delta p}{\rho}} \tag{3-51}$$

Equations 3-50 and 3-51 will give either the pressure drop necessary to produce a given jet velocity or the velocity of the jet when the pressure drop is known.

An equation for the discharge from an orifice in a tank can be obtained by combining Eq. 3-48 and Eq. 3-49 with Eq. 3-51 to give

$$Q = C_c A_o \sqrt{\frac{2\Delta p}{\rho}} \tag{3-52}$$

Example Problem 3-11: The height of water above the center of this 1-in. diameter orifice is 3 ft, as shown. Find the discharge through the orifice.

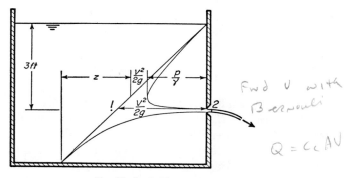

Ex. Prob. 3-11

Solution:

The coefficient of contraction is obtained from Fig. 3-24 for $\alpha = 90$ degrees and $d/D = 0$.

$$Q = C_c A_0 \sqrt{\frac{2\Delta p}{\rho}} = 0.61\,(0.00545) \sqrt{\frac{2\Delta p}{1.94}}$$

$$\Delta p = p_1 - p_2 = \gamma h_1 - 0 = 62.4 \times 3 = 187.2 \text{ psf}$$

$$Q = 0.046 \text{ cfs}$$

The pressure distribution on the side of the tank in Example Problem 3-11 can be studied by way of contrast with the fluid distribution on the end plate of the pipe in Fig. 3-20. In Fig. 3-20, the pressure upstream is considered to be uniform across the section because the fluid is a gas. In Example Problem 3-11, however, the fluid is a liquid with atmospheric pressure at the free surface and, hence, the pressure along the left wall is hydrostatic, which results in a triangular pressure distribution on the left side of the tank, as explained in Chapter 2. On the right side, the distribution is modified from hydrostatic in the vicinity of the orifice, because the fluid has acquired a velocity. The total head, see Eq. 3-38, remains the same, however, as shown in the figure—being a summation (taken horizontally for convenience) of the velocity head $V^2/2g$, the pressure head p/γ, and the elevation head z.

In Eq. 3-52, it is assumed that the velocity of the jet is uniform at the **vena contracta**—a situation which is essentially true for sharp edged orifices, but not exactly true for rounded orifices, in which the velocity along the boundary is reduced slightly, due to drag. In this case, a coefficient of velocity C_v is used to correct the velocity term

$$\sqrt{\frac{2\Delta p}{\rho}}$$

so that
$$V_j = C_v\sqrt{\frac{2\Delta p}{\rho}} \tag{3-53}$$

and
$$Q = C_c C_v A_o\sqrt{\frac{2\Delta p}{\rho}} \tag{3-54}$$

Under the ideal conditions of a sharp-edged orifice the boundary drag approaches zero and $C_v \approx 1.0$. If a rounded orifice is considered, C_v may decrease to $C_v = 0.98$. Usually, however, the degree of accuracy of the problem does not warrant the refinement of considering C_v to be less than 1.0. The boundary drag which causes C_v to be less than 1.0 can also be represented by simply a head loss, as shown in Chapter 7 on Flow in Closed Conduits.

Orifices of various types have various contraction coefficients C_c and velocity coefficients C_v. Figure 3-21 shows several different types of orifices in the side of a tank. Figure 3-21(a) is a sharp-edged orifice in which C_c is 0.61 (see also Fig. 3-25). The streamlines must turn around the sharp edge and, in so doing, the inertia of the fluid forces the flow to contract to an area less than that of the orifice. Along the outer edge of the jet the pressure is atmospheric (zero gage pressure) and, hence, by the Bernoulli equation, the velocity is the same over the entire outer surface of the jet. Inward from the outer edge of the jet, an increasing pressure is encountered.

which causes a normal acceleration of the flow outward to change the direction of the velocity. At the **vena contracta,** the streamlines become straight and parallel, the internal pressure goes to zero, and the normal acceleration also goes to zero. Consequently, there is no further change in the nature of the jet, except from outside forces, such as gravity or drag on

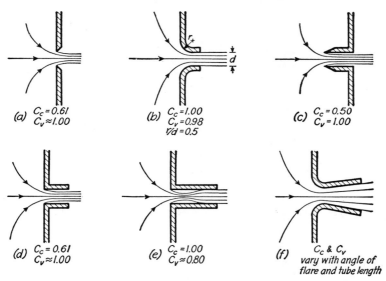

Fig. 3-21. Types of orifice.

the outer edge of the jet. Therefore, the pressure is atmospheric inside the jets as well as outside, and there is no further curvature of the streamlines caused by internal pressure.

From upstream to downstream the streamlines are seen to converge. The continuity equation shows that this convergence results in an increased velocity in the direction of flow, and the Bernoulli equation, in turn, shows that when the velocity is increased the pressure is decreased. This increasing velocity and decreasing pressure also occur inside the jet up to the **vena contracta.** The greater pressure upstream from the orifice is to be expected also from the momentum equation, since the velocity (and hence the momentum flux) is being changed in direction as well as magnitude.

Figure 3-22 shows the influence of rounding the edge of the orifice—which is a sort of streamlining—as illustrated in Fig. 3-21(b). Here, it can be noted, that as the ratio r/d, where r is the radius of curvature and d is the minimum diameter of the orifice, varies from zero to say 0.1, the product $C_c C_v$ varies from 0.61 to about 0.80. The equation of this line is

$$C_c C_v = 0.61 + 1.86 \frac{r}{d}$$

In other words, rounding the orifice so that the radius of curvature is one-tenth the diameter of the orifice causes an increase in discharge of 30 percent through the orifice (see Eq. 3-54.) When the rounding of the entrance is

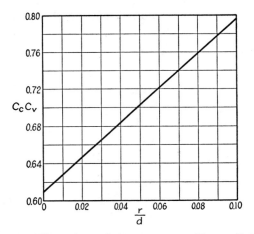

Fig. 3-22. Effect of rounded entrance on orifice coefficients.

increased so that r/d is about 0.5, the product $C_c C_v$ becomes nearly 1.0. It cannot reach 1.0 exactly, however, because C_v decreases with increasing r/d, as shown in the following paragraph.

When the edge of the orifice is sharp—that is, when $r/d = 0$—the contraction coefficient C_c is a minimum, and the velocity coefficient C_v approaches a maximum of 1.0, since the boundary beyond the beginning of the orifice offers only slight resistance to the flow. As r/d is increased, however, C_c increases and C_v decreases slightly (due to boundary resistance and the tendency toward separation along the rounded portion), until at about $r/d = 0.5$ the C_c reaches a maximum of 1.0 and C_v decreases slightly. Despite the fact that C_c will not increase further beyond 1.0, C_v will continue to decrease, due to increased length of the boundary offering resistance, as r/d is increased beyond 0.5.

Whereas the rounded orifice in Fig. 3-21 (b) increases C_c, the reentrant or Borda tube orifice in Fig. 3-21 (c) decreases C_c to 0.5. In this case, the outer streamlines must take a path even more circuitous than in Fig. 3-21 (a), with smaller radii of curvature, so that C_c is reduced to 0.5. This has been proved both mathematically and experimentally to be the minimum value of C_c.

By placing a ring, as in Fig. 3-21 (d), downstream from a sharp-edged orifice, the contraction coefficient is not changed from that for Fig. 3-21 (a), where $C_c = 0.61$, because the air has ready access to the jet and the surrounding pressure is maintained at atmospheric. As the length of the tube or ring is increased, however, the jet is not freely ventilated, and a negative ambient pressure surrounding the jet increases with the length of the tube until finally, at a length of approximately $3d$ or greater, the jet fills the tube, the Δp becomes a maximum, and the discharge is increased appreciably, see Fig. 3-21 (e), because C_c is increased to 1.0. The velocity coefficient C_v, however, is decreased because of the separation and consequent energy loss, together with boundary resistance.

For the sharp-edged orifice with the short tube, the discharge cannot be increased as much as the Bernoulli equation indicates. This results from the energy which is lost in the eddies and turbulence caused by the separation zone immediately downstream from the sharp-edged entrance, which is largely responsible for the decrease in C_v. As the edge is rounded more and more, however, the separation zone becomes smaller and smaller until the only source of loss is the drag along the boundary of the tube.

Both the sharp edge and the rounded edge orifices have a nonuniform velocity distribution across the end of the tube. Hence, the velocity coefficient C_v is less than 1.0, which is primarily responsible for a reduced discharge in Eq. 3-54.

By rounding the orifice and using an expanding tube, as shown in Fig. 3-21 (f), the largest possible discharge is obtained for a given throat size. In this case, the orifice is rounded to the shape of the free jet, as shown in Fig. 3-21 (a), from the orifice to the *vena contracta*, so that both C_c and C_v approach 1.0. Furthermore, if the pressure at the outlet of the expanding tube is atmospheric, then by the Bernoulli equation and the continuity equation the pressure in the throat is less than atmospheric. Hence, the Δp actually causing the flow is greater than for the other cases in Fig. 3-21. This is the principle of the draft tube used in hydroelectric power plants and the Venturi tube used for measurement of discharge of fluids. These devices are discussed in Chapter 7 on Flow in Closed Conduits.

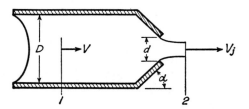

Fig. 3-23. Tapered orifice at end of pipe.

If the orifice at the end of a pipe (commonly known as an end orifice) is arranged at various angles of contraction α, as shown in Fig. 3-23, and the diameter of the pipe D is varied relative to the diameter of the orifice d, the analytical determination* of the coefficient of contraction gives values varying from 0.5 to 1.0, as shown in Fig. 3-24. This figure applies also to two-dimensional flow, for which b and B' correspond to d and D respectively in Fig. 3-23.

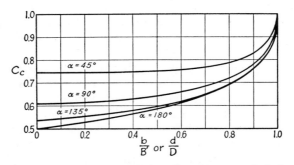

Fig. 3-24. Contraction coefficient for orifice at end of pipe.

Example Problem 3-12: A 2-in. diameter nozzle with a 45° contraction is placed on the end of a 6-in. diameter pipe line, see Fig. 3-23. If the upstream pressure at section 1 is 20 psi, what is the discharge through the orifice?

Solution:

Use Eq. 3-54, $Q = C_c C_v A_0 \sqrt{2\Delta p/\rho}$

$$d/D = 1/3$$

From Fig. 3-24 $C_c = 0.748$

Assume $C_v = 1.0$

$$\Delta p = p_1 - p_2 = (20)(144) - 0 = 2880 \text{ psf}$$

Substituting into the Equation

$$Q = (0.748)(1.0)(0.0218)(\sqrt{2 \times 2880/1.94} = \underline{\underline{0.887}} \text{ cfs}$$

The **cross-sectional shape of the orifice** has considerable effect upon the cross-sectional shape of the jet issuing from it, but very little influence upon the coefficient of contraction C_c. Of the various shapes of orifices, the same cross-sectional shape is retained along the jet only with the circular orifice. With other orifice shapes, such as triangular and rectangular, the jet develops fins or vanes which change in both shape and size along the jet, in accordance with the corners and unsymmetrical features of the orifice and the influence of surface energy. Despite the strange shapes which develop in the jets, the contraction coefficients for these orifice shapes are approximately the same as for the circular orifice.

* See Chapter 10 on Flow and Fluid Measurement for experimental determination of the coefficients for an end orifice.

For two-dimensional orifices, such as slots, the coefficient of contraction is approximately the same as that for the circle.

The foregoing developments are applicable for flow of any fluid in a pipe or from a tank wherein the pressure drop Δp is used. As pointed out in Chapter 2, however, the pressure below the free surface of a liquid is γh in which $h = (p/\gamma) + z$ is the piezometric head. In this way, Eq. 3-50 and Eq. 3-51 can be modified to

$$V_j = \sqrt{\frac{2\gamma\Delta h}{\rho}} = \sqrt{2g\Delta h} \qquad (3\text{-}55)$$

and is applicable to Fig. 3-25, in which $p_2 = 0$, $V_1 = 0$, and $h_2 = z_2$, which means that the entire head h_1 is converted to velocity head. This basic equation stems from the relationship between velocity and head which was

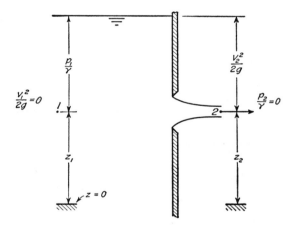

Fig. 3-25. Orifice in side of tank.

developed first by Torricelli in about 1645, and is simply a special adaptation of the Bernoulli equation which was developed later.

Flow under a **sluice gate,** which is a special two-dimensional type of orifice as shown in Fig. 3-26, is readily analysed by the continuity, energy, and momentum equations. Because the flow is contracting rapidly, and hence is essentially irrotational, the pressure distribution along the bottom of the channel can be determined indirectly by drawing a flow net, or directly by using piezometers connected along the bottom. In either case, the broken-line curve represents the pressure head along the bottom, and the velocity head is the distance from the broken line curve to the total head line. In the same manner, the pressure and velocity distribution can be determined along any other streamline in the flow or across the streamlines, since the flow is irrotational. This is illustrated by the distribution of the pressure head and the velocity along the upstream face of the gate and across the flow just below the gate, as shown in Fig. 3-26(a). Refer to the discussion of Example Problem 3-11 for an explanation of this distribution.

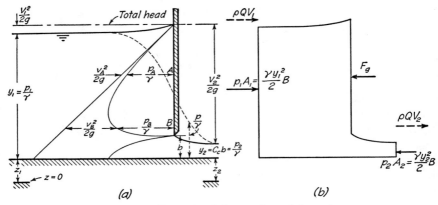

Fig. 3-26. Irrotational flow under a sluice gate.

By using the continuity equation (together with the proper contraction coefficient) and the energy or Bernoulli equation, it is possible to utilize a sluice gate as a measuring device as well as a discharge control structure. The contraction coefficient C_c is approximately 0.61 for all values of b/y_1 from zero to about 0.2. For progressively larger values of b/y_1, C_c increases rapidly because the upper streamlines are required to deviate progressively less from straight lines. This is illustrated in Fig. 3-25, where the two-dimensional flow of a slot can be cut in half longitudinally to represent a sluice gate. The curve for $\alpha = 90$ degrees and $B' = y_1$ in the abscissa must be used.

The momentum equation (as well as an integration of the pressure distribution) can be used to determine the total force against the sluice gate. By using velocity and depth of flow, as determined by the Bernoulli and continuity equations, at both sections 1 and 2, an unbalanced free-body diagram, Fig. 3-26(b), can be drawn to represent the forces involved. Equating the forces in the free-body diagram of Fig. 3-26(b) results in the following equation

$$\frac{\gamma y_1^2 B}{2} - \frac{\gamma y_2^2 B}{2} - F_g = \rho Q(V_2 - V_1)$$

Solving for the force on the gate F_g gives

$$F_g = \frac{\gamma y_1^2 B}{2} - \frac{\gamma y_2^2 B}{2} - \rho Q(V_2 - V_1)$$

For a vertical sluice gate with values of b/y_1 between 0 and 0.2, C_c for the sluice gate can be taken as 0.61, and the downstream depth can be evaluated as follows

$$y_2 = C_c b$$

The velocities can be evaluated from the continuity equation,

$$Q = V_1 y_1 B = V_2 y_2 B$$

Once the upstream depth y_1, the gate opening b, and the width of the gate B are known, the force on the gate can be determined.

By applying the continuity equation and the energy equation across the sluice gate, a general equation can be derived for calculating the unit discharge q for which $q = Q/B$. The discharge equation for the **sluice gate** is the following

$$q = \frac{C_c C_v b}{\sqrt{1 - \left(C_c \dfrac{b}{y_1}\right)^2}} \sqrt{2g(y_1 - y_2)} \tag{3-57}$$

in which C_c is 0.61 for large y_1/b values and usually $C_v \approx 1.0$.

A **weir** involves flow of a liquid with a free surface over the top of a plate (in contrast to flow **under** a plate or sluice gate). Because the flow is contracting in the direction of motion, it can be considered irrotational, and the

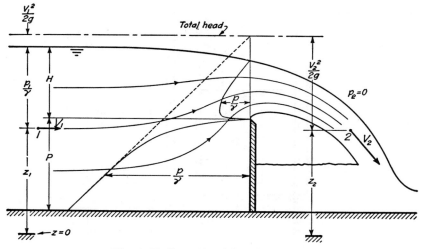

Fig. 3-27. Irrotational flow over a weir.

Bernoulli equation is applicable throughout the flow, as shown in Fig. 3-27. From section 1 to section 2, the flow is gradually accelerated from V_1 to V_2. At section 1 the pressure head is p_1/γ, but at section 2 the pressure is atmospheric—the pressure head having been converted to velocity head—and hence $p_2 = 0$. Across the streamlines directly above the weir crest, the internal pressure varies from atmospheric at the crest to a maximum inside the nappe, and then back to atmospheric at the free surface. Along the upstream face of the weir, the pressure gradually decreases from hydrostatic at the bottom to zero at the crest. The pressure and velocity distributions can be determined either experimentally or by use of the flow net.

An equation for flow over a weir can be derived by assuming that the weir is a type of orifice flow with a free surface. Hence

$$Q = CA\sqrt{2gH} \tag{3-58}$$

in which A is representative of the flow area and, for convenience, is taken as the product of the head H on the weir and width of the flow channel B. The coefficient C is the product of a coefficient of discharge C_d and the fraction $2/3$, which is explained in connection with critical flow in Chapter 8 on Flow in Open Channels. Therefore, Eq. 3-58 can be written for discharge per unit width $q = Q/B$ in which B is also the length of the weir.

$$q = C\sqrt{2g}H^{3/2} \tag{3-59}$$

The variation of the coefficient with the relative height of the weir, and general consideration of weirs of various types and shapes, which are used as devices for measuring discharge, are discussed in Chapter 10 on Flow and Fluid Measurements.

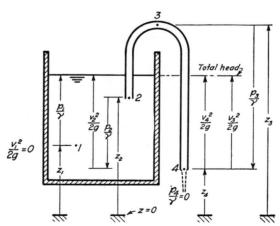

Fig. 3-28. Flow in a siphon.

The Bernoulli equation has direct application to the true *siphon* shown in Fig. 3-28. Since the flow in the container has zero velocity, the total head line throughout both the container and the siphon coincides with the water surface in the container. At point 1, $V_1^2/2g = 0$ and the total head H_T is $(p_1/\gamma) + z_1$. At point 2, however, the velocity head must be taken below the total head line, so that the pressure is

$$\frac{p_2}{\gamma} = H_T - \left(\frac{V_2^2}{2g} + z_2\right) \tag{3-60}$$

The sum of the velocity head and the elevation head is greater than the total head H_T. Therefore, the pressure in the tube must be negative. The greatest negative pressure is at point 3, where z is largest. If the sum of the velocity head and elevation head is too great at this point, the liquid will vaporize and cause the siphon to break or vapor lock and lose its prime. The negative pressure head p/γ at points 2 and 3 are indicated in Fig. 3-28 by a dimension line pointing downward. The dimension lines for the positive heads are pointing upward.

Cavitation is a phenomenon which is encountered in certain types of liquid flow. It may result in serious damage to hydraulic structures, such as turbine runners, gate piers, and spillways, and hence should be studied at this point, so that the student will have an understanding of the basic principles involved.

As the velocity head is increased at a local point, the pressure head decreases by a corresponding amount. For a liquid, however, there is a limit to this decrease in pressure—the limit being the vapor pressure of the liquid. If the pressure is decreased to vapor pressure, the liquid will vaporize and boil. For a given liquid, the vapor pressure depends upon the temperature. For example, at 60°F the vapor pressure of water is 0.256 psia, while at 212°F the vapor pressure of water is 14.7 psia (standard atmospheric pressure). Cavitation is related frequently to the *cavitation number* $\frac{p_v}{\rho V^2}$, as developed in Chapter 1 by dimensional analysis. As the liquid vaporizes, vast numbers of small vapor cavities form, which are carried downstream with the flow. When these cavities reach a region of greater pressure, they suddenly collapse as the vapor condenses again into a liquid. This collapse is accompanied by intense pressures which are caused by the water surrounding the cavity rushing in to fill the cavity where its momentum is converted into pressure from the impact of the walls of the cavity as they meet. Since these pressures reach many tens of thousands of psi (measurements indicate pressures as great as 100,000 psi) any boundary in the immediate vicinity of the collapse will likewise be subjected to repeated, intense localized pressures—the magnitude being dependent upon (among other factors) the distance of the boundary from the point of collapse. Such intense pressures upon a boundary, even though over a very small area for a very short period of time, eventually cause the boundary to fail by pitting—apparently due to fatigue from the repeated pounding which may be many hundreds of times per second.

Since the engineer is frequently confronted with the problem of eliminating, or at least reducing, the pitting of an object, it is well to list the various steps which might be taken to accomplish this:

1. Pitting can best be controlled by *eliminating* or reducing cavity formation (cavitation) by:
   a. Increasing the ambient or surrounding pressure so that the reduction in pressure, caused by the local increase in velocity, will not result in the local pressure being decreased to the vapor pressure of the liquid. This can be accomplished either with an increase of pressure over the entire system, or with a local increase in pressure by dropping the region affected to a lower elevation, so as to increase the hydrostatic pressure.
   b. Decreasing the magnitude of the local velocity, which in turn increases the local pressure. This usually can be accomplished by streamlining the flow passage or the boundary.

2. If the cavity formation (cavitation) cannot be eliminated, then the boundary can be *protected* against pitting by:
   a. Removing the boundary from the region of the collapsing cavities, or
   b. Making the boundary of tough or resilient material, such as stainless steel or rubber, or
   c. Covering the boundary with a tough material, such as stainless steel or rubber.

Thus far, the Bernoulli equation has been applied to problems where no *energy is added to or subtracted from the system.* However, the equation can be used advantageously either when energy is added to the system by means of a pump, or when energy is subtracted by a turbine. Figure 3-29 shows a pump being used to force water over a hill to a turbine below. For simplicity, the assumption is made that there is no energy lost due to boundary drag. In practical pipe line problems, however, this drag cannot be neglected.

As the water enters the intake side of the pump in Fig. 3-29, the pressure head is reduced by the amount of the velocity head. On the discharge side of the pump, however, the total head line is increased by E_{Mp}, so that a positive pressure is maintained in the pipe throughout. Just over the crest of the hill, the pipe size has been decreased, which makes the velocity head greater. At the turbine, energy has been subtracted so that the total head line is reduced by E_{Mt}. The Bernoulli equation must then be written as

$$\frac{V_1^2}{2g} + \frac{p_1}{\gamma} + z_1 + E_{Mp} = \frac{V_2^2}{2g} + \frac{p_2}{\gamma} + z_2 \quad (for\ the\ pump) \quad (3\text{-}61)$$

$$\frac{V_3^2}{2g} + \frac{p_3}{\gamma} + z_3 - E_{Mt} = \frac{V_4^2}{2g} + \frac{p_4}{\gamma} + z_4 \quad (for\ the\ turbine) \quad (3\text{-}62)$$

Points 1 and 3 are on the intake side of the pump and the turbine, and points 2 and 4 are on the discharge or outlet side of the pump and the turbine, as shown in Fig. 3-29.

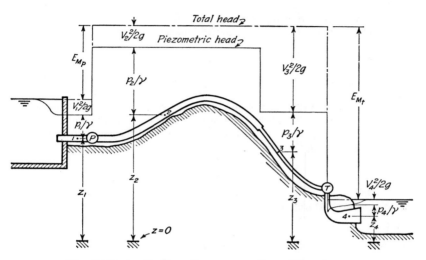

Fig. 3-29. Application of energy equation with no losses.

The draft tube downstream from the turbine is an expanding section which is designed to reduce the velocity and convert velocity head to piezometric head, and thereby increase the head drop across the turbine so that it will operate most efficiently. The jet which is discharged from the turbine through the draft tube into the reservoir is finally completely diffused in the surrounding water of the reservoir. Hence, the final velocity goes to zero, with no recovery of velocity head to increase the piezometric head beyond that at the outlet of the draft tube.

The **horsepower output** sometimes called the water horsepower of the pump is determined by the following equation.

$$HP_p = \frac{Q\gamma E_{Mp}}{550} \quad (3\text{-}63)$$

Substitute E_{Mt} for E_{Mp} to find the input horsepower HP_t, sometimes called the water horsepower, of the turbine.

The foregoing equation for horsepower HP is easily explained when it is realized that E_{Mp} represents the ft-lb of energy which the pump adds to each pound of fluid. Analysis of Q reveals that $Q\gamma$ equals pounds per second of fluid flowing, so that $Q\gamma E_{Mp}$ represents ft-lb/sec, which is power. The figure of 550 ft-lb/sec per horsepower converts the units of the equation to horsepower.

Example Problem 3-13: Referring to Fig. 3-29, assume that the pipe line is 12 in. in diameter and that the pipe line extends only to the top of the hill, where it ends in a tank whose water surface elevation is 2500 ft. The water surface in the reservoir at the pump is at 2000 ft. What horsepower would be required to pump a flow of 2 cfs to the tank on the hill?

Solution:

In applying the Bernoulli equation to this problem, it may be remembered that no energy losses are considered, and so this only illustrates the principle of adding energy to the flow system. In the practical problem, there would be energy losses in the pipe line which would have to be considered.

Applying Eq. 3-61 between the water surface of the reservoir and the water surface of the tank

$$\frac{V_1^2}{2g} + \frac{p_1}{\gamma} + z_1 + E_{Mp} = \frac{V_2^2}{2g} + \frac{p_2}{\gamma} + z_2$$

At the water surfaces, $V_1 = 0$, $p_1 = 0$

and $\qquad\qquad V_2 = 0$, $p_2 = 0$

so that $\qquad\qquad z_2 - z_1 = E_{Mp} = 500$ ft

and $\qquad\qquad HP_p = \dfrac{Q\gamma E_{Mp}}{550} = \dfrac{2 \times 62.4 \times 500}{550}$

$$HP_p = \underline{\underline{113.4 \text{ hp}}}$$

Thus far in this chapter, emphasis has been placed upon the application of the Bernoulli or energy equation to the flow of a fluid which is confined between boundaries. It can be applied also, however, to certain parts of the flow around a **submerged object.** Although detailed discussion of this is given in the chapter on submerged objects, brief consideration of flow about the upstream side of a sphere is now given as an example.

Figure 3-30 is a **sphere,** with streamlines and the diagrams of the pressure and velocity distribution for irrotational flow. Along the centerline upstream from the sphere, the streamlines are parallel—thus indicating that both the velocity and pressure are unchanging in the direction of flow. As the flow approaches the sphere, however, the streamlines near the centerline must diverge, so that they can pass around the sphere, and hence the velocity is decreased with a resulting increase in the pressure—since, by the Bernoulli equation, the sum of the velocity head and the pressure head must remain a constant for a given elevation z. In fact, at the point S the velocity has been reduced to zero and hence stagnation pressure results.

Progressing from the point S along the face of the sphere, the stream-lines gradually converge again, and the velocity increases with a corresponding decrease in pressure. At a point approximately 30 degrees from

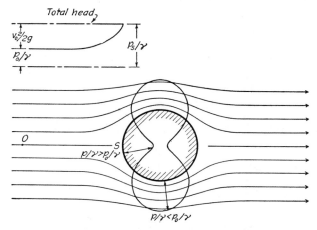

Fig. 3-30. Irrotational flow around a sphere.

the point S around the sphere, the velocity has increased so that the pressure goes to p_o and the head is $\dfrac{V_o^2}{2g} + \dfrac{p_o}{\gamma}$, as it is upstream. Progressing

farther around the sphere, the streamlines continue to converge, with the velocity head becoming even greater than the total head. Consequently, the pressure head must become less than p_o so that the total head (the algebraic sum of the velocity head and the pressure head) remains a constant at all points along the face of the sphere. The minimum pressure near the top of the sphere is approximately $1.25 V_o^2/2g$ less than p_o/γ.

In the flow pattern around the sphere of Fig. 3-30, the elevation head z is ignored, because the variation of pressure head from point to point is usually large compared with the variation of elevation. If the variation of elevation is significant, however, the piezometric head $(p/\gamma) + z$ must be used in place of simply p/γ.

The fact that stagnation pressure develops at the center of the fore-part of a sphere has led to development of the Pitot tube for the measurement of velocity of flow. The Pitot tube, developed by the French scientist Henri Pitot in 1732, has been used widely to measure the air speed of aircraft and the velocity of water flowing in pipes and open channels.

The design characteristics and use of the Pitot tube are discussed in greater detail in the chapter on Flow and Fluid Measurements, and further discussion of flow around submerged objects is also given in a later chapter.

- SUMMARY

1. The three basic equations in fluid dynamics are:
   a. Continuity equation
   b. Momentum equation
   c. Energy equation

2. A *streamline* is an imaginary line within the flow for which the tangent at any point is the time average of the direction of motion at that point. A *streamtube* is a tube of fluid bounded by a group of streamlines which enclose the flow.

3. *Acceleration* is the rate of change of velocity with respect to either time or distance.

4. *Steady flow* has no variation of velocity at a point with respect to time. *Uniform flow* has no variation of velocity at a given time with respect to distance.

5. A *flow net* is a system of streamlines and orthogonal lines which show the idealized flow pattern for a given system of boundaries.

6. *Separation* of flow occurs in a real fluid at changes in boundary shape because of either the inertia of the fluid or the velocity distribution of the flow near the boundary.

7. *Irrotational flow,* which is represented by the flow net, occurs in an ideal fluid when each fluid element in the flow system has no *net* rotation about its own mass center.

8. The *continuity equation* stems from the law of conservation of mass and states that the flux of mass along a streamtube is a constant. It involves the velocity to the first power and is written for mass flux or discharge as

$$G = \rho_1 A_1 V_1 = \rho_2 A_2 V_2 \tag{3-14}$$

and for volume flux or discharge as

$$Q = A_1 V_1 = A_2 V_2 \tag{3-16}$$

9. The *momentum equation* which contains vector quantities is derived from Newton's second law and relates the change in momentum flux to the force which causes this change. It contains the velocity to the second power, and for the x-direction it is

$$\Sigma F_x = \rho Q \Delta V_x = \rho Q (V_2 - V_1)_x \tag{3-18}$$

The momentum equation is used to solve problems involving deflected jets, jet propulsion, flow transitions and bends, propeller thrust, and the hydraulic jump.

10. The *energy equation* involves scalar quantities and is derived from Newton's second law of motion. It is based upon the work-energy principle and can be written as:

$$\frac{V_1^2}{2g} + \frac{p_1}{\gamma} + z_1 + E_M + E_H = \frac{V_2^2}{2g} + \frac{p_2}{\gamma} + z_2 + H_L \qquad (3\text{-}44)$$

The energy equation is used to solve problems involving the relative magnitude of the terms in the equation for such phenomena as: jets issuing from an orifice, jet trajectory, flow under a gate, flow over a weir, siphons, transition flow in pipes and open channels, flow around submerged objects, and flow associated with pumps and turbines.

11. The basic integral equations for the correction coefficients are:

$$\text{Continuity} \quad 1 = \frac{1}{A} \int_A \frac{v}{V} \, dA$$

$$\text{Momentum } K_m = \frac{1}{A} \int_A \left(\frac{v}{V}\right)^2 dA \qquad (3\text{-}24)$$

$$\text{Energy} \quad K_e = \frac{1}{A} \int_A \left(\frac{v}{V}\right)^3 dA \qquad (3\text{-}42)$$

References

1. Binder, R. C., *Fluid Mechanics*, 3rd ed. Englewood Cliffs, N. J.: Prentice-Hall, Inc., 1955.

2. Jaeger, Charles, *Engineering Fluid Mechanics*. London-Glasgow: Blackie & Son, Ltd., 1956.

3. Rouse, Hunter, *Elementary Mechanics of Fluids*. New York: John Wiley & Sons, Inc., 1946.

4. Rouse, Hunter, *Fluid Mechanics for Hydraulic Engineers*. New York: McGraw-Hill Book Co., Inc., 1938.

5. Shoder, Ernest W. and Francis M. Dawson, *Hydraulics*. New York: McGraw-Hill Book Co., Inc., 1934.

6. Streeter, Victor L., *Fluid Mechanics*, 2nd ed. New York: McGraw-Hill Book Co., Inc., 1958.

Problems

3-1. In steady flow, if the velocity varies linearly along a straight line between two points 12 in. apart, what is the acceleration at each of the points A and B if the velocity at A is 2 fps and the velocity at B is 38 fps?

3-2. At point A on a streamline, the velocity is 10 fps and the rate of increase of velocity along the streamline is 3 fps/ft. The radius of curvature of the streamline at point A is 2 ft. What is the total acceleration at point A, and what is the angle between the acceleration and the streamline for steady flow?

3-3. For steady flow in a forced vortex, the velocity varies according to the equation $v = Cr$, where r is in ft. At a radius of 3 in., the velocity is 3 fps. What is the acceleration at a radius of 2 in.? At a radius of 6 in.?

3-4. If horizontal flow of uniform depth occurs radially outward from a point, what is the acceleration at a radial distance of 3 ft from the source, if the velocity 1 ft from the source is 10 fps?

3-5. A body is moving in a circular path at 30 fps. If the radius of the circular path is 30 ft, what is the normal acceleration?

3-6. A body moving in a straight line is decreasing speed linearly with respect to distance. At point A the velocity is 50 fps and at point B, 40 ft away from A, the velocity is 5 fps. What is the total acceleration at A, at B, and at a point C located midway between A and B?

3-7. A body moving from A to B along the circumference of a circle of radius 20 ft is increasing its speed at a rate of 0.5 fps per ft. At point A, the velocity is 8 fps in a direction due east. What is the total acceleration of the body at point B, if it is going due west at B?

3-8. A boat traveling due east at a velocity of 10 knots is crossing a river which flows due south at a velocity of 5 knots. What is the velocity of the water relative to the boat?

3-9. An airplane traveling 300 mph is going due north. A 100 mph tail wind is coming from a direction 45° south of east. What is the velocity of the air relative to the plane.

3-10. A ball is thrown out the window of a moving car at right angles to the direction of motion. If the ball's velocity relative to the car is 50 fps and the car's velocity is 80 fps, what is the ball's speed with respect to the ground, and what angle does its path make with the path of the car?

3-11. A car traveling 50 mph due east has a tail wind of 70 mph. If the wind is coming from the direction 30° north of west, what is the velocity of the car relative to the wind?

3-12. An airplane is flying 400 mph due east. A headwind blowing from the north 45° east has a velocity of 100 mph. What is the velocity of the wind relative to the plane? Give both magnitude and direction.

3-13. A lawn sprinkler rotates in a horizontal plane at 60 rpm. The distance from the center of rotation to the sprinkler nozzle is 9 in. If the nozzle at an angle of 45° to the rotating arm is also in the horizontal plane of the rotating arm, what is the absolute velocity of the issuing jet of water, if the jet velocity relative to the nozzle is 40 fps?

3-14. A man rows a boat 3 miles upstream and 3 miles back to his starting place in 3 hours. If he can row a boat 5 mph in still water, how fast is the stream flowing?

3-15. If the flow net in Fig. 3-6 is two-dimensional flow, and the velocity of the uniform flow in the outlet section is 20 fps, what is the maximum velocity as the flow enters the outlet, and where does the maximum velocity occur?

3-16. Construct a flow net for unconfined irrotational flow around a cylinder.

3-17. If the velocity in the straight pipe leading to the elbow in Fig. 3-8(b) is uniform and equal to 10 fps, find the velocity on (a) the inside of the bend, and (b) the outside of the bend.

3-18. Draw the flow net for flow past the front of a flat plate 1 in. thick, when the front of the plate is half a cylinder 1 in. in diameter.

3-19. A 5-in. diameter pipe is connected to a 3-in. diameter pipe. If the velocity in the 3-in. pipe is 10 fps, find Q and the velocity in the 5-in. pipe.

3-20. A square conduit 4 in. on a side discharges into a 3-in. diameter pipe. If both flow full, what is the relationship between the velocities in the two pipes?

3-21. The average velocity in a 3-in. diameter pipe is 20 fps. Find the discharge (a) in cfs, (b) in gpm, (c) in lb/sec, and (d) in gpd.

3-22. A concrete-lined canal carries 500 cfs at an average velocity of 4 fps. If the bottom width is 6 ft and the slope of the sides is 1H:2V, what is the depth of water in the canal?

3-23. The depth of flow in a rectangular canal 12 ft wide is 6 ft. If the depth changes to 4 ft and the width changes to 15 ft at some point downstream, what is the relationship between the velocities at the two points, if steady flow exists?

3-24. Derive an expression for the relationship between the velocities at any two sections of a variable size circular pipe line carrying steady flow.

3-25. A 4-in. diameter pipe is flowing full at an average velocity of 20 fps. It discharges into a swimming pool which is 30 ft long by 15 ft wide. If the pool has a horizontal bottom, how long will it take to fill the pool to a depth of 7 ft? At what rate will the water surface rise in the pool? Give answer in inches per hour.

3-26. A jet of water issuing from a kitchen sink faucet fills a quart bottle in 10 seconds. If the diameter of the faucet outlet is $\frac{1}{2}$ in., what is the velocity of the jet as it leaves the faucet?

3-27. A fire nozzle often has a diameter of $1\frac{1}{8}$ in. According to many fire codes, a fire hose should deliver at least 300 gpm. What is the velocity of the jet issuing under such conditions?

3-28. A pipe line must carry 500 gpm. If the velocity in the pipe must not exceed 10 fps, what is the minimum size of pipe which can be used?

3-29. A canal has a bottom width of 10 ft and side slopes of 2H:1V. The canal must carry 200 cfs, and the maximum average velocity must not exceed 2 fps, to avoid scouring the earth lining of the canal. What is the minimum allowable depth of flow in the canal?

3-30. A 6-in. diameter pipe carries 93 lb of oil per second. If the specific gravity of the oil is 0.90, what is the discharge in the pipe (**a**) in gpm, and (**b**) in cfs?

3-31. A 2-in. diameter water pipe carries 100 gpm. If the average velocity in a 1-in. diameter pipe is the same as in the 2-in., what is the discharge in the 1-in. pipe?

3-32. Pipe *A*, a 12-in. diameter pipe, branches into two pipes *B* and *C*. Pipe *B* is 6 in. in diameter and pipe *C* is 3 in. in diameter. If the flow in *A* divides so that $\frac{2}{3}$ of the flow goes into pipe *B* and $\frac{1}{3}$ in pipe *C*, what is the discharge in pipe *A*, if the velocity must not exceed 10 fps in any of the pipes?

3-33. What is the smallest size of square conduit necessary to carry 100 cfs at a maximum velocity of 8 fps?

3-34. In a laboratory test, water flowing through a pipe 6 in. in diameter is dumped into a tank set on some platform scales. If 1,200 lb accumulates in 20 seconds, what is the average velocity in the pipe?

3-35. Water flows 18 in. deep in a glass-walled flume which is 2 ft wide. If average concentration of sediment in the water is 0.2% and 300 lb of sediment pool which is 30 ft long, 20 ft wide, and 5 ft deep, if the flume empties into the pool?

3-36. Water flowing in a rectangular channel 10 ft wide is 8 ft deep. If the average concentration of sediment in the water is 0.2% and 300 lb of sediment pass a given section in one minute, what is the average velocity in the channel?

3-37. The discharge in a pipe line is 2 cfs. At one point, the pipe line changes from one diameter pipe to another. If the velocity in the small pipe is 12 fps and the ratio of the velocities is 6 to 1, what are the diameters of the two pipes?

3-38. A 3-in. diameter hose is connected to a 1-in. diameter nozzle. If the hose is carrying 200 gpm, what is the average velocity in the hose and the velocity of the nozzle jet?

3-39. A rectangular canal 20 ft wide is designed to flow at a maximum average velocity of 10 fps with a corresponding discharge of 1500 cfs. How deep is the water in the canal at maximum flow?

3-40. A trapezoidal canal is 6 ft wide at the base, the side slopes are 1H:1V, the depth is 6 ft and the average velocity is 5 fps. What is the discharge?

3-41. A 2-in. diameter horizontal jet of water with a velocity of 75 fps strikes a vertical wall at 90° to the wall. What is the force exerted on the wall?

3-42. A 3-in. jet has a velocity of 40 fps. The jet of water strikes a stationary blade which diverts the water through an angle of 140°. Find the force of the jet on the curved blade.

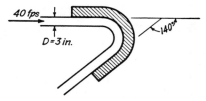

Prob. 3-42

3-43. A horizontal 45° reducing elbow (24-in. diameter to 12-in. diameter) carries 30 cfs. If the pressure in the 12-in. pipe equals 10 psi, what is the force exerted on the elbow? Give the magnitude and direction of the force.

3-44. A 90° elbow in a 10-in. horizontal pipe line is encased in a concrete block which will withstand a maximum force of one ton in any one direction. The relationship between discharge and pressure varies linearly in the following way:

$$Q = 1 \quad \text{cfs} \qquad p = 30 \text{ psi}$$
$$Q = 0.5 \text{ cfs} \qquad p = 45 \text{ psi}$$
$$Q = 1.5 \text{ cfs} \qquad p = 15 \text{ psi}$$

Is the elbow installed safely for the condition of no flow in the pipe? If not, determine what minimum flow must exist in order for the installation to be just barely safe.

3-45. A 90° elbow located in a horizontal 12-in. diameter pipe line is carrying water at the rate of 10 cfs. The pressure in the 12-in. pipe is 10 psi. What is the force exerted on the elbow?

Prob. 3-45   **Prob. 3-46**

3-46. The discharge in a 3-in. diameter free jet is 3 cfs. If the jet strikes a stationary blade which deflects the jet through an angle of 75°, what is the force on the blade?

3-47. The velocity distribution in a 4-in. pipe is parabolic and is represented by the following equation

$$v = J(r_0^2 - r^2)$$

Determine the momentum flux coefficient K_m.

3-48. Solve Example Problem 3-4 when the blade is moving at 30 fps in the opposite direction of the jet.

3-49. A 2-in. jet of water with a velocity of 60 fps strikes a vertical wall at an angle of 45°. What is the force exerted upon the wall?

3-50. A horizontal air jet at 40°F has a 4-in. diameter and a velocity of 200 fps. The jet strikes a vertical plate moving directly into the jet at a velocity of 100 fps. What is the force of the jet on the plate?

3-51. A 2-in. jet of water traveling at a velocity of 50 fps strikes a blade which turns the jet through an angle of 180°. If the blade is traveling at a velocity of 20 fps directly away from the jet, what is the force of the jet on the moving blade?

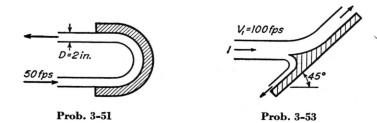

Prob. 3-51   **Prob. 3-53**

3-52. A 2-in. circular jet of water traveling at a rate of 50 ft/sec strikes a flat stationary plate inclined at 60° to the path of the horizontal jet. Determine the force required to hold the plate in position.

3-53. The stationary divider blade divides the free jet so that $Q/3$ goes down. What is the force exerted by the jet on the blade if $Q = 10$ cfs of water?

3-54. A jet issuing from a 3-in. diameter nozzle has a velocity of 50 fps. If the nozzle is attached to a horizontal 6-in. diameter pipe, what is the pressure in the 6-in. pipe if water is flowing?

3-55. A 2-in. diameter siphon is draining a fairly large tank of gasoline (sp. gr. = 0.73). The highest point in the siphon is 13 ft above the surface of the gasoline in the tank, and the discharge end of the siphon is 13 ft below the surface. What is the discharge, and what is the pressure at the highest point in terms of mercury vacuum?

3-56. A 2-in. sharp-edged orifice exists in the horizontal bottom of a large tank. The water in the tank is 12 ft deep. Find the discharge through the orifice and the velocity in the jet 6 ft below the orifice.

3-57. Referring to Problem 3-47, what is the value of the kinetic energy correction coefficient K_e?

3-58. Referring to Fig. 3-19, the following data were obtained:

$$Q = 3 \text{ cfs}, \qquad D_1 = 3 \text{ inches}, \qquad V_{y1} = 2V_{z1}.$$

Find the elevation of the jet at point 3, the diameter of the jet at point 3, and the distance from point 1 to point 4, assuming that the elevations of the two points are the same.

3-59. An end orifice as shown in Fig. 3-20 has a diameter of 3 in. and a contraction coefficient as shown in Fig. 3-21(a). If the orifice is on the end of a horizontal 6-in. pipe and the pressure in the 6-in. pipe is 30 psi, find the discharge.

3-60. A jet issues from a Borda tube as illustrated in Fig. 3-21(c). If the tube diameter is 4 in. and the tube is in the side of a large tank, find the discharge through the tube when the height of water above the centerline of the tube is 8 ft. What is the velocity and direction of the jet after it has fallen 3 ft? How far horizontally is the jet from the tank at that point?

3-61. Derive the equation for flow under a sluice gate as shown in Fig. 3-26.

$$q = \frac{C_c}{\sqrt{1 + C_c \dfrac{b}{y_1}}} \, b\sqrt{2gy_1}$$

3-62. A siphon like the one in Fig. 3-28 has the following dimensions: Water surface elevation = 10 ft, elevation of point 2 = 8 ft, elevation of point 3 = 15 ft, elevation of point 4 = 0 ft. If the siphon is 2 in. in diameter, what is the water discharge? What is the pressure at point 3?

3-63. If the vapor pressure of water is 1.5 psia, what is the maximum discharge which can be obtained by lowering point 4, if point 3 remains at the same elevation as in Problem 3-62?

3-64. Water flows under a sluice gate so that the upstream depth is 6 ft and the depth after contraction is 1 ft at point A. A hydraulic jump occurs immediately beyond point A. What is the discharge per foot of width and what is the depth downstream from the jump?

3-65. A sluice gate in a rectangular channel is the same width as the channel which is 6 ft. The gate opening is 1.5 ft and the coefficient of contraction for the gate is 0.61. The average velocity upstream from the gate is 4 fps. What is the discharge through the gate and the depth upstream from the gate?

3-66. If the coefficient of friction between the gate and the slot in Problem 3-65 is 0.40, what vertical force will be required to start the gate to move upward if the gate weighs 500 lbs?

3-67. A sharp-edged orifice 1 in. in diameter discharges a free jet of water. If the orifice is in a vertical plane and the jet strikes a floor 5 ft below the center of the orifice at a horizontal distance of 10 ft from the orifice, what is the orifice discharge, and how deep is the water above the center of the orifice?

3-68. A short tube as in Fig. 3-21(e) in the vertical side of a large tank discharges 30 gpm when the water is 3 ft above the opening. What diameter is the short tube? Neglect resistance so that $C_v = 1.00$.

3-69. A sharp-edged slot in the horizontal bottom of a tank is 12 in. long and $\frac{1}{2}$ in. wide. What is the discharge in the free jet issuing from the slot if the water is 10 ft deep in the tank which is 12-in. square?

3-70. A flat orifice plate is welded to the end of a 6-in. horizontal pipe. If the sharp-edged orifice in the plate is 4 in. in diameter, what is the discharge through the orifice if the pressure in the 6-in. pipe is 10 psi?

3-71. A sluice gate 4 ft wide is in a 4 ft rectangular channel. If the upstream depth is 6 ft and the downstream depth is 5 ft, what is the flow through the gate when the bottom of the gate is 1 ft above the floor of the channel? Neglect head losses.

3-72. A $\frac{1}{2}$-inch pipe 2 ft long is connected at 90° to the horizontal bottom of a large open tank. The entrance to the pipe is rounded and the depth of water in the tank is 1 ft. What is the discharge through the $\frac{1}{2}$-in. pipe if it discharges freely into the air? What is the pressure in the pipe 18 in. above the lower end? Neglect head losses.

3-73. The sharp-edged orifice is 1-in. in diameter. $p_1 = 35$ psi, $p_2 = 15$ psi. What is the discharge through the orifice?

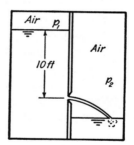

Prob. 3-73

3-74. A nozzle with a $1\frac{1}{8}$-in. outlet is connected to a 3-in. diameter fire hose. The pressure behind the nozzle is 80 psi. Two firemen handling the nozzle hold the nozzle 4 ft off the ground. What is the maximum distance that the firemen can stand back and still get the water on the roof which is 64 ft high? At what angle will they be holding the nozzle?

3-75. A nozzle is fixed so that it is inclined at an angle of 45° with the horizontal, and the velocity of the 2-in. jet leaving the nozzle is 100 fps. What is the maximum height wall that this jet will go over? How far away from the nozzle must the wall be in order for the jet to go over?

3-76. If the discharge is 15 cfs and the diameter at A is 12 in., what is the diameter at B? $z_A = 8$ ft. $z_B = 2$ ft. The specific gravity of the fluid in the gage is 2.50. Water is flowing.

3-77. For purposes of flow measurement, a Venturi meter having a 3-in. throat is installed in a 6-in. water pipe. If a differential mercury manometer connected between piezometer openings in the pipe and the throat reads 1.00 ft, what rate of flow is indicated?

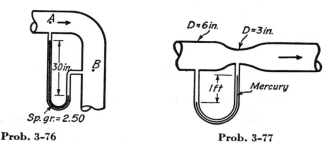

Prob. 3-76   **Prob. 3-77**

3-78. A 45° reducing elbow at the end of a horizontal pipe line is 8 in. in diameter at the upstream end and 4 in. at the other. Determine the magnitude and direction of the force required to hold it in position when a discharge of 2 cfs of water is flowing through it into the air.

3-79. Using the figure for Problem 3-76, compute the gage reading in inches if: elevation of A = 30 ft, elevation of B = 10 ft, diameter at A = 10 in., diameter at B = 6 in., discharge = 3 cfs, fluid in the gage is mercury.

3-80. If D_A = 6 in., D_B = 3 in., and V_A = 5.68 ft/sec, find the magnitude in inches and the direction of the reading on the mercury manometer.

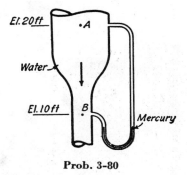

Prob. 3-80

3-81. A free jet of water is being discharged by the nozzle as shown. If the diameter of the jet at the nozzle is 3 in., find the output horsepower of the pump.

3-82. A 2-in. diameter nozzle discharges a 2-in. diameter jet at a height of 10 ft above a large horizontal glass plate which diverts the jet through a 90° angle. The jet velocity at the nozzle is 25 fps and the jet is vertically downward. If the plate will break when acted upon by a dynamic force of 40 lb, will the jet of water break the glass?

3-83. A free jet of water at point A has a 2-in. diameter and is directed at an angle of 45° with the horizontal. If the discharge is 1.09 cfs, where will the centerline of the jet strike the wall? Neglect air friction.

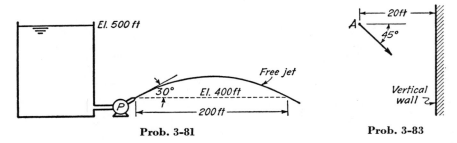

Prob. 3-81   Prob. 3-83

3-84. If the combined weight of the people and the raft is 300 lb, how high above the whale are they supported? The diameter of the jet at the whale is 6 in. and the velocity is 100 fps. Assume the jet deflects 90°.

Prob. 3-84

3-85. An 18-in. vertical pipe leading into a turbine carries a discharge of 40 cfs. A pressure gage on the 18-in. pipe reads 50 psi at a point 10 ft above the center of the turbine. The turbine draft tube discharges below the surface of a pool. The water surface in the pool is 10 ft below the center of the turbine. If the turbine is 75% efficient, what is the output horsepower of the turbine?

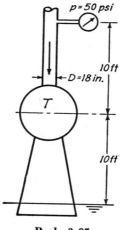

Prob. 3-85

3-86. What is the discharge in the pipe line if oil (sp. gr. = 0.8) is flowing as shown. Neglect viscous effects.

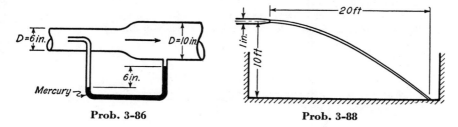

Prob. 3-86   **Prob. 3-88**

3-87. In Problem 3-81, if, instead of the 200 ft horizontal distance, the jet went up to a maximum height of 40 ft, what would the pump horsepower be?

3-88. A garden hose with 1 in. diameter is held horizontally while filling a swimming pool at the rate of 0.05 cfs. When the pool is empty, the jet strikes the floor 20 ft horizontally from the end of the nozzle and 10 ft below the nozzle. Calculate the force of the water exerted on the nozzle.

3-89. A nozzle at elevation 100 ft is shooting a 3-in. jet of water vertically into the air. The nozzle is on the end of a 6-in. pipe. If a pressure gage at elevation 95 on the 6-in. pipe reads 30 psi, what is the discharge and how high will the jet rise?

3-90. A free jet of water 2 in. in diameter is discharged from a nozzle at an angle of 60° up from the horizontal. If the discharge is 980 gpm, what is the maximum distance that the nozzle can be placed from a building and still get water into a window which is 60 ft above the nozzle?

3-91. The pump is pumping 1 cfs of water. What is the input horsepower to the pump if the efficiency of the pump is 80%?

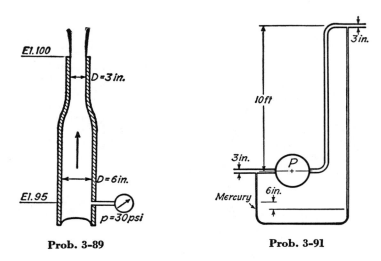

Prob. 3-89   **Prob. 3-91**

3-92. A lawn sprinkler with a horizontal arm, as shown, is held stationary by a force P applied perpendicular to the arm 3 in. from the center of rotation. If the pressure in the $\frac{3}{4}$-in. pipe is 50 psi, what is the required force P? If P were removed, how fast would the arm rotate (a) neglecting friction in the rotating arm, and (b) if the torque required to overcome friction is 4-in-lb.

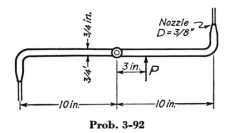

Prob. 3-92

3-93. For water purification purposes, a chlorine solution ($\rho = 1.98$ slugs/cf) is pumped from an open tank into a pipe line. To obtain efficient mixing, the solution should enter the pipe line at a velocity of 30 fps. The quantity required is 10 cu ft per hour. The pressure in the pipe line where the chlorine enters is 100 psi, and the solution enters the pipe line at a point 35 ft above the free surface of the reservoir from which it is pumped. Assuming a pump efficiency of 80% and frictionless flow, what is the horsepower input required by the pump which is pumping the chlorine?

3-94. A stream of water (25 cfs) must be conveyed through a pipe line from canal A across a depression, the bottom of which is 50 ft below the water surface of canal A, and up the hill to be discharged into canal B, the water surface of which is 40 ft above that of canal A. Assuming an overall operating efficiency of

55%, determine (a) the size (HP) pump required. (b) If the pipe is 2 ft in diameter, what is the maximum elevation at which the pump can be placed without developing a pressure less than 10 in. of mercury vacuum?

3-95. A vertical plate hinged by a frictionless hinge at the bottom is hit on each side by a jet from nozzles located equidistant from the plate but on opposite sides. If the nozzles have the same cross section but are inclined with the horizontal at angles of 30° and 60° respectively, and if the jets do not reach their maximum height before hitting the plate, what relationship must exist between the two velocities at the nozzles in order to keep the plate in vertical equilibrium? Assume that the jets strike the plate at the same elevation.

3-96. A transition in a pipe line is connected by a differential manometer. If the manometer reads 12 in. of mercury, what is the discharge through the transition?

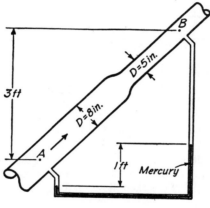

Prob. 3-96

3-97. In Problem 3-96, if the pressure at point A is 10 psi, find the pressure in psi at point B.

3-98. A vertical free jet of water has a velocity of 30 ft per second at a point where the diameter is 2 in. Find the diameter of the jet 10 ft above point A.

3-99. A jet of water directed vertically upward rises to a height of 40 ft. If the diameter of the nozzle with a contraction coefficient of 1.0 is 3 in., what is the discharge of the jet?

3-100. Using the figure for Problem 3-86, the following data apply to this problem: Small diameter = 4 in., large diameter = 9 in., Q = 2.0 cfs of oil (S = 0.85), the gage fluid has a specific gravity of 4.0. Find the gage reading in inches.

3-101. A pipe line carrying water goes from elevation 320 ft to elevation 400 ft. At elevation 320 ft the diameter is 6 in. and at 400 ft the velocity of flow is 8 ft/sec. If the pressure at the higher point is 20 psi less than the pressure at the 320 ft elevation, what is the discharge and what is the pipe diameter at the 400 ft elevation? Neglect losses.

3-102. A 30-HP pump is pumping water against a total head of 30 ft. Assuming 100% efficiency, what is the flow through the pump? Neglect losses.

3-103. The inlet side to a turbine is a 12-in. diameter pipe and the outlet side is an 18-in. pipe. If the pressure at the inlet is 100 psi and the pressure on the outlet side is 12 in. of mercury vacuum, what horsepower is the turbine producing when the flow through the turbine is 30 cfs and the efficiency is 85%?

3-104. If the flow is steady in this system and the pressure at point A is 5.2 psi, find the pressure at point B.

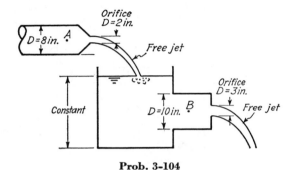

Prob. 3-104

3-105. A pipe transition from A to B involves the following data: $Q = 5$cfs, $V_B = 5V_A$, static pressure at $A = 30$ psi, and a Pitot tube when inserted at point A gives a reading of 31 psi. Find the following: V_A, V_B, D_A, D_B, and p_B.

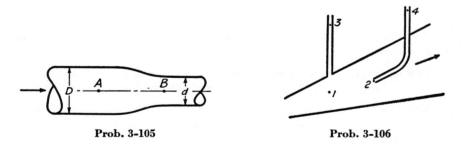

Prob. 3-105   **Prob. 3-106**

3-106. The following information applies to the figure shown: $D_1 = 6$ in., $D_2 = 12$ in. Elevations of the various points are: 1–20 ft, 2–50 ft, 3–60 ft, 4–100 ft. When the Pitot tube is removed, what is the pressure at point 2? Water flowing.

3-107. A 90° elbow in a horizontal pipe line changes diameter from a 12-in. pipe to a 10-in. pipe. The pressure in the 12-in. pipe equals 10 psi, and the pressure in the 10-in. pipe is 8 psi. Find the force on the elbow.

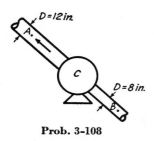

Prob. 3-108

3-108. The following applies to the figure: $p_A = 30$ psi, $p_B = 100$ psi, $Q = 5$ cfs. Elevation of $A = 220$ ft, elevation of $B = 100$ ft. Is the machine at C a turbine or a pump? What is the water horsepower associated with this machine?

3-109. A tank on wheels sits on a horizontal plane. A 3-in. sharp-edged orifice in the back end of the tank discharges water under a constant 3-ft head. Assuming that the tank and water weigh 500 lb and that there is no friction in the wheels, what is the acceleration of the tank?

3-110. If a liquid discharges through a vertical orifice in the side of a tank, prove that the propulsive force on the tank is evaluated by this equation:

$$F = 2\gamma h A_j$$

where h is the head on the orifice in ft and A_j is the area of the jet at the *vena contracta*.

3-111. In Fig. 3-27, assume that V_2 is 10 fps and that the angle between V_2 and the horizontal is 60°. Also assume that the sheet of water at point 2 is 6 in. thick and that it remains essentially constant in cross-section as it is deflected into the horizontal direction. Estimate the depth of water between the weir and the jet. Assume two-dimensional flow.

3-112. A pump draws water from reservoir A and discharges it into reservoir B. The difference in the elevations of the two water surfaces is 100 ft. If the centerline of the 6-in. suction pipe of the pump is located 20 ft above the surface of reservoir A, what is the maximum discharge which the. pump will produce if the vapor pressure of the water is 26 in. of mercury vacuum? What horsepower would be required for the maximum discharge?

3-113. If the uniform velocity of water approaching the sphere in Fig. 3-30 is 40 fps and the pressure at point 0 is 6 psi, what are the minimum and maximum pressures on the boundary of the sphere, and where are they located? Analyze the problem using Fig. 3-30 as the flow pattern in a horizontal plane.

3-114. If $h_1 = 8$ in. and $h_2 = 12$ in., calculate the pump horsepower required for the conditions shown. Water flowing.

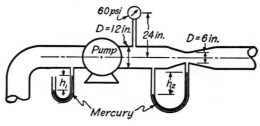

Prob. 3-114

3-115. A vertical jet of water issuing from a 1-in. nozzle at point B rises to a height of 75 ft. What is the discharge and the pressure at point A?

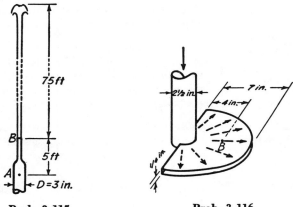

Prob. 3-115   Prob. 3-116

3-116. A discharge of 0.2 cfs is flowing in the $2\frac{1}{2}$-in. pipe. If the flow between the plates discharges into the air, what is the pressure at point B? What is the pressure gradient at B? Water flowing.

3-117. Find the discharge in the pipe if water is flowing.

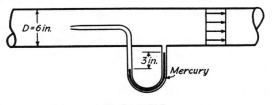

Prob. 3-117

chapter 4

Fluid

Compressibility

In the modern age of supersonic velocities, guided missiles, and rarified gases, the subject of compressible flow takes on special significance. There are other more commonplace phenomena which also illustrate the influence of compressibility. The change in the pitch of the sound caused by a vehicle on the highway as it approaches and passes the listener is observed frequently—as is the lapse in time between the flash of lightning and the clap of thunder. The basic principles for understanding these and other phenomena are developed in this chapter.

As stated in Chapter 1, all forms of matter, whether gas, liquid, or solid, are compressible. Because of the close molecular spacing of liquids and solids, they may sometimes be considered to be incompressible when compared with a gas, which generally has a wide molecular spacing. Furthermore, under certain conditions, even a gas is considered incompressible— as was done throughout Chapter 3. The present chapter is intended to discuss the various factors involved in deciding whether fluid compressibility is an important variable, and how to include it in problems of fluid mechanics when it is important.

In Chapter 1, the bulk modulus of elasticity E was stated to be

$$E = \rho \frac{dp}{d\rho} = -\tilde{V} \frac{dp}{d\tilde{V}} \qquad (1\text{-}4/1\text{-}5)$$

For an **isothermal process,** p/γ = constant, and the foregoing equation can be written as

$$E = \rho \frac{dp}{d\rho} = p \qquad (1\text{-}8)$$

and for an **adiabatic process** p/γ^k = constant, hence the foregoing equation can be written as

$$E = \frac{dp}{dp/kp} = kp \qquad (1\text{-}9)$$

For various fluids, there is a rather wide range of values for the bulk modulus of elasticity, as shown in Tables V and VI. For example, air is approximately 20,000 times more compressible than water.

There is no sharp line of demarcation between a flow system in which compressibility is unimportant and one in which compressibility is important. In general, if the density (or volume) changes are appreciable, compressibility must be considered, and if the density changes are insignificant, it need not be considered. In Chapter 3, the changes in density were well within the category of being relatively small.

There are three types of problems in which the **changes in density are** *important.* These can be classified as follows:

1. Fluid systems which are subjected to sudden (and sometimes repeated) **accelerations.** Typical problems involving such accelerations are those associated with quick-acting valves, propagation of **elastic waves** such as explosions, and rapid and repeated vibrations such as those associated with cavitation, noise reduction, and various problems in acoustics.

2. Fluid systems in which **very large velocities** are prevalent. Such velocities are those approaching or exceeding the velocity of sound. Typical problems involving very large velocities are those associated with high-speed aircraft, large velocity flow through an opening such as a nozzle, jet propulsion, steam and gas turbines, airplane propellers, and projectiles.

3. Gas systems in which there are appreciable variations in density due to the **great heights** involved. Most problems involving this type of system are meteorological. One of them is the determination of the variation in atmospheric pressure with elevation, which was treated in Chapter 2. In this text, no further consideration is given to this type of system.

A problem involving the first type of fluid system is the simple elastic wave. To analyse this wave, however, both the continuity equation and the momentum equation are needed. These will now be considered. Problems of the second type which are considered in this chapter involve the energy equation in addition to the continuity and momentum equations. In the remainder of the chapter, consideration is given to compressible flow through contracting and expanding sections of closed conduits without boundary resistance.

• CONTINUITY EQUATION

In Chapter 3, the continuity equation for **weight rate of flow** (weight discharge) is given as

$$W = \gamma_1 A_1 V_1 = \gamma_2 A_2 V_2 \tag{3-15}$$

in which W is the weight of fluid per unit of time passing a section;

    γ is the specific weight of the fluid;

    A is the cross-sectional area of the stream tube.

Equation 3-15 can be expressed in differential form as

$$d(\gamma A V) = 0 \tag{4-1}$$

In other words, there is no change in the product $\gamma A V$ from one section to the next.

For **mass rate of flow** (mass discharge) G, the equation of continuity is

$$G = \rho_1 A_1 V_1 = \rho_2 A_2 V_2 \tag{3-14}$$

in which ρ is the mass density in mass per unit volume;

    G is the mass discharge in mass per unit time which is passing a section of a stream tube.

Equation 3-14 can be expressed in differential form as

$$d(\rho A V) = 0 \tag{4-2}$$

Because this chapter is concerned with compressible fluids, the equation cannot be simplified, as done in Chapter 3 for incompressible fluids.

Example Problem 4-1: Calculate the specific weight, the weight rate of flow, and the discharge in cfs of air at sections 1 and 2 in a pipe line, if at section 1 the diameter, temperature, pressure, and velocity are 6 in., 90°F, 30 psia, and 150 ft per sec, and at section 2 the diameter, pressure, and temperature are 4 in., 20 psia, and 50°F, respectively.

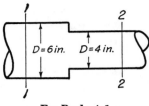

Ex. Prob. 4-1

Solution:

A flow diagram should be constructed and the given and unknown quantities should be shown on the diagram, as follows:

Given: $T_1 = 90°F$ $T_2 = 50°F$

        $p_1 = 30$ psia $p_2 = 20$ psia

        $V_1 = 150$ fps

Find: γ_1 γ_2

        W W

        Q_1 Q_2

Solving for γ

$$\gamma_1 = \frac{(30)(144)}{(53.3)(550)} = \underline{0.147 \text{ lb/cu ft}}, \qquad \gamma_2 = \frac{(20)(144)}{(53.3)(510)} = \underline{0.106 \text{ lb/cu ft}}$$

Solving for W

$$W = \frac{\pi}{4}\left(\frac{6}{12}\right)^2 (0.147)(150) = \underline{4.32 \text{ lb/sec.}}$$

Solving for Q

$$V_2 = \frac{W}{A_2 \gamma_2} = \frac{4.32}{(0.087)(0.106)} = 468 \text{ fps}$$

$$Q_1 = A_1 V_1 = (0.196)(150) = \underline{29.4 \text{ cfs}}$$

$$Q_2 = A_2 V_2 = (0.087)(468) = \underline{40.6 \text{ cfs}}$$

• MOMENTUM EQUATION

The momentum equation, as developed in Chapter 3, has application not only to incompressible fluids, but also to compressible fluids, because it equates the change in momentum flux to the force required to cause this change. The momentum flux is the product of the mass flux ρVA and the velocity V. By the continuity equation, Eq. 3-14, the mass flux is a constant. Hence, if the density varies in compressible flow along a stream tube, the product of the velocity V and the area A must vary inversely with density, so that the final product ρVA is a constant from point to point.

The momentum equation can be written in two convenient forms for the x direction. Either

$$\Sigma F_x = (\rho A V V_x)_2 - (\rho A V V_x)_1 \qquad (3\text{-}18)$$

for the total net force F_x, or dividing by the area A of a conduit with a constant cross section

$$\frac{\Sigma F_x}{A} = p_x = V\Delta(\rho V_x) \qquad (4\text{-}3)$$

for the unit net force or pressure p_x in the x direction.

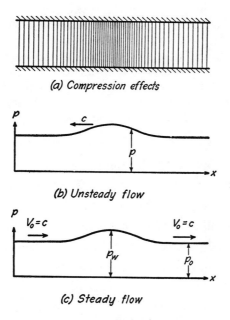

(a) Compression effects

(b) Unsteady flow

(c) Steady flow

Fig. 4-1. Small elastic wave.

With both the continuity and momentum equations available for compressible flow, an analysis can now be made of the simple elastic wave mentioned earlier in this chapter as an important example of the first type of fluid system (see Fig. 4-1).

• ELASTIC WAVE

When the pressure at a point in a fluid is changed gradually, the new pressure is spread throughout the fluid so quickly (i.e., at the speed of sound) that there is no significant pressure gradient involved. When the pressure is changed suddenly with respect to distance at a given instant, or with respect to time at a given plane, however, the elastic or compressibility properties of the fluid take on great significance. Actually, any change in pressure causes an elastic wave reflecting this change to travel through the fluid at a velocity or celerity of c in much the same way that a gravity wave travels on a water surface. Fig. 4-1(b) shows such a wave of pressure traveling from right to left. This wave of momentary pressure-increase (where the lines are close together) causes localized compression of the molecules and, hence, movement of the fluid takes place. At a point, the movement involved in the passage of an elastic wave is unsteady flow, see Fig. 4-1(b). As described in Chapter 3, however, unsteady flow can be made steady simply by applying to the entire flow system, a velocity V_o to the right which is equal to c. By this means the wave is held steady, as shown in Fig. 4-1(c). When $V < c$, the pressure wave will move to the left at the speed $c - V$, and when $V > c$ the pressure wave will move to the right at the speed $V - c$. The flow from left to right is steady at the velocity c, but within the standing wave of pressure increase, there is a corresponding increase in density and a decrease in velocity at the same point. Hence, the continuity equation is satisfied, and

$$\rho_o A_o V_o = \rho_w A_w V_w = \text{constant}$$

in which ρ_o is the density of the approaching flow;

A_o is the area of the approaching flow;

V_o is the velocity of the approaching flow (c the wave celerity);

ρ_w is the density of the fluid within the elastic wave;

V_w is the velocity of the fluid within the elastic wave;

A_w is the area of flow at the wave.

For simplicity of analysis, assume that the wave occurs within a stream tube of constant cross-sectional area A, so that $A_o = A_w$. The remaining part of the continuity equation can then be written in differential form as

$$d(\rho V) = \rho \, dV + V d\rho = 0 \tag{4-4}$$

which shows that there is no change in the product of density and velocity from section to section along the stream tube.

In the simple elastic wave, there is a constant movement of mass flow ρV from section to section, but the velocity changes as the flow passes through the wave, which results in a change of momentum flux. This change must be equal to the change in force or total pressure across the sections. In equation form, for unit values of momentum flux per unit area and force per unit area (which is pressure), this becomes

$$\rho_o A V_o \Delta V = F = A(p_w - p_o) = -A\Delta p \tag{4-5}$$

which can be divided by the area A to obtain unit values, and the differential form becomes

$$\rho_0 V_0 dV = -dp \tag{4-6}$$

Equations 4-4 and 4-6 can be combined to eliminate dV and yield

$$\frac{V\,d\rho}{\rho} = \frac{d\rho}{\rho_0 V_0} \tag{4-7}$$

which can be rearranged as

$$V^2 = \frac{dp}{d\rho} \tag{4-8}$$

Since the velocity is the celerity c, Eq. 4-8 becomes

$$c^2 = \frac{dp}{d\rho} \tag{4-9}$$

or

$$c = \sqrt{\frac{dp}{d\rho}} = \sqrt{\frac{dp/dx}{d\rho/dx}} \tag{4-10}$$

in which x is the direction of flow. This equation has particular significance in showing how the celerity of a small elastic wave varies with the relative gradients of pressure and density.

In Chapter 1, it was shown that

$$E = \rho\frac{dp}{d\rho} \tag{1-5}$$

in which E is the bulk modulus of elasticity. This equation can be combined with Eq. 4-10 to yield the general equation

$$c = \sqrt{\frac{E}{\rho}} \tag{4-11}$$

for the celerity or ***speed of an elastic wave.***

Equation 4-11 shows that the celerity of a simple elastic wave is dependent only upon the elastic properties and the density of the fluid in question. Therefore, elastic waves of various magnitudes and frequencies can travel through a fluid and be entirely independent of each other. Sound is composed of elastic waves of this nature which travel with the celerity c. Hence, c is known as the velocity of sound. Of interest is the fact that, although there is considerable variation in the bulk modulus of elasticity among the various fluids, there is much less variation of the velocity of sound c. For example, there is about a 20,000-fold variation of E between air and water and little more than a four fold variation of c.

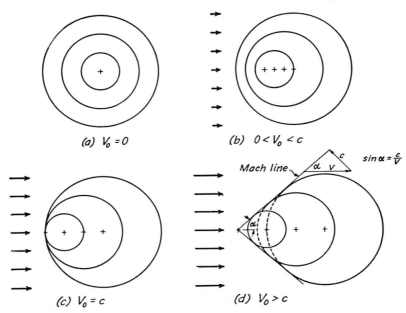

(a) $V_0 = 0$

(b) $0 < V_0 < c$

$\sin \alpha = \frac{c}{V}$

Mach line

(c) $V_0 = c$

(d) $V_0 > c$

Fig. 4-2. Relation between ambient velocity and speed of sound.

It should be pointed out that Eq. 4-11 gives the celerity of a **small** elastic wave—that is, the increase in density within the wave is small compared with the ambient density ρ. Waves of large magnitude have a celerity exceeding that of Eq. 4-11, as do large gravity waves discussed in Chapter 8.

The foregoing analysis of a small and simple elastic wave can be applied to the analysis of a pressure pulse at a point. If a series of elastic pressure waves are radiating from a line source, as shown in Fig. 4-2, each wave will progress radially outward with a velocity of c. If there is no motion of the fluid, the waves will be represented by a series of concentric circular rings (cylinders) shown in Fig. 4-2(a) with a radius of ct, where t is the time since the impulse. When the fluid is moving to the right at a velocity less than the speed of sound $(0 < V_o < c)$, see Fig. 4-2(b), each *ring wave* is carried downstream by this velocity. The upstream portion of the wave progresses upstream more slowly than for Fig. 4-2(a)—the velocity now being $(c - V_o)$. If the velocity of the fluid is increased to the speed of sound $(V_o = c)$, see Fig. 4-2(c), the upstream portion of each wave remains stationary as in Fig. 4-1(c)—so that the wave rings are tangent at the upstream side and the downstream portion is moving downstream at a speed of $(V + c)$. In this case, the pressure wave is unable to move upstream against the oncoming flow. Finally, when the surrounding flow has a velocity greater than the speed of sound $(V_o > c)$, see Fig. 4-2(d), the waves are swept downstream on the upstream side as well as on the downstream side.

When $V_o > c$, a line of tangency to each wave ring can be drawn with an angle α between the tangent line and the direction of flow. This line is known as the Mach line, and the angle α as the Mach angle. The angle α can be seen to depend upon the relative magnitudes of V_o and c. In fact, the sine of the angle has particular significance

$$\sin \alpha = \frac{c}{V_o} = \frac{1}{\mathrm{Ma}} \tag{4-12}$$

in which Ma is the Mach number. The Mach number is a dimensionless parameter $\mathrm{Ma} = V_o/c$, which has special importance in the study of phenomena involving velocities of movement which approach, equal, or exceed the speed of sound. In Fig. 4-2, the Mach numbers are: $\mathrm{Ma} = 0$ when $V_o = 0$, $0 < \mathrm{Ma} < 1.0$ when $0 < V_o < c$, $\mathrm{Ma} = 1.0$ when $V_o = c$, and $\mathrm{Ma} > 1.0$ when $V_o > c$. The phenomena discussed in Chapter 3 involved Mach numbers considerably less than one, in which case the flow pattern is affected on all sides of a disturbance. As the Mach number increases, however, the changes in flow pattern are confined to a region closer and closer to the disturbance, and when $\mathrm{Ma} > 1.0$, all influence is confined to the interior of the angle 2α.

The Mach number is given more extensive discussion in later chapters—particularly in the chapters on Flow in Closed Conduits, Flow Around Submerged Objects, and Similitude. At present, it is sufficient to point out that, for *adiabatic flow* (where there is no heat exchange), $E = kp$, see Eq. 1-9, and hence

$$c = \sqrt{\frac{E}{\rho}} = \sqrt{\frac{kp}{\rho}} = \sqrt{kgRT} \tag{4-13a}$$

and for *isothermal flow*, $E = p$ (see Eq. 1-8), so that

$$c = \sqrt{\frac{E}{\rho}} = \sqrt{\frac{p}{\rho}} = \sqrt{gRT} \tag{4-13b}$$

which shows that, although the speed of sound varies with both pressure and density, the net effect for a given gas is reflected by absolute temperature alone. The other variables (k, g, and R) are constant.

In Eq. 4-13(b), it is assumed that the temperature remains constant during the passing of a wave. For this to be true, it would be necessary to have some of the heat taken out while the wave is present, and then have it added again after the wave passes. Such transfer of heat is not actually possible, because the wave passes so quickly. Consequently, Eq. 4-13(b) differs considerably from experimental data. In Eq. 4-13(a), on the other hand, it is assumed that the heat content (entropy) remains constant, which means the process is assumed to be adiabatic. Experimental data agree very closely with Eq. 4-13(a), which verifies the adiabatic assumption.

From the foregoing analysis, it can be concluded that the sonic velocity depends only upon the characteristics of the fluid, and is not influenced in any way by the frequency or length of the wave. In other words, all waves (including those which are audible) in a given fluid at a given temperature move with the same speed.

There are many applications of the analysis of the speed of sound in various media. These include the sonic determination of the ocean depth and the height of clouds, and the location of lightning and of an explosion (or any noise) by sound ranging. Each of these applications utilizes the fact that a definite amount of time is required for the sound to travel from the source to the receiver.

Example Problem 4-2: Measuring the shockwave angle α is an accurate means of measuring supersonic velocities for sharp nosed objects such as projectiles. Referring to Fig. 4-2(d), estimate the velocity and Mach number of a projectile which reflects the given shockwave in air at a pressure of 13.0 psia and at a temperature of 40°F.

Solution:

Referring to Eq. 4-12

$$\sin \alpha = \frac{c}{V_0} = \frac{1}{\mathrm{Ma}}$$

and, assuming a frictionless process

$$c = \sqrt{kgRT} = \sqrt{(1.4)(32.2)(53.3)(500)}$$

$$= 1095 \text{ fps}$$

The angle α in Fig. 4-2(d) measures 42.5°; hence

$$V_0 = \frac{c}{\sin 42.5°} = \frac{1095}{0.675} = 1625 \text{ fps}$$

and

$$\mathrm{Ma} = \frac{V_0}{c} = \frac{1625}{1095} = 1.48$$

• ENERGY EQUATION

Although the momentum equation is completely independent of compressibility effects, the energy equation is very much dependent upon variable density. Furthermore, it is important to know whether the process involves heat exchange or not and, if so, then how and to what extent. In the following paragraphs are treated these and other aspects of the energy equation, as applied to compressible flow.

In Chapter 3, the energy equation was developed for the limited case of incompressible fluids. In this case, certain important aspects of the complete energy equation were neglected which must be included in the consideration of compressible fluids. The total energy (per unit volume or per unit weight) at a point in a compressible system must include the following*:

1. Kinetic energy.

2. Potential energy.

3. Molecular energy.

In a word equation, this can be expressed as

$$\text{Total energy} = \text{Kinetic energy} + \text{Potential energy} + \text{Molecular energy} \qquad (4\text{-}14)$$

Using symbols, this is

$$E_t = E_k + E_e + E_m \qquad (4\text{-}15)$$

According to the fundamental law of physics on conservation of energy, the total energy at one point in a system, plus the energy added E_a (or subtracted E_s) between two points, must be equal to the total energy at the second point in the same system. That is

$$(E_k + E_e + E_m)_1 + E_a = (E_k + E_e + E_m)_2 \qquad (4\text{-}16)$$

This equation shows that energy can be changed from one form to another on the two sides of the equation, but that the total must be the same. In other words, energy can be neither created nor destroyed.

In Chapter 3, heat energy (which is related directly to the molecular energy) was considered only indirectly as an energy "loss" when fluid resistance changed kinetic and potential energy to heat energy. For compressible fluids, however, heat (or molecular) energy must be included as a specific part of the equation.

The kinetic energy consists not only of energy of straight translation and rotation, but also energy of turbulence which ultimately will be converted (through internal shear) into heat energy. The energy of straight

---

* Electrical, chemical, and atomic energy are not considered here.

translation and rotation can be converted directly into mechanical work, but the turbulent energy is converted into mechanical work only after it as been converted into heat energy.

The potential energy in a compressible fluid exists because of the elevation of the fluid. This type of energy can be converted directly into mechanical energy and work.

The molecular energy of a compressible fluid is caused by the activity of the molecules, which increases as the temperature of the fluid increases. In a gas, this molecular activity also creates the pressure—which is that part of molecular energy usually converted into mechanical energy. In thermodynamics, the molecular energy is known as the enthalpy $En = p/\gamma + I$, in which I is that part of the molecular energy other than the pressure energy p/γ.

Equation 4-16 can now be written in terms of energy per unit weight as

$$\left(\frac{V^2}{2g} + z + \text{En}\right)_1 + (E_a - E_s) = \left(\frac{V^2}{2g} + z + \text{En}\right)_2 \qquad (4\text{-}17)$$

in which $V^2/2g$ is the kinetic energy per unit weight of fluid;

z is the elevation or potential energy per unit weight;

En is the enthalpy or molecular energy per unit weight.

The foregoing equations can be applied to the line flow diagram in Fig. 4-3. Flow is entering from the left at A, with a certain amount of the three types of energy considered. At point B, the flow passes through a

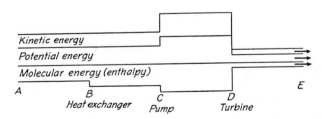

Fig. 4-3. Flow diagram for the energy equation.

heat exchanger in which heat is added to (or is subtracted from), the flow, which increases (or decreases) the molecular energy or enthalpy. At point C, a pump adds energy to the system. This addition can result in an increase in either kinetic energy or potential energy or both—and, in the case of a compressible fluid, it can increase the molecular energy or enthalpy by an increase in pressure energy. The turbine at point D has an effect which is just opposite to that of the pump. As a consequence, all three types of energy may be reduced to very small amounts.

Consideration is now given to the application of the energy equation to the two flow processes involving no heat exchange (adiabatic) and no temperature change (isothermal).

• ISOTHERMAL FLOW

Under conditions of isothermal flow, the **temperature** *does not change from one point to another* and, hence, both I and p/γ remain constant—that is, $I_1 = I_2$ and $(p/\gamma)_1 = (p/\gamma)_2$. This means that an increase (or decrease) in pressure p from point 1 to point 2 is accompanied by a corresponding increase (or decrease) in specific weight γ. Furthermore, it can be assumed that the change in elevation $(z_2 - z_1)$ is small compared with the other changes. Hence, Eq. 4-17 becomes

$$\frac{V_2^2}{2g} - \frac{V_1^2}{2g} = E_a \qquad (4\text{-}18)$$

which shows that energy E_a (usually heat energy) must be added to the system if there is to be an increase in kinetic energy, and yet no change in temperature (isothermal flow). This is discussed further and then applied in Chapter 7 on Flow in Closed Conduits.

• ADIABATIC FLOW

An adiabatic process is one in which there is **no heat** *added or subtracted from point to point* within the system. In this case, Eq. 4-17 becomes

$$\frac{V_2^2}{2g} - \frac{V_1^2}{2g} = \text{En}_1 - \text{En}_2 \qquad (4\text{-}19)$$

where the elevation (potential energy) difference is again assumed to be small compared with the differences of other types of energy.

It is known from the fundamentals of thermodynamics that

$$\text{En}_1 - \text{En}_2 = \frac{k}{k-1}\left(\frac{p_1}{\gamma_1} - \frac{p_2}{\gamma_2}\right) = \frac{k}{k-1}R(T_1 - T_2) \qquad (4\text{-}20)$$

in which k is the adiabatic exponent—a ratio of the specific heat at constant pressure c_p to the specific heat at constant volume c_v;

R is the gas constant;

T is the absolute temperature.

Equation 4-19 now becomes

$$\frac{V_2^2}{2g} - \frac{V_1^2}{2g} = \frac{k}{k-1}\left(\frac{p_1}{\gamma_1} - \frac{p_2}{\gamma_2}\right) = \frac{k}{k-1}R(T_1 - T_2) \tag{4-21}$$

The second part of this equation is general, but the third part is restricted to perfect gases (no shear). From Eq. 1-6, the adiabatic relation between p and γ is

$$\frac{p_1}{\gamma_1^k} = \frac{p_2}{\gamma_2^k} = \text{constant} \tag{4-22}$$

Equations 4-21 and 4-22 can be arranged in terms of γ_1, p_1, and p_2 to eliminate γ_2

$$\frac{V_2^2}{2g} - \frac{V_1^2}{2g} = \frac{p_1}{\gamma_1}\frac{k}{k-1}\left[1 - \left(\frac{p_2}{p_1}\right)^{(k-1)/k}\right] \tag{4-23}$$

The foregoing equations have application to various specific phenomena. Common examples are discussed in some detail.

• STAGNATION PRESSURE

One application of the foregoing principles is to the computation of stagnation pressure for compressible flow—for example, at the upstream side of an object such as that in Fig. 3-30. At the point of stagnation, the velocity head is converted to pressure head. Therefore, $V_2 = 0$. This conversion occurs quickly as the fluid moves around the object. For this reason, *there is no appreciable heat exchange and the process can be considered **adiabatic.*** Hence, Eq. 4-23 can be rearranged as

$$\frac{p_2}{p_1} = \left[1 + \frac{k-1}{2}\left(\frac{V_1^2}{kp_1/\rho_1}\right)\right]^{k/(k-1)} \tag{4-24}$$

whicn may be expanded by the binomial theorem to

$$\frac{p_2}{p_1} = 1 + \frac{k}{2}\left(\frac{V_1}{c_1}\right)^2 + \frac{k}{8}\left(\frac{V_1}{c_1}\right)^4 + \frac{k(2-k)}{48}\left(\frac{V_1}{c_1}\right)^6 + \cdots \tag{4-25}$$

where c^2 from Eq. 4-13(a) is substituted for kp/ρ. Rearrangement of Eq. 4-25 gives

$$\frac{p_2 - p_1}{\dfrac{\rho_1 V_1^2}{2}} = 1 + \frac{1}{4}\left(\frac{V_1}{c_1}\right)^2 + \frac{2-k}{24}\left(\frac{V_1}{c_1}\right)^4 + \cdots \tag{4-26}$$

It may be seen from the foregoing that, when the approach velocity V_1 is small compared with the velocity of sound, Eq. 4-26 reduces to

$$\frac{p_2 - p_1}{\dfrac{\rho_1 V_1^2}{2}} = 1 \tag{4-27}$$

which is the same as for an incompressible fluid—see Chapter 3. The ratio V_1/c_1 is the Mach number which has been mentioned previously in this chapter. Using this terminology, Eq. 4-26 becomes

$$\frac{p_2 - p_1}{\dfrac{\rho_1 V_1^2}{2}} = 1 + \frac{1}{4}\,\mathrm{Ma}^2 + \frac{2-k}{24}\,\mathrm{Ma}^4 + \cdots \tag{4-28}$$

Comparison of Eqs. 4-26 and 4-27 shows that, as the approach velocity becomes large, the stagnation pressure becomes greater for compressible fluids (with adiabatic flow) than for incompressible fluids. In fact, for a

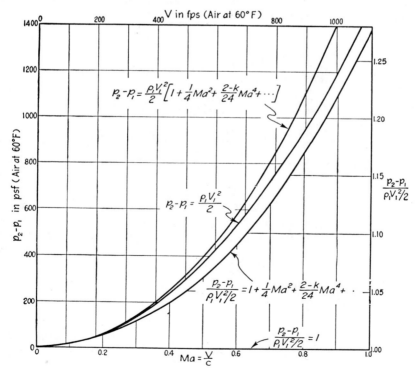

Fig. 4-4. Influence of compressibility on stagnation pressure p_2 for air.

velocity half the speed of sound (Ma = 0.5) in air, the stagnation pressure
is 6 percent greater and, when sonic velocity is reached, the pressure is
28 percent greater. Of considerable importance is the fact that, if air is
considered incompressible instead of compressible, the velocity must be as
great as 300 fps before there is an error of as much as 2 percent from using
Eq. 4-27 which ignores secondary terms, instead of Eq. 4-26. This is shown
graphically in Fig. 4-4.

The assumption of adiabatic flow and the resulting analysis are satisfactory unless Ma > 1.0, in which case a more extensive theoretical treatment must be used (see Chapter 9.)

The foregoing development has direct application to the Pitot tube,
which is used for the measurement of velocities of a gas, provided the velocities are not greater than the speed of sound. The use of the Pitot tube is
discussed in greater detail in Chapter 10 on Flow Measurement.

Example Problem 4-3: A jet airplane is propelled at a velocity of 600 mph
through air in which the atmospheric pressure is 12 psia. The temperature is
10°F and the wind velocity is negligible. Calculate the pressure intensity, the
temperature, the density of the air at the stagnation point on the nose of the jet,
and determine its Mach number.

Solution:

$$\gamma_1 = \frac{p_1}{RT} = \frac{(12.0)(144)}{(53.3)(470)} = 0.069 \text{ lb/cu ft}$$

$$\rho_1 = \frac{\gamma_1}{g} = \frac{0.069}{32.2} = 0.00214 \text{ slug/cu ft}$$

$$c_1 = \sqrt{kgRT} = \sqrt{(1.4)(32.2)(53.3)(470)} = \underline{1062 \text{ fps}}$$

Applying Eq. 4-25

$$\frac{p_2}{p_1} = 1 + \frac{k}{2}\left(\frac{V_1}{c_1}\right)^2 + \frac{k}{8}\left(\frac{V_1}{c_1}\right)^4 + \cdots$$

$$\frac{p_2}{p_1} = 1 + \frac{1.4}{2}\left(\frac{880}{1062}\right)^2 + \frac{1.4}{8}\left(\frac{880}{1062}\right)^4 + \cdots$$

$$p_2 = 1.562\, p_1 = \underline{18.78 \text{ psia}}$$

Applying Eq. 4-21

$$\frac{V_1^2}{2g} = \frac{k}{k-1} R(T_2 - T_1)$$

$$T_2 = \frac{(k-1)(V_1^2)}{kR2g} + T_1 = \frac{(1.4-1)(880)^2}{(1.4)(53.3)(64.4)} + 470$$

$$= 64.5 + 470 = 534.5°F \text{ abs.}, \ T_2 = \underline{74.5°F}$$

$$\rho_2 = \frac{p_2}{gRT_2} = \frac{(18.78)(144)}{(32.2)(53.3)(534.5)} = \underline{\underline{0.00294 \text{ slug/cu ft}}}$$

$$\text{Ma} = \frac{V_1}{c_1} = \frac{880}{1062} = \underline{\underline{0.827}}$$

• VARIATION OF VELOCITY WITH AREA

With incompressible flow along a stream tube, the velocity simply varies inversely with the area, since $Q = A_1V_1 = A_2V_2$. When the flow is compressible, however, the variation of velocity with area is complicated by the influence of the density ρ, since $\rho_1A_1V_1 = \rho_2A_2V_2$ (see Eq. 3-14). The influence of ρ is now considered.

The continuity equation for incompressible flow can be differentiated to give

$$\frac{dA}{A} = -\frac{dV}{V} \tag{4-29}$$

and for compressible flow

$$\frac{dA}{A} = -\frac{dV}{V} - \frac{d\rho}{\rho} \tag{4-30}$$

From the energy equation, Eq. 4-21, it can be seen that

$$\frac{V^2}{2} + \frac{k}{k-1}\left(\frac{p}{\rho}\right) = \text{constant} \tag{4-31}$$

which can be differentiated to yield

$$V \, dV + \frac{k}{k-1}\left(\frac{dp}{\rho} - \frac{p \, d\rho}{\rho^2}\right) = 0 \tag{4-32}$$

By assuming adiabatic flow, and thereby combining $c = \sqrt{dp/d\rho} = \sqrt{k\,p/\rho}$ and Eq. 4-32, the following significant equation can be obtained

$$\frac{dA}{A} = -\left[1 - \left(\frac{V}{c}\right)^2\right]\frac{dV}{V} = -\left[1 - \text{Ma}^2\right]\frac{dV}{V} \tag{4-33}$$

The speed of sound is $c = \sqrt{gkRT}$, see Eq. 4-13(a), which can be differentiated, rearranged, and combined with the Mach number to give

$$\frac{dT}{T} = 2\left[\frac{dV}{V} - \frac{d(\text{Ma})}{\text{Ma}}\right] \tag{4-34}$$

and Eq. 4-21 can be written in differential form as

$$\frac{V\,dV}{g} + \frac{k}{k-1}R\,dT = 0 \tag{4-35}$$

or multiplying by g, dividing by V^2, and multiplying the second term by $\frac{T}{T}$ gives

$$\frac{dV}{V} + \left(\frac{1}{(k-1)\mathrm{Ma}^2}\right)\frac{dT}{T} = 0 \tag{4-36}$$

Combining Eqs. 4-36, 4-34, and 4-33 yields

$$\frac{dA}{A} + \frac{(1-\mathrm{Ma}^2)d(\mathrm{Ma})}{\mathrm{Ma}\left[\left(\dfrac{k-1}{2}\right)\mathrm{Ma}^2 + 1\right]} = 0 \tag{4-37}$$

which can be integrated to give

$$\frac{A}{A_c} = \frac{1}{\mathrm{Ma}}\left[\frac{2 + (k-1)\mathrm{Ma}^2}{2 + (k-1)}\right]^{(k+1)/2\,(k-1)} \tag{4-38}$$

in which A_c is the critical minimum area shown in Fig. 4-5 where Eq. 4-38 is plotted.

Equation 4-33 for compressible adiabatic flow is particularly significant when it is compared with Eq. 4-29 for incompressible flow. These curves differ only by the term $(1 - \mathrm{Ma}^2)$, which enters because of the variable density. This term shows that for small velocities, which give small values of the Mach number Ma, the influence of compressibility is small. As the Mach number becomes larger and approaches 1.0, however, the influence of compressibility becomes increasingly important, as can be seen by the increasing separation of the two curves in Fig. 4-5. For example, if incompressibility is assumed, (the lower curve in Fig. 4-5) an error of about 30 percent will result at $V = 550$ fps (approximately Ma $= 0.5$), as compared with the upper curve in Fig. 4-5 for compressible flow.

In Fig. 4-5, the scales for A and V apply to both curves for air at 60°F. A discharge of 500 cfs was selected simply to permit comparison of compressible and incompressible flow. The scales for the dimensionless parameters Ma and A/A_c are used with the upper curve, Eq. 4-38, which is general for air under dry adiabatic conditions, $k = 1.4$.

When $V < c$, the gradient dA/dV is negative and the area decreases with increasing velocity. However, at Ma $= 1.0$, $V = c$ and the gradient is zero, which shows the area has reached a minimum. This means that *sonic velocity is reached only at the minimum cross-sectional area*—a fact which is utilized for the measurement of flow of a gas with an orifice. This is discussed in greater detail in Chapter 10 on Flow Measurement. If $V > c$, then Ma > 1.0, the gradient dA/dV is positive, and the area must increase with increasing velocity. In other words, ***supersonic velocities are reached only in an expanding section downstream from a minimum section in which*** $V = c.$

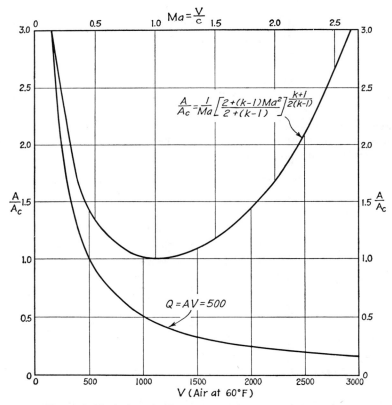

Fig. 4-5. Variation of velocity with area for compressible and incompressible flow of air.

• CONVERGING FLOW

Further information regarding the fundamentals of compressible flow can be learned from a detailed consideration of one-dimensional flow in a converging passage or stream tube such as the nozzle shown in Fig. 4-6. Since $V_1 = 0$, the energy equation (Eq. 4-23) becomes

$$\frac{V_2^2}{2g} = \frac{p_1}{\gamma_1}\left(\frac{k}{k-1}\right)\left[1 - \left(\frac{p_2}{p_1}\right)^{(k-1)/k}\right] \tag{4-39}$$

or, since this is an adiabatic process,

$$\frac{p_1}{\gamma_1^k} = \frac{p_2}{\gamma_2^k} \tag{4-40}$$

so that Eq. 4-39 can be written as

$$\frac{V_2^2}{2g} = \frac{p_2}{\gamma_2} \frac{k}{k-1} \left[\left(\frac{p_1}{p_2}\right)^{(k-1)/k} - 1 \right] \tag{4-41}$$

which can be rearranged in dimensionless form as

$$\left(\frac{V_2}{c_2}\right)^2 = \frac{2}{k-1} \left[\left(\frac{p_1}{p_2}\right)^{(k-1)/k} - 1 \right] \tag{4-42}$$

in which $c_2 = \sqrt{kp_2/\rho_2}$ is the speed of sound in the jet. From this equation, it can be seen that the velocity of the jet V_2 relative to the speed of sound c_2 is dependent only upon the pressure ratio p_1/p_2.

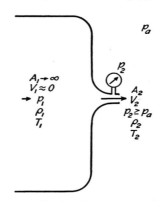

Fig. 4-6. Compressible flow from a nozzle in a large tank.

Equation 4-39 can be combined with the continuity equation

$$W = A_2 V_2 \gamma_2$$

and Eq. 4-40 to yield

$$\frac{W}{A_2} = V_2 \gamma_2 = \sqrt{\frac{2gk}{k-1} (p_1 \gamma_1) \left[\left(\frac{p_2}{p_1}\right)^{2/k} - \left(\frac{p_2}{p_1}\right)^{(k+1)/k} \right]} \tag{4-43}$$

This equation is plotted as the curved line in Fig. 4-7 (a), from which it can be seen that, as the pressure p_2 decreases from $p_2 = p_1$, the rate of flow per unit area W/A_2 increases until it reaches a maximum at $p_2 = 0.53p_1$. As the pressure is decreased further to $p_2 < 0.53p_1$, the rate of discharge theoretically decreases, as shown by the broken portion of the curved line.

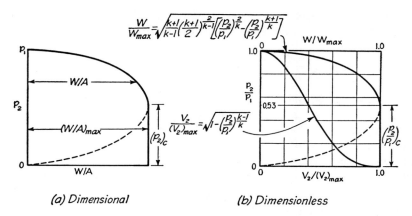

Fig. 4-7. Discharge and velocity diagrams for converging compressible flow of air.

Actually, however, the discharge remains at the maximum value throughout $p_2 < 0.53 p_1$, because the velocity V_2 has reached the speed of sound and a pressure wave cannot travel upstream to influence the discharge. For the same reason, the pressure in the jet can be equal to or greater than the ambient pressure p_a.

The ratio p_2/p_1 for the point of maximum discharge can be determined by differentiating Eq. 4-43 and setting the result equal to zero. This gives

$$\left(\frac{p_2}{p_1}\right)_c = \left(\frac{2}{k+1}\right)^{k/(k-1)} \tag{4-44}$$

which shows that the critical pressure ratio depends only upon the adiabatic exponent k. This is approximately 0.53 for $k = 1.4$. The subscript in Eq. 4-44 indicates the critical condition at which the velocity of flow equals the speed of sound.

The maximum rate of flow can be determined from Eq. 4-43 by setting

$$\frac{p_2}{p_1} = \left(\frac{p_2}{p_1}\right) = \left(\frac{2}{k+1}\right)^{k/(k-1)} \tag{4-45}$$

which yields

$$\left(\frac{W}{A_2}\right)_{max} = \sqrt{\frac{2gk}{k-1}(p_1\gamma_1)\left(\frac{2}{k+1}\right)^{2/(k-1)}\frac{k-1}{(k+1)}} \tag{4-46}$$

and shows that the **maximum flow** *depends only upon the pressure p_1 and the specific weight γ_1* in the tank upstream. This fact is used as an important and convenient means for measuring discharge as discussed in Chapter 10 on Flow Measurement.

Equation 4-43 can be arranged in significant dimensionless form, upon dividing by Eq. 4-46 for the maximum discharge, to yield

$$\frac{W}{W_{max}} = \sqrt{\frac{k+1}{k-1}\left(\frac{k+1}{2}\right)^{2/(k-1)}\left[\left(\frac{p_2}{p_1}\right)^{2/k} - \left(\frac{p_2}{p_1}\right)^{(k+1)/k}\right]} \quad (4\text{-}47)$$

This equation is plotted in Fig. 4-7(b), where the Mach number Ma is also used as an ordinate scale based upon Eq. 4-42.

Further study of Fig. 4-7 shows that, since the discharge is not reduced to less than the maximum for $p_2/p_1 < (p_2/p_1)_c$, it is logical to reason that p_2/p_1 does not become less than $(p_2/p_1)_c$. Experiment has verified this reasoning.

Example Problem 4-4: Calculate the temperature, pressure, velocity, and weight rate of flow in a jet of nitrogen which discharges through a 1-in. convergent nozzle into standard air from a large tank, such as shown in Fig. 4-6, in which the temperature and pressure are 40°F and 50 psia, respectively.

Solution:

$$\gamma_1 = \frac{p_1}{RT_1} = \frac{(50)(144)}{(55.1)(500)} = 0.262 \text{ lb/cu ft}$$

$$\left(\frac{p_2}{p_1}\right)_c = \left(\frac{2.0}{2.4}\right)^{1.4/0.4} = 0.528$$

$$\frac{p_{at}}{p_1} = \frac{14.7}{50} = 0.294 < 0.528$$

Therefore

$$p_2 = (0.528)(50) = \underline{26.4 \text{ psia}}$$

$$\gamma_2 = \left(\frac{p_2}{p_1}\right)^{1/k}\gamma_1 = (0.528)^{1/1.4}(0.262) = 0.166 \text{ lb/cu ft}$$

$$T_2 = T_1\left(\frac{p_2}{p_1}\right)^{k-1/k} = (500)(0.528)^{0.4/1.4} = \underline{416°F \text{ abs}}$$

or

$$T_2 = \frac{p_2}{R\gamma_2} = \frac{(26.4)(144)}{(55.1)(0.166)} = \underline{416°F \text{ abs}}$$

$$V_2 = c_2 = \sqrt{kgRT_2} = \sqrt{(1.4)(32.2)(55.1)(416)} = \underline{1018 \text{ fps}}$$

$$W = \gamma_2 A_2 V_2 = (0.166)(0.0054)(1018) = \underline{0.910 \text{ lb/sec}}$$

• DIVERGING FLOW

In the foregoing section, consideration was given only to flow converging and issuing from a nozzle at the point of minimum cross section. Furthermore, the discharge and the velocity were found to reach a maximum

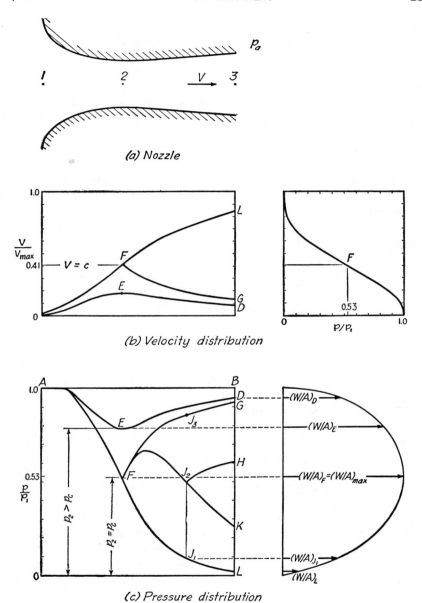

(a) Nozzle

(b) Velocity distribution

(c) Pressure distribution

Fig. 4-8. Distribution of velocity and pressure through a contracting and expanding nozzle with compressible flow.

(which could not be exceeded) when the critical pressure ratio was reached. In the section on variation of velocity with area, it was shown that the velocity reached the speed of sound at the minimum cross-sectional area and, if the velocity is to be greater than the speed of sound, there must be an increase in area. In other words, the speed of sound cannot be exceeded in the throat of a nozzle and, therefore, a diverging or expanding section must be added downstream in order to obtain velocities greater than the speed of sound.

The foregoing facts can be used to initiate an analysis of the flow through a Venturi-type conduit such as the Laval nozzle shown in Fig. 4-8, in which it is possible to obtain velocities greater than the speed of sound in the expanding section of the nozzle. An understanding of the various possible phenomena can be obtained by simply considering the variation of the ambient pressure p_a downstream from the nozzle, together with the following general form of Eq. 4-43.

$$\frac{W}{A} = \sqrt{\frac{2gk}{k-1}(p_1\gamma_1)\left[\left(\frac{p}{p_1}\right)^{2/k} - \left(\frac{p}{p_1}\right)^{(k+1)/k}\right]} \qquad (4\text{-}48)$$

The various curves in Fig. 4-8 represent the following form of Eq. 4-39

$$\frac{p}{p_1} = \left[1 - \frac{k-1}{k}\frac{\rho_1}{p_1}\frac{V^2}{2}\right]^{k/(k-1)} \qquad (4\text{-}49)$$

in which the velocity can be determined from the continuity equation (Eq. 3-15.)

When the ambient pressure p_a downstream is equal to the pressure p_1 upstream, there obviously can be no flow and $p/p_1 = 1.0$ throughout the nozzle, as well as downstream, as represented by the horizontal line AB. As p_a/p_1 is reduced, however, the pressure p_3/p_1 at section 3 is likewise reduced. The influence of this pressure drop can move upstream, because the velocity of flow is less than the speed of sound, which causes flow to take place with a greater drop in pressure at section 2. The pressure distribution is represented by AED and the unit discharge is $(W/A)_D$ at the exit and $(W/A)_E$ at the throat, as shown on the discharge diagram.

When the ambient pressure ratio p_a/p_1 is reduced to that at G, the pressure at the throat (section 2) is critical, the velocity is equal to the speed of sound, and the unit discharge W/A is a maximum. Therefore, the pressure distribution follows the curve AFG.

Although the pressure distribution upstream from the throat follows only the one curve AF when sonic velocity exists in the throat, there are two possible curves downstream from the throat—one greater than critical (an adiabatic compression), and the other less than critical (an adiabatic expansion). Which of these curves is applicable depends only upon the relative ambient pressure p_a/p_1. If $(p_a/p_1) < (p_3/p_1)_L$, then the pressure distribution follows AFL and the unit discharge is $(W/A)_L$. Curve AFL represents an adiabatic expansion in the expanding section, and curve AFG an adiabatic compression. In either case, the velocity at the throat is equal to the speed of sound and, hence, the discharge W is a maximum which cannot be exceeded, except by an increase in p_1 upstream.

As can be seen from Fig. 4-8 and Eq. 4-48, the area at section 3 can increase indefinitely, with the result that the pressure will continue to increase along curve FG or to decrease along curve FL, provided only that $p_a = (p_3)_G$ or $p_a \leq (p_3)_L$, respectively. This fact is the basis of supersonic wind tunnel design following the curve AFL—the primary limitation being the tremendous power requirement (as much as many thousands of horse-power) for maintaining a very small ambient pressure downstream.

The reader should make a comparison between the flow phenomena described in this section and those explained for contracting sections in Chapter 8 on Flow in Open Channels.

When $(p_2/p_1)_G > (p_a/p_1) > (p_3/p_1)_L$, it is impossible for flow to take place without the formation of a shockwave and consequent energy loss—as discussed in the following section.

• SHOCK WAVE

In the foregoing section, an analysis was given for the flow in Fig. 4-8, provided the downstream ambient pressure is either $(p_a/p_1) > (p_3/p_1)_G$ or

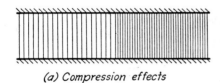

(a) Compression effects

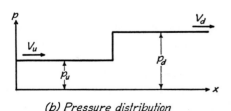

(b) Pressure distribution

Fig. 4-9. Shock wave.

$(p_a/p_1) < (p_3/p_1)_L$. When the ambient pressure is between those at G and L, however, a shock wave is formed which has certain similarities to the hydraulic jump discussed in Chapter 8.

Consider the shockwave as illustrated in Fig. 4-9. Here, a sudden increase in **pressure** is experienced (as compared with the sudden increase in **depth** for the hydraulic jump) as a wave which is moving upstream with the speed of sound. Consequently, if the flow is to be steady, a velocity V_c to the right, which is equal in magnitude but opposite in direction to that of the wave, must be superposed on the entire system. Under steady flow conditions, the velocity upstream from the wave is supersonic $V_u > V_c$, and downstream the velocity is subsonic $V_d < V_c$.

In a theoretical analysis of this phenomenon, the continuity, momentum, and energy equations can be used to advantage. The continuity equation is

$$G = \rho_u A_u V_u = \rho_d A_d V_d \tag{4-50}$$

The momentum equation is

$$p_u A_u - p_d A_d = G V_d - G V_u = \rho_d A_d V_d^2 - \rho_u A_u V_u^2 \tag{4-51}$$

or, since $A_u = A_d$

$$p_u - p_d = \rho_d V_d^2 - \rho_u V_u^2 \tag{4-52}$$

and the energy equation is

$$\frac{k}{k-1}\left(\frac{p_d}{\rho_d}\right) - \frac{k}{k-1}\left(\frac{p_u}{\rho_u}\right) = \frac{V_u^2}{2} - \frac{V_d^2}{2} \tag{4-53}$$

It is assumed that a perfect gas exists ($p = \gamma RT$), boundary resistance is negligible, the specific heats are constant, and the process takes place over a very short distance. These three equations can be combined in two ways:

1. Assuming adiabatic flow in which there is no shockwave, and
2. Assuming non-adiabatic flow in which there is a shockwave.

With adiabatic flow, the adiabatic equation

$$\frac{p_u}{\rho_u^k} = \frac{p_d}{\rho_d^k} \tag{4-54}$$

can be combined with the energy and continuity equations to yield

$$\mathrm{Ma}_u = \sqrt{\frac{2}{k-1}\left[\frac{(p_d/p_u)^{(k-1)/k} - 1}{1 - (p_u/p_d)^{2/k}}\right]} \tag{4-55}$$

which gives the variation of Mach number of approach with the pressure ratio p_d/p_u, as shown in Fig. 4-10. This equation represents the relationship between curves FL (for p_u) and FG (for p_d) in Fig. 4-8 for which no shock wave is formed. Equation 4-55 is used at this point in Fig. 4-8 for comparative purposes only.

If a shockwave forms, the kinetic energy is not all converted to pressure energy as an adiabatic process in accordance with Eq. 4-55. Instead, a considerable portion of the kinetic energy is converted directly to heat energy. As a consequence, the pressure immediately downstream from the shockwave is less than it would be for an adiabatic change. The momentum equation can be used (instead of the adiabatic equation) with the continuity and energy equations to yield

$$\mathrm{Ma}_u = \sqrt{\frac{k+1}{2k}\left(\frac{p_d}{p_u}\right) + \frac{k-1}{2k}} \tag{4-56}$$

which is also plotted in Fig. 4-10. In Fig. 4-8, this equation represents the relationship between the pressure before the shock (p_u in Eq. 4-56 and point J_1 in Fig. 4-8) and the pressure after the shock (p_d in Eq. 4-56 and point J_2 in Fig. 4-8).

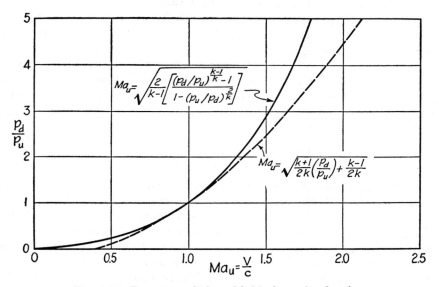

Fig. 4-10. Pressure variation with Mach number for air.

It is significant that the velocity of a pressure wave relative to the speed of sound c_1 in the approaching fluid is dependent only upon the pressure upstream p_u relative to the pressure downstream p_d, in both Eqs. 4-55 and 4-56. The velocity of the wave is the speed of sound in the fluid downstream. Therefore, for a polytropic process Eq. 4-13(a) becomes

$$c_d = \sqrt{n \, \frac{p_d}{\rho_d}} = \sqrt{ngRT_d} \qquad (4\text{-}57)$$

where the number n depends upon the type of process involved ($n = 1.0$ for an isothermal process and $n = k = 1.4$ for an adiabatic process). Consequently, as p_d or T_d increases, c_d also increases and the approach velocity V_u must be increased in order to maintain steady flow. It is only when the magnitude of the pressure wave is infinitesimal ($p_d/p_u = 1$) that the wave velocity downstream equals the speed of sound upstream and $Ma_u = V_u/c_u = 1.0$.

When an adiabatic process is assumed, there is more energy available **downstream from the wave to maintain the downstream pressure than** when the momentum relation is assumed (a nonadiabatic process). There-fore, the relative pressure downstream from the wave for Eq. 4-55 is some-what greater than for Eq. 4-56.

From the foregoing discussion, it may be seen that flow can be caused to follow the adiabatic process by the proper variation of cross-sectional area and downstream pressure. If a shockwave is formed, however, the process is nonadiabatic and the momentum relation is applicable. For a very small wave, the process is adiabatic.

The locus of the point J_2 after the shock in Fig. 4-8 as it moves from F to K can be determined by solving simultaneously Eqs. 4-47, 4-49, and 4-50; and by using Eq. 4-23 between the upstream point of the flow and the beginning of the shock in order to have p_u in terms of p_1. The result is

$$\frac{p_d}{p_1} = \frac{1}{k-1}\left[\frac{4k}{k+1}\left(\frac{p_1}{p_u}\right)^{(k-1)/k} - (k+1)\right] \tag{4-58}$$

which is plotted in Fig. 4-8 as curve FJ_2K.

Since the curve of pressure distribution with the shockwave in the nozzle is AFJ_1J_2H, the ambient pressure p_a is that at H, which must be somewhere between $(p_3)_G$ and $(p_3)_K$, if the foregoing theory is to apply. Under these conditions, a normal compression wave occurs which is at right angles to the flow. Experiment has shown that the behavior of the wave is accurately predicted by the theory and equations thus far pre-sented (using the momentum relation). The pressure downstream from the shockwave develops as the usual adiabatic compression along J_2H, in accordance with Eq. 4-23. The energy available, however, has been de-creased by the amount of $J_3 - J_2$ because of the shockwave, and therefore curve J_2H, is below curve FG. In all cases, the flow downstream from the shock is subsonic and the flow upstream is supersonic.

When the ambient pressure is between K and L, the formation of the shockwave and the general flow pattern are markedly different from those already described. The wave develops obliquely to the flow and separation may occur, so that the flow does not fill the entire cross section. There is no complete and satisfactory mathematical solution for this condition of flow. However, the general process can be understood somewhat from photo-graphs of the flow. As the ambient pressure is increased from that at L in Fig. 4-8, a wave pattern moves upstream into the nozzle, with oblique waves coming from each wall. These oblique waves may extend to the center of the nozzle and cross one another, or they may move to a center wave which is normal to the flow. As p_a approaches the pressure at G, the central normal wave becomes wider relative to the flow section, and the oblique waves gradually disappear.

The discussion thus far has considered only the **compression shockwave.** **An** **expansion shockwave,** *however, is not possible.* In the compression shockwave, there is a loss of available energy, as already discussed, which results in an increase in entropy—the form of energy which remains constant during an adiabatic process. This is both theoretically and physically possible. However, the reverse process, an expansion shockwave, would involve an increase in available energy (a decrease in entropy), which is not physically possible. In open channel flow, the hydraulic jump is irreversible for the same reason, see Chapter 8.

Example Problem 4-5: Air discharges from one tank into another through a convergent-divergent nozzle, such as illustrated in Fig. 4-8(a), at the rate of 3 lb/sec. If the pressure and temperature in the upstream tank are 50 psia and 240°F respectively and the pressure in the downstream tank is 10 psia, calculate (a) the pressure, temperature, and Mach number in the constriction and at the nozzle exit, (b) the required throat diameter, and (c) the nozzle diameter at the exit plane required to achieve full expansion.

Solution:

(a) Assuming that $V_2 = c_2$

$$p_2 = p_1 \left(\frac{2}{k+1}\right)^{k/k-1} = (50)(0.528) = \underline{\underline{26.4 \text{ psia}}}$$

For full expansion the velocity in the divergent section is supersonic, and

$$\frac{T_2}{T_1} = \frac{2}{k+1} \quad \text{or} \quad T_2 = (700)(0.834) = \underline{\underline{584°\text{F abs}}}$$

$$\frac{T_3}{T_2} = \left(\frac{p_3}{p_2}\right)^{(k-1)/k} \quad \text{or} \quad T_3 = (584)\left(\frac{10}{26.4}\right)^{0.286} = \underline{\underline{441°\text{F abs}}}$$

$$V_2 = c_2 = \sqrt{kgRT_2} = \sqrt{(1.4)(32.2)(53.3)(584)} = 1185 \text{ fps}$$

$$\text{Ma}_2 = \frac{V_2}{c_2} = \frac{1185}{1185} = \underline{\underline{1.0}}$$

$$\gamma_2 = \frac{p_2}{RT_2} = \frac{(26.4)(144)}{(53.3)(584)} = 0.122 \text{ lb/cu ft}$$

$$\frac{V_3^2 - V_1^2}{2g} = \frac{p_2}{\gamma_2}\left(\frac{k}{k-1}\right)\left[1 - \left(\frac{p_3}{p_2}\right)^{k-1/k}\right]$$

from which

$$V_3 = 1310 \text{ fps}$$

$$c_3 = \sqrt{kgRT_3} = \sqrt{(1.4)(32.2)(53.3)(441)} = 1028 \text{ fps}$$

$$\text{Ma}_3 = \frac{V_3}{c_3} = \frac{1310}{1028} = \underline{\underline{1.27}}$$

$$W = \gamma_2 A_2 V_2 \quad \text{and} \quad A_2 = \frac{3.0}{(1185)(0.122)} = 0.0208 \text{ sq ft}$$

$$d_2 = \sqrt{\frac{(0.0208)(4)(144)}{\pi}} = \underline{\underline{1.95 \text{ in.}}}$$

(c) $\quad A_3 = \frac{W}{\gamma_3 V_3} = \frac{3.0}{(0.0607)(1310)} = 0.0376$

$$d_3 = \sqrt{\frac{(0.0376)(4)(144)}{\pi}} = \underline{\underline{2.61 \text{ in.}}}$$

- ## SUMMARY

1. *Compressibility is important* if in the fluid system the density (or volume) changes are appreciable. These cases usually involve: sudden accelerations, elastic waves, velocities approaching or exceeding the speed of sound, and great heights.

2. The *continuity equation* for compressible flow is

$$G = \rho_1 A_1 V_1 = \rho_2 A_2 V_2 \tag{3-14}$$

which is expressed in differential form as

$$dG = d(\rho \, A V) = 0 \tag{4-2}$$

3. The *force-momentum flux equation* for compressible flow is

$$\Sigma F_x = (\rho A V V_x)_2 - (\rho A V V_x)_1 \tag{3-18}$$

4. The *celerity* (speed) of a *small elastic wave* varies with the relative gradients of pressure and density and the bulk modulus of elasticity as follows:

$$c = \sqrt{\frac{dp/dx}{d\rho/dx}} = \sqrt{\frac{E}{\rho}} \tag{4-10}$$

5. For *adiabatic flow,*

$$c = \sqrt{\frac{kp}{\rho}} = \sqrt{kgRT} \tag{4-13a}$$

and for *isothermal flow*

$$c = \sqrt{\frac{p}{\rho}} = \sqrt{gRT} \tag{4-13b}$$

which shows that in a given gas the celerity (speed of sound) varies only with the absolute temperature.

6. The *Mach number* at a point in the flow is the ratio of the velocity at the point to the celerity

$$\text{Ma} = \frac{V}{c}$$

7. The *energy equation* for compressible flow is

$$E_t = E_k + E_e + E_m \tag{4-15}$$

8. For *isothermal flow* in which the change in elevation is small,

$$\frac{V_2^2}{2g} - \frac{V_1^2}{2g} = E_a \tag{4-18}$$

which shows that energy must be added or subtracted if there is to be any change in kinetic energy and yet no change in temperature.

9. For *adiabatic flow,* in which the change in elevation is small, the change in kinetic energy is equal to the change in enthalpy. Consequently,

$$\frac{V_2^2}{2g} - \frac{V_1^2}{2g} = \frac{k}{k-1}\left(\frac{p_1}{\gamma_1} - \frac{p_2}{\gamma_2}\right) \tag{4-21}$$

10. *Stagnation pressure* for adiabatic flow is

$$\frac{p_2 - p_1}{\rho V_1^2/2} = 1 + \tfrac{1}{4}\text{Ma}^2 + \frac{2-k}{24}\text{Ma}^4 + \cdots \tag{4-28}$$

11. In a contracting conduit, **sonic** *velocity* is reached in the minimum cross-sectional area, and **supersonic** *velocities* are reached only in an expanding section downstream from the minimum section.

12. The *maximum rate of flow* per unit area from a constriction occurs at the critical condition for which the velocity is sonic

$$\left(\frac{p_2}{p_1}\right)_c = \left(\frac{2}{k+1}\right)^{k/(k-1)} \tag{4-44}$$

and the rate of flow cannot be increased by further reduction of p. When $k = 1.4$, $(p_2/p_1)_c \approx 0.53$.

13. In the expanding section downstream from a sonic nozzle the flow may follow any one of three different distributions of pressure and velocity, see Fig. 4-8c:

a. An *adiabatic compression* occurs if $(p_a/p_1) = (p_3/p_1)_G$ so that the pressure increases and the velocity decreases in the expanding section.

b. An ***adiabatic expansion*** occurs if $(p_a/p_1) \leq (p_3/p_1)_L$ to that the pressure decreases and the velocity increases to supersonic in the expanding section.

c. A ***shock wave***, with consequent energy loss, occurs in the expanding section if

$$(p_3/p_1)_G > (p_a/p_1) > (p_3/p_1)_L$$

14. A ***shock wave*** is a non-adiabatic process in which part of the kinetic energy is converted directly into heat instead of being converted only into pressure energy.

15. The continuity, momentum, and energy equations can be combined to relate the upstream Mach number to the pressures upstream and downstream from the shock wave

$$\text{Ma}_u = \sqrt{\frac{k+1}{2k}\left(\frac{p_d}{p_u}\right) + \frac{k-1}{2k}} \qquad (4\text{-}56)$$

References

1. Binder, R. C., *Fluid Mechanics*, 3rd ed. Englewood Cliffs, N. J.: Prentice-Hall, Inc., 1955.

2. Dodge, Russell A. and Milton J. Thompson, *Fluid Mechanics*. New York: McGraw-Hill Book Co., Inc., 1937.

3. Hunsaker, J. C. and B. G. Rightmire, *Engineering Applications of Fluid Mechanics*. New York: McGraw-Hill Book Co., Inc., 1947.

4. Leipmann and Roshko, *Elements of Gasdynamics*. New York: John Wiley & Sons, Inc., 1957.

5. Prandtl, Ludwig, *Fluid Dynamics*. New York: Hafner Publishing Co., 1952.

6. Rouse, Hunter, *Elementary Mechanics of Fluids*. New York: John Wiley & Sons, Inc., 1946.

7. Shapiro, Ascher H., *The Dynamics and Thermodynamics of Compressible Fluid Flows*. New York: The Ronald Press Co., 1953.

8. Sibert, H. W., *High Speed Aerodynamics*. Englewood Cliffs, N. J.: Prentice-Hall, Inc., 1948.

9. Vennard, John K., *Elementary Fluid Mechanics*, 3rd ed. New York: John Wiley & Sons, Inc., 1954.

Problems

4-1. The bursting pressure for an automobile tire is 60 psia. If the tire has a constant volume of 1400 cu in. and it is pumped up to 28 psi at 60°F, to what temperature must the air in the tire be raised before the tire bursts?

4-2. A tank of compressed air has a volume of 10 cu ft and the tank pressure is 300 psi. If a release valve is opened to the atmosphere until the tank pressure is reduced to 100 psi and the temperature in the tank remains constant, determine the mass of air released. $T = 60°$.

4-3. Find the volume of a tank required to hold 100 lb of air at 140°F when subjected to a pressure of 20 atmospheres.

4-4. Derive the general equation for the velocity of a pressure wave through a fluid. Explain the application of the foregoing equation to the isothermal, adiabatic, and isentropic flows.

4-5. Compute the speed of sound in air, water, methane, and hydrogen. Assume standard atmospheric pressure and a temperature of 68°F. Compute the relative speed of sound in the foregoing fluids with respect to water.

4-6. A sonic device used to detect submarines receives a return signal 20 seconds after emmission. What is the distance to the submarine?

4-7. Methane flows through a 2-in. constriction in a 4-in. pipe at the rate of 0.40 lb/sec. The velocities and absolute pressures in the pipe and in the constriction are: $V_1 = 50$ ft/sec, $V_2 = 300$ ft/sec, $p_1 = 30$ psia, and $p_2 = 25$ psia. Determine the temperature in the pipe and the constriction.

4-8. Derive Eq. 4-23.

4-9. Referring to Fig. 4-2(d) taken from a Schlieren photograph, measure the wave angle formed by the projectile, and compute the Mach number and the speed of flight in standard air (59°F and 14.7 psia).

4-10. The Mach angle as measured from the photograph of a bullet has a magnitude of 30°. (a) Estimate the speed of the bullet if the temperature and pressure of the atmosphere are 40°F and 12 psia, respectively. (b) What Mach angle would indicate the same velocity in standard air?

4-11. If the difference between static and stagnation pressure in standard air is 26 in. of mercury, compute the air velocity, assuming (a) the air is incompressible, (b) the air is compressible.

4-12. A Pitot tube is used to measure the velocity of an air stream from a nozzle. The Pitot tube is connected to a differential manometer. The manometer fluid is kerosene at 68°F. The temperature and static pressure in the air stream are 50°F and 12 psia, respectively. When the deflection of the manometer fluid is 30 in., what is the velocity of the stream?

4-13. Referring to Problem 4-12, calculate the velocity in the stream, assuming the air to be incompressible. What error is introduced by assuming the fluid incompressible?

4-14. Calculate the pressure on the nose of a modern plane moving at 500 mph through still air and the Mach number when the atmospheric pressure and air temperature are 10 psia and 30°F, respectively.

4-15. Verify Equations 4-25 and 4-26.

4-16. Oxygen flows through a 12-in. circular duct at a rate of 150 lb/minute. Gages at points 1 and 2 indicate pressures of 70 and 7 psi, respectively. Calculate the Mach numbers of the flow at points 1 and 2. Assume $T = 80°F$.

4-17. A large closed tank discharges carbon dioxide through a small orifice in its side. The pressure and temperature in the tank are 18 psia and 70°F, respectively. Neglecting friction losses, determine the jet velocity of the carbon dioxide leaving the tank, (**a**) assuming incompressible flow, (**b**) assuming adiabatic flow. (**c**) What percent error results, assuming constant density?

4-18. Assuming the same tank and orifice as in Problem 4-17 if the tank pressure is increased to 40 psia and the temperature remains the same, compute the jet velocities, (**a**) assuming incompressible flow, (**b**) assuming adiabatic flow. (**c**) What percent error results, assuming constant density?

4-19. Considering the variation of velocity with area, prove that

$$\frac{dA}{dV} = \frac{A}{V}\left[\text{Ma}^2 - 1 \right]$$

and explain variation of velocity with change in area for the subsonic and supersonic velocities.

4-20. The temperature and pressure in a closed tank are 100 psia and 100°F respectively. Air from this tank discharges through a converging nozzle whose exit area is 1.0 sq in. Assuming various back pressures and isentropic flow, establish a curve relating weight rate of flow and back pressure.

4-21. Air flows through a 3-in. pipe with a weight rate of flow of 2.4 lb/sec. The pressure and the stagnation temperature in the duct are 10 psia and 100°F. Calculate (**a**) the velocity, (**b**) the stagnation pressure, and (**c**) the Mach number in the duct.

4-22. Air flows from a 4-in. pipe to a 2-in. pipe. If the pressure and temperature in the 4-in. pipe are 50 psi and 50°F, respectively, and the pressure in the 2-in. pipe is 25 psi, calculate the velocity and weight rate of flow in each pipe, assuming no losses.

4-23. Derive Eq. 4-43 starting with Eq. 4-39. What ratio of p_2/p_1 maximizes the ratio W/A_2? Verify this mathematically.

4-24. Air discharges from one tank to another through a short converging tube. Beginning with Equation 3-15, prove that

$$W = 0.53 \frac{A_2 p_1}{\sqrt{T_1}}$$

if the pressure in the receiving tank is less than critical pressure. A_2 is the area of the throat, and p_1 and T_1 are the pressure and temperature in the upstream tank.

4-25. Air flows from a large tank through a convergent nozzle into the atmosphere. The nozzle has a tip diameter of 2 in. The temperature in the tank is 240°F. Atmospheric pressure is 13 psi. (a) Calculate the pressure in the tank required to produce acoustic velocity in the jet. (b) What is the magnitude of this velocity? (c) Calculate the weight rate of flow.

4-26. Air flows in a circular duct at a pressure of 10 psi. If the Mach number is 0.5, the weight rate of flow is 0.51 lb/sec, and the diameter of the duct is 1 in., calculate the stagnation temperature in the duct in degrees centigrade.

4-27. Derive the relationship

$$\frac{T_2}{T_1} = \left(\frac{p_2}{p_1}\right)^{(k-1)/k}$$

and specifically state the conditions for which the equation is valid.

4-28. Nitrogen discharges from a closed tank into the atmosphere through a $\frac{1}{2}$ in. orifice. Determine the velocity in the jet of fluid, if the tank pressure and temperature are 115 psi and 40°F respectively, and the atmospheric pressure is 12 psia.

4-29. Air flows through a convergent-divergent nozzle. At the throat section in the nozzle, the pressure is 15 psia and the temperature is 100°F. The pressure at a section just downstream from the throat is 10 psia. Assuming no friction losses and adiabatic flow, calculate the velocity and Mach number at this section.

4-30. Air discharges into the atmosphere through a convergent-divergent nozzle of 1-in. throat diameter. The source of air is a constant-pressure temperature tank in which $p_1 = 200$ psia and $T_1 = 140°F$. (a) Assuming that the divergent reach is 8-in. long, compute the diameter at three equally spaced points between the constriction and the nozzle tip required to insure a linear pressure drop through the divergent part of the tube. (b) Calculate the Mach number at the throat, the three intermediate points and the nozzle tip. Assume full expansion and atmospheric pressure of 30 in. mercury.

4-31. Compare the Mach number of a missile moving through standard air at 2640 ft/sec to that for an airplane moving at 600 mph.

4-32. Derive Eqs. 4-55 and 4-56, verify Fig. 4-10, and explain its use.

chapter 5

Fluid Viscosity
and Turbulence

Fluid viscosity and turbulence are present in every aspect of life which involves fluids. On the one hand, a thin sheet of water flowing down a window pane and the flow of a thick oil, molasses, or honey constitute laminar flow, because of the viscosity of the fluid. Furthermore, the flow of water, blood, or other liquids in small tubes or capillaries is laminar flow. In all cases, laminar flow involves mixing on a molecular scale only. On the other hand, the usual flow of water in a stream or river, the smoke or dust issuing from the stack of an industrial plant, and the usual flow of air over the surface of the ground or around objects (particularly downwind from the objects) are turbulent flow, with mixing taking place on a molar scale.

Both types of flow are illustrated in the same flow system, in the case of smoke from a cigarette in a quiet room. For the first few inches, the smoke plume rising vertically is laminar flow, because the viscosity is controlling. At some point along the plume, however, the flow becomes unstable and breaks up into turbulent flow.

In the foregoing chapter, the assumptions were made that the flow was irrotational and that there was no energy loss in the flow system. When viscosity plays an important part in the flow system, however, the flow may be predominantly rotational and the total energy of the system, as represented by the Bernoulli equation (Eq. 3-38), can be reduced as the kinetic energy is converted to heat through viscous shear.

Care must be taken to distinguish between viscosity as a property of the fluid and turbulence as a property of the flow. Since all fluids possess viscosity at least to some extent, the analysis of most flow systems requires consideration of the viscous forces involved. Therefore, detailed consideration is given to the nature of shear and viscosity.

• DEFINITIONS

Shear is the internal stress within a fluid which resists deformation or change of shape of the fluid during motion. It is a tangential stress (in contrast to pressure, which is a normal stress), and it is a vector quantity. In other words, shear is the stress represented by the unit force which acts over a given area taken parallel to the motion. The shear stress in solids occurs under static conditions, wherein the change in shape or the deformation is proportional to the shear stress caused by a steady force exerted on the solid. The shear within a fluid, however, exists only under dynamic (rather than static) conditions, and the shear stress is proportional to the *rate of deformation.* Shear within plastic materials is both static and dynamic. In certain plastic materials, there is an initial static shear stress required to reach a sort of "yield point", beyond which the plastic behaves somewhat like a liquid (see Fig. 5-1.) Another type of substance is the so-called colloidal liquid, in which a fiber-like or thread-like structure causes it to behave somewhat like a plastic when at rest and a liquid when in motion.

In contrast to solids and plastics, a fluid will move and deform under even the slightest external force, and this deformation is resisted by internal shear only as long as the deformation takes place. In other words, the shear is not proportional to the deformation (as in the case of solids), but rather to the *rate* of deformation—a dynamic condition. Figure 5-1 shows how the shear τ varies with the velocity gradient dv/dy within fluids and plastics. Within a Newtonian fluid (such as water and common oils and gases), the shear is directly proportional to the velocity gradient, and the proportionality coefficient is a constant. Within a non-Newtonian fluid (such as slurries, mud flows, gels, and other suspensions and colloidal liquids), and for small shear-values, the shear increases rapidly with increasing velocity gradient—eventually reaching asymptotically a constant rate of increase. The plastic (see Fig. 5-1) has an initial set which first must be overcome to initiate motion. Once motion is initiated, the shear in a plastic varies directly with the velocity gradient, as it does in the fluid.

Because the external force necessary to produce motion is dynamic rather than static, momentum is being transferred continuously to the fluid from some external source. Within the fluid, this momentum is distributed by the internal shear as a tangential stress. In fact, *shear* can be defined significantly as the *rate of transfer of momentum* (momentum flux) *per unit area.* The full significance of this statement is understood better from consideration of fluid viscosity.

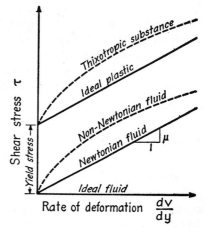

Viscosity is the property of a fluid which causes shear to develop within the fluid to resist deformation of the flow. It is related to the internal cohesiveness of a liquid, although it is a

Fig. 5-1. Variation of shear with velocity gradient.

property which can be measured only while the fluid is in motion. In other words, it is a dynamic property rather than a static one.

For any fluid, there are two terms in fluid mechanics which are useful for expressing the viscosity—the dynamic viscosity represented by μ (mu), and the kinematic viscosity ν (nu). These expressions of viscosity are related by the following equation, see Chapter 1

$$\mu = \rho\nu \quad \text{or} \quad \nu = \frac{\mu}{\rho} \tag{5-1}$$

which expresses the kinematic viscosity as the dynamic viscosity divided by the density.

The cause of the fluid viscosity is not known exactly. However, consideration of existing theories throws considerable light on the subject. For many years, the *kinetic theory of gases* has explained the viscosity of a gas as the result of momentum transfer or exchange due to the activity of the individual gas molecules. Furthermore, the results of numerous experiments which have been made support this theory exceptionally well.

The kinetic theory states that the molecules of a gas are moving over very short distances at great velocities, and that they are continually striking one another and changing their velocities in both magnitude and direction. These collisions result in the transfer of momentum from one molecule to another, so that a shear is created if there is a gradient of momentum concentration ρV. As the temperature is increased, the speed of motion increases, the rate of momentum transfer increases, and hence the viscosity increases.

Perhaps the concept of momentum transfer can be understood better by the use of a simple illustration suggested by Bakhmeteff. Consider two trains of open cars traveling side by side on parallel tracks at velocities of V_A and V_B. The relative velocity is $V_B - V_A = \Delta V$. Further assume that balls are being thrown from train B into the cars of train A. Each ball has a momentum, in the direction the train is moving, of $M V_B$, in which M is the mass of the ball. As each ball comes to rest in train A, there is a drag applied to it to change its velocity from V_B to V_A, so that its momentum is changed from $M V_B$ to $M V_A$. Together with this drag of the train on the ball is an equal and opposite thrust of the ball on train A, which tends to cause train A to change its velocity to that of train B. If balls are also being thrown from train A into train B, a similar thrust is produced on train B, which tends to change its velocity to that of train A. This is a simple example of **momentum exchange**, which tends to produce equal velocities in two bodies. The foregoing description has direct application to the momentum transfer which results from the viscosity of a gas, as explained by the kinetic theory of gases.

Changes in pressure (with temperature remaining constant) usually cause only small changes in viscosity for the common ranges of pressure. The molecules of rarefied gases (extremely low pressures) are very sparse and have much greater travel distances—consequently they possess markedly different properties.

Although the molecules of a **liquid** have a certain amount of activity, the molecules are so closely packed together that their activity seems to be principally a sort of oscillating motion, which is very much restricted in both speed and distance of movement. Viscosity under these conditions appears to be more nearly related to the **cohesive properties** of the molecules than to their kinetic activity. This idea is supported by the fact that the viscosity decreases with temperature. Although pressure generally has little or no influence on the absolute viscosity, extremely great pressures have been found to have marked effects on the viscosity of such liquids as oils.

In **summary**, note that the viscosity of a gas increases with temperature, apparently due to increased activity of the molecules, and that viscosity of a liquid decreases with an increase in temperature, apparently due to a decrease in the cohesive forces associated with the molecules.

• LAMINAR FLOW

If a fluid is flowing relative to a bounding surface and the momentum transfer is by molecular activity only, the flow is known as **laminar flow.** In laminar flow, the forces of viscosity predominate over other forces, such as fluid inertia, as is discussed later in this chapter. Various types of laminar flow are now considered.

• SHEAR EQUATION

To illustrate the action of viscosity and its influence upon the flow system, consider two parallel plates—one fixed and the other moving with a velocity V_o, as shown in Fig. 5-2.

Since the fluid molecules in contact with a given plate must have the same velocity as the plate, the velocity of the fluid varies from zero at the fixed boundary to V_o at the moving boundary. This variation in velocity has been found experimentally to be a straight line and, therefore, the velocity gradient dv/dy is a constant across the flow.

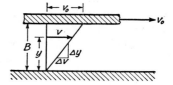

Fig. 5-2. Laminar flow with a moving boundary.

Newton stated that the shear τ at a point within a fluid is proportional to the velocity gradient dv/dy at that point. Mathematically, this is

$$\tau = \mu \frac{dv}{dy} \tag{5-2}$$

in which the proportionality constant μ is the viscosity which causes the deformation and develops the gradient of velocity. From Eq. 5-2, it can be seen that since dv/dy is a constant, the shear is also a constant across the flow section in Fig. 5-2.

Equation 5-2 can be rearranged as

$$\tau = \frac{\mu}{\rho} \frac{d(\rho v)}{dy} = \nu \frac{d(\rho v)}{dy} \tag{5-3}$$

to show that **shear** (*momentum flux per unit area*) *is proportional to the kinematic viscosity* ν *and the gradient of the momentum concentration* ρv. This is a case of laminar flow, because the momentum transfer is by molecular action only.

Example Problem 5-1: The moving plate in Fig. 5-2 which might represent the side of a piston in a cylinder is moving at a velocity of 40 fps. If the spacing B between the two parallel plates is 0.06 inch, what is the kinematic viscosity of the oil in this space between the plates if the shearing stress developed along the stationary wall is 1.0 psf and the specific gravity of the oil is 0.80.

Solution:

From Eq. 5-2

$$\tau = \mu \frac{dv}{dy} = 1.0 = \mu \frac{40}{\dfrac{0.06}{12}}$$

or
$$\mu = 1.25 \times 10^{-4} \frac{\text{lb-sec}}{\text{sq ft}}$$

To solve for kinematic viscosity, use Eq. 5-1 or Eq. 1-10

$$\nu = \frac{\mu}{\rho} = \frac{1.25 \times 10^{-4}}{1.94 \times 0.8} = 8.05 \times 10^{-3} \text{ sq ft/sec}$$

• FLOW BETWEEN TWO PARALLEL PLATES

In connection with lubrication and certain other problems in fluid flow, another simple case is studied—that of flow between two fixed parallel plates. If the flow is assumed to be one-dimensional as shown in Fig. 5-3,

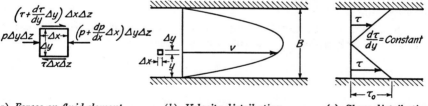

(a) *Forces on fluid element*   (b) *Velocity distribution*   (c) *Shear distribution*

Fig. 5-3. Laminar flow between fixed parallel plates.

the velocity varies from zero at each boundary to a maximum $v_{\max}$ in the center. From Fig. 5-3 and the shear equation $\tau = \mu \, dv/dy$, it can be seen that the velocity gradient (and hence the shear) is a maximum at the boundaries and zero in the center.

By taking a free body of length Δx, width Δz, and thickness Δy, the forces (see enlarged sketch in Fig. 5-3(a)) can be summed in the x direction to give

$$p \, \Delta y \, \Delta z - \left(p + \frac{dp}{dx} \Delta x\right) \Delta y \, \Delta z - \tau \, \Delta x \, \Delta z + \left(\tau + \frac{d\tau}{dy} \Delta y\right) \Delta x \, \Delta z = 0 \quad (5\text{-}4)$$

which reduces to

$$\frac{dp}{dx} \Delta x \, \Delta y \, \Delta z = \frac{d\tau}{dy} \Delta x \, \Delta y \, \Delta z \qquad (5\text{-}5)$$

and finally, upon dividing by the volume $\Delta x \, \Delta y \, \Delta z$

$$\frac{dp}{dx} = \frac{d\tau}{dy} \qquad (5\text{-}6)$$

Equation 5-6 is a very significant equation, aside from its use in the present derivation. It shows that, in the absence of inertia forces, the pressure variation in the direction of flow between parallel plates is equal to the variation of shear in the direction perpendicular to the flow. Note that the presence of one requires the presence of the other. Likewise, Eq. 5-6 shows that a pressure gradient can exist in a given direction without the accompanying acceleration previously studied—provided there is a gradient of shear in a perpendicular direction.

For the flow system involving the moving plate in Fig. 5-2, there is no pressure gradient (except the hydrostatic variation which is taken into account in Eq. 5-13), and consequently $d\tau/dy = dp/dx = 0$ variation in the system. Since the energy for the system is supplied only from the moving plate, there is no pressure gradient supplying any of the energy which is being dissipated in viscous shear. This is illustrated by the journal bearing for which Fig. 5-2 can be considered as a small section which is greatly enlarged.

In solving the problem of flow between parallel plates, Eq. 5-6 can be integrated with respect to y to obtain the shear

$$\tau = \frac{dp}{dx} \, (y + c) \qquad (5\text{-}7)$$

Since $\tau = 0$ when $y = B/2$, this equation becomes

$$\tau = -\frac{dp}{dx} \left(\frac{B}{2} - y \right) \qquad (5\text{-}8)$$

and can be combined with Newton's shear equation to give

$$\frac{dv}{dy} = -\frac{1}{\mu} \frac{dp}{dx} \left(\frac{B}{2} - y \right) \qquad (5\text{-}9)$$

Integration of this equation with respect to y yields

$$v = -\frac{1}{2\mu} \frac{dp}{dx} \, (By - y^2) \qquad (5\text{-}10)$$

in which the constant of integration has been evaluated for the condition that $v = 0$ when $y = 0$.

Equation 5-10, which expresses the *velocity distribution for laminar flow between two parallel plates,* shows that the velocity at a point y varies directly with the pressure gradient dp/dx and inversely with the viscosity μ.

Because Eq. 5-10 is the equation of a parabola, the average velocity between the boundaries is $2/3$ the maximum velocity which occurs at the midpoint between the boundaries. In equation form

$$V = \frac{2}{3} v_{\max} = -\frac{B^2}{12\mu}\frac{dp}{dx} \tag{5-11}$$

Equation 5-11 can be rearranged to solve explicitly for the pressure gradient

$$-\frac{dp}{dx} = \frac{12\mu V}{B^2} \tag{5-12}$$

which shows that, if the distance B between the boundaries and the viscosity remains constant, and if the flow is steady, then the pressure gradient dp/dx is also a constant, and the drop in pressure Δp is the same for a given distance in the direction of flow, regardless of where along the flow this distance is taken. Furthermore, since dp/dx is a constant, Eq. 5-6 and Fig.5-3 show that the gradient of shear across the flow $d\tau/dy$ is also a constant.

If the plates are inclined, gravity forces affect the pressure gradient dp/dx. This effect can be accounted for by replacing the pressure gradient with the gradient of piezometric head

$$\frac{dh}{dx} = \frac{dz}{dx} + \frac{1}{\gamma}\frac{dp}{dx}$$

in which the x-direction is taken parallel to the plates and the z-direction is taken vertically upward. For this more general case, Eq. 5-6 becomes

$$\gamma\frac{dh}{dx} = \frac{d\tau}{dy} \tag{5-13}$$

From the velocity distribution expressed in Eq. 5-10, it is obvious that in the x-y plane the flow is rotational. In the x-z plane, however, the flow is irrotational—a fact which was utilized by Hele-Shaw to demonstrate visually the irrotational flow pattern around various submerged two-dimensional objects.

Example Problem 5-2: Oil is flowing between two parallel plates, as in Fig. 5-3. Piezometers along the plates indicate that the pressure is decreasing in the direction of flow at the rate of 0.5 psf per ft. If the dynamic viscosity is 10^{-3} lbs-sec/sq ft and the spacing of the plates is 2 inches, (a) what is the velocity of the fluid midway between the plates? (b) what is the average velocity?

Solution:

(a) Applying Eq. 5-10

$$v = -\frac{1}{2\mu}\frac{dp}{dx}(By - y^2)$$

Since $-\dfrac{dp}{dx} = \dfrac{0.5}{1.0} = 0.5$ psf/ft

$$v = \frac{1}{2 \times 10^{-3}}(0.5)\left(\frac{1}{6}\frac{1}{12} - \frac{1}{12^2}\right)$$

$$v = \frac{0.25 \times 10^3}{12}\left(\frac{1}{6} - \frac{1}{12}\right)$$

and $\qquad v = \underline{\underline{1.74 \text{ fps}}}$

(b) Part (b) requires the use of Eq. 5-11

$$V = -\frac{B^2}{12\mu}\frac{dp}{dx}$$

Since $-\dfrac{dp}{dx} = 0.5$ psf/ft

$$V = \frac{\left(\frac{1}{6}\right)^2}{(12)10^{-3}}(0.5)$$

and $\qquad V = \underline{\underline{1.16 \text{ fps}}}$

or $\qquad V = \frac{2}{3}v_{max} = \frac{2}{3}(1.74) = \underline{\underline{1.16 \text{ fps}}}$

In connection with certain lubrication problems, the lubricant is forced under pressure through the space between the boundaries, with one boundary also moving. The problem is then a combination of the cases illustrated by Figs. 5-2 and 5-3. Because of the linear nature of the equations describing the flow shown in these figures, the equations can be combined by simple addition to give the velocity distribution and the pressure distribution.

If the oil is being forced into the bearing in the same direction as the boundary is moving, then the equation for velocity in Fig. 5-2, which is $v = V_o y/B$, can be added to Eq. 5-10 to give the velocity distribution for the condition of oil being forced into a bearing when the shaft in the bearing is moving. That is,

$$v = -\frac{1}{2\mu}\frac{dp}{dx}(By - y^2) + V_o\frac{y}{B}$$

If the elevation variation is to be considered, the piezometric head h must be used, and Eq. 5-10 can be written as

$$v = -\frac{\gamma}{2\mu}\frac{dh}{dx}(By - y^2) \tag{5-14}$$

and Eq. 5-12 as

$$h_f = \frac{12\mu VL}{\gamma B^2} \tag{5-15}$$

in which h_f is the loss in piezometric head and is considered positive if the piezometric head decreases in the direction of flow.

The mean velocity is

$$V = \frac{\gamma B^2 h_f}{12\mu L} \tag{5-16}$$

Viscous, free-surface flow down an inclined plane, is a special case corresponding to the lower half of the flow between inclined plates with the pressure gradient equal to zero (see Fig. 5-3).
Equation 5-14 then becomes

$$v = \frac{\gamma S}{2\mu}(2yy_o - y^2) \tag{5-17}$$

Equation 5-11 becomes

$$V = \frac{\gamma S y_o^2}{3\mu} = \frac{\gamma y_o^2 h_f}{3\mu L} \tag{5-18}$$

and Eq. 5-12 becomes

$$S = \frac{3\mu V}{\gamma y_o^2} \tag{5-19}$$

or

$$h_f = 3\frac{\mu VL}{\gamma y_o^2} \tag{5-20}$$

in which S is the slope dz/dx taken as positive when sloping downward in the direction of flow (L/L);

h_f is the loss of total head (L);

L is the length of flow considered (L);

V is the mean velocity of flow (L/T);

μ is the dynamic viscosity (FT/L^2);

γ is the specific weight (F/L^3);

y_0 is the depth of flow or $B/2$ in Fig. 5-3;

x-direction is in the direction of flow;

z-direction is vertical.

• FLOW IN CIRCULAR PIPES

Laminar flow in a circular pipe was studied by both Hagen, a German engineer working with brass pipes, and Poiseuille, a French scientist interested in the flow of blood through veins. Through experiment, they discovered independently, and published in 1839 and 1840 respectively, the variation of pressure drop with velocity. The analytical derivation which expresses this relationship can be developed through the use of Fig. 5-4,

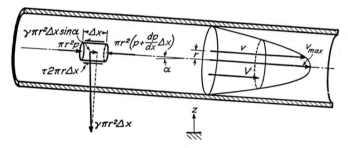

Fig. 5-4. Laminar flow in a pipe.

in which the cylindrical element is considered to be a free body. The forces acting on the free body are the pressure change from one end of the body to the other, the shear on the periphery of the body, and the weight of the body in the direction of flow. These can be summed to give

$$\pi r^2 p - \pi r^2 \left(p + \frac{dp}{dx} \Delta x \right) - \tau 2\pi r \Delta x + \gamma \pi r^2 \Delta x \sin \alpha = 0 \qquad (5\text{-}21)$$

which can be simplified to

$$\tau = \tfrac{1}{2}\gamma r \sin \alpha - \tfrac{1}{2} r \frac{dp}{dx} = -\gamma \frac{r}{2} \frac{dh}{dx} \qquad (5\text{-}22)$$

since $\sin \alpha = -dz/dx$.

The shear equation $\tau = \mu \dfrac{dv}{dy}$ (in which $dy = -dr$), can be introduced so that $-\mu \dfrac{dv}{dr}$ gives

$$\frac{dv}{dr} = \frac{\gamma r}{2\mu} \frac{dh}{dx} \qquad (5\text{-}23)$$

Integrating and solving for v yields the equation for *laminar flow* in a pipe

$$v = -\frac{\gamma}{4\mu}\frac{dh}{dx}(r_o^2 - r^2) \tag{5-24}$$

since $v = o$ at $r = r_o$.

From Eq. 5-24, it can be seen that the velocity distribution for laminar flow in a circular pipe is represented by a paraboloid of revolution. This fact makes it possible to evaluate the mean velocity as half the maximum velocity v_{max} Thus

$$V = \frac{1}{2}v_{max} = -\frac{\gamma D^2}{32\mu}\frac{dh}{dx} = \frac{\gamma D^2}{32\mu}\frac{h_f}{L} \tag{5-25}$$

in which D is the pipe diameter.

Equation 5-25 is the Hagen-Poiseuille equation for the mean velocity. This velocity can be multiplied by the cross-sectional area to yield the discharge. Since dh/dx is a constant for steady, uniform flow, Eq. 5-25 can be solved directly for head loss

$$h_f = 32\frac{\mu VL}{\gamma D^2} \tag{5-26}$$

in which h_f is measured vertically downward from the total head line to give the **head loss** over the length L. To determine changes in pressure along the conduit, the piezometric head loss h_f can be divided into the change in pressure head $\Delta p/\gamma$ and the change in elevation head Δz. Thus

$$h_f = -\left(\frac{\Delta p}{\gamma} + \Delta z\right)$$

and Eq. 5-26 becomes

$$-\Delta p = 32\frac{\mu VL}{D^2} + \Delta(\gamma z) \tag{5-27}$$

Example Problem 5-3: A 2-in. diameter pipe is carrying 45 gallons per minute (gpm) of oil (sp. gr. $= 0.83$) whose dynamic viscosity is 7×10^{-4} lb-sec/sq ft. The flow in the pipe is laminar. Find (a) the head loss in 100 feet of pipe, and (b) the velocity at a point 1/2 in. from the wall.

Solution:

(a) To solve for the head loss, use Eq. 5-26

$$h_f = 32\frac{\mu VL}{\gamma D^2}$$

First convert gpm to cfs. Since

$$1 \text{ cf} = 7.48 \text{ gallons}$$
$$1 \text{ cfs} = 448 \text{ gpm.}$$

In practice   1 cfs is taken as 450 gpm so that

45 gpm = 0.1 cfs

From continuity

$$V = \frac{Q}{A} = \frac{0.1}{0.0218} = 4.59 \text{ fps}$$

By substitution

$$h_f = \frac{32(7 \times 10^{-4})(4.59)(100)}{(62.4)(0.83)\left(\frac{2}{12}\right)^2} = 7.13 \text{ ft}$$

(b) From Eq. 5-24, the velocity 1/2 in. from the wall can be calculated

$$v = -\frac{\gamma}{4\mu}\frac{dh}{dx}(r_o^2 - r^2)$$

From part (a)

$$-\frac{dh}{dx} = \frac{7.13}{100} = 0.0713$$

$$v = \frac{51.7(0.0713)}{4(7 \times 10^{-4})}\left[\left(\frac{1}{12}\right)^2 - \left(\frac{1}{24}\right)^2\right]$$

$$v = 6.84 \text{ fps} \text{ at a point 1/2 in. from the pipe wall.}$$

• FLOW AROUND A SPHERE

Problems involving laminar flow in which the viscous forces predominate are not confined to flow between or within boundaries. Laminar flow may exist around an object such as a dust particle falling in the air or a fine sand grain falling in water. The well known solution to this problem is the one developed and published by Stokes in 1851 for spheres

$$F_D = 3\pi d\mu V \qquad (5\text{-}28)$$

in which

F_D is the drag on the sphere (F);

d is the diameter of the sphere (L);

μ is the dynamic viscosity of the fluid (FT/L^2);

V is the velocity of the fluid relative to the sphere (L/T).

As will be discussed later in greater detail, the drag force F_D is equal to the submerged weight of the sphere if it is falling (or rising in the case of a bubble) through a fluid at constant velocity. In this case the ***"terminal"*** ***fall velocity*** is

$$V = \frac{\Delta \gamma d^2}{18\mu} \tag{5-29}$$

in which $\Delta\gamma$ is the difference between the specific weight of the fluid and the specific weight of the material of which the sphere is made.

Terminal fall velocity is defined as the velocity which a body reaches in falling through a fluid at rest, when the drag on the body is equal to the submerged weight of the body.

• FLOW IN POROUS MEDIA

The problem of laminar flow in porous media has certain similarities to both laminar flow within boundaries and laminar flow around a submerged object. The basic relationship regarding flow through porous media was published in 1856 by Henri Darcy, a French engineer. Darcy discovered experimentally that, for small discharges, the discharge varied directly with head loss h_f over the distance L, as shown in Fig. 5-5. During the century which has followed, the coefficients involved in this relationship have been refined considerably, but the basic variations have remained the same. Darcy's equation is

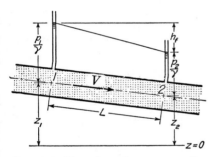

Fig. 5-5. Flow through porous media.

$$h_f = c \frac{\mu V L}{\gamma d^2} \tag{5-30}$$

in which

h_f is the head loss $(LF/F = L)$;

V is the mean velocity of flow (L/T) based on total area;

γ is the specific weight of the fluid flowing (F/L^3);

d is the characteristic grain diameter of the porous material (L);

c is the dimensionless coefficient which describes the porous media by including the size distribution and size of the grains, the porosity, and the orientation and arrangement of the grains. This is commonly known as the coefficient of permeability.

Solving for the velocity directly

$$V = \frac{\gamma d^2}{c\mu} \frac{h_f}{L} = k \frac{h_f}{L} \tag{5-31}$$

in which k is the transmission constant, as commonly used in the Darcy equation.

The similarity between Eq. 5-30 for laminar flow through porous media and Eq. 5-26 for laminar flow through pipes demonstrates that laminar flow through porous media may be considered as occurring through a series of sinuous and nonuniform tubes. The variation in shape of the tubes is apparently characterized by the coefficient 32 in Eq. 5-26 and c in Eq. 5-30.

• SIMILARITY AMONG EQUATIONS FOR LAMINAR FLOW

The similarity among Eqs. 5-15, 5-20, 5-26, and 5-30, is worthy of discussion and special consideration. In each, the head loss or drag varies directly with the velocity, the dynamic viscosity, and the distance over which the loss occurs. Since the flow pattern is dependent upon the nature of the boundaries, the remainder of each equation characterizes the nature of these boundaries.

It is significant also that the density ρ does not enter anywhere in Eqs. 5-15, 5-20, 5-26, and 5-30. In other words, for *laminar flow the forces of inertia, which depend upon the density, are negligible and the forces of viscosity are in complete control.* This fact will be important to recall later in the discussion of turbulent flow.

Since viscosity is a fluid property which, in most cases, does not change appreciably with pressure or location within the region of flow, a number of problems in laminar flow can be solved mathematically. Those problems which have already been considered in this chapter are examples of such mathematical treatment. When the flow is turbulent, however, the factors to be considered are so complicated that rigorous mathematical analysis becomes virtually impossible.

• INSTABILITY OF LAMINAR FLOW

Osborn Reynolds was the first to study laminar flow extensively and to demonstrate the transition from laminar flow to turbulent flow. He constructed an apparatus similar to that shown in Fig. 5-6(a), consisting of a quieting reservoir and a glass tube with a bell-mouthed entrance. A very small tube in the reservoir injected a filament of dye into the flow entering the bellmouth. When the flow was laminar, Reynolds found that the filament passed through the glass tube completely intact as a thread. Furthermore, he found that, provided the specific gravities of the fluid and dye are the same, the filament did not move transversely across the tube. Instead, it remained at the same location relative to the walls throughout the length of the tube, regardless of how close or far it was from the walls initially, see Fig. 5-6(b).

When the rate of flow through the glass tube was increased sufficiently, Reynolds discovered that a *waviness* and *instability* developed and that small parts of the filament tended to break away, see Fig. 5-6(c). Only a small additional increase in flow was necessary to cause the *waviness* to expand transversely into a widespread dispersion of the filament through the flow, see Fig. 5-6(d). This sequence of events demonstrated the transition from stable laminar flow to fully-developed turbulent flow. Turbulent flow is discussed in detail in the next section.

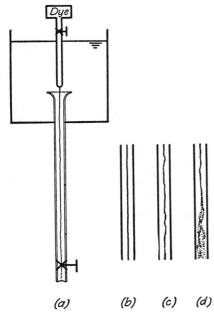

(a)   (b)   (c)   (d)

Fig. 5-6. Reynold's apparatus.

The cause of the waviness and eventual breakdown into turbulent flow can be understood by considering in detail a small portion of the flow within a tube, Fig. 5-7(a). Due to the velocity gradient within the flow, the upper part of the fluid is moving more rapidly than that nearer the boundary. Hence, a shear stress $\tau = \mu \, dv/dy$ is created. As this shear increases, at a point given within the fluid the laminar flow becomes unstable, so that the slightest disturbance, such as that represented by the wave in Fig.

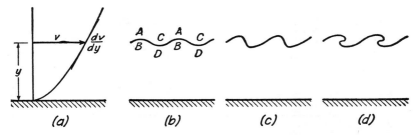

(a)   (b)   (c)   (d)

Fig. 5-7. Development of turbulent flow from instability of laminar flow.

5-7(b), will cause unbalanced forces. Above the crest of a wave (point A), the streamlines are compressed closer together, the velocity is increased (as demonstrated by the continuity equation), and the pressure is reduced (as predicted by the energy equation). On the underside of a wave crest (point B), the pressure is increased. Obviously, with this pressure condition existing, the wave tends to expand vertically in the direction of decreasing pressure, as shown in Fig. 5-7(c). When the undular disturbance or wave reaches upward appreciably, it encounters the higher-velocity flow which bends the crests and eventually causes them to break, as shown in Fig. 5-7(d), and mixing of finite fluid masses has begun. These fluid masses are known as eddies, which are random both in size and orientation — due to the way in which they are formed initially and subsequently change their character.

A criterion has been developed by Rouse for stability of laminar flow at any point within a flow system. He reasoned that the velocity gradient dv/dy, the viscosity μ, the density ρ, and the distance y from a boundary are the variables controlling the stability of laminar flow. He combined these variables into the following dimensionless parameter:

$$\chi = \frac{y^2 \rho \, dv/dy}{\mu} \tag{5-32}$$

With this parameter, Rouse found that, if the magnitude of χ (chi) at any point in the flow exceeds approximately 500, the laminar flow will be unstable and a minor disturbance could cause the laminar flow to break up into turbulent flow.

A study of the variables in the χ-parameter shows that, on the one hand, if the viscosity μ is large, the parameter is small and the flow is stable. On the other hand, if any one or a combination of the distance y, the density ρ, or the velocity gradient dv/dy are large, the parameter is large and the flow tends to become unstable. Figure 5-8 is a plot of Eq. 5-32. Note that

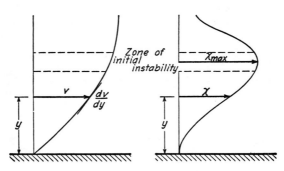

Fig. 5-8. Stability criterion.

the maximum x-value occurs somewhere between the boundary (where dv/dy is a maximum but $y = 0$) and the outer edge of the flow (where dv/dy is a minimum and y is a maximum). In other words, Eq. 5-32 predicts that instability will occur first at some intermediate distance from the boundary. Such prediction is verified by observations of laminar flow in pipes, in which it has been found that initial breakdown of laminar flow occurs between the pipe wall and the centerline of the pipe.

With stable laminar flow, it is possible to introduce any type of disturbance into the flow, and the forces of viscosity will cause the disturbance to be damped and to die out. As the velocity is increased, however, the forces of inertia increase and the instability increases until the viscous forces can no longer damp a disturbance — no matter how minor. It is quite logical, then, that a dimensionless parameter called the Reynolds number, in honor of Osborn Reynolds, should be developed which is a ratio of the forces of inertia to the forces of viscosity within the flow system. The magnitude of the Reynolds number

$$\text{Re} = \frac{VD\rho}{\mu} = \frac{VD}{\nu} \tag{5-33}$$

has been found to be a reliable and important criterion of the stability of laminar flow. In Eq. 5-33

> V is the mean velocity (L/T);
> D is the diameter of the pipe (L);
> ρ is the density of the fluid (FT^2/L^4);
> μ is the dynamic viscosity (FT/L^2);
> ν is the kinematic viscosity (L^2/T).

When $\text{Re} < 2000$ in a circular pipe or tube, it has been found that a disturbance will be damped out, but if $\text{Re} > 2000$, the laminar flow is unstable and a disturbance will probably be magnified, eddies will form, and the laminar flow will break up into turbulent flow. Some experimenters have been able to maintain laminar flow for extremely large Reynolds numbers by allowing the liquid in the reservoir to rest for several days, so that all natural disturbances within the liquid will die out before the experiment is run.

Example Problem 5-4: At what distance from the wall of a circular pipe does the point of maximum instability exist when the flow is laminar?

Solution:

For laminar flow, the velocity gradient is expressed by Eq. 5-23

$$\frac{dv}{dr} = \frac{\gamma r}{2\mu} \frac{dh}{dx}$$

The χ-parameter is expressed by Eq. 5-32

$$\chi = \frac{y^2\rho\ dv/dy}{\mu}$$

Since y and r are measured in opposite directions

$$r = r_0 - y \quad \text{and} \quad -dr = dy$$

so that

$$\chi = \frac{y^2\rho\left(\dfrac{\gamma(r_0 - y)}{2\mu}\right)\dfrac{dh}{dx}}{\mu}$$

or $\qquad\qquad \chi = -\dfrac{\rho\gamma}{2\mu^2}\dfrac{dh}{dx}\,(y^2r_0 - y^3)$

Differentiating with respect to y and setting the differential equal to zero for the point of maximum instability

$$d\chi = \left(\frac{-\rho\gamma}{2\mu^2}\frac{dh}{dx}\right)(2yr_0\ dy - 3y^2\ dy)$$

$$\frac{d\chi}{dy} = 0 = 2r_0 - 3y$$

$$y = \frac{2}{3}r_0 \qquad \text{Point of maximum instability}$$

Example Problem 5-5: A 1-in. diameter pipe carries 0.010 cfs of oil whose kinematic viscosity is 10^{-4} ft^2/sec. Is the flow laminar or turbulent?

Solution:

$$\mathrm{Re} = \frac{VD}{\nu}$$

$$V = \frac{Q}{A} = \frac{0.01}{0.00545} = 1.836 \text{ fps}$$

By substitution

$$\mathrm{Re} = \frac{(1.836)\left(\dfrac{1}{12}\right)}{10^{-4}} = 1530$$

Since $\mathrm{Re} = 1530 < 2000$, the <u>flow is laminar</u>.

• TURBULENT FLOW

Once laminar flow has become unstable, and eddies are formed and spread throughout the flow as described in the preceding section, the instantaneous velocity at a given point fluctuates markedly with time, as shown in Fig. 5-9. The eddies which make up the turbulence are random, both in size and in orientation.

The fluctuations v' take place about a mean or average velocity $\bar{v}$. This action can be visualized graphically by considering the action of an individual eddy. Each eddy is rotating, see Fig. 5-10, and as it moves to the point where the velocity is being measured, it may first increase the

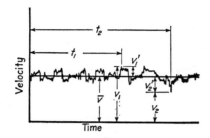

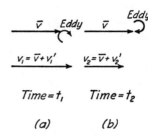

Fig. 5-9. Fluctuations of velocity at a point in turbulent flow.

Fig. 5-10. Velocity fluctuations caused by eddy motions.

mean velocity to cause greater total velocity, as shown in Fig. 5-10(a). An instant later, the other side of the eddy, which is moving in the opposite direction, may move to the point of measurement, where it superposes a reverse or negative velocity to cause a net velocity less than the mean velocity, as shown in Fig. 5-10(b). This continuous and random action of numerous eddies causes the frequent change in velocity at a point with respect to time, as illustrated in Fig. 5-9.

In turbulent flow, the instantaneous velocity varies not only with respect to time at a point, but also with respect to distance at a given time, as shown in Fig. 5-11. In other

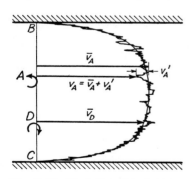

Fig. 5-11. Variation of instantaneous velocity with distance.

words, at a given point the velocity fluctuates from instant to instant; and at a given instant of time the velocity varies from point to point across the flow, as shown in Fig. 5-11. Here again, the single eddy can be considered in order to understand in more detail the action taking place. At point A, the upper part of the eddy is moving in the reverse or negative direction, thereby causing the net velocity to be less than the mean or average velocity. The lower part of the eddy, however, is moving in the positive direction, and causes the net velocity to be greater than the mean velocity. With a random occurrence of eddies (turbulence) across the flow from B to C, a random distribution of instantaneous velocity also occurs.

An eddy not only has rotational velocity, but it also has a velocity of translation which depends upon its origin and its history or experience. It may move for a brief period of time in the general direction of the mean velocity of the flow, and then take off transversely into a region of different velocity. Since the eddy is carrying with it the mean velocity of the region from where it came, the new region will experience a drag force either forward or backward, depending upon whether the new region has a velocity less than or greater than the region from which it came. This interchange of eddies results in a much more effec-

tive mixing action and diffusion of momentum than with laminar flow. Hence, the distribution of mean velocity across the flow is much more uniform, as shown in Fig. 5-12.

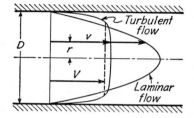

With such action as the foregoing taking place, it is to be expected that considerable shear and energy dissipation would occur. Indeed, each individual eddy gradually dissipates its energy

Fig. 5-12. Velocity profiles for laminar and turbulent flow.

through viscous shear as it blends into its new surroundings and loses its identity. New eddies, however, are continuously being generated, as discussed previously, so that fully-developed turbulence—a completely random occurrence, movement, and dissipation of eddies—is created and maintained throughout the flow system.

Because of the completely random and heterogeneous nature of turbulent flow, it can not be treated in a simple analytical fashion, as in the case of laminar flow. Therefore, it has been necessary to evaluate turbulence in a statistical manner. Although the eddies within a system vary both in speed of rotation and size or scale, they can be characterized by mean or average values of speed and size. These two characteristics are called the intensity of the turbulence and the scale of the turbulence.

The *intensity of the turbulence* is defined as the root-mean-square $\sqrt{\overline{v'^2}}$ of the fluctuations of the instantaneous velocity about the mean velocity. It can be determined from Fig. 5-9 by measuring the velocity fluctuation v' at a given time interval, squaring each value, taking the mean or average of the squares, and then taking the square root of the mean. The determination of the intensity of turbulence can also be accomplished electrically and continuously, as measurements are being taken by use of the hot-wire anemometer in air. The sensing element of this instrument usually consists of a very fine wire (approximately $\frac{1}{10}$ the size of a human hair) about $\frac{1}{8}$ in. long. An electrical current is passed through the wire to heat it, and the air moving past the wire at fluctuating velocities cools it, causing the wire to have a fluctuating resistance which is measured.

The wire is usually calibrated by towing it at various speeds in quiet air. By electrical means, the measurements are converted to root-mean-square (rms-values) of the velocity fluctuation to give the intensity of turbulence.

The **scale of the turbulence** l is usually determined by the use of two sensing elements placed in the flow at varying distances apart. When the elements are very close together, they are within the same eddy, they experience the same velocities of the air, and their readings are closely co-related. As the sensing elements are separated more and more, their readings have less and less correlation—the correlation eventually going to zero as the distance becomes greater. The scale of the turbulence can then be determined by a measure of the distance separating the sensing elements for a certain degree of correlation between their readings, or by the area under the spacing-correlation curve.

Earlier in this chapter, consideration was given to the mixing or diffusion of momentum by the action of molecular viscosity in laminar flow. In turbulent flow, however, diffusion takes place through the action of eddies which are finite groups of molecules—that is, **eddy viscosity.** The use of eddy viscosity was suggested by Boussinesq and, although eddy viscosity has the same dimensions as molecular viscosity, it is a property of the flow, while molecular viscosity is entirely a property of the fluid. For turbulent flow, then, Newton's shear equation, Eq. 5-2, can be modified to

$$\tau = (\mu + \eta)\frac{dv}{dy} \tag{5-34}$$

or

$$\tau = (\nu + \epsilon)\frac{d(\rho v)}{dy} \tag{5-35}$$

in which η (eta) is the dynamic eddy viscosity (a property which is dependent only on the flow and is independent of the fluid);

$\epsilon = \eta/\rho$ is the kinematic eddy viscosity (epsilon).

The eddy viscosity can be related to the intensity and scale of the turbulence by

$$\epsilon = cl\sqrt{\overline{v'^2}} \tag{5-36}$$

in which c is a proportionality constant. This equation shows that ϵ increases with l (which in turn increases with the scale of the flow system) and with $\sqrt{\overline{v'^2}}$ (which increases with the velocity of the flow system).

The eddy viscosity has the disadvantage of varying in magnitude with distance from the boundary, whereas the molecular viscosity is a constant throughout the flow (assuming, of course, the temperature also remains a constant). Eq. 5-35, then, is the completely general form of the shear equation wherein ν predominates with ϵ going to zero for laminar flow, and ϵ predominates with ν becoming insignificant by comparison in turbulent flow.

By making certain assumptions regarding the variation of ϵ and $d(\rho v)/dy$ across the flow, this equation could be used to derive a velocity distribution across the flow in somewhat the same way that Eq. 5-24 was derived for laminar flow. However, other derivations (see Chapter 6) have proved more satisfactory and representative of actual conditions. These derivations are based upon the shear equation proposed by Reynolds in 1895

$$\tau = -\rho v_x' v_y' \tag{5-37}$$

which gives the internal shearing stress due to turbulence. In Eq. 5-37, v_x' and v_y' are the instantaneous velocity fluctuations measured simultaneously in the x and y directions. A detailed study of Eq. 5-37 is helpful in explaining the mechanics of turbulent shear.* Prandtl used this equation and assumed that the change in velocity l (dv/dy) experienced by an eddy moving over a certain distance called the mixing length l was proportional to both v_x' and v_y'. Therefore, he proposed the equation

$$\tau = \rho l^2 \left(\frac{dv}{dy}\right)^2 \tag{5-38}$$

for the shear, where the mixing length l includes the proportionality factor.

The qualitative equation 5-38 becomes a useful equation for evaluating shear by assuming that l is proportional to y.

$$\tau = \rho \kappa^2 y^2 \left(\frac{dv}{dy}\right)^2 \tag{5-39}$$

in which κ (kappa) (the proportionality constant) is the so-called Karman universal constant, which has been determined empirically to be 0.4.

From the foregoing presentation, it is obvious that the phenomenon of turbulent flow is very complex and its analysis is far more difficult than that of laminar flow. Nevertheless, most of the problems in fluid flow involve the complex turbulent motion rather than the simpler laminar motion.

- ## SUMMARY

1. ***Shear*** in fluids is the internal stress within a fluid which resists deformation. It exists only under dynamic conditions of motion and is *proportional to the **rate** of deformation.*

* See Reference 2, Vol. I, pp. 85–89.

2. **Shear** in fluids is the rate of transfer of momentum (momentum flux) per unit area. Hence,

$$\tau = \mu \frac{dv}{dy} = \frac{\mu}{\rho} \frac{d(\rho v)}{dy} \qquad \text{(5-2) and (5-3)}$$

3. **Viscosity** is the basic fluid property which resists deformation and causes shear.

4. The relation between dynamic viscosity μ and kinematic viscosity ν is

$$\nu = \frac{\mu}{\rho} \qquad (5\text{-}1)$$

5. **Viscosity** is related to the *activity of gas molecules* and the *cohesiveness of liquid molecules*. Consequently, viscosity of a gas increases with increasing temperature, whereas viscosity of a liquid decreases with increasing temperature.

6. In **laminar flow** the shear and momentum flux are caused by molecular activity only.

7. **Eq. 5-6**

$$\frac{dp}{dx} = \frac{d\tau}{dy} \qquad (5\text{-}6)$$

shows that the *pressure gradient* in the direction of flow is equal to the *shear gradient* perpendicular to the direction of flow.

8. In **laminar flow** between fixed parallel boundaries the velocity distribution is parabolic.

9. Table 5-1 shows the similarities among the various **equations for laminar flow** in which the head loss or drag varies directly with the velocity, the dynamic viscosity, and the distance over which the loss occurs; and inversely with the specific weight and the cross-sectional area.

10. **Instability** in laminar flow is related to the magnitude of χ in Eq. 5-32.

$$\chi = \frac{y^2 \rho \; dv/dy}{\mu} = \frac{y^2}{\nu} \frac{dv}{dy} \qquad (5\text{-}32)$$

11. **Reynolds number**

$$\text{Re} = \frac{VD\rho}{\mu} = \frac{VD}{\nu} \qquad (5\text{-}33)$$

expresses the ratio of the inertia forces to the viscous forces, and is a criterion for the influence of viscosity in a flow system.

12. *Turbulent flow* is the result of the disintegration of eddies into a random pattern of mixing so that the momentum transfer is by discrete masses of fluid.

$$\tau = (\mu + \eta)\,\frac{dv}{dy} = (\nu + \epsilon)\,\frac{d(\rho v)}{dy} \qquad \text{(5-34) and (5-35)}$$

13. *Turbulence* is commonly described by the intensity $\sqrt{\overline{v'^2}}$ of the turbulence $\sqrt{\overline{v'^2}}$ and the scale of the turbulence l.

SUMMARY OF EQUATIONS FOR LAMINAR FLOW

Flow System	Equation Number	Head Loss or Drag	Equation Number	Mean Velocity	Equation Number	Velocity at a Point
Parallel boundaries	5-15	$h_f = 12\,\dfrac{\mu VL}{\gamma B^2}$	5-16	$V = \dfrac{\gamma B^2}{12\mu}\dfrac{h_f}{L}$	5-14	$v = -\dfrac{\gamma}{2\mu}\dfrac{dh}{dx}(By - y^2)$
Free surface	5-20	$h_f = 3\,\dfrac{\mu VL}{\gamma y_0^2}$	5-18	$V = \dfrac{\gamma y_0^2}{3\mu}\dfrac{h_f}{L}$	5-17	$v = \dfrac{\gamma S}{2\mu}(2yy_o - y^2)$
Circular pipes and tubes	5-26	$h_f = 32\,\dfrac{\mu VL}{\gamma D^2}$	5-25	$V = \dfrac{\gamma D^2}{32\mu}\dfrac{h_f}{L}$	5-24	$v = -\dfrac{\gamma}{4\mu}\dfrac{dh}{dx}(r_o^2 - r^2)$
Porous media	5-30	$h_f = c\,\dfrac{\mu VL}{\gamma d^2}$	5-31	$V = \dfrac{\gamma d^2}{c\mu}\dfrac{h_f}{L}$	....	....
Submerged sphere	....	$h_D = 3\,\dfrac{\mu Vd}{\gamma d^2}$	....	$V = \dfrac{\gamma d^2}{3\mu}\dfrac{h_D}{d}$	....	....

Note: The equation for the drag on a submerged sphere, Eq. 5-28, $F_D = 3\pi\mu Vd$ has been rearranged to put it in the same form as the other laminar flow equations, by setting the total drag force F_D equal to the product of the drag per unit area γh_D and the area of a sphere πd^2 over which this drag acts (see Chapter 9). The term h_D is the drag (analogous to head loss h_f) which occurs over the distance d in the direction of flow.

References

1. Bakhmeteff, B. A., *The Mechanics of Turbulent Flow*. Princeton: Princeton University Press, 1936.

2. Binder, R. C., *Advanced Fluid Mechanics*. Englewood Cliffs, N. J.: Prentice-Hall, Inc., 1958.

3. Goldstein, S., *Modern Developments in Fluid Dynamics*. London: Oxford University Press, 1938.

4. Rouse, Hunter, *Fluid Mechanics for Hydraulic Engineers*. New York: McGraw-Hill Book Co., Inc., 1938.

5. Streeter, Victor L., *Fluid Mechanics*, 2nd ed. New York: McGraw-Hill Book Co., Inc., 1958.

Problems

5-1. Calculate the kinematic viscosity of glycerine at 68°F, using the tabulated data on viscosity.

5-2. Determine the kinematic and dynamic viscosities of water at 40° and 80°F.

5-3. Plot a curve of the kinematic viscosity ν of water versus temperature, in degrees centigrade.

5-4. Plot a curve of the dynamic viscosity μ of water versus temperature in degrees centigrade.

5-5. Plot a curve of dynamic viscosity of water versus kinematic viscosity, between 32°F and 212°F.

5-6. Plot the relationship between dynamic viscosity of air and dynamic viscosity of water, between 32°F and 212°F.

5-7. Plot the relationship between kinematic viscosity of air and kinematic viscosity of water, between 32°F and 212°F.

5-8. Estimate the kinematic viscosity of hydrogen where subjected to a pressure of 200 psia at 100°F.

5-9. Estimate the viscosity of water at 70°F and 10,000 psi.

5-10. If a given fluid has a kinematic viscosity of 0.442 x 10^{-5} sq ft/sec and a dynamic viscosity of 0.838 x 10^{-5} lb-sec/sq ft, calculate the mass density in terms of slug/cu ft. Also determine its specific weight.

5-11. If the fluid of Problem 10 is water, what is the temperature of the water?

5-12. Find the kinematic viscosity of glycerine at a temperature of 68°F.

5-13. Air is compressed adiabatically from 14.7 psia at 40°F to 100 psia. Calculate the kinematic and dynamic viscosities of the air before and after the adiabatic compression.

5-14. Calculate the acoustic velocities before and after compression in Problem 5-13.

5-15. Compute the conversion factor which converts dynamic viscosity in poise to dynamic viscosity in the foot-pound-second system.

5-16. A liquid has a viscosity of 0.98 poise. What is its kinematic viscosity ν in the English system if sp. gr. = 0.8?

5-17. A force of 0.5 lb is required to drag a thin plate at a constant velocity of 1 ft/sec between two fixed plates when the space between these two fixed plates is filled with a liquid. If the area of the plate is 1 sq ft, it travels equidistant from and parallel to the fixed plates, and the distance from the moving plate to the fixed plates is 0.12 in., calculate the viscosity of this liquid. Could this liquid be water and, if so, what is its temperature?

5-18. What force is required to move the plate of Problem 5-17, if the liquid is glycerine at a temperature of 68°F?

5-19. The space between a piston and a piston wall is filled with oil having a dynamic viscosity of 1.0×10^{-3} lb-sec/sq ft. If the diameters of the piston and piston housing are 6.0 and 6.02 in. and the length of the piston is 1.0 ft, calculate the force required to overcome friction at a velocity of 2.0 ft/sec.

5-20. For two-dimensional flow between two parallel plates 12 in. apart, determine the shearing stress at 1-in. intervals from the boundary of the plates to the centerline between the plates, if the velocity varies linearly from zero at the plates to a maximum of 6 ft/sec midway between the two plates.

5-21. Solve Problem 5-20 assuming that the velocity varies parabolically.

5-22. Work Problem 5-20 assuming that, instead of a parabola as given in Problem 5-21, the velocity varies as the quadrant of a circle from zero at the wall to a maximum of 6 ft/sec at the centerline.

5-23. Using the equation $\tau = \mu \, (dv/dy)$, calculate the shearing stress at 1 in. increments from the boundary to the centerline of a pipe 12 in. in diameter, if the velocity distribution has a straight line variation such that the velocity at the wall is zero and is 6 ft/sec at the centerline. (First determine the equation describing the velocity distribution.) The fluid has a viscosity of 3.0×10^{-3} lb sec/sq ft.

5-24. Solve Problem 5-23 assuming that the velocity varies parabolically between the foregoing limits.

5-25. Solve Problem 5-23 assuming that the velocity varies as the quadrant of a circle from zero at the wall to a maximum of 6 ft/sec at the center.

5-26. Using the velocity gradients determined in Problem 5-19, plot the magnitude of the velocity gradient-versus distance from the pipe wall. Where is the velocity gradient a maximum? Verify this mathematically.

5-27. Kerosene at 70°F is flowing between two horizontal parallel plates which are spaced 1 in. apart. Piezometers along the plates indicate that the pressure is decreasing in the direction of flow at the rate of 0.05 psf/ft. (a) What is the velocity midway between the plates? (b) Find the average velocity.

5-28. From Eq. 5-10, develop Eq. 5-17.

5-29. Assuming that the velocity distribution for laminar, free-surface flow down an inclined plane is parabolic, develop Eq. 5-18.

5-30. A wide sheet of glycerine at 80°F $\frac{1}{2}$ in. thick flows down a 30° incline. Find the discharge per foot of width and the velocity at a point $\frac{1}{4}$ in. from the free surface.

5-31. A thin sheet of water at 50°F flows down a plate of glass inclined at an angle of 85° with the horizontal. If the average velocity is 1 ft/sec, how deep is the water and what is the velocity at the free surface?

5-32. For laminar free-surface flow down an inclined plane, at what depth below the free surface does the average velocity occur?

5-33. Prove that the average velocity for viscous flow in a circular pipe is equal to $\frac{1}{2}$ the maximum velocity.

5-34. SAE 30 oil at 80°F is flowing in a 2-in. pipe at 20 gpm. What is the head loss in 1000 ft of pipe line, and what is the velocity at a point $\frac{1}{4}$ in. from the wall of the pipe?

5-35. A 3-in. pipe line is constructed with a slope of 15°. If 100 gpm of SAE 10 oil at 80°F is flowing in the line, what is the change of pressure in 1000 ft of pipe line? What is the head loss per foot if flow is down?

5-36. A brass pipe with a 1-in. diameter is carrying oil whose viscosity is 4×10^{-4} sq ft/sec. If the discharge is 0.03 cfs, find the velocity (a) at the center of the pipe, (b) at a point $\frac{1}{4}$ in. from the wall of the pipe.

5-37. The Reynold's number for flow of oil through a 2-in. diameter pipe is 1700. If $\nu = 8 \times 10^{-4}$ sq ft/sec, what is the velocity at a point $\frac{1}{4}$ in. from the wall?

5-38. If olive oil at 70°F is flowing in a 2-in. diameter pipe, what is the maximum discharge which can occur and still insure laminar flow?

5-39. A small spherical drop of water falls steadily in air. Neglecting the weight of the air, how long will it take the droplet to fall 1 ft if it has a diameter of 10^{-4} ft? Assume static atmospheric conditions. $T = 60°F$.

5-40. A sphere 1 in. in diameter falls in a bucket of glycerine at 90°F at a constant velocity. If the velocity of the sphere is 0.1 ft/sec, what is the specific gravity of the material in the sphere?

5-41. If a $\frac{1}{2}$-in. sphere falls in a barrel of SAE 10 at 50°F, what is the terminal velocity if the specific gravity of the sphere is 1.1?

5-42. Derive Eq. 5-29.

5-43. A horizontal column of sand 2 ft long and 3 in. in diameter has water flowing through it. If the pressure drop across the column of sand is 3 psi, and the discharge is 1 gallon per minute, what is the transmission constant?

5-44. If the water temperature in Problem 5-43 is 50°F, what is the coefficient of permeability? Use a grain diameter of 2 mm.

5-45. The transmission constant k determined by experiment for a given soil is 0.03 ft/hr when the temperature is 40°F. What is the value of k when the water temperature is raised to 80°F?

5-46. List and briefly discuss the factors which affect the transmission constant.

5-47. A porous layer is confined between two horizontal layers of clay which are 3 ft apart. Tests show that the transmission constant is 0.6 ft/hr at 40°F. What is the flow per unit width of the porous layer if the pressure gradient of piezometric head in the direction of flow is 0.01?

5-48. For a flow of oil (sp. gr. = 0.80) in a 1-in. diameter pipe, the Reynold's number is 1000. If the discharge in the pipe is 20 gpm, is the flow stable at a point located $\frac{1}{4}$ in. from the wall of the pipe?

5-49. In Prob. 5-48, is the flow stable at all points in the cross section?

5-50. What is the maximum discharge of water at 50°F which can be carried by a 1-in. pipe under laminar flow conditions? Is this laminar flow stable at all points in the cross section? Why?

5-51. If kerosene at 68°F is flowing in a 1-in. pipe, what is the Reynolds number of the flow if the maximum value of the χ parameter is 500?

5-52. If 200 gpm of SAE 30 oil at 100°F flows in a 3-in. pipe, is the flow laminar or turbulent?

5-53. Thirty gpm of water at 60°F flows in a 1-in. pipe. Is the flow laminar or turbulent?

5-54. The Reynolds number for glycerine flowing in a 2-in. pipe at 80°F is 180. What is the discharge?

5-55. The following velocities were measured at 1 second intervals. Find the mean velocity and the root mean square of the velocity fluctuation. Also plot the variation of velocity with time. The instantaneous velocity measurements in ft/sec were: 3.49, 3.32, 3.69, 3.75, 3.51, 3.27, 3.42, 3.69, 3.80, 3.53, 3.39, 3.48, 3.72, 3.61, 3.51, 3.33, 3.49, 3.67, 3.78, 3.47, 3.59, 3.32, 3.67, 3.77, 3.50, 3.29, 3.46, 3.73, 3.62, 3.37, 3.48, 3.68, 3.45, 3.31, 3.64, 3.79, 3.34, 3.59, 3.44.

5-56. Assuming that the equation for the velocity in turbulent flow in an 8-in. pipe is $v = 15 + 3.0 \log_{10}(y/D)$, find the mixing length, eddy viscosity and velocity at a point 2-in. from the wall. Water is flowing at 60°F.

5-57. If the shear varies linearly from zero at the center to a maximum at the boundary in a circular pipe, what equation would represent the velocity distribution across a pipe cross section, if the dynamic eddy viscosity were assumed to be constant everywhere in the cross section for turbulent flow?

5-58. At 70°F, what is the ratio of: (a) The dynamic viscosities of water and air? (b) The kinematic viscosities of water and air? Explain the significance of the difference between these two ratios.

5-59. The flow of oil between two parallel plates spaced at $B = 0.05$ in. is 10 gal/min per foot. The kinematic viscosity of the oil is 5×10^{-3} sq ft/sec and the specific gravity is 0.85. Find: (a) Velocity half way between the plates; (b) Velocity at $y = B/4$; (c) Shearing stress at the boundary; (d) Shearing stress at $y = B/4$.

5-60. A vertical cylinder with a 0.42-in. inside diameter is full of oil with a viscosity of 10^{-3} sq ft/sec. A piston 8 in. long with a diameter of 0.36 in. is moved down the cylinder at a velocity of 0.5 fps. The clearance between the piston and cylinder is 0.03 in. What is the velocity distribution in the clearance space, where is the maximum velocity in each direction, and where in the clearance space is the velocity equal to zero?

5-61. Derive the Reynolds number, on the assumption that it is the ratio between a representative inertial reaction and a representative viscous reaction.

5-62. List some examples of laminar flow in the events of daily living.

5-63. The absolute viscosity of a fluid is a property which:
(a) is a function of the motion of the fluid;
(b) is independent of pressure and temperature;
(c) can only be measured when the fluid is in motion;
(d) is a function of the shearing stress in the fluid;
(e) is none of these.
Check one or more of the statements above.

5-64. The Reynolds number:
(a) has a dimension of length;
(b) is dimensionless;
(c) is only applied to pipe flow;
(d) is a parameter which gives an indication of the type of flow which is occurring;
(e) of 2,000 is always the boundary between laminar and turbulent flow.
Check one or more of the statements above.

5-65. The Chi (χ) parameter:
(a) is always zero at the boundary;
(b) is a measure of the stability of the flow at a point in laminar flow;
(c) is a maximum at the center of the pipe for flow in a circular pipe;
(d) does not vary with the Reynolds number;
(e) is not valid if the flow is in a rectangular conduit.
Check one or more of the statements above.

5-66. The eddy viscosity:
(a) is a constant if the temperature remains the same;
(b) is greater for laminar flow than for turbulent flow;
(c) is principally a function of the kinematic viscosity;
(d) is mainly a function of the type of turbulence involved;
(e) is none of these
Check one or more of the statements above.

5-67. The mixing length:
(a) is a maximum near the boundary;
(b) is a rough measure of the size of the turbulent eddies in the flow;
(c) is dimensionless;
(d) is easily determined by direct measurement;
(e) is none of these.
Check one or more of the statements above.

chapter 6

Fluid

Resistance

The resistance to flow is a subject which is arising continually in the field of fluid mechanics. For example: pipe lines require pumps which can drive a fluid through a pipe against various types of resistance; liquids will flow in open channels, through the action of gravity, with the expenditure of potential energy to overcome the resistance to flow; and various objects, such as airplanes, automobiles, raindrops, and particles of sand and silt, move through fluids—but their motion results in an expenditure of energy, which is converted into heat, to overcome the resistance to flow of the fluid around these objects. In short, almost any flow system involves some aspect of resistance to fluid motion. The basic principles of resistance which are developed in this chapter have direct application to these problems and to others which are discussed in detail in this and subsequent chapters.

• CAUSE OF FLUID RESISTANCE

As a first step in the study of fluid resistance, the cause of such resistance is considered.

Basically, fluid resistance may be divided into two types:

1. *Shear* resistance.

2. *Pressure* resistance.

The *shear* resistance, a *tangential stress,* is caused by the fluid viscosity and takes place along a boundary of the flow in the general direction of local motion. *Pressure* resistance, a *normal stress,* is caused by acceleration of the fluid, which results in a decrease in pressure from the upstream side to the downstream side of an object. All fluid resistance can be divided into one or a combination of these two categories. In some problems, the resistance is obviously due only to the shear at the surface of the boundary and in the direction of motion, whereas in other problems the resistance is clearly due only to the pressure decrease in the direction of motion on the upstream and downstream sides of the flow boundary. Frequently, however, problems in fluid resistance involve both shear resistance and pressure resistance to varying degrees.

• SHEAR RESISTANCE

In the foregoing chapter, it was observed that, when a fluid flows past a boundary, the molecules of fluid in immediate contact with the boundary have zero velocity (i.e., the same relative velocity as that of the boundary), and that a shear is set up within the fluid and also at the boundary. This boundary shear results in a drag force transmitted from the fluid to the

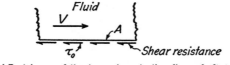

(a) Resistance of the boundary to the flow of fluid

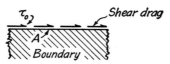

(b) Drag of fluid on the surface of the boundary

Fig. 6-1. Shear resistance and shear drag.

boundary, and the boundary in turn transmits to the fluid a force, equal in magnitude but opposite in direction, which offers resistance to the flow of the fluid. The resistance or drag is associated only with the shear at the boundary, a tangential stress, and hence is called *shear resistance,* or *shear drag.* The direction of the drag on the boundary is the same as the direction of motion, while the total surface resistance of the boundary against the flow is in the opposite direction and is the product of the unit shear τ_0 and the area A over which this unit shear is acting. This is shown in Fig. 6-1.

When the fluid is a liquid, there are molecules of the liquid in contact with the boundary at all points on its surface. When the fluid is a gas, however, the number of molecules in contact is greatly reduced. Consequently, for gases under atmospheric pressure or greater, the fluid molecules are in sufficient number per unit area to warrant the assumption that the velocity of the fluid at the boundary is the same as the velocity of the boundary. As the number of molecules per unit area of boundary is decreased, however, they are less influenced by the boundary and, for rarefied gases, their mean velocity can be somewhat different from that of the boundary, which results in a slip of the fluid past the boundary. This slip takes place when the number of molecules in the immediate vicinity of a unit area of the boundary is small, either because of a *decrease in pressure* of the gas, or because of an *increase in area* of the boundary per unit volume of the gas. Examples of such flow are movement of an object in the outer rarefied atmosphere, flow of a gas under reduced pressure in a pipe line, and flow of a gas in fine-grained porous media. This phenomenon is known as the Klinkenberger effect.

• PRESSURE RESISTANCE

When flow takes place past a boundary which is not parallel to the flow, a drag on the object and a resistance to flow are created which depend upon the *shape or form* of the boundary. The shape of the boundary causes a deflection of the streamlines and local acceleration of the fluid. Consequently, there occurs a decrease in pressure, a normal stress, from the upstream to the

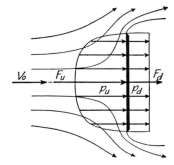

Fig. 6-2. Pressure drag with separation.

downstream sides of the object or boundary. The summation, over the surface, of this difference in pressure results in a pressure drag on the boundary and a pressure resistance against the fluid. For the disc in Fig. 6-2, the total drag force is the algebraic sum of the pressure forces on the two sides of the disc. As can be seen from this figure, the pressure on the upstream side is essentially all positive, whereas on the downstream side the pressure is entirely negative, so that the corresponding forces F_u and F_d both contribute to the drag of the fluid on the disc.

Pressure drag and pressure resistance can occur with either laminar flow or turbulent flow. With laminar flow, however, the flow pattern is markedly different from that for turbulent flow. In either case, the flow around an object may separate from the object on the downstream side. This separation is discussed in greater detail later in this chapter.

• GENERAL DRAG EQUATION

Actual determination of the drag of a flowing fluid upon an object can be done best by a combination of:

1. Consideration of the variables involved;
2. Use of dimensional analysis to arrange the variables in simplified and significant dimensionless parameters; and
3. Use of experimental data to establish the numerical relationships between these parameters over a wide range.

The variables involved are those describing the **geometry**, the **flow**, and the **fluid**. Hence

$$F_D = \phi_1(L, \text{sf}, e, V_0, \rho, \mu) \tag{6-1}$$

in which L and sf describe the geometry, and are the characteristic length and shape factor (dimensionless) of the boundary over which the shear drag and pressure drag are present;

e is the roughness of the boundary;

F_D and V_0 describe the flow, and are the total drag and ambient velocity of the flow;

ρ and μ describe the fluid, and are the density and the dynamic viscosity.

As illustrated in Chapter 1, one repeating variable can be selected from each of these groups and the pi-theorem applied. Choosing L, V_0, and ρ as the repeating independent variables and applying the pi-theorem yields

$$\frac{F_D}{L^2 \rho V_0^2} = \phi_2\left(\frac{V_0 L \rho}{\mu}, \text{sf}, \frac{e}{L}\right) \tag{6-2}$$

which can be arranged for the sake of convenience as

$$\frac{F_D}{A\rho V_0^2/2} = C_D = \phi_3\left(\text{Re, sf}, \frac{e}{L}\right) \tag{6-3}$$

since L^2 is an area and can be set equal to A, and the stagnation pressure can be obtained by dividing by 2 to yield $\rho V_0^2/2$. The first parameter is commonly known as the drag coefficient and can be arranged to give the **general drag equation**

$$F_D = \frac{C_D A \rho V_0^2}{2} \tag{6-4}$$

in which F_D is the drag on any boundary having the area A, the ambient velocity V_0 of the flow, and the density ρ of the fluid.

Equation 6-3 shows that the drag coefficient C_D depends upon the viscous effects relative to the inertial effects (as contained in the Reynolds number Re), the shape factor of the boundary sf and the relative roughness e/L. Since inertial effects are also included in the term for stagnation pressure, it is logical to expect that at large Re-values the viscous effects become relatively very small and the drag coefficient C_D becomes independent of Re. Indeed, this expectation is supported by both theory and experiment. At small Re-values, C_D is found to be inversely proportional to Re and, as Re increases, C_D becomes more and more independent of Re. This is demonstrated later in this chapter for plane boundaries, and in Chapter 9 for flow around submerged objects.

The area A is selected differently for different boundary conditions. For plane boundaries, it is the area of the plane parallel to the flow, and for submerged objects, it is the projected area normal to the flow.

In effect, the product $A\rho V_0^2/2$ gives the drag force which would result from the stagnation pressure $\rho V_0^2/2$ being applied over the area A, and the drag coefficient C_D acts as a correction factor which adjusts this stagnation force to the actual drag found for the particular boundary conditions (shape of object experiencing the drag) and flow conditions (represented by the Reynolds number).

In many problems, a combination of shear resistance and pressure resistance is involved. Furthermore, with the same geometric conditions, shear resistance (a tangential stress) can predominate for one condition of flow (small Re), and pressure resistance (a normal stress) can predominate under another condition of flow (large Re). The reasons for this are presented in the following sections.

Example Problem 6-1: What are the physical dimensions of the drag coefficient C_D of Eq. 6-4?

Solution:

$$C_D \sim \frac{F_D}{A\rho V_0^2}$$

the dimensions of which are

$$\frac{F}{L^2 \dfrac{FT^2}{L^4} \dfrac{L^2}{T^2}} = 1$$

Therefore C_D is dimensionless.

• BOUNDARY LAYER THEORY

In years past, the science of hydraulics was largely empirical. The bulk results of hydraulic phenomena were measured and design criteria were established without actually understanding the significant details and physical processes involved in the phenomena. Although many scientists, such as Osborne Reynolds, made major contributions to the science, it was not until Prandtl proposed his boundary layer theory early in this century that a really satisfactory explanation was at hand for certain important physical processes. From Prandtl's hypothesis, it became possible to understand and visualize the detailed mechanics of the process and cause of both surface resistance and form resistance.

Prandtl reasoned that, in a flow system, the influence of the boundary is confined to a relatively thin layer of fluid near the boundary. His theory is that near a boundary a layer of fluid is decelerated due to the resistance to flow which is offered by the shear at the boundary, see Fig. 6-3. This relatively thin layer is called the boundary layer. It can develop either in laminar flow or in turbulent flow. Because this layer acts somewhat like a submerged object within the fluid, it causes the influence of the boundary to be extended even beyond the limits of the boundary layer itself. As will be shown in the following sections, this boundary layer has an important influence on flow in uniform conduits (both closed and open), on expanding flow, on contracting flow, on flow around bends, and on flow around submerged objects. Although outside the boundary layer the flow is normally irrotational, within the boundary layer (at any given section) the nonuniform distribution of velocity results in rotational flow—which prevents the use of the energy equation in the analysis of flow within the boundary layer.

The thickness of the boundary layer is represented by δ as shown in Fig. 6-3. The resistance to flow which is caused by the boundary nearby produces a reduced velocity and, hence, a reduced momentum flux of the fluid moving past the boundary. By means of the momentum-flux equation, it is possible to relate the change in momentum flux to the total resistance which the boundary exerts on the fluid to cause this reduction in velocity. Such a relationship, together with the continuity equation and certain assumptions, permits the development of a mathematical expression for the variation in velocity within the boundary layer.

Consider a very thin plate within the flow, as shown in Fig. 6-3. The velocity of the fluid approaching the boundary is assumed to be V_0 and uniform throughout. Furthermore, it is assumed that the acceleration of the fluid normal to the boundary is insignificant compared with the acceleration of the fluid along the boundary. As the fluid reaches the boundary,

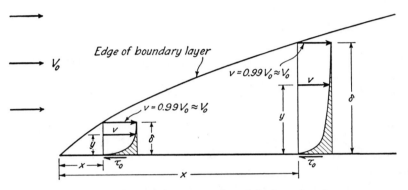

Fig. 6-3. Schematic representation of the boundary layer (vertical scale greatly enlarged).

the shear τ_0 resists the flow and retards the motion of the fluid. Hence, the velocity at the boundary is zero and, at some distance δ from the boundary, the velocity is nearly V_0. Strictly speaking, the velocity at the edge of the boundary layer, where $y = \delta$, is slightly less than V_0 because, theoretically, the velocity goes to V_0 only at infinity. For practical reasons, however, the edge of the boundary layer has been defined arbitrarily as being where the local velocity v is nearly equal to the ambient velocity V_0, say $v/V_0 = 0.99$.

This boundary layer, which is developing on the flat plate, can be contrasted with the boundary layer in a pipe. On the one hand, the boundary layer over the flat plate expands indefinitely (provided the depth of flow and the length of plate are infinite), so that δ becomes greater as it progresses downstream, and the pressure remains constant throughout the field of motion. On the other hand, the boundary layer in a pipe has a constant thickness equal to the pipe radius (once the flow is established), and the pressure is uniform across the flow, but decreases in the direction of motion.

• BOUNDARY LAYER EQUATION

By means of the momentum-flux equation, the reduction in velocity and momentum flux represented by the shaded area can be set equal to the total shear which has accumulated over the distance x from the leading edge to the point of consideration.

In the development of the boundary layer equations for the flat plate, see Fig. 6-3, certain simplifying assumptions are first necessary. These assumptions include:

1. Steady flow ($\partial v / \partial t = 0$).

2. Outside the boundary layer the velocity is V_0 throughout.

3. The thickness δ is very small compared with the distance x.

4. The pressure remains constant throughout the flow, both within and outside the boundary layer. In other words $dp/dy = 0 = dp/dx$.

5. The acceleration of the fluid normal to the boundary is negligible, compared with the acceleration of the fluid parallel with the boundary.

The fluid which is approaching the boundary in Fig. 6-3 has a momentum flux of $(\rho V)(V)$ per unit area of cross section. At a given point x along the boundary, however, the velocity has been reduced from V_0 to v at distance y from the boundary. The total reduction in momentum flux is represented by the shaded area, since ρ is constant throughout. In order for equilibrium to exist, this reduction in momentum flux per unit width must equal the total resistance F_D/B caused by the shear τ_0 which has accumulated from the leading edge to the point x. Therefore,

$$\frac{F_D}{B} = \int_0^x \tau_0 \, dx \tag{6-5}$$

in which B is the width of the plate.

The flux of mass passing through the boundary layer per unit width at the point x is $\int_0^\delta \rho v \, dy$, and the change in velocity at any point y is $(V_0 - v)$. Therefore, the change in momentum flux is $\int_0^\delta \rho (V_0 - v) v \, dy$, which can be equated to the resistance equation, Eq. 6-5, to give

$$\frac{F_D}{B} = \int_0^x \tau_0 \, dx = \int_0^\delta \rho (V_0 - v) v \, dy \tag{6-6}$$

which must be integrated to obtain the drag on a flat plate. Note that such integration will produce an equation having the form of Eq. 6-4.

Equation 6-6 shows that either the variation of the boundary shear τ_0 with the distance x, or the variation of the velocity v with the distance y from the boundary, must be known before the integral can be evaluated. These variations depend upon whether the flow in the boundary layer is laminar or turbulent, as shown in the following sections.

• LAMINAR BOUNDARY LAYER

In the previous chapter, the shear equation $\tau = \mu\, dv/dy$ was presented, and it was shown that the velocity distribution for laminar flow between flat plates is parabolic. The velocity distribution in a laminar boundary layer has also been experimentally found to be approximately parabolic, and hence

$$\frac{v}{V_0} = 2\frac{y}{\delta} - \frac{y^2}{\delta^2} \qquad (6\text{-}7)$$

as shown in Fig. 6-4.

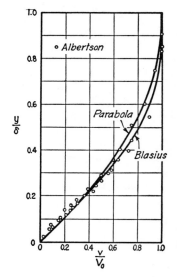

Inserting Eq. 6-7 for the parabolic velocity distribution into Eq. 6-6, integrating, and using the coefficient from the more exact (but more complicated) solution of Blasius (based upon experimental data) results in an equation for the ***thickness of the laminar boundary layer***

$$\frac{\delta}{x} = \frac{5.2}{\sqrt{\dfrac{V_0 x}{\nu}}} = \frac{5.2}{\mathrm{Re}^{1/2}} \qquad (6\text{-}8)$$

Fig. 6-4. Velocity distribution in a laminar boundary layer.

in which Re is the Reynolds number of the boundary layer wherein the length term is x. From this equation, it can be seen that the thickness varies inversely with the square root of the velocity, and directly as the square root of the viscosity and the length of the boundary. The combined effect of these variables is represented in the two dimensionless parameters: the relative thickness of the boundary layer δ/x and the Reynolds number Re. This variation is contrasted later with that for the turbulent boundary layer.

Example Problem 6-2: A thin rectangular plate is towed through water whose temperature is 40°F at a uniform velocity of 1.0 fps.

(a) At what distance downstream along the plate will laminar flow cease to exist?

(b) What is the thickness of the laminar boundary layer at this point?

Solution:

Assuming an upper limit for laminar flow at Re = 4 × 10⁵

(a)
$$\mathrm{Re} = \frac{V_0 x}{\nu}$$

and
$$x = \frac{(4 \times 10^5)(1.664)}{(10^5)(1.0)} = 6.65 \text{ ft}$$

(b) Utilizing Eq. 6-8 with the Blasius coefficient

$$\frac{\delta}{x} = \frac{5.2}{\mathrm{Re}^{1/2}}$$

and
$$\delta = \frac{(5.2)(6.65)}{(4 \times 10^5)^{1/2}} = 0.055 \text{ ft}$$

By substituting the expression for the parabolic velocity distribution into the shear equation

$$\tau_0 = \mu \left(\frac{dv}{dy}\right)_{y=0} \tag{6-9}$$

The boundary shear τ_0 at $y = 0$ is

$$\tau_0 = 2\frac{\mu V_0}{\delta} \tag{6-10}$$

which can be combined with Eq. 6-8 and the more accurate Blasius coefficient to give

$$\tau_0 = 0.33 \sqrt{\frac{\rho \mu V_0^3}{x}} \quad (Laminar \ boundary \ layer) \tag{6-11}$$

Equation 6-11 can be rearranged significantly to give the **local drag coefficient** for laminar flow

$$c_D = \frac{\tau_0}{\rho V_0^2/2} = \frac{0.66}{\mathrm{Re}^{1/2}} \tag{6-12}$$

to show how the boundary shear τ_0 relative to the stagnation pressure $\rho V_0^2/2$ varies inversely with the square root of the Reynolds number Re.

To determine the total drag on a flat plate, the first part of Eq. 6-6 can be solved by combining it with Eq. 6-11 and integrating. The last part of Eq. 6-6 can be solved by inserting the Blasius distribution of the velocity and integrating. The resulting equation is

$$(\tau_0)_{\text{ave}} = \frac{F_D}{BL} = \frac{1.33}{\sqrt{\dfrac{V_0 L}{\nu}}} \frac{\rho V_0^2}{2} \tag{6-13}$$

This equation in which x is now L can be compared with Eq. 6-4 to see that the area A is the product BL of the boundary width and the length of the boundary from the leading edge, and that the **mean drag coefficient** is

$$C_D = \frac{1.33}{\sqrt{\dfrac{V_0 L}{\nu}}} = \frac{1.33}{\text{Re}^{1/2}} \quad \left(\begin{array}{l}\textit{Laminar boundary layer over a}\\ \textit{flat plate, } 10^4 < \text{Re} < 6 \times 10^5\end{array}\right) \tag{6-14}$$

The lower limit of Re $\approx 10^4$ for this equation is explained in connection with Eq. 6-15 and the upper limit Re $\approx 6 \times 10^5$ has been determined experimentally as shown in Fig. 6-9.

From Eqs. 6-8, 6-11, and 6-14 the following **significant facts** can be observed regarding the **laminar boundary layer:**

1. **The thickness δ of the laminar boundary layer** (although relatively very thin) increases with the square root of the length of boundary L and the viscosity ν, and it decreases with the square root of the velocity. These variations are all combined into a single more significant relationship by use of the Reynolds number Re and the relative thickness δ/x of the boundary layer.

2. The **boundary shear** τ_0 at any point along the boundary increases with the square root of the density ρ and the viscosity μ, and it decreases with the square root of the distance L of the point from the leading edge. The shear also increases as the square root of the velocity V_0 cubed.

3. The **boundary shear** τ_0 relative to the stagnation pressure $\rho V_0^2/2$ decreases with the square root of the boundary-layer Reynolds number Re $= V_0 L/\nu$.

4. The **mean drag coefficient** C_D can be considered as the ratio of the mean shear over the length L to the stagnation pressure $\rho V_0^2/2$. Like the local drag coefficient c_D, the mean drag coefficient C_D decreases with the square root of the Reynolds number Re $= V_0 L/\nu$.

Recently, Kuo has shown that, for very small Reynolds numbers (Re $< 10^3$), the Prandtl theory of the boundary layer and the Blasius derivation for laminar flow are not applicable. By considering accelerations

in the transverse as well as the longitudinal directions, he has derived
a more complete relationship for the drag coefficient

$$C_D = \frac{1.33}{\mathrm{Re}^{1/2}} + \frac{4.12}{\mathrm{Re}} \tag{6-15}$$

in which the relative importance of the last two terms depends upon the
relative magnitude of the Reynolds number. When $\mathrm{Re} > 10^4$, the Blasius
equation Eq. 6-14 is applicable, but Eq. 6-15 is applicable for the entire
laminar boundary layer in which $\mathrm{Re} < 6 \times 10^5$.

• TURBULENT BOUNDARY LAYER

As the laminar boundary layer continues downstream from the leading
edge of a flat plate, it expands in thickness until it becomes unstable. In
other words, the magnitude of the χ-parameter becomes so large, at its
maximum point within the flow, that eddies develop and turbulence sets in.
Once this turbulence is initiated, it quickly spreads throughout the thickness of the boundary layer until the laminar boundary layer has been
changed to a turbulent boundary layer. The region over which this change
takes place is known as the transition region, see Fig. 6-5.

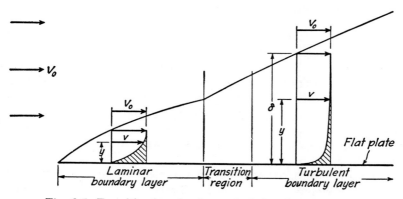

Fig. 6-5. Transition from laminar to turbulent boundary layer
(vertical scale greatly enlarged).

The point L-distance from the leading edge at which the transition
begins varies considerably, depending upon the conditions of the ambient
flow and the boundary. Consequently, the transition from laminar flow
to turbulent flow can take place over a rather large range of Reynolds
number. When the ambient flow is turbulent and /or the boundary is rough,

the transition begins at a small Re-value, say Re = 10^5. On the other hand, when the ambient flow is free of turbulence and internal shear, and the boundary is smooth, the Re-value can become much larger before there is sufficient instability developed to cause turbulent flow, say Re = 10^6. Laminar flow in a pipe will damp a disturbance immediately downstream until the critical value of Reynolds number of about Re = 2000 is reached. Laminar flow past a flat plate, however, does not possess such innate stability. In fact, a disturbance such as that caused by unusual roughness near the leading edge, or a bar or a ribbon placed in the flow upstream from the leading edge, will induce a turbulent boundary layer and the disturbance will not be damped in the flow downstream. Instead, the flow will remain turbulent throughout, as shown in Fig. 6-6.

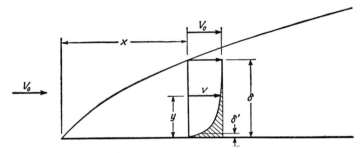

Fig. 6-6. Turbulent boundary layer (vertical scale greatly enlarged).

The turbulent boundary layer over a smooth flat plate has certain similarities to the laminar boundary layer, and yet it is quite different in certain other important respects. Immediately adjacent to the boundary, there is a thin layer or film in which the flow is laminar because of its proximity to the boundary. This thin layer is known as the *laminar sublayer*, with a thickness of δ'. As y increases from zero to δ' and on to δ, there is a transition from this laminar flow to the turbulent flow. Hence, like δ, the sublayer thickness δ' must be defined somewhat arbitrarily, as shown in Chapter 7.

Turbulent flow, as pointed out in Chapter 5, possesses much greater mixing ability than does laminar flow. This characteristic is reflected in the shape of the curve of velocity distribution. In Fig. 6-5, the more effective mixing in the turbulent boundary layer results in a much more uniform velocity distribution than in the laminar boundary layer. This same characteristic of the turbulent boundary layer causes the thickness δ to increase more rapidly with the distance x. Likewise, the velocity gradient dv/dy at the boundary is much greater, so that the shear at the boundary is greater with a turbulent boundary layer than with a laminar boundary layer.

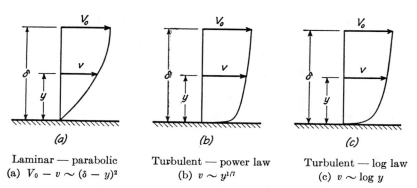

Laminar — parabolic
(a) $V_0 - v \sim (\delta - y)^2$

Turbulent — power law
(b) $v \sim y^{1/7}$

Turbulent — log law
(c) $v \sim \log y$

Fig. 6-7. Velocity distribution in various boundary layers.

Whereas the velocity distribution in the laminar sublayer is approximately parabolic, see Fig. 6-7(a), the velocity distribution in the turbulent boundary layer has been found to follow approximately either the one-seventh power of the distance y from the boundary, Fig. 6-7(b), or the log of this distance, Fig. 6-7(c). A comparison of these velocity variations is shown in Fig. 6-8

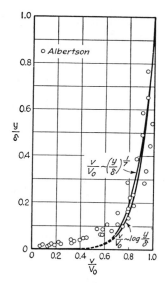

Fig. 6-8. Velocity distribution in turbulent boundary layer.

For small Reynolds numbers with turbulent flow, the power law applies with considerable accuracy. As the Reynolds number increases, the exponent gradually decreases to $\frac{1}{8}$, $\frac{1}{9}$, $\frac{1}{10}$, etc. This variation in the exponent can be expressed more conveniently, however, by means of the log-law for all moderate to large Reynolds numbers, see Fig. 6-9.

To develop the equations for the turbulent boundary layer, it is necessary to return to the general boundary layer equation, Eq. 6-6, and solve it for both the power-law and the log-law velocity distributions. The equation has been solved for the power law to yield the following relationships for a *smooth boundary*

$$\frac{\delta}{x} = \frac{0.38}{\mathrm{Re}^{1/5}} \qquad \left(\begin{array}{c} \textit{Turbulent boundary layer—} \\ \textit{one-seventh power law} \\ 2 \times 10^5 < \mathrm{Re} < 2 \times 10^7 \end{array}\right) \qquad (6\text{-}16)$$

$$C_D = \frac{0.074}{\mathrm{Re}^{1/5}} \qquad\qquad\qquad\qquad\qquad\qquad\qquad\qquad (6\text{-}17)$$

which are reasonably applicable for the turbulent boundary layer when $\mathrm{Re} < 2 \times 10^7$, see Fig. 6-9.

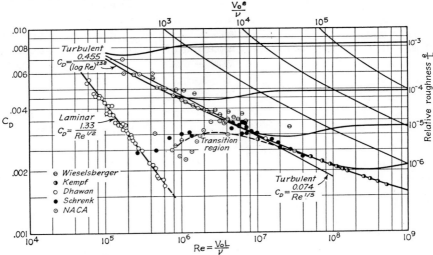

Fig. 6-9. Variation of drag coefficient with Reynolds number for a flat plate parallel to the flow.

The log distribution of the velocity stems from the shear equation for turbulent flow, Eq. 5-39 from Chapter 5,

$$\tau = \rho \kappa^2 y^2 \left(\frac{dv}{dy}\right)^2 \qquad (5\text{-}39)$$

which can be rearranged to give

$$\frac{dv}{dy} = 2.5 \frac{\sqrt{\tau_0/\rho}}{y} \qquad (6\text{-}18)$$

in which $\sqrt{\tau_0/\rho}$ is the shear velocity and the Karman dimensionless proportionability coefficient κ has been determined empirically as 0.4, on the basis of experimental evaluation. This coefficient has been considered to be a universal constant for all turbulent flow, as long as the assumptions which have been made are satisfied. The empirical determination of $\kappa = 0.4$ is based upon many measurements of velocity distribution. The Karman coefficient varies from values less than 0.4 to values considerably in excess of 0.4 when the basic assumptions are not satisfied, such as when the fluid is not homogeneous or the flow deviates from strictly steady, uniform flow.

Equation 6-18 can be integrated and rearranged to yield

$$\frac{v}{\sqrt{\tau_0/\rho}} = 2.5 \ln y + c \tag{6-19}$$

in which c is the constant of integration which can be determined by use of the condition that, for a logarithmic distribution of the velocity, the velocity goes to zero at a fictitious distance y' above the boundary. That is, $v = 0$ when $y = y'$ and Eq. 6-19 becomes

$$\frac{v}{\sqrt{\tau_0/\rho}} = 2.5 \ln \frac{y}{y'} = 5.75 \log \frac{y}{y'} \tag{6-20}$$

which is a **general equation** *for the velocity distribution in a turbulent boundary layer*.

It now remains to evaluate y' in terms of known and useable quantities. Experiments by Nikuradse have shown that for a smooth boundary $y' \sim \nu/\sqrt{\tau_0/\rho}$, which can be substituted in Eq. 6-20 to obtain a general equation of the velocity distribution in a turbulent boundary layer with a smooth boundary

$$\frac{v}{\sqrt{\tau_0/\rho}} = 2.5 \ln \frac{y\sqrt{\tau_0/\rho}}{\nu} + \beta \tag{6-21}$$

in which β (beta) involves the constant of proportionality.

Equation 6-21 can now be employed to develop a general resistance equation by using experimental data to evaluate the numerical constants. The mathematics of this development, however, is beyond the scope of this text. The Karman-Schoenherr equation for a **smooth boundary**

$$\frac{1}{\sqrt{C_D}} = 4.13 \log (Re\ C_D) \qquad \begin{pmatrix} Turbulent\ boundary \\ layer—log\ law \end{pmatrix} \tag{6-22}$$

is based upon the foregoing principles, and represents the data very well for a wide range of Reynolds number Re where C_D is the mean drag coefficient used in Eq. 6-4.

Another form of log equation has been developed by Schlichting for a *smooth boundary*

$$C_D = \frac{0.455}{(\log \text{Re})^{2.58}} \quad \left(\begin{array}{l} Turbulent\ Boundary\ Layer - \\ log\ law\ 10^6 < \text{Re} < 10^9 \end{array} \right) \quad (6\text{-}23)$$

which in some respects is simpler and easier to use, since it does not involve the product $\text{Re}C_D$. This equation is applicable over the range of Reynolds number $10^6 < \text{Re} < 10^9$.

Figure 6-9 shows that the laminar flow with a flat plate does not become unstable at any particular or unique value of Reynolds number, as it does for flow in pipes, see Chapter 7. Instead, as stated previously, the transition from laminar to turbulent flow can begin over a considerable range of Reynolds number—say $10^5 < \text{Re} < 10^6$. Furthermore, by means of a disturbance upstream, the boundary layer can be made turbulent throughout the length of the boundary.

The data for the laminar boundary layer follow Eq. 6-14 reasonably well. Likewise, Eq. 6-17 based on the power law follows the data for $2 \times 10^5 < \text{Re} < 2 \times 10^7$. For the larger values of Reynolds number, however, it is necessary to use equations based upon the log law, such as Eq. 6-22 and 6-23. The e/L curves above the data in Fig. 6-9 are for rough boundaries and are discussed later in this chapter.

The *simplifying assumptions* which were made in the development of the boundary layer equations are now considered. The assumption of steady flow remains applicable—and since the flow in most problems wherein the boundary layer equations are needed is steady, the assumption of steady flow is applicable to most problems. When the velocity of flow is changed very slowly, the boundary layer equations still describe the flow with reasonable accuracy. The velocity outside the boundary layer can usually be considered to be a constant, although for certain problems, such as flow in pipes, certain adjustments must be made. The assumption that the thickness δ is small compared with the length L is amply justified by Eqs. 6-8 and 6-16, in which it is seen that δ is a small fraction of L. It is at very small Re-values that this assumption is no longer valid and that Eq. 6-15 applies.

The assumption that the pressure distribution is constant both inside and outside the boundary layer is applicable only for the special case of a boundary layer developing on a boundary submerged in a large expanse of fluid surrounding it on both sides, and to flow in open channels. Flow in pipes experiences a pressure drop in the direction of motion.

Acceleration of the flow in the direction normal to the boundary is negligible compared with the tangential acceleration throughout the flow, except in the immediate vicinity of the leading edge, where the normal acceleration is appreciable (see comments regarding Eq. 6-15.) This is confined to a relatively short distance, however, and has little or no influence on the flow some distance downstream.

Example Problem 6-3: A smooth rectangular plate 2 ft wide and 20 ft long is held stationary in a stream of water which is traveling at a uniform velocity of 20 fps in the direction of the long axis of the plate, and the water temperature is 60°F. (a) Determine the drag force on the plate. (b) What is the thickness of the boundary layer at the trailing edge of the plate.

Solution:

(a) $\quad \text{Re} = \dfrac{V_0 L}{\nu} = \dfrac{(20)(20)}{1.217 \times 10^{-5}} = 3.29 \times 10^7$

From Fig. 6-9, $C_D = 0.0025$

$$F_D = C_D A \frac{\rho V_0^2}{2} = \frac{(0.0025)(2 \times 2 \times 20)(1.94)(20)^2}{2} = \underline{\underline{77.6 \text{ lb}}}$$

The area is based on 2 sides of the plate.

(b) Using Eq. 6-16

$$\frac{\delta}{L} = \frac{0.38}{\text{Re}^{1/5}}$$

$$\delta = \frac{0.38 \times 20}{(3.29 \times 10^7)^{1/5}} = \underline{\underline{0.24 \text{ ft}}}$$

• INFLUENCE OF FLUID COMPRESSIBILITY

Although theory and experimental data are quite limited with regard to a plane boundary in supersonic flow, they both indicate there is a marked decrease in the drag coefficient with increasing Mach number. Fig. 6-10 is based upon data taken at the California Institute of Technology. It can be seen that, as Ma increases from zero to 5, the magnitude of C_D decreases 50 percent for $\text{Re} = 10^7$. For laminar flow, the theoretical reduction is considerably less—being about 15 percent throughout when $\text{Ma} = 4$.

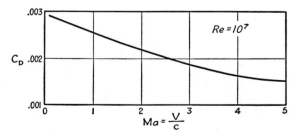

Fig. 6-10. Influence of Mach number on drag coefficient for turbulent flow and a smooth boundary.

• BOUNDARY ROUGHNESS

In the foregoing section, the turbulent boundary layer was explained and the equations were developed for flow over a boundary that is hydrodynamically smooth. It is possible, however, for the boundary to be *hydrodynamically rough*, in which case the flow immediately adjacent to the boundary in the region of the laminar sublayer is markedly different from the flow in the boundary layer beyond this sublayer. This difference is explained in some detail.

Although the flow is laminar within the laminar sublayer, the remainder of the boundary layer is turbulent, see Fig. 6-11(a), and the eddies from the turbulent region are continually impinging upon and penetrating to some extent into the laminar region. As a result, the velocity within the laminar sublayer is not entirely steady. Instead, there is a random surging of the flow, resulting in a fluctuation of the velocity caused by the influence of the eddies from the turbulent region. As the Reynolds number is increased, the influence of the eddies from the turbulent region is increased, which causes the thickness of the laminar sublayer to decrease. Eventually, the laminar sublayer becomes so thin the boundary roughness projects through it, see Fig. 6-11(b), so that the viscous influence of the laminar sublayer is no longer in effect. Once the laminar sublayer is destroyed, the relative roughness and the forces of inertia (instead of the combination of viscous and inertial forces) control the resistance to flow which is offered by the boundary.

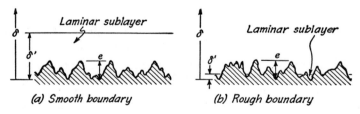

(a) Smooth boundary   *(b) Rough boundary*

Fig. 6-11. Definition of smooth and rough boundaries.

Regardless of how smooth a boundary surface may appear to the naked eye, this surface will have irregularities or roughness which can be seen under the microscope. Therefore, if the Reynolds number is increased sufficiently, the laminar sublayer becomes so thin that the roughness projects through it and the *boundary becomes rough.* In other words, under conditions of small Reynolds number, the laminar sublayer may be sufficiently thick to cover the boundary roughnesses (somewhat like a wind-blown snowfall covers irregularities on the ground surface), and the boundary is hydrodynamically smooth. Under conditions of larger Reynolds number, however, the laminar sublayer may be so thin that the same boundary will be hydrodynamically rough.

When the boundary is rough, the Reynolds number is no longer important—which means that the viscous forces are insignificant compared with the forces of inertia. Hence, for a given relative roughness e/L, the drag coefficient C_D becomes a constant, see Fig. 6-9, and the drag varies only with the square of the mean ambient velocity V_0, as seen in Eq. 6-4.

From the foregoing concept of boundary roughness, together with the marked increase of resistance for a rough boundary as compared with a smooth boundary, has developed the idea of the **admissible roughness.** The admissible roughness is the maximum roughness which can be permitted for a given flow condition and yet have the boundary remain smooth (i.e., the laminar sublayer is sufficiently thick to cover the roughness) so that the resistance or drag is a minimum. This idea is used extensively in the design of airplane wings and ships, and is responsible in some cases for the increased speed which an airplane is able to attain when it has been waxed to cover and smooth the irregularities.

In order for the roughness parameter e to have quantitative significance, it is necessary to have a standard of roughness which is fixed, reproducible, and representative (at least approximately) of natural roughness. The work of Nikuradse, who used uniform sand grains cemented inside pipes, has resulted in a standard which is fixed and is representative of many natural roughnesses, but it has not been reproducible and does not represent adequately certain natural roughnesses. Despite the inadequate features of the Nikuradse roughness, however, it is the only generally-accepted roughness standard and, until a better one is developed, it will continue to fill, at least in part, an important need.

The plane boundary curves in Fig. 6-9 for various relative roughness values are based upon the Nikuradse standard of roughness, in which the absolute roughness e is the size of uniform sand grains which he cemented to the inside of pipes. Since the data of Nikuradse are for pipes instead of plane boundaries, the curves are only schematic, and hence the data are not plotted. Detailed consideration of roughness applied to specific problems is given in the following chapters.

• SEPARATION

Thus far, the boundary layer development has been studied only for steady uniform flow past a plane boundary, where the boundary is parallel to the direction of motion. In this case, the general form of the velocity profile does not change from section to section along the flow. However, when the boundary is very angular, such as it is with a disc, a zone of separation is established downstream from the object. The point of separation is clearly at the edge of the disc and it remains there regardless of the fluid or flow characteristics. Downstream from the separation zone, the turbulent and eddying flow is called a wake. This is discussed in more detail in Chapter 9.

Although these two limiting conditions of flow have no separation in the case of the plane boundary, on the one hand, and a large zone of separation with a definite point of separation in the case of the disc, on the other hand, there are other conditions of flow past boundaries of intermediate shape, for which separation may or may not occur. Furthermore, if separation does occur, the point of separation may not be fixed but, instead, may depend upon the characteristics of the fluid or the flow. This situation is now studied for a gradually expanding boundary—in contrast to the boundary which suddenly contracts and then expands, such as the disc, and the plane boundary, which neither contracts nor expands.

Figure 6-12 can be considered to represent flow around a segment of a large radius cylinder or an airplane wing. In Fig. 6-12(a), the boundary is expanding slightly and the velocity near the boundary is reduced, so that the velocity gradient dv/dy at the boundary is markedly reduced. However, dv/dy does not reach zero or become a negative quantity. In Fig. 6-12(b), on the other hand, the boundary expands more rapidly and the

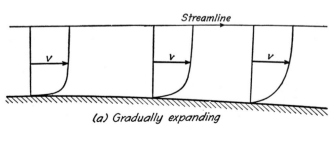

(a) Gradually expanding

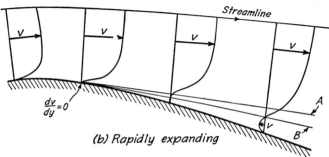

(b) Rapidly expanding

Fig. 6-12. Separation of flow from an expanding boundary.

velocity gradient at the boundary not only goes to zero, but also reverses itself to become negative. In other words, the continuity equation is satisfied since the fluid reverses its direction of flow and moves upstream. The point of separation occurs where $dv/dy = 0$ at the boundary.

The points where the velocity is zero are connected by line B in Fig. 6-12(b), which is the line of zero velocity. Above this line, the velocity is to the right, and below it the velocity is to the left, so that a pattern of secondary flow is established.

As the fluid moves past the boundary, the continuity equation must be satisfied. In other words, for any quantity of flow upstream in the zone of separation, there must be an equal quantity of flow downstream, which is in addition to the quantity entering from the left outside the separation zone. Line A connects the points from the velocity profiles above which the discharge ($\int v \, dy$) is the same as that entering from the left, and below which the rotating secondary flow results in a net discharge of zero.

Although separation occurs at the point where $dv/dy = 0$, the location of this point along the boundary profile is somewhat indefinite—depending upon the fluid and flow conditions. The boundary layer, for example, is one important factor in the location of the point of separation. If the boundary layer is laminar, the velocity gradient dv/dy at the boundary is relatively small and, therefore, the boundary does not need to expand appreciably before the velocity gradient at the boundary has reached zero, and even become negative. When the boundary layer is turbulent, however, the mixing is much greater and the velocity profile entering from the left is more nearly uniform. This makes the velocity gradient at the boundary much greater and, therefore, the boundary can expand from the flow much more before separation occurs. The location of the point of separation along the boundary has a significant effect upon the pressure against the boundary downstream from the separation point. Consequently, it has a marked influence on the pressure drag. This is discussed in greater detail in Chapter 9 on Flow around Submerged Objects.

The upper boundary of the separation zone for a laminar boundary layer is approximately a straight line which is tangent to the boundary at the point of separation. Once the laminar flow leaves the proximity of the boundary and the Reynolds number of the flow is increased, the flow may become unstable and change to turbulent flow. Because turbulent flow involves turbulent mixing, the outer boundary of the separation zone may turn back to the solid boundary at an angle of 10 to 30 degrees with the tangent line, and may even reattach itself to the boundary—thereby causing the separation zone to be an isolated bubble in the flow. Conditions causing such a bubble are a laminar boundary layer at the upstream end with a turbulent boundary layer downstream. Whether the flow will become turbulent and reattach itself to the boundary depends upon the ambient turbulence, the Reynolds number, and the geometry of the boundary. Increasing the ambient turbulence or the Reynolds number decreases the stability of the laminar flow and hence increases the tendency to change to turbulent flow.

The foregoing indicates the great importance of the nature of the boundary layer and the velocity distribution in the phenomenon of separation. Furthermore, it indicates that, for a turbulent boundary layer, the laminar sublayer may also play an important part in separation. Indeed, if the laminar sublayer is relatively thick in a turbulent boundary layer, the separation will occur further upstream than under conditions of greater velocity and thinner laminar sublayer. As the laminar sublayer becomes still thinner and eventually is destroyed, due either to increased Reynolds number (and ambient turbulence) or to increased relative roughness, the point of separation moves increasingly downstream, because the velocity gradient dv/dy at the boundary increases.

The foregoing considerations are also of great importance in connection with the shear drag and the pressure drag which a boundary will experience. The shear drag depends upon the boundary shear τ_0 in the direction of motion, whereas the pressure drag depends upon the summation of normal pressures over the surface of the boundary. Therefore, on the one hand the shear drag in Fig. 6-12(b) is a minimum if the point of separation is as far to the left as possible, so that the velocity gradient is negative over a large portion of the boundary. On the other hand, the pressure drag is a minimum if the point of separation is as far to the right as possible so that the streamlines will expand, the velocity will decrease, and the normal pressure (a back pressure) on the downstream side of the object will be a maximum. For one boundary (assuming the flow to be from left to right), the shear drag may be predominant, and therefore the point of separation should be as far to the left as possible, whereas for another boundary the pressure drag may be predominant, and the point of separation should be to the right as much as possible to hold the total drag to a minimum. If the drag needs to be a maximum, as in the case of parachutes, sea anchors, and baffles, the foregoing principles apply in a reverse sense.

The problem then remains to control the location of the point of separation. This can be done in several different ways. Basically, the velocity gradient in the approach section along the boundary must be as small as possible to hold the point of separation to the left, and as large as possible to hold the point of separation to the right. This velocity gradient at the boundary is a minimum when the flow is laminar. To increase the velocity gradient, the turbulent mixing must be increased in the proximity of the boundary. In other words, the laminar boundary layer or the laminar sublayer (of a turbulent boundary layer) must be made turbulent. This has been done successfully by placing a disturbance near the boundary in the approach section. The disturbance may consist of a greatly increased surface roughness (such as a coating of sand or gravel, or other projections), a wire stretched across the boundary, or a band or strip placed slightly above the boundary. Increasing the general level of turbulence in the ambient fluid is also effective. This can be accomplished by disturbances such as bars or a lattice upstream. Each of the foregoing tends to increase the mixing near the boundary, which removes the slowly moving fluid from near the boundary and, hence, moves the point of separation downstream. The point of separation can also be controlled to a very great extent by changing the radius of curvature of the boundary.

Although the velocity near the boundary can be increased by increased turbulence, it can also be increased by injecting high-velocity fluid at the boundary and in the direction of flow, or by draining or sucking away the slowly-moving fluid. This has been done quite successfully, not only in the laboratory, but also in the design of aircraft wings. To protect the laminar boundary layer or the laminar sublayer, however, the ambient turbulence should be kept to a minimum and the boundary should be as smooth as possible.

The concepts and ideas presented in this chapter have widespread application to problems in fluid mechanics. In fact, it is seldom that a problem in fluid flow arises that is not affected at least to some extent by the boundary layer concept, by roughness, and by separation. Streamlining of an object or a boundary, for example, is intended to reduce the pressure drag of the object, both by reducing the energy loss due to turbulence and by increasing the pressure on the downstream side of the object. The detailed application of these subjects is given in the following chapters. Further discussion of the relative roles played by shear resistance and pressure resistance, as functions of shape of the boundary and the Reynolds number, is given in Chapter 9 on Flow around Submerged Objects.

• SUMMARY

1. *Resistance* is the force transmitted from a boundary to the fluid and *drag* is the force of the fluid on the boundary.

2. Resistance (or drag) can be divided into:
    a. *Shear* resistance (or drag).
    b. *Pressure* resistance (or drag).

3. *Shear* is the force per unit area *tangential* to the boundary, and *pressure* is the force per unit area *normal* to the boundary.

4. The **general drag (or resistance) equation is:**

$$F_D = \frac{C_D A \rho V_0^2}{2} \tag{6-4}$$

in which

$$C_D = \frac{F_D}{A \rho V_0^2/2} = \phi\left(\text{Re, sf,}\ \frac{e}{L}\right) \tag{6-3}$$

5. The **boundary layer theory** for a flat plate is based upon the following assumptions:
   a. The influence of the boundary on the fluid is confined to a relatively thin layer (the boundary layer) of fluid near the boundary, which is small compared with the length of the boundary.
   b. The acceleration of the fluid normal to the boundary is negligible compared with the acceleration of the fluid parallel to the boundary.
   c. The pressure is a constant throughout the boundary layer.
   d. The flow is steady.
   e. Within the boundary layer the accumulated resistance to flow is equal to the reduction in momentum flux.

6. The **boundary layer equation** can be presented in differential form as

$$\frac{F_D}{B} = \int_0^x \tau_0 \, dx = \int_0^\sigma \rho(V_0 - v)v \, dy \tag{6-6}$$

7. For the **laminar boundary layer,** the following equations apply

$$\frac{\delta}{x} = \frac{5.2}{\text{Re}^{1/2}} \tag{6-8}$$

$$C_D = \frac{1.33}{\text{Re}^{1/2}} + \frac{4.12}{\text{Re}} \tag{6-15}$$

8. Immediately adjacent to a smooth boundary associated with a turbulent boundary layer, there is a thin film of essentially laminar flow known as the **laminar sublayer.**

9. For a **turbulent boundary layer,** there are two limiting types of boundaries:
   a. A **smooth boundary** for which the roughness elements are covered with the laminar sublayer so that the roughness has no influence on the flow within the boundary layer.

    b. A **rough boundary** for which the laminar sublayer is destroyed and the roughness elements are contributing directly to the turbulence.

10. For the **turbulent boundary layer** over a **smooth boundary** the equations are based upon

    a. The **one-seventh power law** of velocity distribution, which has application over the range $2 \times 10^5 < \text{Re} < 2 \times 10^7$

$$\frac{\delta}{x} = \frac{0.38}{\text{Re}^{1/5}} \tag{6-16}$$

$$C_D = \frac{0.074}{\text{Re}^{1/5}} \tag{6-17}$$

    b. **The log-law** of velocity distribution, which has application over the range $2 \times 10^6 < \text{Re} < 2 \times 10^9$

$$C_D = \frac{0.455}{(\log \text{Re})^{2.58}} \tag{6-23}$$

11. For the **turbulent boundary layer** along a **rough boundary** the drag coefficient varies only with **relative roughness** e/L and not with Reynolds Number.

12. For **supersonic flow** along a smooth boundary the drag coefficient decreases with increasing Mach number

    a. With **laminar flow** the decrease is about 15 percent at $\text{Ma} = 4$.

    b. With **turbulent flow** for which $\text{Re} = 10^7$, the decrease is about 50 percent at $\text{Ma} = 5$.

13. **Separation** occurs in the flow at a point along a boundary where the streamlines are spreading and the velocity gradient dv/dy is zero.

References

1. Dodge, Russell A. and Milton J. Thompson, *Fluid Mechanics*. New York: McGraw-Hill Book Co., Inc., 1937.

2. Goldstein, S., *Modern Developments in Fluid Dynamics*. London: Oxford University Press, 1938.

3. Prandtl, Ludwig, *Fluid Dynamics*. New York: Hafner Publishing Co., 1952.

4. Prandtl, Ludwig, and O. G. Tietjens, *Fundamentals of Hydro-and Aeromechanics*. New York: McGraw-Hill Book Co., Inc., 1934.

5. Rouse, Hunter, *Elementary Mechanics of Fluids*. New York: John Wiley & Sons, Inc., 1946.

6. Rouse, Hunter, *Fluid Mechanics for Hydraulic Engineers*. New York: McGraw-Hill Book Co., Inc., 1938.

7. Schlichting, Hermann, *Boundary Layer Theory*. New York: McGraw-Hill Book Co., Inc., 1955.

8. Streeter, Victor L., *Fluid Mechanics*, 2nd ed. New York: McGraw-Hill Book Co., Inc., 1958.

Problems

6-1. The Stoke's equation, Eq. 5-28, is a special case for laminar flow of the general drag equation, $F_D = C_D A \rho V_0^2 / 2$. Determine C_D for the special case of laminar flow.

6-2. A thin rectangular plate is towed through a large body of SAE 10 oil at 60°F at a velocity of 1 ft/sec. What is the thickness of the boundary layer 1 ft from the leading edge of the plate.

6-3. A thin rectangular plate is towed through water at 70°F at a velocity of 20 ft/sec. Assuming that the boundary layer is turbulent from the leading edge, (a) what is the thickness of the boundary layer 3 ft from the leading edge? (b) At the section where $x = 3$ ft, what is the velocity at a point where $y = \delta/2$?

6-4. A submerged thin plate 1 ft square is towed through a body of glycerine at a temperature of 80°F. The velocity of the plate is 2 ft/sec and the direction of the relative velocity between the plate and the glycerine is parallel to the plate. (a) Find the force on the plate, (b) Find the maximum thickness of the boundary layer for this case.

6-5. Water at 70°F and flowing at a uniform velocity of 10 ft/sec flows along a smooth flat plate parallel to the plate. How far from the leading edge of the plate does the boundary layer change from laminar to turbulent flow. What is the thickness of the boundary layer at this point?

6-6. In turbulent flow near the earth, the velocity of the wind is measured at two points as listed

$$y = 5 \text{ ft} \qquad v = 6 \text{ ft/sec}$$
$$y = 15 \text{ ft} \qquad v = 8 \text{ ft/sec}$$

Find: (a) the boundary shear; (b) the velocity at a point 25 ft above the ground; (●) the equation for the velocity distribution. Assume $T = 60°F$.

6-7. A rectangular plate 8 ft by 2 ft is submerged in a uniform flow of air whose temperature is 80°F at atmospheric pressure. If the flow is parallel to the 8 ft axis of the plate, what is the drag on the plate if the relative roughness $e/L = 10^{-4}$ and the air velocity is 500 fps?

6-8. A smooth rectangular plate 3 ft wide and 15 ft long is held stationary in a stream of water which is traveling at a uniform velocity of 25 ft/sec in the direction of the long axis of the plate. The water temperature is 40°F. (a) Determine the drag force on the plate. (b) What is the thickness of the boundary layer at the trailing edge?

6-9. A rectangular plate 2 in. by 6 in. is towed through a tank of glycerine at a velocity of 10.0 ft/sec. The temperature is 68°F and the direction of the motion is along the 6-in. length. Find the force on the plate and the maximum thickness of the laminar boundary layer.

6-10. An oil whose specific gravity is 0.80 and whose kinematic viscosity is 3×10^{-3} sq ft/sec is flowing past a flat plate at a uniform velocity of 2 ft/sec. Find: (a) the point along the plate where Re = 10,000; (b) the maximum value of the chi (χ) parameter in the boundary layer at the point where Re = 10,000; (c) at the approximate section where the laminar boundary layer breaks down, what is the maximum value of the chi parameter in this section? What is the value of χ at $y = \delta/2$?

6-11. The boundary layer:
(a) is a layer of laminar flow next to the boundary;
(b) is a layer of fluid near the boundary in which the velocity distribution curve is generally parabolic;
(c) is a layer of the flow near the boundary in which the velocity is smaller than the ambient velocity approaching the boundary;
(d) thickness is constant along the boundary;
(e) none of these.
Check one or more of these statements.

6-12. Separation:
(a) is not very important in hydraulic design problems;
(b) does not occur in the case of boundaries experiencing pressure drag;
(c) occurs at points of abrupt change in the direction of the flow;
(d) cannot occur in laminar flow;
(e) none of these.
Check one or more of these statements.

6-13. An approximate critical Reynolds number for the transition from a laminar boundary layer to a turbulent boundary layer is: (a) 2,000, (b) 5,000,000, (c) 500,000, (d) 50,000, (e) none of these. Check one or more of these answers.

6-14. The thickness of the laminar boundary layer varies with (a) x, (b) x^{-1}, (c) $x^{-1/2}$, (d) $x^{1/2}$, (e) x^2. Check one or more of these answers.

6-15. The velocity distribution in a turbulent boundary layer is (a) parabolic, (b) linear, (c) logarithmic, (d) a function of $y^{1/5}$, (e) none of these. Check one or more of these answers.

6-16. The thickness of the turbulent boundary layer varies as (a) $x^{4/5}$, (b) $x^{-1/5}$, (c) $x^{2.58}$, (d) x^2, (e) none of these. Check one or more of these answers.

6-17. The general equation for the drag coefficient is:

(a) $C_D = \dfrac{F_D}{A}$ (b) $C_D = \dfrac{A\rho V_0^2/2}{F_D}$

(c) $C_D = \dfrac{F_D}{\rho V_0^2/2}$ (d) $C_D = \dfrac{\rho V_0^2/2}{A}$

(e) $C_D = \dfrac{F_D}{\rho V_0^2/2A}$

Check one or more of these equations.

6-18. The boundary layer varies in the direction of flow in the following sequence:
(a) Laminar boundary layer, transition zone, turbulent boundary layer.
(b) Transition zone, turbulent boundary layer, laminar boundary layer.
(c) Turbulent boundary layer, transition region, laminar boundary layer.
(d) None of these.

Check one or more of these answers.

6-19. A flat plate 3 ft long is set parallel to a flow of air at a temperature of 100°F. The ambient velocity is 30 ft/sec. (a) Find the mean drag coefficient. (b) Find the maximum thickness of boundary layer. (c) Find the total drag per unit width of the plate if the air covers both sides.

6-20. As a first approximation, assume a linear distribution of velocity across the laminar boundary layer. (a) Find an expression for the drag coefficient for a flat plate parallel to flow. (b) Find an expression for the thickness of the boundary layer formed on the plate.

6-21. Wind velocities, recorded simultaneously at a weather station, at distances of 12 ft and 30 ft above the ground are 4.88 mph and 8.28 mph respectively. If the air temperature is 60°F (a) Calculate the shear velocity $\sqrt{\tau_0/\rho}$, the boundary shear τ_0, and the distance y'. (b) Develop the velocity distribution equation and plot the velocity profile of the wind from y' to 100 ft above the earth.

6-22. Develop the equation

$$\tau_0 = \rho \left[\frac{v_2 - v_1}{5.75 \log y_2/y_1} \right]^2$$

for the shear on a boundary if v_1 and v_2 are the velocities in the fluid flowing adjacent to the boundary at distances y_1 and y_2 above the boundary.

6-23. Water is flowing in a wide channel at depth y_0. If the velocities are 1.75 and 2.00 fps at distances of 0.4 and 0.8 ft above the bed calculate the shear stress on the channel bed. The water temperature is 50°F.

6-24. Determine the boundary-layer thickness at one-third points on a smooth plate 9-ft long when:

$$\text{(a)} \quad \frac{V_oL}{\nu} = 100,000 \quad \text{(b)} \quad \frac{V_oL}{\nu} = 1,000,000 \quad \text{(c)} \quad \frac{V_oL}{\nu} = 10,000,000$$

and the flow is assumed to be completely turbulent.

6-25. The boundary layer on a flat plate is completely turbulent. Determine the values of C_D by means of the 1/7-power law formula, and by the log law as developed by Schlichting, for Re $= 10^7$ and 10^9. If Schlicting's formula is assumed correct, what percent change in C_D is obtained when the 1/7-power law formula is used?

6-26. An aircraft wing has a length of 40 ft and an average chord length of 8 ft. Neglecting the thickness of the wing, what would be the shear resistance on the wing while flying at a barometric pressure of 20 in. of mercury and an air temperature of $-40°F$ if the air speed is: **(a)** 100 mph, and **(b)** 500 mph?

chapter 7

Flow in
Closed Conduits

Closed conduits include all types of pipes and tubes having many different shapes and sizes. Industry is employing increasing numbers of closed conduits to transport fluids used in processing and fluids being processed. Air conditioning (heating and ventilating) systems use conduits of various shapes. Waterworks involve extensive piping systems, both large and small. Oil and gas pipe lines supply users thousands of miles away from the well. Various other industries are using pipe lines to transport their product as raw material, during processing, and as the finished product. Indeed, pipe line engineering has become very important to the progress of civilization.

Flow in closed conduits involves a combination of:

1. Steady or unsteady flow;

2. Uniform or non-uniform flow;

3. Laminar or turbulent flow.

258

If flow in a conduit is steady, then at any given point in the flow there is no variation in velocity with respect to time, i.e., $\partial v/\partial t = 0$. If there is variation of velocity with time, then the flow is unsteady, i.e., $\partial v/\partial t \neq 0$. Flow at a constant discharge in a pipe is steady flow. If a tank under varying head is being emptied, the flow in the outlet pipe is unsteady.

When flow is uniform, there is no change at a given time in either the magnitude or the direction of the velocity with respect to distance along a streamline, i.e., $\partial v/\partial s = 0 = \partial v/\partial n$. When the flow is nonuniform, the velocity along a streamline varies either in magnitude, $\partial v/\partial s \neq 0$, or in direction, $\partial v/\partial n \neq 0$, or both. In closed conduits, the flow is uniform if the conduit is straight and the walls are parallel. If the conduit has a bend, an inner obstruction, diverging or converging walls, or any other feature which causes the streamlines to curve, the velocity magnitude and/or direction changes, and the flow is nonuniform. Flow of a gas which is expanding due to decreasing pressure in the direction of flow results in an increase in the magnitude of the velocity along a streamline; hence, the flow is nonuniform.

In laminar flow (see Chapter 5), the mixing of fluid in one region with that in another region across the flow is accomplished only by the extremely slow process of molecular activity or diffusion. Turbulent flow, on the other hand, has finite fluid masses which rotate and move about as eddies, making the mixing process much more rapid. With turbulent flow, the boundary may be either hydrodynamically smooth or hydrodynamically rough.

The solution of various engineering problems involving the foregoing types of flow is considered in the remainder of this chapter.

In some industrial processes, non-Newtonian fluids are transported through pipes. The viscosity μ for non-Newtonian fluids is dependent upon dv/dy, see Fig. 5-1. These flow systems include various types of plastic flow, such as plug flow in which a central core moves as a body that is lubricated by a thin film near the wall; and thixotropic fluids, such as drilling mud used in drilling oil wells. Thixotropic fluids change from the solid to the fluid state as a result of mechanical agitation, and problems involving these fluids are not considered in this text.

• ESTABLISHMENT OF FLOW

The flow of a fluid along the wall of a pipe has many similarities to the flow past a plane boundary, as discussed in the preceding chapter. In fact, much of the original boundary layer theory and experimental data were developed first for the flow in a pipe.

The boundary layer which develops in a pipe is illustrated in Fig. 7-1. At the streamlined entrance to the pipe, the boundary layer develops in much the same manner as it does for a plane boundary. When the thickness

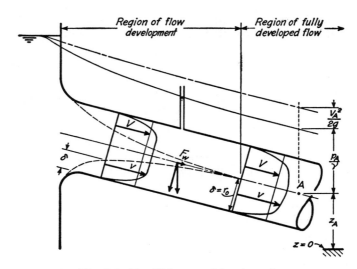

Fig. 7-1. Establishment of flow in a pipe.

of the boundary layer δ becomes equal to the radius of the pipe at the downstream end of the region of flow development, however, it can expand no farther, and the boundary layer proceeds downstream unchanged in its distributions of velocity and turbulence throughout the region of fully-developed flow. The length of pipe x required for the region of flow development has been found to be approximately

$$\frac{x}{D} = 0.06 \text{ Re} \tag{7-1}$$

for *laminar flow*, and

$$\frac{x}{D} = 0.7 \text{ Re}^{1/4} \tag{7-2}$$

for *turbulent flow* throughout. Experimental data, however, have shown that the length given by Eq. 7-2 is too short if part of the boundary layer is laminar.

In the previous chapter, it was found that the change in momentum flux within the boundary layer was equal to the accumulated shear from the leading edge. In the region of flow establishment within the pipe, this same principle applies, but, in addition, the piezometric head is decreasing in the direction of flow and the velocity of the undisturbed core is increasing (in order to satisfy the continuity equation $Q = AV$), which complicates the analysis of the flow in this region.

Example Problem 7-1: Calculate the distance required to establish uniform flow in a 12 in. pipe assuming a temperature and velocity of 68°F and 4.8 fps respectively if (a) the liquid is SAE 30 oil, (b) if the liquid is water.

Solution:

(a) $\text{Re} = \dfrac{(4.8)(1)\ 1.785}{9.2 \times 10^{-3}} = 931$

flow is laminar

$$x = 0.06\ \text{Re}\ D = (0.06)(931)(1) = \underline{\underline{55.9\ \text{ft}}}$$

(b) $\text{Re} = \dfrac{(4.8)(1)(10)^5}{1.09} = 441{,}000$

flow is turbulent

$$x = 0.7\ \text{Re}^{1/4}\ D = 0.7(441{,}000)^{1/4}\ (1) = \underline{\underline{18.1\ \text{ft}}}$$

The relatively short distance is required to establish uniform turbulent flow because of the larger intensity of turbulent mixing.

• STEADY, UNIFORM FLOW

Once the flow has become established, there is no further change in momentum flux and the shear at the boundary does not change from point to point along the pipe. This is illustrated in Fig. 7-2, where the boundary shear τ_0 is the shear resistance the boundary exerts on the fluid.

To analyse the flow in Fig. 7-2, consider a free body of length L of the fluid. Over this distance, the piezometric head changes by the amount h_f. This change in head is caused by the shear τ_0 acting over the cylindrical inner surface of the pipe, producing a total resisting force equal to $\tau_0(\pi D L)$.

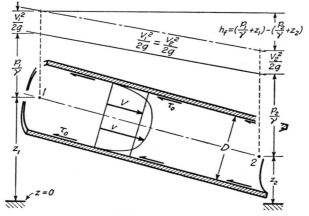

Fig. 7-2. Established flow in a pipe.

It can be equated to the net force acting on the ends of the cylinder, which is caused by the combination of the change in pressure and the component of the weight of the fluid acting in the direction of motion. This combination is equal to the decrease in piezometric head h_f over the distance L. Now, the unbalanced force on the ends of the cylinder can be expressed as $\gamma h_f \dfrac{\pi D^2}{4}$, which when equated to the force of shear resistance yields

$$\tau_0(\pi DL) = \gamma h_f\left(\frac{\pi D^2}{4}\right) \tag{7-3}$$

which may be rearranged and divided by the velocity head $V^2/2g$ to give

$$4\frac{\tau_0/\gamma}{V^2/2g} = \frac{h_f}{L}\frac{D}{V^2/2g} \tag{7-4}$$

The left term in the equation is equal to the Darcy-Weisbach resistance coefficient f, which is very similar to the resistance coefficient C_D in Eq. 6-4. Equation 7-4 can now be rearranged as

$$h_f = f\frac{L}{D}\frac{V^2}{2g} \tag{7-5}$$

This equation is used to determine that part of the total loss H_L in the Bernoulli equation which is caused by the shear resistance at the boundary

$$\frac{V_1^2}{2g} + \frac{p_1}{\gamma} + z_1 = \frac{V_2^2}{2g} + \frac{p_2}{\gamma} + z_2 + H_L \tag{7-6}$$

in which H_L is the sum of the losses caused both by shear drag h_f and pressure drag h_L (pressure drag in a pipe is discussed later in this chapter). That is

$$H_L = h_f + h_L \tag{7-7}$$

It now remains to determine the magnitude of the coefficient f.

As in the case of the resistance coefficient C_D in Chapter 6, the Darcy-Weisbach resistance coefficient f depends upon the Reynolds number Re and the relative roughness e/D. That is,

$$f = \phi_1\left(\text{Re}, \frac{e}{D}\right) \tag{7-8}$$

For laminar flow, and for turbulent flow with a smooth boundary, the relative roughness e/D is unimportant and, for a rough boundary, the Reynolds number Re is unimportant. Hence

$$f = \phi_2(\text{Re}) \quad (Hydrodynamically\ smooth\ boundary) \tag{7-9}$$

$$f = \phi_3\left(\frac{e}{D}\right) \quad (Hydrodynamically\ rough\ boundary) \tag{7-10}$$

• LAMINAR FLOW

For **laminar flow,** the Poiseuille equation, Eq. 5-26, may be rearranged in the form of Eq. 7-5 so that

$$f = \frac{64}{Re} \quad (Laminar\ flow,\ Re < 2000) \tag{7-11}$$

which shows that the resistance coefficient f varies inversely with the Reynolds number Re—that is, if the Reynolds number is doubled the coefficient f is cut in half when the flow is laminar. As shown in Fig. 7-3, Eq. 7-11 is applicable for Reynolds numbers up to 2000.

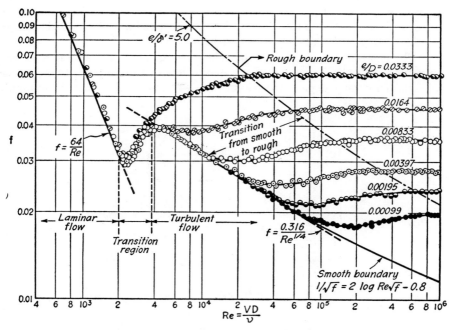

Fig. 7-3. Variation of resistance coefficient with
Reynolds number and relative roughness.

• TURBULENT FLOW

Many investigators have studied **turbulent flow** in pipes. For turbulent flow in a smooth pipe, both the power law and the log law of the velocity distribution have been used as a basis to solve for the resistance coefficient f.

Blasius studied the data of Saph and Schoder and developed the following empiral equation

$$f = \frac{0.316}{\text{Re}^{1/4}} \quad \left(\begin{array}{c} \textit{Turbulent flow, smooth boundary} \\ 3000 < \text{Re} < 100{,}000 \end{array} \right) \tag{7-12}$$

for turbulent flow with smooth boundaries. This equation shows that the coefficient f varies inversely with the $\frac{1}{4}$-th power of the Reynolds number. In other words, the Reynolds number must be increased 16 times in order to cut the resistance coefficient in half.

More recent data obtained for larger Reynolds numbers proved the Blasius equation for smooth boundaries to be applicable only for Reynolds numbers between 3000 and 100,000 as shown in Fig. 7-3. For Reynolds numbers greater than 100,000, the log law of the velocity distribution for smooth boundaries has proved useful, just as it was for flow past a flat plate at larger Reynolds numbers, see Chapter 6.

The logarithmic distribution of velocity in most of the boundary layer, together with the linear velocity distribution in the laminar sublayer, is shown in a greatly expanded scale in Fig. 7-4. The thickness of the laminar

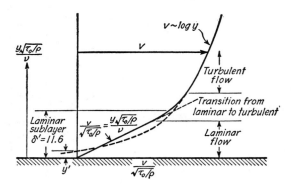

Fig. 7-4. Velocity distribution near a smooth boundary.

sublayer δ' has been defined as that point where these two curves cross. Because the parabolic distribution of velocity used in the Poiseuille equation and the laminar boundary layer is so nearly a straight line for small distances from the boundary, the distribution within the laminar sublayer is approximated by a straight line without appreciable error. Hence, the velocity gradient within the laminar sublayer becomes v/y and the shear equation $\tau = \mu \, dv/dy$ becomes

$$\tau = \tau_0 = \mu \, \frac{v}{y} \tag{7-13}$$

which may be rearranged for later use as

$$\frac{v}{\sqrt{\tau_0/\rho}} = \frac{y\sqrt{\tau_0/\rho}}{\nu} \tag{7-14}$$

The logarithmic distribution of velocity outside the laminar sublayer is expressed by Eq. 6-20. This equation can be solved for the particular case of a smooth boundary to obtain Eq. 6-21. The constant β in this equation has been determined experimentally to be 5.5. Therefore, converting to the base 10 for the log, Eq. 6-21 becomes

$$\frac{v}{\sqrt{\tau_0/\rho}} = 5.75 \log \frac{y\sqrt{\tau_0/\rho}}{\nu} + 5.5 \quad \left(\begin{matrix} Turbulent\ boundary\ layer, \\ smooth\ boundary \end{matrix}\right) \tag{7-15}$$

which is also plotted in Fig. 7-4.

Equation 7-14 and Eq. 7-15 may be solved simultaneously to yield

$$\delta' = 11.6 \frac{\nu}{\sqrt{\tau_0/\rho}} \tag{7-16}$$

which is a logical definition of the thickness of the laminar sublayer. Despite its logic, however, the definition is arbitrary, since the point is actually within the transition region between laminar flow and turbulent flow—the transition actually extending to some distance on each side of 11.6.

- BOUNDARY ROUGHNESS

When the laminar sublayer is destroyed, the viscous forces are no longer effective and y' in Eq. 6-20 must be evaluated in terms of the absolute roughness e. Again using the Nikuradse experiments, it has been found that y' varies directly with e—in fact, $y' = e/30$. By converting to log base 10, Eq. 6-20 becomes

$$\frac{v}{\sqrt{\tau_0/\rho}} = 5.75 \log \frac{y}{e} + 8.5 \quad \left(\begin{matrix} Turbulent\ boundary\ layer, \\ rough\ boundary \end{matrix}\right) \tag{7-17}$$

Equations 7-15 and 7-17 can be substituted in the equation $Q = \int v\,dA$, integrated, and the constants adjusted slightly to fit the data best in order to determine the exact functions expressed in Eqs. 7-9 and 7-10.

$$\frac{1}{\sqrt{f}} = 2 \log \text{Re} \sqrt{f} - 0.8 \quad \left(\begin{matrix} Turbulent\ boundary\ layer, \\ smooth\ boundary \end{matrix}\right) \tag{7-18}$$

and

$$\frac{1}{\sqrt{f}} = 2 \log \frac{D}{e} + 1.14 \quad \left(\begin{matrix} Turbulent\ boundary\ layer, \\ rough\ boundary \end{matrix}\right) \tag{7-19}$$

Equations 7-11, 7-12, 7-18, and 7-19 have been plotted in Fig. 7-3 where together with the data of Nikuradse, they illustrate many significant facts. Figure 7-3 shows not only the variation of the coefficient f with respect to Reynolds number for laminar flow, Eq. 7-11 (the Poiseuille equation), but it also shows the transition from laminar flow to turbulent flow, where f must be evaluated empirically, as it suddenly increases by about 30 percent. In other words, for a given Reynolds number, considerably less power is required to move a fluid through a pipe as laminar flow than as turbulent flow. Unfortunately, however, from the practical standpoint this is seldom possible to accomplish, except in the case of transport of certain viscous fluids, or by the use of very small pipes or tubes.

With turbulent flow, considerable advantage is gained by maintaining a smooth boundary. For example, when $Re = 5 \times 10^4$, it is possible to cut the coefficient f at least in half by the use of a smooth boundary, as compared with a rough boundary of $e/D = 0.0167$.

Figure 7-3 also shows that, when the boundary is smooth, the relative roughness e/D plays no part in the flow system. Hence, f is dependent only on the Reynolds number Re, and Eqs. 7-9, 7-12, and 7-18 are applicable. When the boundary is rough, on the other hand, the effect of Reynolds number is insignificant and f varies only with the relative roughness e/D. Under these conditions, Eqs. 7-10, and 7-19 are applicable.

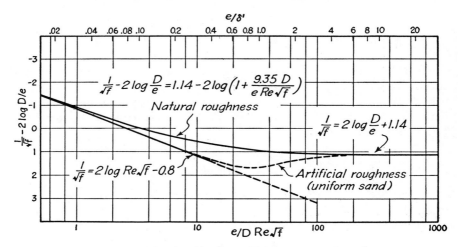

Fig. 7-5. Transition function for artificial and natural roughness.

In the region between the completely smooth boundary and the completely rough boundary is a transition zone, in which both Reynolds number Re and relative roughness e/D are important, and Eq. 7-8 is applicable. Although an analytical solution for this complicated transition region is not possible, it can be visualized in Fig. 7-5, which is a plot of the transition function for both artificially-roughened pipe (Nikuradse sand-grain roughness) and pipe having a natural or commercial roughness determined from various types of pipes.

Figure 7-5 does not include the region of laminar flow, but it does bring onto one curve all data for the rough boundaries, and it distinguishes clearly between artificial and natural roughness.

Colebrook and White have developed the following semi-empirical equation for the transition from smooth to rough pipe, by simply combining Eq. 7-18 and Eq. 7-19.

$$\frac{1}{\sqrt{f}} = 1.14 - 2 \log \left(\frac{e}{D} + \frac{9.35}{\mathrm{Re}\sqrt{f}} \right) \tag{7-20}$$

Consequently, for smooth pipe (where the roughness is ineffective), this equation reduces to Eq. 7-18, and for rough pipe (where Re is ineffective), it reduces to Eq. 7-19.

The magnitude of the natural roughness has been evaluated according to its equivalent roughness, based on the uniform sand roughness of Nikuradse. In other words, the unknown roughness for each experiment with a pipe having natural roughness was found by setting it equal to that value of the Nikuradse roughness e which gave the same f-value for a given pipe diameter in Eq. 7-19.

The relative thickness of the laminar sublayer e/δ' is another important part of Fig. 7-5. In this figure, the abscissa $\mathrm{Re}\sqrt{f}\, e/D$ may be simplified to yield e/δ' by combining with Eq. 7-16 as follows

$$\frac{\mathrm{Re}\ \sqrt{f}}{D/e} = \frac{VD\ (\sqrt{8\tau_0/\rho})\ e}{\nu VD} = \sqrt{8}\ \frac{e\ \sqrt{\tau_0/\rho}}{\nu} = 32.8\ \frac{e}{\delta'} \tag{7-21}$$

The numerator of the first term can be expressed as

$$\mathrm{Re}\ \sqrt{f} = \frac{D^{3/2}}{\nu} \sqrt{\frac{2gh_f}{L}} \tag{7-22}$$

which can be used for the solution of problems where V and Q are unknown.

As pipes become older, the roughness increases due to corrosion, incrustations, and tuberculation. Colebrook and White have suggested that, for water in steel pipes, this increase in roughness is approximately linear, so that the following equation applies

$$e = e_0 + \alpha t \tag{7-23}$$

in which

    e is the roughness at age t;

    e_0 is the initial roughness;

    α is the coefficient to be determined by experiment (a typical variation in α is 0.0002 to 0.007);

    t is the time pipe has been in use, in years.

The increase in e depends principally upon the chemical quality of the water. Therefore, α varies from one geographical region to another.

The relative thickness of the laminar sublayer e/δ' is given as the abscissa in the upper part of Fig. 7-5. Here it shows that when the thickness of the laminar sublayer is several times as great as the roughness height e (say $e/\delta' \leq \frac{1}{5}$), the boundary is smooth for the uniform roughness. Likewise, when the roughness e is several times as great as δ' (say, $e/\delta' \geq 5$), the boundary is rough. In the region between, where $\frac{1}{5} < e/\delta' < 5$, is the transition from a smooth boundary to a rough boundary. With commercial or natural roughness, the boundary is rough when $e/\delta' > 5$, but the transition from smooth to rough is initiated at much smaller values—say at approximately $e/\delta' = 0.05$—due to the disturbance caused by an infrequently-occurring but exceptionally large roughness starting the breakdown of the laminar sublayer.

Since the commercial type of pipe usually encountered in design work has a natural roughness which is not uniform in size, the transition curve for natural roughness in Fig. 7-5 is the one which is most applicable for most design problems. Therefore, this transition curve is used in Fig. 7-6, which is designed specifically for speed and ease of computation of nearly any pipe problem involving steady, uniform flow. For further simplification of the solution of steady flow problems, refer to the nomograph Pipe Resistance Diagram inside the back cover.

The roughness for various pipe materials and inside coatings is given in the table in Fig. 7-6. *The average of the range of e should be utilized, unless additional information gives reason to use the smaller or larger parts of the range.* From Fig. 7-6, it may be seen that a rather large error in the estimate of e results in a smaller error in f. For example, if an e is selected which gives an e/D value of 0.001 when it should have been 0.0005—a 100 percent error in e/D—, the error in the f-value is less than 15 percent.

Table 7-1. VALUES OF ROUGHNESS e FOR VARIOUS CONDUIT MATERIALS

	(e in ft)
Glass, drawn brass, copper, and lead	Smooth
Commercial steel or wrought iron	0.0001–0.0003
Asphalted cast iron	0.0002–0.0006
Galvanized iron	0.0002–0.0008
Cast iron	0.0006–0.003
Wood stave	0.0004–0.002
Concrete	0.001 –0.01
Riveted steel	0.003 –0.03
Corrugated metal pipe	0.1 –0.2
Large tunnels, concrete or steel lined	0.002 –0.004
Blasted-rock tunnels	1.0 –2.0

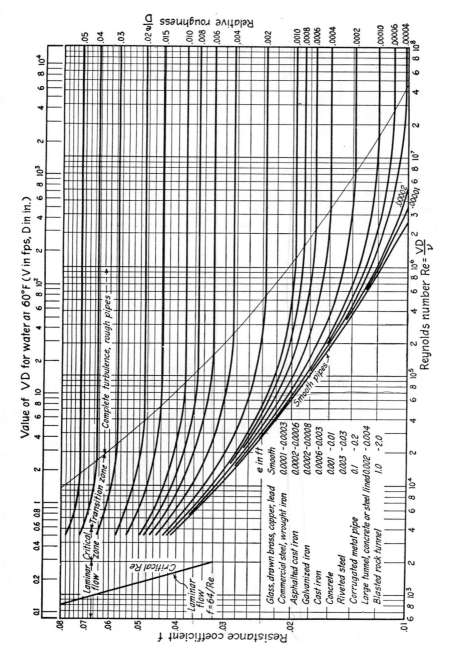

Fig. 7-6. Moody resistance diagram for uniform flow in conduits.

• SOLUTION OF PROBLEMS

There are **three types of problems** which occur most frequently in the design of pipe lines and pipe systems. Usually, the head loss h_f is needed for a given length of pipe L, velocity V or discharge Q, pipe diameter D, roughness e, and fluid viscosity ν. For this problem, the solution is direct. From Reynolds number Re and relative roughness e/D, the resistance coefficient f can be found and the head loss solved from the equation

$$h_f = f\frac{L}{D}\frac{V^2}{2g} \qquad (7\text{-}24)$$

When V or Q is needed, however, Re cannot be computed directly. Hence, another parameter Re $\sqrt{f}$, see Eq. 7-22, is used which does not involve either V or Q. By use of Re $\sqrt{f}$ and e/D the f-value can be determined directly from Eq. 7-20, and V or Q can be computed by

$$V = \sqrt{\frac{2gh_fD}{fL}} \qquad (7\text{-}25)$$

or

$$AV = Q = \sqrt{\frac{\pi^2 gh_f D^5}{8fL}} \qquad (7\text{-}26)$$

By assuming the pipe is rough with a given value of e/D, f can be determined from Fig. 7-6. After computing V from Eq. 7-25, the value of f can be checked by calculating Re, to see if the pipe has the roughness assumed. If the pipe is not rough as assumed but is in the transition zone, adjust the answer by using a new value of f, from the Re vs. f curves.

When the diameter D of a pipe is the unknown, none of the scales in Fig. 7-6 permit direct solution for D. However, a trial-and-error method can be used which is reasonably rapid. The f-value is estimated (on most problems this is between 0.02 and 0.04), and a trial D value is found from either

$$D = f\frac{L}{h_f}\frac{V^2}{2g} \qquad (7\text{-}27)$$

or

$$D^5 = f\frac{8LQ^2}{\pi^2 gh_f} \qquad (7\text{-}28)$$

depending upon whether V or Q is known. The trial D is used to compute both Re and e/D, then f is found from Fig. 7-6. This new f value is now used as the new estimate of f, and the process is repeated. Usually, not more than three trials are necessary to obtain good agreement between the estimated f value and the computed f value. Since the D value obtained by this process is seldom an exact pipe size, the next larger size of pipe is usually selected.

The three foregoing types of problems and methods of solution can be summarized as follows:

Table 7-2. OUTLINE OF METHODS OF SOLUTION FOR PIPE PROBLEMS

Known	Needed	Method of Solution
(1) V or Q, D, L, e, and ν	h_f	Use Re and e/D scales to obtain f from Fig. 7–6, then use $$h_f = f\,\frac{L}{D}\frac{V^2}{2g}$$
(2) h_f, D, L, e, and ν	V or Q	Compute e/D and Re $\sqrt{f}$, see Eq. 7–22, then determine f directly from Eq. 7–20, or use the e/D-curves in Fig. 7–6 to determine f by trial-and-error procedure. Then use $$V = \sqrt{\frac{2gh_f D}{fL}} \quad \text{or}$$ $$Q = \sqrt{\frac{\pi^2 g h_f D^5}{8fL}}$$
(3) V or Q, h_f, L, e, and ν	D	Trial-and-error solution. Estimate f and solve for trial D by $$D = f\,\frac{L}{h_f}\frac{V^2}{2g} \quad \text{or}$$ $$D^5 = f\,\frac{8LQ^2}{\pi^2 g h_f}\,.$$ With trial D compute Re and e/D as in (1) to find new estimate of f from Fig. 7–6. Repeat process until calculated f-value agrees with the estimated f-value.

In practical field problems, the exact value of the roughness e and the temperature are often not known. However, by assuming a reasonable temperature and choosing a value of e from Fig. 7-6, a satisfactory answer can be obtained for most pipe line problems. The foregoing methods of solution can be used for any fluid when there is steady, uniform flow in a pipe or tube.

Example Problem 7-2: Find the head loss in ft per 1000 ft in a 9 in. commercial steel pipe which is carrying 900 gpm of kerosene at 68°F.

Solution:

$$Q = \frac{900}{450} = 2.0 \text{ cfs}$$

$$V = \frac{2.0}{0.442} = 4.52 \text{ fps}$$

$$\text{Re} = \frac{VD\rho}{\mu} = \frac{(4.52)(0.75)(1.57)}{4.0 \times 10^{-5}} = 133,000$$

$$e/D = \frac{0.0002}{0.75} = 0.000267$$

Based upon Re, e/D and Fig. 7-6 $f = 0.019$
and

$$h_f = \frac{(0.019)(1000)(4.52)^2}{(0.75)(64.4)} = 8.05 \text{ ft}/1000 \text{ ft}$$

or referring to the pipe resistance nomogram which is inside the back cover

$$h_f = 8.0 \text{ ft}/1000 \text{ ft}$$

Example Problem 7-3: Determine the diameter of a cast iron pipe and the average velocity in the pipe if it must carry 10 cfs of water at a temperature of 40°F with a head loss not to exceed 20 ft per 1000 ft.

Solution:

Assume $f = 0.018$

From Eq. 7-24 or Table 7-2

$$D^5 = \frac{f8LQ^2}{\pi^2 g h_f} = \frac{(0.018)(8)(1000)(100)}{\pi^2(32.2)(20)} = 2.26$$

$$D = 1.177 = 1.18 \text{ ft}$$

$$V = \frac{10}{1.10} = 9.1 \text{ fps}$$

$$\text{Re} = \frac{(9.1)(1.18)(10)^5}{1.664} = 645,000$$

$$e/D = \frac{0.0012}{1.18} = 0.00102$$

From Fig. 7-6

Corresponding $f = 0.02$

Computing a new D based on $f = 0.02$

$$D^5 = \frac{0.02(8)(1000)(100)}{\pi^2(32.2)(20)} = 2.51$$

$$D = 1.21 \text{ ft}$$

$$V = \frac{10}{1.15} = 8.7 \text{ fps}$$

$$\text{Re} = \frac{(8.7)(1.21)(10)^5}{1.664} = 632{,}000$$

$$e/D = \frac{0.0012}{1.21} = 0.000992$$

From Fig. 7-6

$$f = 0.0195 \approx 0.02$$

and

$$D = \underline{\underline{1.21 \text{ ft}}} \qquad\qquad V = \underline{\underline{8.7 \text{ fps}}}$$

The next larger standard size pipe would be selected.

Although the Darcy-Weisbach equation and its accompanying graph Fig. 7-6 is the best rational equation for solving pipe flow problems, various empirical equations for the flow of water in pipes have been developed from data taken in the laboratory and in the field. Perhaps the best known and most extensively used of these is the Hazen and Williams formula

$$V = 1.32C_1R^{0.63}S^{0.54} \tag{7-29}$$

in which C_1 is the Hazen and Williams discharge coefficient;

> R is the hydraulic radius, i.e., the area A divided by the wetted perimeter P (the wetted perimeter for a pipe flowing full is simply the circumference);
>
> S is the slope of the energy line or the hydraulic gradient h_f/L.

This equation is widely used in waterworks design and is most applicable for pipes of 2-in. or larger and velocities less than 10 fps. The principal advantage of this equation is that the coefficient C_1 does not involve Reynolds number, and all problems have direct solutions. This is also a disadvantage, however, since temperature and viscosity variations are ignored. These are at times very important and ignoring them can cause serious error.

Table 7-3. HAZEN AND WILLIAMS COEFFICIENTS FOR FLOW IN PIPES

Description of Pipe	C_1 value
Extremely smooth and straight...................	140
Very smooth......................................	130
Smooth wood and wood stave.....................	120
New riveted steel................................	110
Vitrified...	110
Old riveted steel................................	100
Old cast iron....................................	95
Old pipes in bad condition.......................	60 to 80
Small pipes badly tuberculated...................	40 to 50

Other empirical equations have been developed by various experimenters, but the Hazen-Williams equation is the only one presented, because of
its extensive application in the field of water supply. This equation is
utilized in practice because it is easily adapted to use with a nomograph.
However, a separate nomograph is necessary for each value of C_1 and, since
Fig. 7-6 and the Pipe Resistance Diagram give results for all values of pipe
roughness, they are the only graphs presented in this text. They are used
because they have the best theoretical and rational development, and they
are the most versatile graphs known by the authors for solving all types
of pipe line problems involving uniform flow.

• CONDUITS OF NONCIRCULAR CROSS SECTION

Pipes having a noncircular cross section but a simple geometric shape,
such as a rectangle, a trapezoid, or an ellipse which does not differ markedly
from circular, can be solved by Fig. 7-6 if the hydraulic radius $R = A/P$
$= D/4$ for a circular pipe is used. The Reynolds number then becomes
$\mathrm{Re} = 4RV/\nu$ and the head loss equation becomes

$$h_f = f \frac{L}{4R} \frac{V^2}{2g} \tag{7-30}$$

For turbulent flow, this use of hydraulic radius gives reasonably accurate
results. For laminar flow, however, it gives increasingly inaccurate results
as the shape of conduit differs more and more from circular.

• COMPRESSIBLE FLUIDS

Flow of compressible fluids in a conduit involves not only change in
pressure in the downstream direction, but also a change in both the density
of the fluid and the velocity of flow, see Eq. 4-2. Consequently, the energy
equation must be written in the differential form.

$$\frac{d(V^2)}{2g} + \frac{dp}{\gamma} + dz + dE_s = 0 \tag{7-31}$$

in which dz may be neglected for conditions of small change in altitude.
The gradient of energy subtracted dE_s is

$$\frac{dE_s}{dL} = \frac{fV^2}{D2g} \tag{7-32}$$

which may be combined with Eq. 7-31 to yield

$$\frac{V\,dV}{g} + \frac{dp}{\gamma} + \frac{fV^2}{D2g}\,dL = 0 \tag{7-33}$$

or

$$2\frac{dV}{V} + 2\frac{dp}{\rho V^2} + f\frac{dL}{D} = 0 \tag{7-34}$$

Equation 7-34 is the basic energy equation for flow of compressible fluids in pipes of constant cross section and small changes in elevation dz. If the change in elevation is appreciable, the term $dz = \sin \alpha\, dL$ must be included. This equation has application to both isothermal flow and adiabatic flow.

• ISOTHERMAL FLOW

If the temperature of the fluid remains constant along the length of conduit considered, the flow is isothermal. This situation prevails for long natural gas lines, for example. Under this condition, $p/\rho = gRT$ is a constant, and dynamic viscosity μ is also considered a constant. Consequently, the Reynolds number

$$\mathrm{Re} = \frac{VD}{\nu} = \frac{DW}{\mu g A} \tag{7-35}$$

is a constant, which means that the resistance coefficient f is also a constant.

The equation of continuity, Eq. 4-2, and the equation of state, Eq. 1-2 can be combined to yield

$$\rho V^2 = \frac{p_1}{p\rho_1}\frac{W^2}{g^2 A^2} \tag{7-36}$$

which can be substituted into the pressure term of Eq. 7-34, and the equation can then be arranged in integral form between sections 1 and 2 as

$$2\int_1^2 \frac{dV}{V} + \frac{2\rho_1 g^2 A^2}{p_1 W^2}\int_1^2 p\,dp + \frac{f}{D}\int_1^2 dL = 0 \tag{7-37}$$

Integration yields

$$p_1^2 - p_2^2 = \rho_1 V_1^2 p_1 \left(2\ln\frac{V_2}{V_1} + f\frac{L}{D}\right) \tag{7-38}$$

Since $\rho_1 V_1 = \rho_2 V_2$, Eq. 7-38 can be written in terms of p_1/p_2 by substitution and rearrangement as

$$p_1^2 - p_2^2 = \rho_1 V_1^2 p_1 \left(2\ln\frac{p_1}{p_2} + f\frac{L}{D}\right) \quad (Short\ pipes) \tag{7-39}$$

Equation 7-39 is a *general equation for* **isothermal flow in pipes.** If the pipe is long, the logarithmic term can be neglected, because it is small compared with the $f(L/D)$ term. Then Eq. 7-39 can be arranged as

$$p_1^2 - p_2^2 = f \frac{L}{D} \rho_1 V_1^2 p_1 \tag{7-40}$$

or

$$p_1 - p_2 = \left(\frac{2}{1 + p_2/p_1}\right) f \frac{L}{D} \frac{\rho_1 V_1^2}{2} \quad (Long\ pipes) \tag{7-41}$$

which, when divided by γ, is the same as Eq. 7-24 for incompressible fluids—except for the term involving the ratio p_2/p_1. These equations can be used to solve problems of isothermal flow in long pipes (Eq. 7-40) or short pipes (Eq. 7-39).

Example Problem 7-4: If 1 lb per sec of air at a temperature of 80°F flows in a 4-in. smooth horizontal pipe line and the pressures at points 1 and 2 are 60 and 40 psia respectively, calculate the distance between points 1 and 2.

Solution:

$$\text{Re} = \frac{DW}{\mu g A} = \frac{(4/12)(1)}{(3.85)10^{-7}(32.2)(0.097)} = 278{,}000$$

$$\gamma_1 = \frac{p_1}{RT_1} = \frac{(60)(144)}{(53.3)(540)} = 0.305 \text{ lb/ft}^3$$

$$\rho_1 = \frac{\gamma_1}{g} = 0.00934$$

$$V_1 = W/\gamma_1 A = 38.2 \text{ fps}$$

Based upon Re, and Fig. 7-6

$$f = 0.0148$$

Assuming a long pipe, Eq. 7-40

$$p_1^2 - p_2^2 = f \frac{L}{D} \rho_1 V_1^2 p_1$$

$$[(60)(144)]^2 - [(40)(144)]^2 = (0.0148) \frac{(L)}{(4/12)} (0.00934)(38.2)^2(60)(144)$$

$$L = \underline{\underline{7960 \text{ ft}}}$$

The magnitude of L justifies the use of Eq. 7-40.

Equation 7-39 can be expressed dimensionlessly in terms of the Mach number Ma which is

$$\text{Ma}_1 = \frac{V_1}{c_1} = \frac{V_1}{\sqrt{kp_1/\rho_1}} \tag{7-42}$$

The resulting dimensionless equation is

$$1 - \left(\frac{p_2}{p_1}\right)^2 = k\,\text{Ma}_1^2\left(2\ln\frac{p_1}{p_2} + f\frac{L}{D}\right) \qquad (7\text{-}43)$$

which is plotted as solid lines in Fig. 7-7 for p_2/p_1 as a function of $f(L/D)$ and the Mach number—with $k = 1.4$. The theoretical drop in pressure along the conduit with incompressible flow is represented by a dot-dash straight line for each Mach number Ma_1. Note that the assumptions of incompressible flow is most accurate for small Ma_1 values and small $f(L/D)$ values. As either of these variables increases, the deviation also increases.

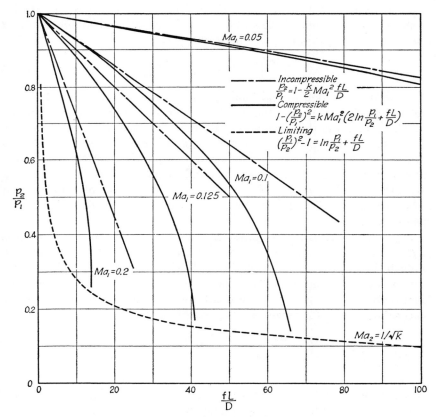

Fig. 7-7. Isothermal flow in pipes.

The continuity equation and the equation of state can be combined with Eq. 7-34 to yield

$$\frac{dp}{p}\frac{1}{k\,\text{Ma}^2} - \frac{dp}{p} + f\frac{dL}{2D} = 0$$

which can be solved for the pressure gradient as a function of Mach number

$$\frac{d(fL/D)}{dp/p} = 2\left(1 - \frac{1}{k\,\text{Ma}^2}\right) \tag{7-44}$$

Equation 7-44 shows that, when $k\,\text{Ma}^2 = 1.0$ or $\text{Ma} = 1/\sqrt{k}$, the gradient goes to zero on the locus of points represented by the broken line in Fig. 7-7. This limiting condition applies to the local Mach number at the end of the pipe. When $\text{Ma} < 1/\sqrt{k}$, the flow is subsonic and the pressure *decreases* in the downstream direction, but when $\text{Ma} > 1/\sqrt{k}$, the flow is supersonic and the pressure *increases* in the downstream direction. Evidently $\text{Ma} = 1/\sqrt{k}$ represents a limit for continuous isothermal flow.

Since $p_1/p_2 = \text{Ma}_2/\text{Ma}_1$, Eq. 7-43 can be expressed in terms of Ma_2 at the end of the pipe. When $\text{Ma}_2^2 = 1/k$ is substituted in the resulting equation, the following relationship is obtained

$$\left(\frac{p_1}{p_2}\right)^2 - 1 = \ln\frac{p_1}{p_2} + f\frac{L}{D} \tag{7-45}$$

which is the equation of the limiting curve represented in Fig. 7-7 as a broken line. For a given initial Mach number Ma_1 of the flow, the broken line also represents the minimum pressure, below which the subsonic flow cannot take place, and a maximum length of pipe (for given f and D) permissible for the given flow. If the length of pipe is increased, the flow is choked by resistance of the conduit boundary and the Mach number Ma_1 will automatically decrease. The reader should note the similarity between the curves in Fig. 7-7 and the backwater curves in Chapter 8 on flow in open channels.

Supersonic velocities in pipes are confined only to short lengths $(L/D < 50)$. Under these conditions, the resistance coefficient f has been found to be about half that for the small Mach numbers. For long pipes there is sufficient length for the velocity profile to become established, the Mach numbers are usually quite small, and the resistance coefficient f is nearly the same as for flow of incompressible fluids.

• ADIABATIC FLOW

If a pipe is insulated, the total energy in the flow remains a constant throughout its length. This type of flow has certain similarities to adiabatic flow as discussed in Chapter 4, although in reality it lies between truly adiabatic flow and isothermal flow—approaching the isothermal flow at great pressures—because the heat of resistance is retained in the pipe. With insulation, the temperature and viscosity change causes the Reynolds number and the resistance coefficient f also to change along the length of the pipe. If the actual variation of these factors were to be included in Eq. 7-34, its integration would be extremely difficult. Consequently, it is necessary to assume that f remains constant and that the flow is adiabatic. Under these conditions, the continuity equation and equation of state combine to yield

$$V = \frac{W}{A\gamma} = \frac{W}{A\gamma_1}\left(\frac{p_1}{p}\right)^{1/k} \tag{7-46}$$

Equation 7-46 can be substituted into Eq. 7-34, rearranged, and integrated as

$$f\frac{L}{D} = \frac{2}{k}\ln\frac{p_2}{p_1} - \frac{2k}{k+1}\left(\frac{g\gamma_1 A^2}{W^2 p_1^{1/k}}\right)(p_2^{(k+1)/k} - p_1^{(k+1)/k}) \tag{7-47}$$

in which variation of elevation again has been assumed negligible. Equation 7-47 can be expressed in terms of the Mach number Ma_1 at section 1

$$f\frac{L}{D} = \frac{2}{k}\ln\frac{p_2}{p_1} - \frac{2}{(k+1)\,Ma_1^2}\left[\left(\frac{p_2}{p_1}\right)^{(k+1)/k} - 1\right] \tag{7-48}$$

With Eq. 7-48 and Eq. 7-47, it is possible to solve directly for length L if f, D, Ma_1, p_2, p_1, and k are known, or directly for discharge W if f, D, L, p_2, p_1, and k are known.

Figure 7-8 is a graphical representation of Eq. 7-48 with p_2/p_1 a function of $f(L/D)$ for different Ma_1 values. As in Fig. 7-7 for isothermal flow, the straight lines represent the pressure change for the flow of an incompressible fluid, and the broken line represents the limiting condition for maximum length of pipe and minimum outlet pressure possible, if the given W is to be maintained at the initial Mach number Ma_1. If the pipe were longer, the initial conditions would be changed to a smaller Mach number. The broken line represents $Ma_2 = 1.0$ at the outlet. With subsonic flow, the pressure decreases and the velocity and Mach number increase in the downstream direction, whereas with supersonic flow (short tubes), the pressure increases and the velocity and Mach number decrease in the downstream direction. In both cases the flow goes toward $Ma = 1.0$ at the outlet.

If the ambient pressure at the outlet is greater than that indicated in Fig. 7-8, the initial Ma_1 will be less than indicated. Reducing the pressure at the outlet will increase Ma_1 only to the point of $Ma = 1.0$ at the outlet. If the pipe is too long for reduction of Mach numbers to just $Ma = 1.0$ at the outlet, a shock wave will form in the pipe, downstream from which the flow will be subsonic and $Ma = 1.0$ at the outlet. Again, there is considerable similarity to the backwater curves and the hydraulic jump, as discussed in Chapter 8.

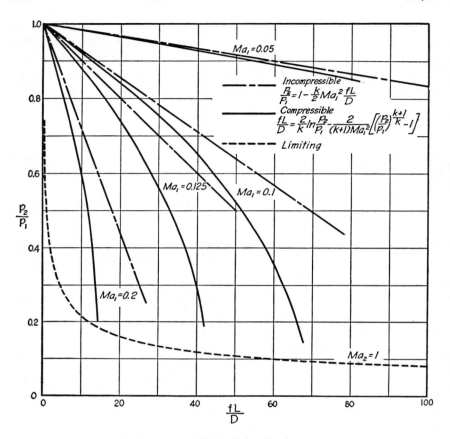

Fig. 7-8. Adiabatic flow in pipes.

• STEADY, NONUNIFORM FLOW

Nonuniform flow exists when either the magnitude or direction of the velocity varies with distance along a streamline. Tangential acceleration occurs if the velocity is changed in magnitude, and normal acceleration occurs if the velocity is changed in direction.

These changes in velocity result in a change in momentum flux, a vector, and the change in momentum flux is accomplished only by pressures against the fluid that are in addition to the pressures associated with uniform flow. When such changes in velocity occur, zones of separation and secondary flow frequently result, and this increases the shear and the turbulence at the expense of the piezometric head. Hence, a head loss h_L results. Since the foregoing changes in velocity and the resulting head losses are caused by nonuniform distribution of pressures on the boundary, the losses are termed *form losses due to pressure resistance and associated increases in shear resistance.*

In Eq. 7-5, the loss is expressed as a factor $f(L/D)$ multiplied by the velocity head $V^2/2g$. Thus, $f(L/D)$ may be thought of as the number of velocity heads which are lost due to shear resistance. Form losses h_L can also be expressed as C_L times the velocity head $V^2/2g$, where C_L can be thought of as the number of velocity heads lost due to the form of the conduit—a form loss. That is

$$h_L = C_L \frac{V^2}{2g} \qquad (7\text{-}49)$$

in which C_L is called the form loss coefficient. It now remains to determine the magnitude and variation of the loss coefficient C_L for various conduit boundary forms which cause changes in either the magnitude or the direction of the velocity. Like the values of f for boundary resistance, almost all values of C_L must be determined by experimentation rather than by theoretical analysis. Although shear resistance has been found to be dependent entirely upon the Reynolds number and relative roughness, pressure resistance, which is primarily responsible for form loss, is dependent upon Reynolds number and roughness only to a minor extent, the shape or form of the boundary being the major factor.

These form losses are sometimes called "minor" losses. Such a term represents the true situation literally when the pipe line is relatively long and the shear loss coefficient $f(L/D)$ is large by comparison with C_L. For shorter pipe lines, however, the form losses caused by pressure resistance may be of major importance instead of minor importance. The so-called minor losses are summarized in Tables 7-4, 7-5, and 7-6, and a detailed discussion is given of each of these losses.

Table 7-4. LOSS COEFFICIENTS FOR MINOR LOSSES

Bends......................	See Figs. 7–10, 7–11, and 7–12
Sudden expansions...........	$C_L = \left(1 - \dfrac{A_1}{A_2}\right)^2$ or Fig. 7–20
Gradual Expansions..........	See Fig. 7–16
Laterals....................	See Fig. 7–18
Sudden Contractions.........	$C_L = \left(\dfrac{1}{C_c} - 1\right)^2$ or Fig. 7–20
Entrances...................	See Fig. 7–21

Table 7-5. Loss Coefficients for Commercial Pipe Fittings

Fitting	C_L
Globe valve, fully open	10
Angle valve, fully open	5
Swing check valve, fully open	2.5
Closed return bend	2.2
Tee, through side outlet	1.8
Short radius elbow	0.9
Medium radius elbow	0.8
Long radius elbow	0.6
45-degree elbow	0.4
Gate valve, fully open	0.2
Gate valve, 3/4 open	1
Gate valve, 1/2 open	5.6
Gate valve, 1/4 open	24

• CONDUIT BENDS

When the flow in a pipe encounters a bend, the outside wall of the bend must exert an additional pressure against the flow in order to turn it. On the inside of the bend, the pressure is reduced to less than the pressure without a bend. In fact, the flow may separate so that a separation zone forms in the vicinity of A, see Fig. 7-9. The total force acting against the

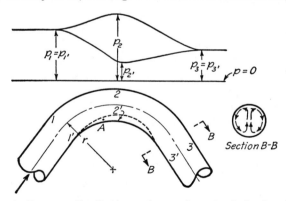

Fig. 7-9. Pressure distribution and secondary circulation in a bend.

flow in excess of the pressure without a bend is equal to the change in momentum flux in that direction. A typical pressure distribution for a bend is shown in Fig. 7-9. Because the pressure difference between the inside and the outside of the bend is so great, the bend or elbow has been used successfully as a flow measuring device by installing piezometers on the inside and outside of the bend. This idea is presented in greater detail in the chapter on Flow Measurement.

The flow pattern of secondary circulation set up in a bend is shown in section B-B of Fig. 7-9. Upstream from the bend, the fluid in the central part of the pipe has a greater velocity than that near the walls of the pipe. As the flow enters the bend, this high-velocity core has a greater momentum and, hence, is more difficult to turn than the slower moving fluid surrounding the core. Consequently, the streamlines of the high-velocity core turn on a larger radius than the bend and the core strikes the outside of the bend as the slower fluid is forced along the walls to the inside, see section *B–B* of Fig. 7-9. This double spiral of secondary flow carries downstream for a distance as great as 50 to 100 diameters before it is finally damped and normal flow is again established. This circular motion causes increased boundary shear and, hence, the gradient of piezometric head dh/dx is increased until established flow is again obtained.

Another source of head loss in the bend is the separation zone which occurs and the resulting increase in turbulence which is created at the expense of the piezometric head. The separation zone also decreases the cross-sectional area of the flow, which forces a local increase in velocity—again at the expense of the piezometric head.

From the foregoing discussion, it is evident that, in general, if either the strength of the secondary circulation or the size of the separation zone

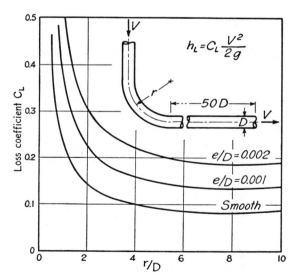

Fig. 7-10. Loss coefficients for 90-degree pipe bends.

are decreased, the head loss will be decreased. Likewise, if either of these are increased, the head loss will increase. Figure 7-10 shows the variation of the head loss coefficient C_L with the relative radius of curvature r/D of the bend and relative roughness e/D of the pipe walls. This plot shows that the coefficient C_L is decreased only slightly when r/D values are greater than about 6, whereas C_L increases very rapidly as the r/D values are reduced to about 2 and less.

As the radius of curvature r/D goes to zero, the bend becomes a miter bend, as shown in Fig. 7-11(a). Under these conditions, the separation zones which occur are a maximum in size and the local increase in velocity is also a maximum, so that the loss coefficient C_L becomes as large as 1.2.

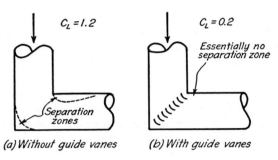

(a) Without guide vanes   *(b) With guide vanes*

Fig. 7-11. Miter bends.

Pipe bends of less than 90 degrees have C_L values which are also less, as shown in Fig. 7-12. However, the angle α must be less than 45 degrees to reduce C_L appreciably. For the larger r/D values, this is due at least in part to the fact that the loss caused by separation is negligible and, once the spiraling secondary flow is established, the increase in loss with α is quite gradual—so that at $\alpha = 180$ degrees C_L is only about 50 percent greater than for the 90-degree bend. As the r/D values become smaller, however, that part of the loss caused by separation is greatly increased. In fact, for the miter bend where $\alpha = 0$, the loss is caused principally by separation.

The loss in miter bends can be reduced greatly by the use of guide vanes, which guide the fluid around the bend, as shown in Fig. 7-11(b), because they practically eliminate both the separation zone and the secondary circulation. Guide vanes reduce C_L from 1.2 to about 0.2, which is in the same order of magnitude as the loss in a bend where $r/D = 2.0$. The mitered bend with guide vanes, however, takes much less space and usually is less expensive to make.

Since the ratio of the centerline velocity v_{max} to the mean velocity V for laminar flow in pipes is so much greater than for turbulent flow, the loss coefficient for laminar flow in slight bends is greater than for turbulent flow. This is because the higher-velocity core of laminar flow causes more secondary circulation. For sharp bends, however, C_L is less for laminar flow than

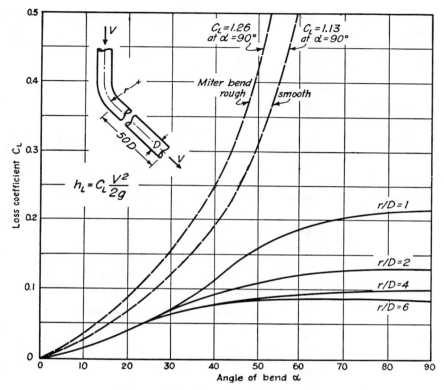

Fig. 7-12. Loss coefficient for pipe bends of 90-degrees and less.

for turbulent flow. In Fig. 7-10, for turbulent flow, the increased C_L values for increased roughness may be caused by this same effect of the v_{max}/V ratio.

The secondary flow, and hence the loss coefficient, is reduced appreciably in rectangular ducts by making them several times as wide as deep. The losses obtained by Wirt are given in the following table, where B and D are the width and depth of the duct, and r is the radius of curvature. A similar spreading of the flow is used in the design of draft tubes for hydro-power plants.

Table 7-6. Loss Coefficients C_L for Bends in Rectangular Ducts

	$B/D = 6$			$B/D = 3$		
r/D	5/3	1	2/3	5/3	1	2/3
C_L	0.09	0.16	0.38	0.15	0.22	0.55

The influence of Reynolds number on C_L has been found to be small compared to the other influences discussed in the foregoing paragraphs.

Loss coefficients due to standard threaded pipe bends or elbows are listed in Table 7-5 for pipe fittings.

• CONDUIT EXPANSIONS

Whereas the study of flow in a bend is concerned chiefly with changing the direction of the velocity, the study of flow in an expansion is concerned chiefly with changing the *magnitude* of the velocity. In either case, however, the momentum flux is changed and there are corresponding forces associated with the phenomenon.

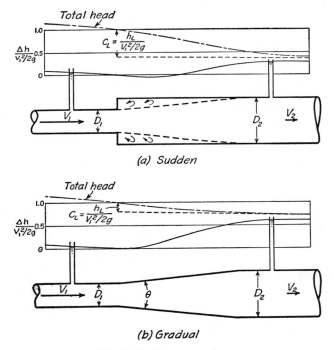

(a) Sudden

(b) Gradual

Fig. 7-13. Pipe expansions.

Flow in a sudden expansion is shown in Fig. 7-13(a), where the jet downstream is diffused transversely until it fills completely the downstream pipe. Surrounding the jet is a separation zone with the associated turbulence and energy loss, as reflected by the rather sudden decrease in total head.

By means of the continuity, momentum, and energy equations, and the assumption of uniform pressure across any given section, it is possible to solve theoretically for a loss coefficient for a sudden expansion. As the flow expands in the downstream direction, the pressure increases to supply the retarding force necessary to decrease the momentum flux. On the basis of the force-momentum flux equation, the force plus momentum flux at section 1 must equal the force plus momentum flux at section 2. An unbalanced free-body diagram, Fig. 7-14, is used to illustrate the application of this equation to the problem of a sudden expansion. The average pressure at section 1 is $\bar{p}_1$ and the average pressure at section 2 is $\bar{p}_2$. The

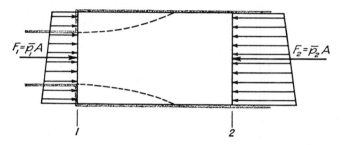

Fig. 7-14. Pressure diagram for a sudden expansion.

momentum flux at the two sections is $\rho Q V_1$ to the right and $\rho Q V_2$ also to the right. In equation form this now becomes

$$F_1 + K_m \rho Q V_1 = F_2 + K_m \rho Q V_2 \tag{7-50}$$

Substituting $F = \bar{p}A$, dividing by γ, and transposing terms

$$\frac{A(p_2 - p_1)}{\gamma} = K_m Q \frac{(V_1 - V_2)}{g} \tag{7-51}$$

$$\frac{\pi D_2^2}{4} \frac{(p_2 - p_1)}{\gamma} = \frac{K_m Q (V_1 - V_2)}{g} \tag{7-52}$$

$$\frac{p_2 - p_1}{\gamma} = K_m \frac{V_2}{g} (V_1 - V_2) \tag{7-53}$$

in which K_m is the momentum coefficient to correct for the nonuniform velocity distribution.

The Bernoulli or energy equation is

$$K_{e1} \frac{V_1^2}{2g} + \frac{p_1}{\gamma} + z_1 = K_{e2} \frac{V_2^2}{2g} + \frac{p_2}{\gamma} + z_2 + H_L \tag{7-54}$$

in which K_e is the energy coefficient to correct for a nonuniform velocity distribution.

Equation 7-54 can be combined with Eq. 7-53 to eliminate p_1 and p_2, if z_1 and z_2 are assumed to be equal, to obtain

$$C_L = \frac{h_L}{V_1^2/2g} = K_e - 2K_m \left(\frac{A_1}{A_2}\right) + (2K_m - K_e) \left(\frac{A_1}{A_2}\right)^2 \tag{7-55}$$

in which C_L is the loss coefficient. Assuming a velocity distribution following the $\frac{1}{7}$-power law, Cermak has shown that $K_m = 1.022$ and $K_e = 1.058$. When those values are substituted into Eq. 7-55

$$C_L = 1.058 - 2.044 \left(\frac{A_1}{A_2}\right) + 0.986 \left(\frac{A_1}{A_2}\right)^2 \tag{7-56}$$

and if $K_m = 1.0 = K_e$ is assumed,

$$C_L = \left(1 - \frac{A_1}{A_2}\right)^2 \tag{7-57}$$

Since no assumptions have been made regarding the shape of cross section, it is understandable that experimental results indicate the area ratio A_1/A_2 can be used with reasonable accuracy for noncircular expansions.

Both Eq. 7-56 and Eq. 7-57 are plotted in Fig. 7-16, together with the curves from experimental data for various angles of expansion θ. In the derivation of these equations, no assumptions have been made regarding the angle or the shape of the cross section. Instead, it has been assumed that the conversion of velocity head to piezometric head in the expansion is represented entirely by the momentum relationship, which requires that an increase in pressure downstream from the expansion is necessary to reduce the momentum of the flow from the smaller conduit upstream. Laboratory experiments have shown that the theoretical equations represent the experimental curves particularly well for $\theta > 60$ deg.

However, for gradual expansions, see Fig. 7-13(b), the angle plays a very important role, as shown in Fig. 7-16. As θ becomes smaller than 60 deg, there is increasing recovery of velocity head, because of reduced separation along the walls of the expansion.

When θ approaches an angle of zero, the resistance to flow is due entirely to shear resistance, and Eq. 7-5 is applicable. For the other extreme, a sudden expansion of $\theta \geq 60$ deg, the loss expressed in Eq. 7-56 is due entirely to pressure resistance. A transition from one type of resistance to another can be explained as follows.

For small angles of θ, the velocity gradient dv/dy at the walls of the expansion is positive, so that the shear against the boundary is positive and to the right, as shown in Figs. 7-13 and 7-15. Although this condition prevails as the angle θ increases, the velocity dv/dy at the boundary decreases with a corresponding decrease in boundary shear τ_0 until the angle of θ is reached where $dv/dy = 0$ and $\tau_0 = 0$, see Fig. 7-15(b). A further increase in the angle θ causes dv/dy to be reversed and become negative, so that the shear against the boundary is to the left, as shown in Fig. 7-15(c). With this type of flow, the boundary is expanding more rapidly than the fluid can follow, and a separation zone is formed.

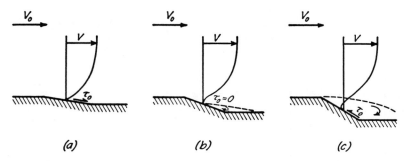

Fig. 7-15. Velocity distribution, shear, and separations in a gradual expansion.

As discussed in the previous chapter, the initial formation of a separation zone is influenced to a considerable extent by the Reynolds number and the relative roughness, because of their influence on the velocity profile in the approach section upstream from the boundary divergence or expansion. If the velocity profile is relatively uniform, the angle θ may be considerably larger before separation occurs than for a velocity profile which is markedly nonuniform. In general, C_L has been found to decrease slightly with increasing Reynolds number, because the larger Re, the more uniform the velocity distribution.

The pronounced influence of Reynolds number is evident in Fig. 7-16. In fact, at Re $= 10^4$ the experimental curve most nearly fits Eq. 7-56, while at Re $= 1.5 \times 10^5$ it most nearly fits Eq. 7-57. The cause of this variation is the difference in the separation zone due to the changing laminar sublayer and the distribution of velocity across the approaching flow. At larger Reynolds numbers, the velocity distribution is more nearly uniform and the thickness of the laminar sublayer is reduced. Consequently, the momentum flux near the boundary is increased, and the size of the separation zone is decreased. With a smaller separation zone, the conversion of the velocity head to piezometric head is more complete, K_m and K_e approach 1.0, and C_L is less.

A special method which is used to insure high-velocity flow at the boundary is to induce spiral flow. Centrifugal force keeps the high-velocity

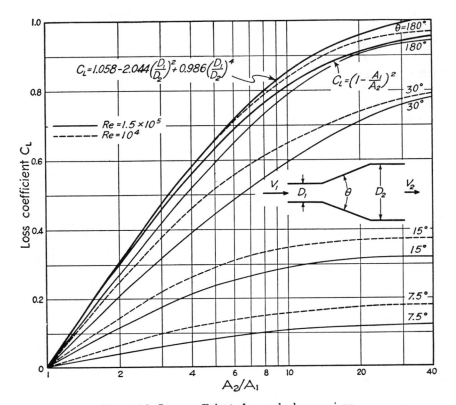

Fig. 7-16. Loss coefficients for gradual expansions.

flow at the outside and also helps the flow to follow the expanding boundary. Consequently, the efficiency is increased.

The length of the expansion, as reflected in the area ratio A_2/A_1, is also significant for angles of θ between about 20 and 120 deg. This, like the Reynolds number and the relative roughness, apparently is due to the effect on the initiation of and nature of the separation zone. Further research is necessary, however, before this phenomenon can be explained completely.

In an effort to reduce the loss in an expansion, trumpet-shaped expansions have been tested by Gibson for the conditions of $dv/dt = $ constant, and $d(V^2)/dx = $ constant. The latter design, where the rate of change of velocity head with respect to distance remained constant, was found to be 20 to 60 percent more efficient than the gradual expansion with straight sides—the reduction being greatest for the shortest expansions.

Guide vanes are also very effective in reducing the loss in gradual expansions.

• CONDUIT LATERALS

If several laterals are supplied from a main pipe line, the discharge of the main line decreases at each lateral by the amount of flow supplied to the lateral. If the main pipe has a constant cross section, the streamlines expand at each lateral, as the discharge and velocity are reduced and nonuniform flow results. Hence, along those expanding streamlines which continue downstream, the energy equation shows that an *increase in pressure head* will occur. On the other hand, there are losses along the lateral (due to form resistance and boundary resistance) which will cause a *decrease in pressure head.*

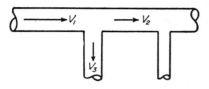

Fig. 7-17. Pipe laterals.

These two conditions which cause a change in piezometric head tend to counteract each other so that under some conditions the net change in piezometric head will be an increase, and under other conditions the piezometric head will decrease in the direction of flow.

The flow with laterals cannot be analysed exactly by either the energy equation or the momentum equation. However, if boundary resistance is negligible, the flow is very similar in some respects to a sudden expansion. Therefore, the loss h_L caused by a lateral can be written

$$h_L = C_L \frac{(V_1 - V_2)^2}{2g} \qquad (7\text{-}58)$$

The light lines in Fig. 7-18 represent Eq. 7-58 and the heavy lines the experimental data. Comparison shows that, for small discharges in the lateral, there is an apparent gain in energy. This is due to the tendency of the lateral to draw off the low-velocity fluid around the boundary of the pipe and leave the higher-velocity fluid to progress on downstream. The loss of head due to boundary resistance in the main line is steadily decreasing in the downstream direction, since the velocity is decreasing. Although fully-established flow may not occur when the laterals are close together, Eq. 7-5 can be written in differential form as

$$dh_f = f \frac{V^2}{2g} \frac{dx}{D} \qquad (7\text{-}59)$$

which can be integrated if the variation of f, V, and D with x is known.

If the pipe diameter D and the resistance coefficient f do not vary with x, and the fluid can be assumed to leave the main line uniformly, then

$$\frac{dv}{dx} = \frac{V_2 - V_1}{L} \qquad (7\text{-}60)$$

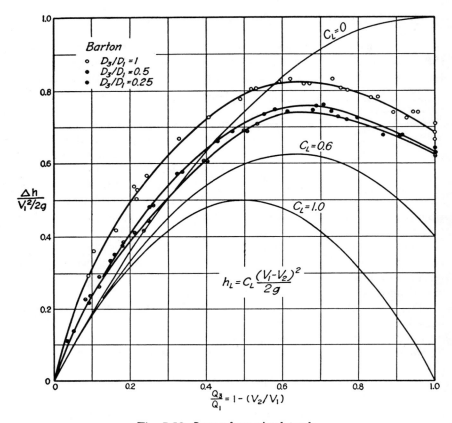

Fig. 7-18. Losses from pipe laterals.

in which L is the length between points 1 and 2. Now Eq. 7-59 can be integrated as

$$h_f = \frac{1}{3}\frac{fL}{D2g}\left(\frac{V_1^3 - V_2^3}{V_1 - V_2}\right) \tag{7-61}$$

From this equation, it can be seen that, if there is zero velocity at the downstream end so that $V_2 = 0$, then Eq. 7-61 becomes

$$h_f = \frac{1}{3}f\frac{L}{D}\frac{V_1^2}{2g} \tag{7-62}$$

which shows that, under these conditions, the loss h_f is one-third the loss which would occur in the same pipe without the lateral flow.

If the variation of f, V, and D are too complex to integrate Eq. 7-59, then graphical integration may be necessary.

• CONDUIT CONTRACTIONS

The flow in a reducer results in a separation zone and ***vena contracta*** downstream from the contraction, as shown in Fig. 7-19, which is quite similar to the flow in Fig. 3-21(e) for an orifice, as described in Chapter 3. This type of flow pattern again illustrates the use of the continuity, momentum, and energy equations. The continuity equation applies at all three sections. Between section 1 and the contraction c, the energy or Bernoulli equation is applicable with virtually no head loss, because the

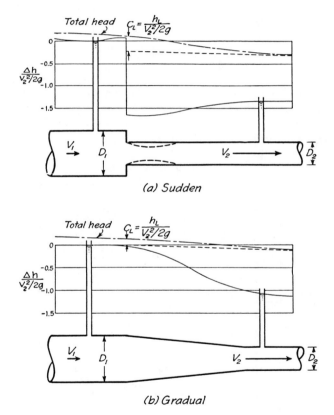

(a) Sudden

(b) Gradual

Fig. 7-19. Pipe contractions.

streamlines are converging from section 1 to the vena contracta at section c. Between the vena contracta and section 2, however, the flow is expanding in much the same way as it does for the expansions shown in Fig. 7-13. Therefore, the momentum equation is applicable between the vena contracta and section 2.

The foregoing discussion can be expressed in equation form by applying the continuity equation written as

$$C_c A_2 V_c = A_2 V_2 \tag{7-63}$$

in which C_c is the coefficient of contraction $C_c = A_c/A_2$;

A_c is the area of the contracted jet at the vena contracta;

V_c is the velocity of the jet at the vena contracta.

The head loss h_L then becomes

$$h_L = \frac{(V_c - V_2)^2}{2g} = \left(\frac{V_c}{V_2} - 1\right)^2 \frac{V_2^2}{2g} \tag{7-64}$$

$$= \left(\frac{1}{C_c} - 1\right)^2 \frac{V_2^2}{2g} = C_L \frac{V_2^2}{2g} \tag{7-65}$$

Using the data of Weisbach, Eq. 7-65 can be plotted as shown in Fig. 7-20. This figure shows that, when A_1 is very great and $A_2/A_1 \to 0$, the loss coefficient C_L is 0.5, as it is for the orifice shown in Fig. 3-21(e). In other words, half a velocity head is lost.

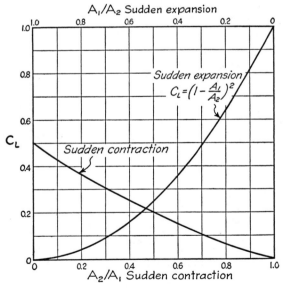

Fig. 7-20. Loss coefficients for sudden expansions and sudden contractions.

Gradual contractions, like gradual expansions, reduce the size of the separation zones, and hence reduce the head loss. This can be seen by comparing Fig. 7-19(a) with Fig. 7-19(b). In the sudden contraction, the piezometric head line dips markedly and, although there is considerable recovery, as shown by the momentum principle, it does not rise to the level of the piezometric head line for the gradual contraction, which has dropped gradually with no marked dip. In both the expansions and contractions, the head lost in excess of the standard boundary resistance is because of the separation zones and the resulting diffusion of a jet through shear and turbulent mixing. Hence, the loss takes place downstream from the sudden change in section. In fact, for the sudden contraction the loss is principally downstream from the vena contracta where the jet is expanding through turbulent mixing.

• CONDUIT ENTRANCES

Various shapes of pipe entrances are considered, from the viewpoint of the distribution of velocity head, piezometric head, separation zones, and head losses. There are three basic shapes of entrances to pipes, as shown in Fig. 7-21: the rounded entrance, the square entrance and the re-entrant entrance. The rounded entrance permits the flow to contract gradually and without separation zones. Hence, the loss coefficient C_L is very small, 0.05 or less.

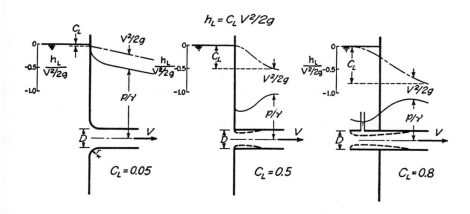

$$h_L = C_L V^2/2g$$

(a) Rounded entrance   (b) Square entrance   (c) Re-entrant entrance

Fig. 7-21. Pipe entrances.

As the radius of curvature of the rounded entrance is decreased to zero, a separation zone always appears, the area of flow is decreased, and the entrance eventually becomes a square-edged entrance. At the same time,

C_L increases to 0.5. In this case, there is the separation zone surrounding the contracted jet and the expanding jet downstream, where the head loss occurs as it does for the sudden contraction. In fact, this is the limiting case for a sudden contraction where $A_2/A_1 = 0$.

The re-entrant entrance is even less streamlined than the square entrance, and hence the separation zone surrounding the jet is larger, the jet is smaller, and the loss h_L is larger. In this case, C_L is 0.8—that is, 0.8 of a velocity head is lost.

• CONDUIT OUTLETS

At the outlet of a pipe, the entire velocity head of exit is lost—which is the limiting case for a sudden enlargement, where $A_1/A_2 = 0$. Therefore, if this loss is to be reduced, the exit velocity head must be decreased. This can be done by the use of a gradual expansion, as shown in Fig. 7-22(b). Although $C_L = 1.0$ in each case, the exit velocity is reduced appreciably in Fig. 7-22(b)—especially the velocity head, since this varies with the square

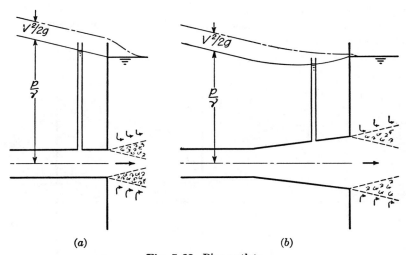

(a)   (b)

Fig. 7-22. Pipe outlets.

of the velocity. For example, if the diameter is doubled in the expanding outlet, the area will be increased four times and the velocity head will be one-sixteenth as great as the velocity head without the expansion. This principle is used successfully in a specialized way with draft tubes for hydro-power plants.

• PIPE FITTINGS

Each of the foregoing shapes, bends, expansions, contractions, entrances, and outlets of conduit transitions and accompanying flow systems, is quite simple and easily analyzed compared with actual fittings found in pipe line work. Such fittings have shapes convenient principally for their manufacture and use. The loss coefficients for these are given in Table 7-5.

In these fittings, there are complex combinations of separation zones and secondary circulation, like the individual cases already presented, which cause the head loss. Globe valves, even fully open, have a very tortuous path, with contractions and expansions for the fluid to follow— hence, $C_L = 10$. Gate valves fully open, on the other hand, have a clear passage for the fluid which is about the same size and shape as the pipe, but with recesses which cause some loss—hence, $C_L = 0.2$. As the gate valve is closed, the fluid passage is blocked due to the gate barrier moving directly across the flow, a jet of increasing velocity issues under the gate and expands downstream somewhat like a sudden expansion, and C_L increases to very great values. When any valve is closed completely, C_L theoretically goes to infinity—i.e., the flow is stopped and C_L no longer has any meaning.

• EQUIVALENT LENGTH FOR FORM LOSSES

For ease in computation, the concept of equivalent length for form losses has been developed. This approach simply states that the losses caused by form or shape of passages may be expressed in terms of an equivalent loss due to the boundary or shear resistance caused by an "equivalent" length of pipe of the same diameter. In other words C_L is set equal to $f(L/D)$ and the equivalent length may be found as

$$L = \frac{C_L}{f} D \qquad (7\text{-}66)$$

If f is assumed to be an average value of 0.02, it may be seen that the equivalent length for a gate valve fully open would be $(0.2/0.02)D = 10D$, and for a globe valve fully open it would be $(10/0.02)D = 500D$. Therefore, the equivalent lengths of a gate valve and globe valve respectively are about $10D$ and $500D$. If long pipe lines are under consideration, these losses become "minor" and may be neglected. For example, if a one-percent error is permissible, the pipe line must be $100(10D) = 1000D$ for the gate valve, and $100(500D) = 50,000D$ for the globe valve, before the valve loss may be neglected or ignored.

• COMPOUND PIPELINES

The principles presented in the foregoing sections of this chapter may be used in combination to solve problems involving compound pipe lines. Figure 7-23 is an example of a compound pipe line which involves an en-

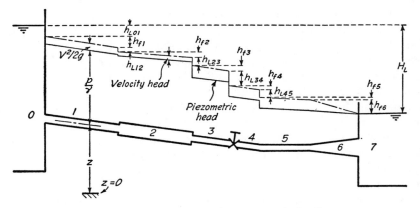

Fig. 7-23. Energy diagram for compound pipe lines.

trance, a sudden expansion, a sudden contraction, a valve, a bend, a gradual expansion, an outlet, and flow in pipes of different diameters. Each of these items involves a head loss. The straight pipe involves shear resistance, and each of the others involves both shear and pressure resistance to make up the form losses.

Bernoulli's equation may be written for any reach of pipe as

$$\frac{V_a^2}{2g} + \frac{p_a}{\gamma} + z_a = \frac{V_b^2}{2g} + \frac{p_b}{\gamma} + z_b + H_L \tag{7-67}$$

in which H_L includes both shear losses and form losses between sections a and b. If the upstream reservoir is chosen as a and the downstream reservoir as b, then H_L is the sum of all the losses indicated in Fig. 7-23. In other words,

H_L	$=$	h_{L01}	$+$	h_{f1}	$+$	h_{L12}	$+$	h_{f2}	$+$	h_{L23}	$+$	h_{f3}
Total loss		Entrance loss		Pipe loss		Expansion loss		Pipe loss		Contraction loss		Pipe loss

	$+$	h_{L34}	$+$	h_{f4}	$+$	h_{L45}	$+$	h_{f5}	$+$	h_{L57}	$+$	h_{L67}
		Valve loss		Pipe loss		Bend loss		Pipe loss		Gradual expansion loss		Exit loss

which are the entrance loss, the sudden expansion loss, the sudden contraction loss, the valve loss, the bend loss, the gradual expansion loss, the exit loss, and the straight pipe losses for sections 1, 2, 3, 4, and 5. Each of these losses must be determined in accordance with the methods already developed, and then added together to get H_L. These losses have been determined for conditions of uniform flow upstream and downstream for a considerable distance. If one loss is close to another (say closer than $50D$), the loss of the two in combination is frequently less than the sum of the two losses individually. Pumps and turbines may also be added to the system, as in Chapter 3.

Example Problem 7-5: Referring to Fig. 7-23 assume that at the end of pipe 2 the water discharges into the atmosphere. If the elevation of the water surface in the reservoir is 100 ft, the elevation of pipe 2 at the outfall is 10 ft, pipes 1 and 2 have diameters of 12 in. and 24 in. and lengths of 500 ft and 800 ft respectively, compute the discharge assuming a value of $f = 0.02$ for both pipes.

Solution:

Write the Bernoulli equation between the water surface in the reservoir and the free outfall.

$$\frac{V_0^2}{2g} + \frac{P_0}{\gamma} + z_0 = \frac{V_2^2}{2g} + \frac{P_2}{\gamma} + z_2 + H_L$$

or

$$0 + 0 + 100 = \frac{V_2^2}{2g} + 0 + 10 + H_L$$

and

$$90 = H_L + \frac{V_2^2}{2g}$$

$$H_L = h_{L01} + h_{f1} + h_{L12} + h_{f2}$$

$$H_L = C_{L1}\frac{V_1^2}{2g} + \frac{f_1 L_1 V_1^2}{D_1 2g} + C_{L2}\frac{V_1^2}{2g} + \frac{f_2 L_2 V_2^2}{D_2 2g}$$

From Fig. 7-21 and Fig. 7-20

$$C_{L1} = 0.50$$

$$C_{L2} = 0.56$$

$$H_L = 0.50\frac{V_1^2}{2g} + \frac{0.02(500)V_1^2}{(1)2g} + 0.56\frac{V_2^2}{2g} + \frac{0.02(800)V_2^2}{(2)2g}$$

Using the continuity equation

$$V_1 A_1 = V_2 A_2$$

from which

$$\frac{V_1^2}{2g} = \left(\frac{A_2}{A_1}\right)^2 \frac{V_2^2}{2g} = \frac{16V_2^2}{2g}$$

then

$$90 = \frac{V_2^2}{2g} + (0.50)(16)\frac{V_2^2}{2g} + (0.02)(500)(16)\frac{V_2^2}{2g}$$

$$+ 8.95\frac{V_2^2}{2g} + \frac{(0.02)(800)}{(2)}\frac{V_2^2}{2g}$$

$$90 = \frac{V_2^2}{2g}(1 + 8 + 160 + 8.95 + 8)$$

$$\frac{V_2^2}{2g} = \frac{90}{185.95} = 0.484$$

$$V_2 = 5.58 \text{ fps}$$

$$Q = (5.58)(3.14) = \underline{\underline{17.5 \text{ cfs}}}$$

• PIPE SIZE OF MINIMUM COST

In the design of a pipe line, consideration must be given to the initial cost of the pipe and pumping equipment, the interest on the initial investment, the life of the pipe, and the salvage value of the pipe. From this information, the annual cost of the pipe and pumping equipment may be determined. Consideration must be given also to the operating expenses for pumping and maintenance. The annual cost of the investment will be increased if the pipe size is increased. However, the annual expense of operting will decrease as the pipe size is increased, because of the reduced head loss. It is the combination of these two costs that determines the total cost of a pipe line. As shown in Fig. 7-24, there is a minimum cost which determines the pipe size that is least expensive.

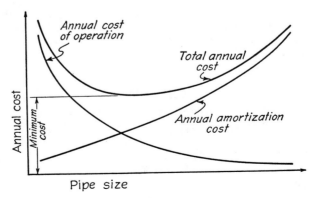

Fig. 7-24. Determination of pipe size for minimum cost.

• DIVIDED FLOW IN PIPES

Frequently, problems of divided flow in pipe lines are encountered. These problems include looping pipes, branching pipes, lateral pipes, and pipe networks; and they may be associated with additions to old systems or with the design of entirely new systems. In any case, however, the form losses are usually so small they are considered minor, and hence ignored. By means of this simplification, and careful consideration of the continuity equation and the piezometric head line, the problems may be solved. Some solutions are direct and others are indirect, using various methods of trial-and-error.

• LOOPING PIPES

The problem of looping pipes is a simple and common one. It frequently arises in connection with increasing the capacity of a pipe line. Figure 7-25 shows a schematic representation of a looping pipe system.

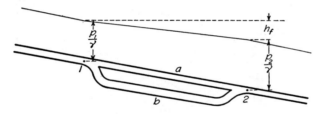

Fig. 7-25. Looping pipes.

The continuity equation for the looping system of Fig. 7-25 is

$$Q_1 = Q_a + Q_b = Q_2$$

The energy equation written from point 1 to point 2 is

$$\frac{V_1^2}{2g} + \frac{p_1}{\gamma} + z_1 = \frac{V_2^2}{2g} + \frac{p_2}{\gamma} + z_2 + h_f$$

Since the energy equation deals with the energy per unit weight of fluid, the total head (that is, the total energy per unit weight) at point 1 and at point 2 does not depend upon whether the unit weight of fluid goes through pipe *a* or through pipe *b*. Therefore, each unit weight of fluid going from 1 to 2 loses the same amount of energy, irrespective of the path it travels. For this reason, the following equation applies to the head loss in the energy equation

$$h_f = f_a \frac{L_a}{D_a} \frac{V_a^2}{2g} = f_b \frac{L_b}{D_b} \frac{V_b^2}{2g}$$

and either pipe (not both at the same time) can be used to determine the head loss from 1 to 2.

- ## BRANCHING PIPES

Pipe systems which branch, but do not rejoin as the looping pipes do, differ from the looping pipes in that the piezometric heads at the downstream ends of the branches may not be equal. Figure 7-26 illustrates a branching pipe system. From this figure, it may be seen that the flow into the junction must equal the flow out of the junction. Furthermore, the

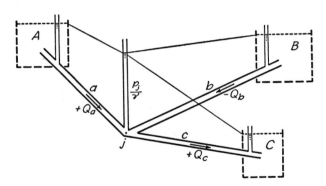

Fig. 7-26. Branching pipes.

piezometric head at the junction is common for all three pipes. The three piezometric readings at A, B, and C may be considered as the water surface elevations in three reservoirs, as shown by broken lines, since the velocity head is considered as insignificant in these problems when compared with the head losses due to boundary resistance.

There are three different flow conditions for the continuity equation, any one of which may be applicable for a given problem. Each flow system depends upon the slope of the hydraulic grade line, as follows:

1. Flow from pipe a into pipe b and pipe c, so that the piezometric line for pipe b slopes downward to the right and $Q_a = Q_b + Q_c$.
2. Flow from pipes a and b into pipe c, so that the piezometric head line for pipe b slopes downward to the left and $Q_a + Q_b = Q_c$.

3. Flow from pipe a into pipe c, with no flow in pipe b, so that the piezometric head line for pipe b is horizontal and $Q_a = Q_c$ while $Q_b = 0$.

The solution of a specific problem will depend not only upon the continuity equation and the piezometric head lines, but also upon the variables which are known and those which are unknown. The variables to be considered are:

1. Pipe sizes and lengths.
2. The piezometric head at the ends of each of the pipes, which includes the elevation of the piezometric head lines at the junction and the water surface elevations in each reservoir.
3. The discharges Q_a, Q_b, and Q_c in the three lines.

If the pipe sizes and lengths, the flow in one line, and the water surface elevations in two reservoirs are known, the solution to the problem is direct, by the use of the pipe resistance diagram to determine the piezometric head at the junction. The solution of most other problems is indirect, by assuming values for the discharge or the piezometric head at the junction. When the water surface elevation in each of the three reservoirs is given, together with the pipe sizes and lengths, the piezometric head at the junction can be assumed as equal to the water surface in reservoir B to determine whether flow is into or out of reservoir B. New assumptions can then be made to complete the solution. It is helpful to plot each computed discharge combination as in Fig. 7-27, so as to interpolate or extrapolate, to determine the values of Q. In Fig. 7-27, Q_b is negative when flow in b is toward the junction.

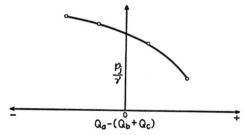

Fig. 7-27. Distribution of discharge in branching pipes.

• PIPE NETWORKS

Pipe lines are frequently branching and looping in a complex network fashion. This is particularly true in connection with city water distribution systems. Due to the complexity of these problems, the controlled trial-and-error solutions proposed for a simple looping or branching system are difficult and time-consuming to use. Perhaps the simplest solution and the one most widely used is that developed by Cross. The Hardy Cross solution provides systematically for a correction to be made in the assumed flow or assumed head loss. Although it is not possible to discuss it in detail, the basic elements of the Cross solution are given.

In the Cross solution of pipe networks, the three conditions for looping pipes must be satisfied:

1. The flow into a single junction must equal the discharge out of this junction.

2. The sum of the head losses around one branch of each circuit must equal the sum of the head losses around the other branch of the same circuit.

3. The correct relation between head loss and discharge in each pipe must exist at all times.

To simplify the computations, the head loss h_f due to boundary resistance in a given pipe line of the network is written in the form

$$h_f = KQ^n \qquad (7\text{-}70)$$

in which K depends upon the fluid viscosity and the pipe size, length, and roughness. When the pipe is rough, $n = 2$, and when it is smooth, $n < 2$—being $n = 1.85$ for the Hazen Williams formula. If the form losses are significant, they should be converted to an equivalent length of pipe and added to the length of the particular pipe line under consideration. Once this equation is determined for each pipe, the following sequence of computation is followed:

1. Assume the most reasonable flow in each pipe line which satisfies condition 1—the continuity equation at each junction.

2. Compute the loss of head in each pipe line by Eq. 7-70, to satisfy condition 3.

3. With due attention to direction of head drop, determine the total head loss for each branch of a circuit (i.e., $h_f = KQ^n$) to determine whether condition 2 is satisfied.

4. Compute for each circuit the quantity nKQ^{n-1}, without regard to direction.

5. Compute an adjusted or counterbalancing flow for each circuit—i.e., to make $KQ^n = 0$. The counterbalancing flow Q is computed by

$$\Delta Q = \frac{\Sigma\, KQ^n \text{ (with regard to direction of flow)}}{\Sigma\, nKQ^{n-1} \text{ (without regard to direction of flow)}} \qquad (7\text{-}71)$$

6. Revise the assumed flows and repeat the process until the accuracy desired is obtained.

   Usually not more than three trials are necessary, except for very complex systems.

The basic value of the Cross solution lies in the selection of the counter-balancing flow ΔQ. This may be shown in the following development. For any pipe line

$$Q = Q_0 - \Delta Q \qquad (7\text{-}72)$$

in which Q_0 is the assumed flow and Q is the correct flow. For each pipe then,

$$h_f = KQ^n = K(Q_0 - \Delta Q)^n = K(Q_0^n - nQ_0^{n-1}\Delta Q \ldots) \qquad (7\text{-}73)$$

where the remaining terms of the series may be ignored if ΔQ is small compared with Q_0. Now for a single loop, with ΔQ the same for all pipes,

$$\Sigma\, h_f = \Sigma\, KQ^n = \Sigma\, (KQ_0^n - \Delta Q K n Q_0^{n-1}) = 0 \qquad (7\text{-}74)$$

Therefore,

$$\Delta Q = \frac{\Sigma\, KQ_0^n}{\Sigma\, K n Q_0^{n-1}} \qquad (7\text{-}75)$$

in which the denominator is taken without regard to direction of flow. The assumption in Eq. 7-75 that the terms of the series in Eq. 7-73 beyond the first two may be neglected is not entirely accurate if ΔQ is of the same order of magnitude as Q_0. However, this error is quickly corrected by subsequent adjustments which bring rapid convergence. Methods have been developed for solving the simultaneous equations involved. Among the various methods are the graphical, the electric network analyser, and the hydraulic model. Complex pipe network problems also can be solved on an electronic computer with a very great saving of time and computational energy.

• UNSTEADY, UNIFORM FLOW

Most problems in fluid mechanics which involve unsteady flow are extremely difficult, if not impossible, to solve analytically. There are certain problems, however, for which analytical or graphical solutions have been developed that are either exact or sufficiently exact to be used with considerable success. Unsteady flow in uniform, closed conduits involves acceleration, either positive or negative, of the flow. This acceleration involves forces, the magnitude of which are reflected in the magnitude of the acceleration.

• FLOW FROM A RESERVOIR

One of the simplest problems involving unsteady flow is the discharge from a reservoir under varying head, as shown in Fig. 7-28. If the valve is opened suddenly at the downstream end of the pipe, the rate of discharge will rapidly increase, until steady flow is reached for the head in the reservoir.

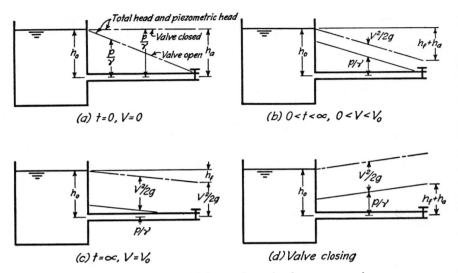

Fig. 7-28. Flow establishment in a pipe from a reservoir.

Before the valve in Fig. 7-28(a) is opened, the total head line in the pipe is the piezometric head line at the same level as h_0, and the velocity head is zero. If the valve is opened instantly, the velocity in the pipe is still zero, but the piezometric head line drops to the valve elevation at the outlet. This piezometric head drop h_a over a distance L is a gradient of piezometric head h_a/L, which acts as a driving force per unit weight or acceleration head to accelerate the fluid in the pipe line. At this instant, the rate of acceleration is a maximum because the gradient of piezometric head h_a/L is a maximum. As the velocity increases, the boundary loss h_f increases and the gradient decreases, so that the acceleration also decreases. If the pipe line is long enough so that the form losses and the issuing velocity head at the end of the pipe are of minor importance when compared to h_0, then the decrease in acceleration gradient can be illustrated graphically, as shown in Fig. 7-28(b). The Newtonian equation of motion for flow in the pipe becomes

$$\frac{h_a}{L} = \frac{1}{g}\frac{dV}{dt} \qquad (7\text{-}76)$$

in which h_a/L is the gradient of piezometric head;

$\dfrac{1}{g}$ is the unit mass per unit weight;

$\dfrac{dV}{dt}$ is the acceleration.

In terms of the original piezometric head, the value of h_a is the original piezometric head h_0 minus the head loss h_f. Equation 7-76 can then be written in the form

$$\frac{L}{g}\frac{dV}{dt} = h_0 - f\frac{L}{D}\frac{V^2}{2g} \tag{7-77}$$

If f is considered as a constant, then the variables can be separated and Eq. 7-77 can be integrated to give the relationship between velocity and time.

$$t = \frac{2D}{f} \int \frac{dV}{V_0^2 - V^2} \tag{7-78}$$

On evaluation of the integral, Eq. 7-78 becomes

$$t = \frac{LV_0}{2gh_0} \ln \frac{1 + V/V_0}{1 - V/V_0} \tag{7-79}$$

in which t is the time after the valve is opened and flow begins;

V is the velocity at any time t;

V_0 is the equilibrium velocity;

L is the length of pipe;

h_0 is the difference in piezometric head between the two ends of the pipe.

From this equation, it may be seen that the time t required to establish a given velocity ratio V/V_0 varies directly with the length of pipe L and the established velocity V_0, and inversely with the head h_0 on the pipe. Figure 7-29 is a plot of Eq. 7-79 which shows that, theoretically, equilibrium veloc-

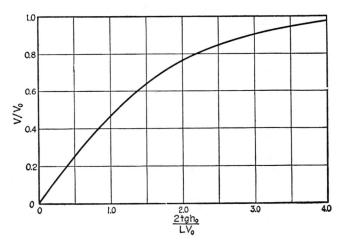

Fig. 7-29. Time of flow establishment.

ity V_0 is reached only after a great length of time. However, from the practical viewpoint, the velocity V is within about one percent of the equilibrium velocity V_0 very soon after the valve is opened.

The problems of the time required for flow establishment and the rate of valve closure are of importance in connection with a number of problems in fluid mechanics, such as the frequency of operation of a hydraulic ram, the evaluation of water hammer, and the design of surge tanks.

Once the flow in Fig. 7-28(c) is established, the reservoir will gradually empty and h_0 will decrease. Since this causes a variation of velocity with time, this condition of flow is unsteady and requires a special analysis. Consider the tank with an orifice in the bottom, as shown in Fig. 7-30. During the incremental time interval dt, the head h in the reservoir falls an incremental distance dh. This causes a decrease in volume of the fluid by $A\ dh = (\pi/4)D^2\ dh$ which is equal to the volume that has flowed out of the orifice $Q\ dt$. Hence

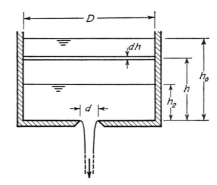

Fig. 7-30. Discharge from a reservoir with a falling head.

$$-\frac{\pi}{4} D^2\ dh = Q\ dt \qquad (7\text{-}80)$$

The rate of discharge at any time may be computed by a modification of Eq. 3-54 to include the entire piezometric head

$$Q = C_c C_v A_0 \sqrt{2gh} \qquad (7\text{-}81)$$

in which $A_0 = \pi d^2/4$. Eq. 7-81 may be substituted into Eq. 7-80 to give

$$dt = \frac{D^2}{d^2 C_c C_v \sqrt{2gh}}\ dh \qquad (7\text{-}82)$$

which can be integrated to yield

$$t = \frac{2D^2}{C_c C_v d^2 \sqrt{2g}}\ (h_0^{1/2} - h_2^{1/2}) \qquad \left(\begin{array}{l}\textit{Time for tank to empty}\\ \textit{from } h_0 \textit{ to } h_2\end{array}\right) \qquad (7\text{-}83)$$

in which $t = 0$ when $h = h_0$. This equation is applicable as long as the diameter of the orifice d is relatively small as compared with the size of the tank D. Otherwise, Eq. 7-77 must be employed and developed for the particular problem in question. Furthermore, C_c is usually assumed to be constant, although it changes slightly as the depth in the tank approaches the same order of magnitude as the diameter of the orifice. C_v is generally assumed to be about 1.0.

Example Problem 7-6: The tank shown in Fig. 7-30 has a diameter $D = 10$ ft and it has an orifice in its base with diameter $D_o = 2$ in. which has a contraction coefficient $C_c = 0.61$. Compute the time required to lower the water surface in the tank 4 ft if the initial depth is 8 ft. Assume $C_v = 1.0$.

Solution:

$$Q \, dt = -A \, dh$$

$$Q = 0.61 \frac{\pi}{4} \left(\frac{2}{12}\right)^2 \sqrt{2 \, gh} = 0.107 \, h^{1/2}$$

$$dt = \frac{-A \, dh}{Q} = -735 \frac{dh}{h^{1/2}}$$

$$t_2 - t_1 = -735 \int_8^4 \frac{dh}{h^{1/2}} = 1220 \text{ sec} \quad \text{or} \quad \underline{\underline{20 \text{ min } 20 \text{ sec}}}$$

• OSCILLATING FLOW

The pendulating or oscillating flow of a liquid in a U tube is another example of unsteady flow. If resistance to flow is ignored, Eq. 7-77 can be employed. The general case of a U tube of uniform diameter is shown in Fig. 7-31, where the length of the fluid column is L and the displacement of the fluid at a given time is s. The legs of the U tube are at angles α and β with the vertical.

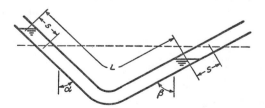

Fig. 7-31. Oscillations in a U-tube.

The vertical distance between the two liquid surfaces is the piezometric head which causes acceleration, since the ambient pressure on each surface is the same and unchanging with either time or distance. This vertical distance is

$$z = s \, (\cos \alpha + \cos \beta) \tag{7-84}$$

Hence, Eq. 7-76 can be written

$$\frac{dV}{dt} = \frac{d^2s}{dt^2} = -\frac{g}{L} s \, (\cos \alpha + \cos \beta) \tag{7-85}$$

If $V = 0$ and $t = 0$ when $s = s_0$, then this equation can be integrated further to give

$$s = s_0 \cos \left(t \sqrt{\frac{g \, (\cos \alpha + \cos \beta)}{L}} \right) \tag{7-86}$$

and the period T becomes

$$T = 2\pi \sqrt{\frac{L}{g \, (\cos \alpha + \cos \beta)}} \tag{7-87}$$

When the legs of the U tube are vertical, Eqs. 7-86 and 7-87 become

$$z = z_0 \cos \left(t \sqrt{\frac{2g}{L}} \right) \tag{7-88}$$

and

$$T = 2\pi \sqrt{\frac{L}{2g}} \tag{7-89}$$

which shows that the period of oscillation in a vertical U tube is the same as that of a pendulum half as long, $L/2$, as the column of liquid in the U tube.

In the foregoing development, the assumption was made that there was no boundary resistance to flow in the U tube. Under certain conditions this assumption is valid, since the forces of resistance may be small in comparison to other forces involved. As a matter of fact, however, there is always some resistance which, for a complete solution of the problem, requires consideration.

● SUMMARY

1. Flow in closed conduits involves a combination of:
    a. *Steady* or *unsteady* flow
    b. *Uniform* or *nonuniform* flow
    c. *Laminar* or *turbulent* flow

2. At the upstream end of a pipe there is a *region of flow development* in which the boundary layer is developing. In this region the greater boundary shear and any initial separation cause a greater rate of energy 'oss than in the *region of fully developed flow* downstream.

3. The **length** *of the region of flow development* is approximately

$$\frac{x}{D} = 0.06 \text{ Re} \qquad (\textit{for laminar flow}) \qquad (7\text{-}1)$$

and

$$\frac{x}{D} = 0.7 \text{ Re}^{1/4} \qquad (\textit{for turbulent flow}) \qquad (7\text{-}2)$$

4. The basic equation for **boundary resistance** *to flow in closed conduits* is

$$h_f = f \frac{L}{D} \frac{V^2}{2g} \qquad (7\text{-}5)$$

5. The **resistance coefficient** f depends upon the influence of viscosity and boundary roughness

$$f = \phi_1 \left(\text{Re}, \frac{e}{D} \right) \qquad (7\text{-}8)$$

6. For **laminar flow**

$$f = \frac{64}{\text{Re}} \qquad (\text{Re} < 2000) \qquad (7\text{-}11)$$

7. For **turbulent flow** with a **smooth boundary**.

$$f = \frac{0.316}{\text{Re}^{1/4}} \qquad (3000 < \text{Re} < 100{,}000) \qquad (7\text{-}12)$$

$$\frac{1}{\sqrt{f}} = 2 \log \text{Re}\sqrt{f} - 0.8 \qquad (\text{Re} > 50{,}000) \qquad (7\text{-}18)$$

8. For **turbulent flow** with a **rough boundary**.

$$\frac{1}{\sqrt{f}} = 2 \log \frac{D}{e} + 1.14 \qquad (7\text{-}19)$$

9. The *thickness of the* **laminar sublayer** for turbulent flow with a smooth boundary is

$$\delta' = \frac{11.6\nu}{\sqrt{\tau_0/\rho}} \qquad (7\text{-}16)$$

10. For pipes, the boundary is smooth if $e/\delta' < \frac{1}{5}$ and it is rough if $e/\delta' > 5$. Between these approximate values the boundary is in transition between smooth and rough.

11. For the solution of **pipe problems of boundary resistance,** use the methods outlined in Table 7-2.

12. For *isothermal flow* of a gas in a pipe line, the resistance coefficient f remains essentially a constant despite the fact that density decreases with pressure drop downstream.

13. The equation for the **pressure drop** in a pipe with *isothermal flow* of a gas is Eq. 7-43 rearranged as

$$f\frac{L}{D} = \frac{1}{k\mathrm{Ma}_1^2}\left[1 - \left(\frac{p_2}{p_1}\right)^2\right] + 2\ln\frac{p_2}{p_1}$$

. and for **adiabatic flow** is Eq. 7-48 rearranged as

$$f\frac{L}{D} = \frac{2}{(k+1)\mathrm{Ma}_1^2}\left[1 - \left(\frac{p_2}{p_1}\right)^{(k+1)/k}\right] + \frac{2}{k}\ln\frac{p_2}{p_1}$$

14. **Nonuniform flow** is involved for any flow transition which causes a change of velocity (either magnitude or direction) with respect to distance. Transitions cause *form resistance* which is composed principally of *pressure* resistance but also of some *shear* resistance. The resulting losses are known as **minor losses** C_L which are tabulated in Table 7-4, Table 7-5, and Table 7-6; and plotted in Fig. 7-10, 11, 12, 16, 18, and 20.

15. The **losses** due to both **boundary** *resistance* and **form** *resistance* must be used singly or in combination to solve problems involving *compound pipe lines, determination of minimum cost pipe lines, divided flow in pipes, branching pipes, and pipe networks.*

16. Some problems of **unsteady flow** are sufficiently simple to be solved either analytically or graphically. These include establishment of flow in a pipe line, emptying a reservoir, and oscillating flow.

References

1. Binder, R. C., *Fluid Mechanics*, 3rd ed. Englewood Cliffs, N. J.: Prentice-Hall, Inc., 1955.

2. Binder, R. C., *Advanced Fluid Mechanics*. Englewood Cliffs, N. J.: Prentice-Hall, Inc., 1958, Vols. I & II.

3. Daugherty, R. L. and A. C. Ingersoll, *Fluid Mechanics*. New York: McGraw-Hill Book Co., Inc., 1954.

4. Francis, J. R. D., *Fluid Mechanics*. London: Edward Arnold, Ltd., 1958.

5. Jaeger, Charles, *Engineering Fluid Mechanics*. London: Blackie & Son, Ltd., 1956.

6. King, Horace W. and Ernest F. Brater, *Handbook of Hydraulics*, 4th ed., New York: McGraw-Hill Book Co., Inc., 1954.

7. McNown, John S., "Surges and Water Hammer", Chapter VII of *Engineering Hydraulics*, ed. Hunter Rouse, New York: John Wiley & Sons, Inc., 1950.

8. Parmakian, John, *Water Hammer Analysis*. Englewood Cliffs, N. J.: Prentice-Hall, Inc., 1955.

9. Prandtl, Ludwig and O. G. Tietjens, *Applied Hydro-and Aeromechanics*. New York: McGraw-Hill Book Co., Inc., 1934.

10. Rich, George R., *Hydraulic Transients*. New York: McGraw-Hill Book Co., Inc., 1951.

11. Rouse, Hunter, *Elementary Mechanics of Fluids*. New York: John Wiley & Sons, Inc., 1946.

12. Streeter, Victor L., *Fluid Mechanics*, 2nd ed. New York: McGraw-Hill Book Co., Inc., 1958.

13. Streeter, Victor L., "Steady Flow in Pipes and Conduits", Chapter VI of *Engineering Hydraulics*, ed. Hunter Rouse, New York: John Wiley & Sons, Inc., 1950.

14. Hydraulic Institute, *Pipe Friction Manual*. New York: 1954.

PROBLEMS

7-1. Thirty gallons per minute of SAE 30 oil at 70°F flows from a reservoir through a rounded entrance into a 2-in. diameter pipe. What distance from the entrance of the pipe is required to establish uniform flow?

7-2. Thirty gallons per minute of water at 60°F flows from a reservoir through a rounded entrance into a 2-in. diameter pipe. What distance from the entrance of the pipe is required to establish uniform flow?

7-3. Assuming that laminar and turbulent flow can both occur at a Reynolds number of 10,000 for the same fluid in the same pipe line, what is the relative distance required to establish uniform flow for laminar flow, as compared to the distance required to establish uniform flow for turbulent flow?

7-4. The shear stress at the wall of a 4-in. pipe carrying water is 0.015 lb/sq ft. What is the head loss in 100 ft of straight pipe?

7-5. A 6-in. pipe on a 10% slope has a decrease of pressure in the direction of flow of 2 psi in 100 ft of pipe line. The flow is down the 10% slope. What is the shear stress at the pipe boundary? Water flowing.

7-6. A 1-in. smooth pipe line carries 20 gpm of SAE 10 oil at 50°F. What is the head loss in 1000 ft of pipe line? What horsepower would be required to pump the oil through the 1000 ft of pipe line if it sloped upward on a 3% grade?

7-7. Answer Problem 7-6 if water at 50°F instead of oil is flowing in the pipe line.

7-8. If 2 cfs of water at 60°F flow in a 6-in. pipe, what is the head loss in 100 ft of pipe if the pipe has a uniform roughness of $e = 0.01$ ft?

7-9. If an artificially roughened pipe has a Nikuradse sand grain roughness so that $e/D = 0.00397$, what is the maximum water discharge which can flow in the pipe and still have the pipe act as a smooth pipe? The pipe diameter is 3 in. and the water temperature is 70°F. What is the thickness of the laminar sublayer at this maximum flow?

7-10. For a Reynolds number of 5×10^6 what is the percent error involved by using Eq. 7-12 instead of Eq. 7-18 for finding the resistance coefficient for a smooth pipe?

7-11. The water discharge in a 2-in. brass pipe is 60 gpm. If the water temperature is 60°F, what is the shear at the boundary in pounds per square foot? What is the velocity at the center of the pipe and at a point $\frac{1}{2}$ in. from the pipe wall?

7-12. Water at 80°F is flowing in a 6-in. pipe at a rate of 3 cfs. If the relative roughness $e/D = 0.01$, what is the head loss per 100 ft, and at what distance from the pipe wall does the average velocity occur? What is the velocity at the centerline of the pipe?

7-13. From Fig. 7-5, determine the relationship between the absolute roughness e and the laminar sublayer (**a**) at the point where the boundary becomes rough, (**b**) at the point where a boundary with artificial roughness ceases to act as a smooth boundary, (**c**) at the point where a boundary of natural roughness ceases to act as a smooth boundary. (**d**) Explain the variation between (b) and (c).

7-14. Water at 40°F is flowing in a smooth 6-in. pipe line with an average velocity of 10 ft/sec. What is the velocity 2 in. from the wall of the pipe? What is the thickness of the laminar sublayer?

7-15. If the pipe in Problem 7-14 were rough, with a value of $e = 0.006'$ ft, what would the velocity be in 2 in. from the wall? Find δ'.

7-16. A straight horizontal brass pipe leads from a tank of glycerine whose temperature is 80°F. If the pipe line is 2 in. in diameter and 8 ft long, is uniform flow established in the line? The pipe discharges 30 gpm into the air at the end.

7-17. A straight horizontal brass pipe with a rounded entrance leads from a tank of water at 80°F. If the pipe line is 3 in. in diameter and 10 ft long, is uniform flow established in the line when the pipe discharges 200 gpm into the air?

7-18. Water at 70°F is flowing in a 6-in. average galvanized iron pipe line at the rate of 3 cfs. What horsepower will be required to lift this water from an elevation of 100 ft through 3000 ft of straight pipe to an elevation of 200 ft.

7-19. A 12-in. cast iron pipe line carries water at 40°F. If the maximum allowable change of pressure in 1000 feet of horizontal pipe is 10 psi, what is the maximum allowable discharge in the pipe line?

7-20. Go to the library and find at least ten formulas involving head losses in pipe lines. List the author of the equation and try to list some of the limitations of each equation.

7-21. A water discharge of 2 cfs is flowing in a new galvanized iron pipe whose diameter is 6 in. If the water temperature is 40°F, what is the head loss in 1000 ft of pipe line?

7-22. With water flowing in a 12-in. new steel pipe, the pressure loss in 100 ft of pipe is found to be 3 psi. If the water temperature is 50°F, what is the approximate flow in the horizontal pipe line?

7-23. If the flow of SAE 30 oil in a 2-in. brass pipe is 50 gpm at a temperature of 90°F, find the following: (a) head loss per 100 ft; (b) τ_0; (c) velocity $\frac{1}{2}$ in. from the wall.

7-24. A 2-in. galvanized pipe had a value of $e = 0.0004$ ft when it was new. The pipe was installed as a drain for an open water tank. The pipe discharged freely into the air 202 ft below the surface of the reservoir and was 1200 ft long. The water tank was 20 ft square and 4 ft deep. How long did it take to drain the water tank (a) when the pipe was new, (b) when the pipe was 30 years old if the value of α was 0.001? Assume a water temperature of 80°F.

7-25. If 2 cfs of water at 60°F must be pumped through 1000 ft of straight horizontal pipe line 6 in. in diameter, determine the horsepower required for the following types of pipe: (a) best commercial steel pipe, (b) worst commercial steel pipe, (c) best cast iron pipe, and (d) worst cast iron pipe.

7-26. In Problem 7-25, what is the cost of operating the pump for 1 year for each type of pipe installation if the pump operates 8 hours per day and the cost of power is $3\frac{1}{2}$ cents per kilowatt hour.

7-27. A 4-in. pipe line is designed to carry 500 gpm. The pipe is wrought iron and carries SAE 10 oil at 100°F. Assuming an average value for e, what is the head loss in 1000 ft of pipe line?

7-28. An 8-in. commercial steel pipe carries water at 40°F from reservoir A and discharges it 10 ft below the surface of reservoir B. The length of the pipe is 4000 ft and the water surface elevations in the reservoirs are A, 3272 ft and B, 3098 ft. What is the discharge in the pipe? Neglect minor losses.

7-29. A 6-in. pipe line is installed in a given city with an expected life of 60 years. The pipe is asphalted cast iron and the maximum head loss must never exceed 10 ft per 1000 ft, for a water temperature of 40°F. Based on these requirements, what is the maximum discharge which a designer should predict for this pipe line?

7-30. If $h_f = 20$ ft in 1500 ft, $D = 4$ in., $e = 0.002$ ft, and $T = 70°F$, find the discharge of water.

7-31. For fire requirements in a certain level district of a town, a commercial steel pipe must always be able to supply at least 400 gpm with maximum pressure drop of 10 psi in 1000 ft. If the minimum water temperature is 35°F, what size pipe should be installed? Explain your answer.

7-32. Given: $Q = 1$ cfs, $L = 3000$ ft, change of piezometric head $= 50$ ft, $T = 60°F$, $e = 0.0007$ ft. Find D. Water is flowing.

7-33. Crude oil with a viscosity of 3×10^{-4} sq ft/sec is flowing in a glass tube of $\frac{1}{4}$-in. diameter. If the average velocity in the tube is 3 ft/sec, what is the pressure drop in psi in 100 ft of straight tubing? Sp. gr. $= 0.93$.

7-34. A cast iron pipe 6 in. in diameter is carrying 3 cfs of water at 70°F. Find the head loss in (**a**) feet of water per foot, (**b**) psi per foot.

7-35. Water at 40°F is pumped from a lake through 3000 ft of riveted steel 12-in. pipe to a storage reservoir. The pressure in the 12-in. suction side of the pump is 15 in. of mercury vacuum, while the pressure on the 12-in. discharge side of the pump is 50 psi. The input horsepower to the motor is measured as 35 kilowatts. If the overall efficiency of the pump and the motor is 75%, (**a**) what is the discharge through the pump? (**b**) What is the head loss through the pipe line? (**c**) What is the difference in elevation between the water surface in the lake and the water surface in the storage reservoir? The pipe line discharges below the water surface of the reservoir. Neglect minor losses.

7-36. An engineer is asked to design a swimming pool 40 ft x 20 ft x 6 ft. For water supply, he decides to lay 1800 ft of 2-in. copper pipe from a nearby water main to the pool. The pipe will discharge freely into the pool during the entire filling process. The average pressure at the water main is 80 psi and the difference in elevation between the two ends of the copper pipe is 30 ft, with the discharge end being higher than the intake end. Assuming that 10% of the loss in the pipe line will be due to transitions, fittings, and valves, how long (in hours) will it take to fill the swimming pool? Assume a water temperature of 70°F.

7-37. The flow in a 1-in. circular pipe is laminar and the discharge is 30 gpm. What is the viscosity of the fluid if the Reynolds number is 1500? Is the flow stable or unstable? Explain.

7-38. Liquid having a viscosity of 1 stoke flows through a pipe at a rate of 47.6 gpm. If the head loss per foot of length is found to be 0.015 ft/ft, what is the diameter of the pipe?

7-39. If the Reynolds number for liquid flowing through a 1-in. pipe is at the critical value, what is the velocity of flow if the viscosity of the fluid is 0.032 stoke?

7-40. A very smooth riveted steel pipe line 24 in. in diameter and 4000 ft long connects reservoirs A and B. Reservoir A has a water surface elevation of 4230 ft, while that of B is 3910 ft. If 50 cfs of water at 50°F flow in the line, which is fitted with a turbine discharging below the surface of reservoir B, how much horsepower will the turbine put out if its efficiency is 85%?

7-41. In a straight section of 2-in. pipeline 500 ft long, running from elevation 130 ft to elevation 90 ft, the flow of linseed oil is 30 gpm. If the flow is downward and the temperature is 90°F, calculate the pressure at elevation 90 ft when the pressure at elevation 130 ft is 40 psi.

7-42. The Reynolds number for flow of oil through a 2-in. diameter pipe is 1700. If $\nu = 8 \times 10^{-4}$ sq ft/sec, what is the velocity at a point $\frac{1}{4}$ in. from the wall?

7-43. A very smooth 12-in. pipe is run 5000 ft between two reservoirs. The pipe is connected below the surface of each reservoir and the water surface in reservoir A is 200 ft above the water surface in reservoir B. What is the flow in the 12-in. pipe?

7-44. If the pipe in Problem 7-43 gets old and in very bad condition, what is the percent decrease in the flow that the pipe will now carry? In the design of a pipe system, what discharge coefficient should be used, the one for new or old pipe? Why?

7-45. Two pressure measurements are made on a horizontal 12-in. pipe line of old cast iron pipe. If the pressure measurements were 2000 ft apart and one was 60 psi while the other was 37 psi, what was the flow in the pipe line in cfs?

7-46. If 40 ft of head are available to force a required discharge of 2 cfs through an old riveted steel pipe, what is the maximum length the 6-in. pipe line can be before the discharge begins to fall below the required amount?

7-47. A new riveted steel pipe line 15 in. in diameter runs 3000 ft in a straight line from A to B. If the elevations of the two points are $A = 300$ ft and $B = 100$ ft, what is the difference in pressure $p_A - p_B$ in psi if the flow in the line is (a) 1 cfs, (b) 5 cfs.

7-48. The pressure at point A in a horizontal pipe line with a Hazen Williams discharge coefficient of 100 is 100 psi. At point B 1000 feet from A, for fire protection purposes, the pressure must be at least 80 psi and the flow must be at least 800 gpm. According to calculations, what is the minimum size pipe which could be installed? What size would you probably install? Why?

7-49. Part of the campus water supply for the University of Peshawar in Pakistan consists of a deep well. The water is pumped from the well into a tank. The water surface in the tank is 36 ft above the ground. A 6-in. steel pipe takes water from the tank over to a proposed hydraulics laboratory. If the pipe line is 900 ft long and the water temperature is 60°F, how long will it take to put 4 ft of water in the sump which is 35 ft by 25 ft. The flow from the end of the pipe is free flow and the end of the pipe is 6 ft below the ground elevation at the tank.

7-50. In a laboratory experiment on a horizontal $\frac{3}{4}$-in. brass pipe, a student read the following pressures at points along the pipe where the flow was uniform. Water temperature = 50°F. Find the discharge.

Point	Distance in ft along pipe	Pressure in psi
1	0	33.9
2	3	33.0
3	7	31.8
4	12	30.3

7-51. Water in a 6-in. pipe with an e value of 0.083 in. has a velocity of 5 fps at a point 2 in. from the pipe wall. Find the shear stress at the wall and the discharge in the pipe.

7-52. The Reynolds number for water at 60°F flowing in a 6-in. welded steel pipe is 200,000. Find the discharge and the head loss which would occur in 500 ft of straight pipe.

7-53. A pipe line carries oil whose viscosity is 5.9×10^{-4} sq ft/sec. The average velocity of the oil is not to exceed 3 ft/sec. What is the maximum discharge which can be carried under these conditions and still have stable laminar flow? What is the diameter of the pipe.

7-54. What horsepower must be supplied to pump 2.0 cfs of water from the lower to the upper reservoir? What is the maximum dependable flow rate which can be pumped through this system, if maximum vacuum pressure allowed is 21 in. of mercury? What pump horsepower would be required? Assume the flow is in the completely rough stage for old cast iron pipe. $e = 0.003$ ft.

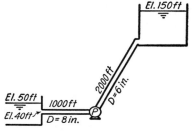

Prob. 7-54

7-55. Water is being pumped through a 3-ft riveted steel pipe for which e has been found to be 0.01 ft. If the highest point in the conduit is 75 ft above the pump at a section 3,000 ft away, what minimum discharge pressure must be maintained at the pump, during a flow of 60 cfs at 50°F, to prevent the pressure intensity from becoming less than atmospheric at the point of maximum elevation.

7-56. Benzene (sp. gr. = 0.9) is pumped through a long smooth 6-in. pipe at 50°F. Point 1, elevation 50 ft and point 2, elevation 100 ft, represent only a 1000 ft length along this long pipeline. What horsepower is required to pump 1.5 cfs of benzene from points 1 to 2.

7-57. A turbine is located at the end of an 18-in. riveted steel pipe line. The pipe line is 2 miles long and connects to a reservoir whose water surface elevation is 5200 ft. The turbine is at elevation 5000. The discharge pipe from the turbine is 24-in. riveted steel pipe, and discharges below the water surface of a reservoir at elevation 4950. The discharge pipe is 2000 ft long. If the input turbine horsepower is 100 hp, what is the discharge in the system?

7-58. Given: $h_f = 30$ ft, $L = 1000$ ft, $D = 8$ in., $T = 70°F$. Compare the discharges for new riveted steel pipe using the Darcy Weisbach and the Hazen Williams equations. Make the same comparison for old riveted steel pipe.

7-59. By the Hazen Williams formula, calculate the head loss in 2000 ft of old cast iron pipe, if the pipe is 12 in. in diameter and is flowing full with a discharge of 7 cfs.

7-60. A vitrified drainage tile is flowing half full. If the tile is 6 in. in diameter and is laid on a slope of 1 ft per 1000 ft, what is the discharge in the pipe?

7-61. If the maximum allowable pressure drop in 1000 ft of horizontal 12-in. pipe carrying 9 cfs when flowing full is 15 psi, what type of pipe, according to the Hazen Williams formula, should be used?

7-62. From the equation $p_1V_1 = p_2V_2$, develop Eq. 7-39 from Eq. 7-38.

7-63. Natural gas is pumped at the rate of 1000 lb per hour through a horizontal 6-in. cast iron pipe. If the pressure is 60 psia at point A, what is the pressure at B 10 miles downstream from point A? The temperature is 60°F and the viscosity may be assumed to be about 8.0×10^{-5} sq ft/sec. (a) Find the pressure, neglecting any change of density. (b) Assuming isothermal flow, determine the pressure at B. Compare answers.

7-64. A discharge of 3000 cfm of air at 40°F flows in a smooth conduit 20 in. in diameter. A pressure gage at point A reads 8 psi and the barometric pressure is 26 in. of mercury. If f is 0.013, what will a pressure gage read 1000 ft downstream from point A (a) assuming constant density, (b) assuming isothermal flow?

7-65. Air flows in an insulated 2-in. pipe at a temperature of 50°F. If the pressure at point A is 200 psia, what is the weight per minute flowing for a pressure drop of 30 psi in 2000 ft of pipeline which is riveted steel.

7-66. One thousand pounds of air per minute are transported through a horizontal smooth 6-in. pipe. At point A, the pressure is 100 psia and the temperature is 90°F. If the pressure at B downstream from A cannot be less than 90 psia, what is the maximum length of pipe between A and B if the flow is isothermal?

7-67. Referring to Fig. 7-7, compute and plot an incompressible and a compressible flow curve for $Ma_1 = 0.075$, using p_1/p_2 and fL/D as ordinate and abscissa.

7-68. Considering the isothermal flow of air in a 6-in. steel pipe in which $T = 60°F$, $p_1 = 50$ psia, $p_2 = 20$ psia, and $L = 2,000$ ft, (a) compute the error introduced in V_1 by ignoring the logarithmic term in Eq. 7-39. (b) For the foregoing conditions, determine the length of pipe for which, neglecting the logarithmic term, 5% error is introduced in V_1.

7-69. Air moves in a horizontal 8-in. diameter pipe at a temperature of 59°F, 150 psia, and a velocity of 200 fps. If the pipe friction factor is 0.018 and the flow is isothermal, what is the longest pipe that will transmit this flow?

7-70. Methane gas flowing in a 9-in. pipe is metered through a 6-in. sharp-edged orifice. The pressure in the pipe upstream of the orifice is 50 psia and the temperature is 60°F. The orifice manometer shows a pressure difference of 40 in. of water which has a temperature of 75°F. (a) Calculate the weight rate of flow. (b) What percent error is introduced in (a) if the effect of compressibility is neglected?

7-71. How many gate valves wide open would be required to give the same head loss as one globe valve wide open?

7-72. What is the loss of pressure through a short-radius elbow if there is 200 gpm of water flowing through a 2-in. elbow?

7-73. What is the increase in head loss if a 45° bend with $r/D = 6.0$ is replaced by a 45° miter bend without vanes, (a) if pipe is smooth, (b) if pipe is rough?

7-74. Compare the values of C_L for a gradual expansion, using Eqs. 7-56 and 7-57. Make calculations for diameter ratios of $\frac{1}{2}$ and $\frac{1}{3}$.

7-75. A rough pipe has a relative roughness of $e/D = 0.001$. If there is a 90° bend with an r/D ratio of 1.0, what length of straight pipe in terms of diameters will give the same loss as the bend? What would be the equivalent length of straight pipe for a 90° miter bend without vanes?

7-76. A wrought-iron pipe line 300 ft long connects flush to the bottom of two open water tanks whose water surface elevations are 300 ft and 280 ft. If the pipe diameter is 3 in. and the fittings consist of one globe valve wide open, four short radius elbows, two 45° elbows, and three side-outlet tees, what discharge will the 3-in. line carry?

7-77. In Problem 7-76, what equivalent length of straight pipe line would be required to replace all the fittings and give the same discharge through the pipe line?

7-78. Flow of water occurs from reservoir A to reservoir B through 300 ft of 12-in. concrete pipe. If the water temperature is 60°F and the discharge is 10 cfs, what is the water surface elevation in reservoir B?

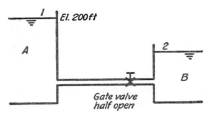

Prob. 7-78

7-79. A 500-ft long brass pipe line joins two reservoirs A and B below the water surface of each reservoir. The 3-in. diameter pipe includes the following fittings: one square entrance, two short radius elbows, one globe valve wide open, and one 45° elbow. If the water surface elevations are $A = 40$ ft, $B = 10$ ft and the water temperature is 40°F, what is the discharge in the 3-in. pipe?

7-80. A 36-in. corrugated pipe extends through an embankment, as shown. What discharge will be flowing in this pipe if the pipe is 100 ft long and the water temperature is 60°F? $e = 0.1$ ft.

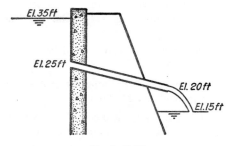

Prob. 7-80

7-81. A riveted steel outlet pipe through a dam is 30 in. in diameter. The inlet has a C_L value of 0.1, for the gate valve $C_L = 0.05$ and for the needle valve wide open, $C_L = 0.15$. The conduit is 200 ft long and the water temperature is 50°F. Estimate the outlet discharge for a depth of 100 ft above the outlet.

7-82. Fargus McFudd has a 1-in. galvanized iron pipe leading from the water main to the faucet where he connects his $\frac{1}{2}$-in. smooth plastic hose, which is 50 ft long. At point A in the 1-in. line, the pressure is 50 psi. Between point A and the nozzle (jet diameter $= \frac{1}{2}$ in.) at the same elevation as point A, there are the following fittings and pipe lengths: four short-radius elbows, one globe valve wide open, one angle valve wide open, one nozzle ($C_L = 8.0$), 100 ft of 1-in. pipe, and 50 ft of hose. Can Fargus get 10 gpm out the end of the hose? Explain why.

7-83. There is a bad leak in a straight horizontal 12-in. concrete pipe line. e for the concrete has a value of 10^{-3} ft. The temperature of the water is 75°F and upstream from the leak the discharge is measured as 8 cfs. Two pressure gages are installed upstream from the leak 2000 feet apart. There are two globe valves wide open and four short-radius elbows between the two gages. What will be the pressure difference between these two gages, in psi? Downstream from the leak, there are two pressure gages installed 2000 ft apart. There are no fittings between these two gages and they show a pressure difference of 22 psi. What is the leak in the pipe in cfs? (*Note:* use the same value of f downstream from the leak as you use upstream from the leak.)

7-84. A 3-in. diameter pipe line leaves a reservoir with a square-edged entrance and ends with a nozzle at point B. The nozzle diameter is 2 in. The pipe is galvanized iron and is 200 ft long. The fittings in the line include one globe valve wide open, one angle valve wide open, and four short-radius elbows. The loss coefficient for the nozzle is 0.5. What is the discharge through the line? Assume that f for the pipe is determined entirely by the pipe roughness. The reservoir water surface is at elevation 300 and the nozzle discharges into the air at elevation 240.

7-85. A discharge of 2 cfs of water is flowing in a 6-in. pipe. A discharge of 1.2 cfs leaves the 6-in. pipe through a 90° lateral which has a diameter of 3 in. What is the head loss h_L at the junction of the main pipe and the lateral?

7-86. In Fig. 7-19(a), $D_1 = 6$ in. and $D_2 = 3$ in. If $V_1 = 10$ fps and the piezometric head at point 1 is 60 ft, what is the piezometric head at point 2?

7-87. If the f value in each case is 0.020, what is the equivalent length of pipe for all of the fittings in Table 7-4?

7-88. A small farm owner wished to sprinkler-irrigate a garden plot with one sprinkler head. He went to an engineer with the following data: The materials represent items which the farmer already has on hand. Q required through sprinkler is 40 gpm. Pressure required in the 1-in. pipe at the sprinkler is 50 psi. Type of pipe: 1 in. diameter of clean galvanized pipe. Length of pipe from pump to sprinkler: 200 ft. Fittings in the line: one globe valve wide open, three 90° elbows, two standard tees. Pressure at the pump: 200 psi. Average temperature: 50°F. Pump elevation: 4990 Sprinkler head elevation: 5050. The engineer studied the problem and told the farmer his plan would not work. Do you agree with the engineer? If so, state how you think the farmer could most easily remedy his problems.

7-89. In Fig. 7-23, the following pipes* and fittings exist:

Pipes	Diameter	Total Length	e Value
1	8 in.	1000 ft	0.0005
2	12 in.	1500 ft	0.008
3, 4, 5	6 in.	500 ft	0.003

Fittings	Type
0-1	Square entrance
1-2	Sudden enlargement
2-3	Sudden contraction
3-4	Gate valve, wide open
4-5	45° elbow
6	Gradual expansion $\theta = 15°$ $A_2/A_1 = 8$

If the difference in water surface elevation between the two reservoirs is 300 ft, find Q.

7-90. Three pipes in series connect two water reservoirs of elevation 600 ft and 200 ft. The pipes are: 10-in. pipe 6000 ft long, 8-in. pipe 3000 ft long, and 6-in. pipe 1000 ft long. Assume that f in all pipes is 0.020. What is the discharge? Neglect minor losses.

7-91. The annual amortization cost of pipe and pump equipment is given by the following equation

$$C = 1000\ D^{3/2} + 200\ D$$

wnere D is the pipe diameter in inches and C is in dollars.
The annual cost in dollars of operation of the system is given by the following equation

$$O = \frac{80,000}{D^{5/2}} + \frac{1000}{D}$$

Find the pipe size which should be installed to give a minimum cost.

7-92. A 12-in. pipe line branches into an 8-in. line 3000 ft long and a 6-in. line 1000 ft long, as shown in Fig. 7-25. At the downstream junction, the pipes again join a 12-in. pipe. If the value of e in all pipes is 0.006 in. and the pipes are completely rough, what is the flow in each pipe when the discharge in the 8-in. pipe is 3 cfs?

7-93. Assuming that the pipe system is the same as in Problem 7-92, what is the flow in each pipe if the flow in the 12-in. pipe is 6 cfs?

* Assume all pipes act as completely rough pipes.

7-94. Water flows in the parallel pipe system as shown, and the following data apply:

4-in. pipe, length = 1000 ft, f = 0.025

6-in. pipe, length = 800 ft, f = 0.022

8-in. pipe, length = 2000 ft, f = 0.018

At A, elevation = 300 ft, pressure = 30 psi

At B, elevation = 100 ft

Average velocity in the 12-in. pipe = 10 fps.

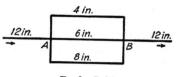

Prob. 7-94

What is the pressure at point B in psi and what is the discharge in each pipe? Neglect minor losses.

7-95. A 10-in. pipe divides into a parallel system at point A, with 3000 ft of 6-in. pipe and 2000 ft of 4-in. pipe. The pipes then join again at point B into a 10-in. pipe. The Hazen Williams coefficients are 120 for the 6-in. pipe and 90 for the 4-in. pipe. At A the elevation is 200 ft and the pressure is 30 psi, and at B the elevation is 50 ft and the pressure is 100 psi. Find the discharge in the 10-in. line.

7-96. A branching pipe system as in Fig. 7-26 has the following dimensions:

Reservoir water surface elevations: B = 700 ft, C = 600 ft.

Diameter of pipes: a = 12 in., b = 8 in., c = 10 in.

Length of pipes: a = 5000 ft, b = 3000 ft, c = 4000 ft.

Pipe friction factors: a = 0.018, b = 0.021, c = 0.020.

The elevation of point j is 650 ft and the discharge going into reservoir C is 6 cfs. What is the discharge in the 8-in. and the 12-in. pipes and what is the elevation of the water surface in reservoir A?

7-97. A turbine is located in the 15-in. line. If the turbine efficiency is 90%, what is the output horsepower of the turbine?

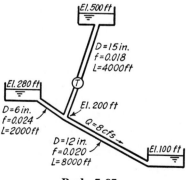

Prob. 7-97

7-98. Three reservoirs as in Fig. 7-26 are joined by three pipes as shown:

Line a diameter = 12 in. of riveted steel pipe, 1000 ft long

Line b diameter = 6 in. of galvanized iron pipe, 2000 ft long

Line c diameter = 8 in. of wrought iron pipe, 1500 ft long.

Reservoir water surface elevation A = 1000 ft, B = 800 ft, C = 700 ft. Elevation of j is 650 ft. If each pipe operates as a completely rough pipe, find the discharge in each pipe.

7-99. Solve Problem 7-98 by the Hazen Williams formula, if the Hazen Williams coefficients are:

C_1 for pipe a = 120

C_1 for pipe b = 100

C_1 for pipe c = 130

7-100. What is the water horsepower of the pump?

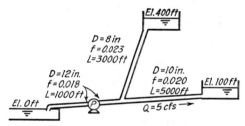

Prob. 7-100

7-101. The discharge in the 12-in. line is 6 cfs. What is the output horsepower of the turbine, if its efficiency is 85%?

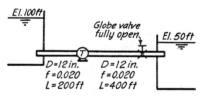

Prob. 7-101

7-102. A 36-in. diameter pipe divides at elevation 500 into three 24-in. pipes. Pipe A is 1 mile long and discharges beneath the surface of a reservoir whose elevation is 400 ft. Pipe B is 3 miles long and discharges beneath the surface of a reservoir whose elevation is 300 ft. Pipe C is 6 miles long and discharges beneath the surface of a reservoir whose elevation is 200 ft. If the discharge in the 36-in. pipe is 100 cfs and all friction factors are 0.020, what is the discharge in each pipe?

7-103. Assuming that the friction factor in all pipes is f = 0.022, solve for the discharge in each pipe.

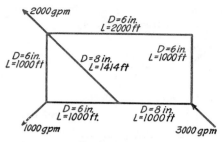

Prob. 7-103

7-104. All pipes have a Hazen Williams coefficient of 100. What is the flow in each pipe?

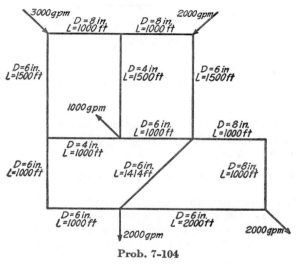

Prob. 7-104

7-105. All pipes between junctions are 1000 ft long and all have a Hazen Williams coefficient of 120. What is the flow in each pipe?

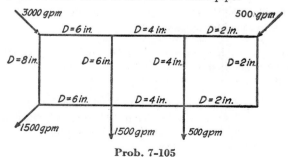

Prob. 7-105

7-106. All pipes between junctions are 1000 ft long and all pipes have a friction factor of 0.020. Find the discharge in each pipe.

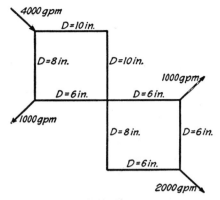

Prob. 7-106

7-107. A 3-in. pipe leads from a reservoir as in Fig. 7-28(a). The pipe is 1000 ft long, $h_0 = 50$ ft and $f = 0.023$. If the valve is opened suddenly, how long will it be before $V = 0.99\ V_0$?

7-108. In Fig. 7-28(a), there is a 2-in. pipe line 2000 ft long. The head $h_0 = 60$ ft and the friction factor $f = 0.025$. How long will it be before $V = 0.99\ V_0$? What is the acceleration when $t = 0.5$ sec, and when $t = 2.0$ sec?

7-109. Solve Problem 7-108 if the head h_0 is changed to 5 ft.

7-110. The following data apply to a figure similar to Fig. 7-28(a). The length of pipe is 100 ft, $h_0 = 30$ ft, diameter = 1 in., and the friction factor = 0.028. At what time does the acceleration equal 3 ft/sec²? What is the time when the acceleration is 0.3 ft/sec²? What is the acceleration and the time when $V = 0.99\ V_0$?

7-111. A sharp-edged orifice as shown in Fig. 7-30 has a diameter of 1 in. If the tank is 3 ft in diameter, how long will it take to drain the water out of the tank from $h_0 = 4$ ft to $h_2 = 1$ ft.

7-112. In Problem 7-111, what is the instantaneous discharge from the orifice, after the tank has drained for 30 seconds?

7-113. Assume that, in Problem 7-111, the orifice is replaced with a short vertical 1-in. pipe which is 6 in. long. If C_v for this short tube is 0.8, how long will it take to drain the tank from $h_0 = 4$ ft to $h_2 = 1.0$ ft?

7-114. A raised water tank has a leak in the bottom which causes the water surface to drop from 16 ft to 15 ft in a period of 12 hours. What will be the elevation of the water surface in the tank after 5 full days, if no water other than the leaking water is either added to or drained from the tank?

7-115. A vertical U-tube is 10 ft long. What is its period of oscillation?

7-116. A U-tube similar to the one in Fig. 7-31 is 30 ft long. The diameter is 3 inches, $\alpha = 30°$ and $\beta = 60°$. What is the period of oscillation? If $s_0 = 6$ in., what is t when $s = 0$?

7-117. In Problem 7-116, find the acceleration when the time equals zero.

7-118. A 4-in. diameter pipe is carrying SAE 30 oil at 70°F. Three Pitot-tube measurements were made at a section where the flow was assumed to be established, and the following data were obtained:

r	
0	11.4 ft.sec
1 in.	8.55 ft.sec
1.5in.	5.0 ft.sec

What is the discharge in the pipe?

7-119. A smooth 4-in. pipe is carrying oil whose kinematic viscosity is 6.5×10^{-4} sq ft/sec. If the discharge changes from 1.5 cfs to 0.15 cfs, what is the change of the maximum velocities in the pipe? Compare the change of maximum velocity with the change of discharge.

7-120. A 12-in. pipe having a parabolic velocity distribution has a shearing stress of 0.15 lb/sq ft at a point 2 in. from the pipe wall. The dynamic viscosity of the fluid is 6×10^{-3} lb-sec/sq ft. Find the discharge in the pipe.

7-121. Water at 60°F is flowing in a 6-in. smooth pipe. The velocity 2 in. from the pipe wall is measured as 10 fps. Find the thickness of the laminar sublayer.

7-122. A 10 hp pump is installed in a horizontal 4 in. pipe line carrying oil. For the oil, μ equals 2×10^{-3} lb-sec/sq ft and the specific gravity equals 0.85. If the pipe carries 0.5 cfs, which discharges into the air at the end of the pipe, what is the shortest distance from the end of the pipe at which the pump can be installed without exceeding a suction head of -20 ft of oil?

7-123. What is the power required to pump 900 gpm of water at 50°F through 500 ft of horizontal 8-in. commercial steel pipe?

7-124. Find the discharge of water at 40°F through a concrete pipe ($e = 0.01$ in.) 12 in. in diameter, if the allowable head loss due to friction is 10 ft of head per 500 ft of length of pipe line.

7-125. It is proposed to pump 900 gpm of water from an irrigation ditch a distance of 2000 ft, to a point in an adjacent field which is 70 ft above the water surface in the ditch. Determine the diameter of commercial steel pipe required, if the friction losses are to be maintained at less than 25% of the change in elevation. Assume a water temperature of 60°F.

7-126. A pump forces 3 cfs of water through an 8-in. pipe from A into a reservoir C. If the pipe is commercial steel pipe, the temperature is 40°F, and the pipe line is 5,000 ft long from A to C, what is the pressure in psi required at A? Neglect all minor losses. What is the pressure in psi at B, midway from A to C?

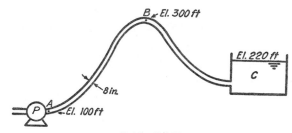

Prob. 7-126

7-127. If 0.5 cfs of water is flowing through the sudden contraction, what is the gage reading h if the pressure in the large pipe is 50 psi?

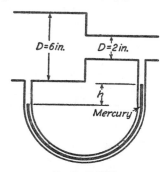

Prob. 7-127

7-128. A 12-in. pipe line branches at A into three lines which join again at B into a 12-in. line. All three pipes are cast iron pipes in very poor condition. The pipes are: (a) a 6-in. pipe, 2000 ft long, (b) an 8-in. pipe, 3500 ft long, (c) a 10-in. pipe, 5000 ft long.

At A the elevation is 300 ft and the pressure is 100 psi, while at B the elevation is 100 ft and the pressure is 150 psi. Assuming that the roughness of the pipe controls f, find the water discharge in the 12-in. line and its direction of flow.

7-129. Neglecting minor losses, find the discharge in each pipe and the horse-power output required by the pump.

7-130. The depth of the tank normal to the sketch is 6 ft. If the square-edged circular orifice has diameter of 4 in., how long will it take the two water surfaces to equalize? What is the velocity of the two water surfaces at the instant shown in the sketch?

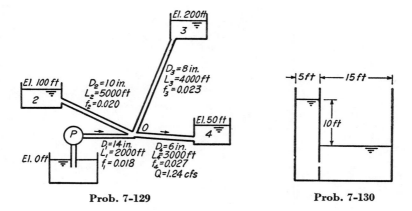

Prob. 7-129   Prob. 7-130

chapter 8

Flow in
Open Channels

Flow in open channels has been nature's way of conveying water on the surface of the earth through rivers and streams since the beginning of time. Furthermore, these streams have constantly been the subject of study by man, as he has been alternately blessed by the life-giving quality of streams under control and plagued by the destructive ability of streams out of control, such as in time of flood.

Man not only has struggled to regulate and control natural stream channels, but he also has built artificial channels to convey water for his own use and benefit. These artificial channels have included canals and flumes for irrigation, flood control, water power, drainage, industry, and domestic water supply. The old Roman aqueducts of 2000 years ago, together with their associated canals, are mute testimony to early achievements made in this field of endeavor.

Open channels include not only those which are completely open overhead, but also closed conduits which are flowing partly full. Examples of such closed conduits are tunnels, storm sewers, sanitary sewers, and various types of pipe lines.

Flow in open channels has certain similarities to flow in closed conduits which are flowing full. The boundary of the channel transmits a shearing force to the flow and converts energy into heat. This energy must be supplied by a gradient of piezometric head dh/dx—just as in a pipe flowing full. For pipes flowing full, however, the piezometric head gradient includes both the gradient of pressure head dp/dx, and the gradient of elevation head dz/dx, whereas for open channels it includes only the gradient of elevation head dz/dx, since the pressure head is everywhere the same on the surface of the stream, and $dp/dx = 0$.

Despite the fact that in this respect flow in open channels is simpler than flow in pipes that are full, the total analysis of flow in open channels has proven more difficult. This is due in large part to the fact that open channels have an infinite variety of shapes and types of boundary, and complex flow patterns, whereas closed conduits are usually rigid and circular in shape, with a simpler flow pattern. The use of a circular shape for pipes has simplified both experimentation and analysis. In fact, the experimental data which have been collected and analyzed for flow in pipes of widely different sizes, roughness, and fluids have resulted in the developments presented in Chapter 7. Experiments with open channels are now under way, both in many laboratories and in the field, in an attempt to advance the understanding of flow in open channels. Fortunately, the knowledge of flow in pipes is contributing much guidance for this development.

As the depth of flow is increased in closed conduits flowing partly full, there is a gradual transition from flow in an open conduit to flow in a closed conduit. Hence, there should exist a single function which can be used to analyze flow in both types of conduit. Indeed, as will be seen later, the Darcy-Weisbach equation (Eq. 7-5) and Fig. 7-6 may be used for both types of flow.

Flow in open channels is usually considered to involve only a liquid with a free surface. However, more fundamentally, this free surface should be considered as an interface between two fluids having different specific weights. The problem most frequently confronting engineers is the interface of water and air, and this chapter deals almost exclusively with that aspect of the problem. However, the flow of water under a gas is simply one type of gravity current wherein the driving force is created by the differential gravitational influence (which is expressed as the specific weight) on the liquid and on the gas.

The same fundamental principles also apply to other gravity currents, such as one liquid flowing horizontally and downward under another liquid of lesser specific weight. Examples of this are the gravity currents of colder water or sediment-laden water flowing from a stream under a reservoir of water which is either warmer or clearer of sediment, so that it has a lesser specific weight. Other examples are streams of cool air flowing down a valley, a hillside, or a mountainside under warmer air, or vice versa. Although these examples are occurring in literally thousands of places every day and creating both river problems and the geology of the future, as well as the weather of today, man has given relatively little attention to them.

• GENERAL TYPES OF FLOW IN OPEN CHANNELS

Because flow in open channels involves a free surface or interface, it has more degrees of freedom than the flow in closed conduits flowing full. This fact results in additional types of flow which, together with the types already encountered in the previous chapters, must be clearly defined and understood. These types include *uniform flow and nonuniform (or varied) flow; steady flow and unsteady flow; laminar flow and turbulent flow; and tranquil flow, rapid flow, and ultrarapid flow.*

Uniform flow in open channels, like that in pipes, depends upon there being no change with distance in either the magnitude or the direction of the velocity along a streamline—that is, both $\partial v/\partial s = 0$ and $\partial v/\partial n = 0$. **Nonuniform flow** in open channels occurs when either $\partial v/\partial s \neq 0$ or $\partial v/\partial n \neq 0$. The particular type of nonuniform flow which occurs when $\partial v/\partial s \neq 0$ in open channels is usually called **varied flow.** An example of uniform flow is flow in a straight canal or flume having a constant depth of flow, a constant slope, and a constant cross section throughout. Obviously, this condition seldom exists exactly. Examples of nonuniform flow, where $\partial v/\partial n \neq 0$, are bends or curving sides of the channel. When $\partial v/\partial s \neq 0$, the flow is varied and occurs when there is a change in depth of flow due either to a change in slope, a barrier or drop, or a change in side or bottom, so that the velocity increases or decreases in the direction of flow.

Steady flow occurs when the velocity at a point does not change with time—that is, $\partial v/\partial t = 0$. When the flow is **unsteady,** $\partial v/\partial t \neq 0$. As in the case of a pipe, unsteady flow is difficult to analyze, unless the change with respect to time is sufficiently slow to permit a step type of analysis. Examples of unsteady flow are traveling surges and flood waves in an open channel.

Whether **laminar flow** or **turbulent flow** exists in an open channel depends upon the Reynolds number Re of the flow, just as it does in pipes. In other words, laminar flow depends upon the viscous forces being predominant compared with the inertial forces, and turbulent flow depends upon the inertial forces being great compared with the forces of viscosity. With laminar flow, the mixing of the fluid is by molecular activity only, whereas with turbulent flow the mixing is by random eddy motion of finite masses of molecules. Laminar flow in open channels occurs very infrequently, except with special liquids such as oils or sediment mixtures.

Turbulent flow may be over a *smooth boundary*, where Reynolds number is relatively small and the laminar sublayer covers the boundary roughness; over a *rough boundary*, where Reynolds number is relatively large and the laminar sublayer is destroyed by the roughness; or over a *transition boundary*, where Reynolds number is intermediate and the boundary is both partly smooth and partly rough.

Unlike laminar and turbulent flow, *tranquil flow* and *rapid flow* exist only with a free surface or interface. The criterion for tranquil and rapid flow is the Froude number $Fr = V/\sqrt{gy}$, which (like the Reynolds number) is a ratio of two types of forces. The Froude number is a ratio of the forces of inertia to the forces of gravity. It is discussed in detail later in this chapter. Therefore, it will suffice at this point to say that when $Fr = 1.0$ the flow is critical, when $Fr < 1.0$ the flow is tranquil, and when $Fr > 1.0$ the flow is rapid. Ultrarapid flow involves slugs or waves superposed over the uniform flow pattern, which makes the flow both nonuniform and unsteady.

In the foregoing paragraphs, it is shown that four classifications are needed to describe completely the type of flow in an open channel:

1. Uniform or nonuniform.

2. Steady or unsteady.

3. Laminar or turbulent.

4. Tranquil or rapid.

One from each of these four types must exist simultaneously.

The energy gradient, and the gradient of the total head are terms which refer to the slope of the total head line—that is, either dH/dx or dH/ds, depending upon the definition being applied, see Fig. 8-2. With uniform flow, all these are the same as the slope of the channel—which is either dz/dx or dz/ds, again depending upon the definition of slope being applied. With nonuniform or varied flow, however, they are not the same.

Uniform flow in an open channel occurs with either a mild, a critical, or a steep slope. With a *mild slope* the flow is tranquil, with a *critical slope* the flow is critical, and with a *steep slope* the flow is rapid. These slopes and the corresponding flow are illustrated in Fig. 8-1 and Fig. 8-2.

Many of the foregoing terms are completely new to the student. Despite this fact, however, it is of utmost importance that each term be understood fully, or considerable difficulty will be experienced in understanding the remainder of this chapter.

• TYPES OF UNIFORM FLOW

Perhaps the problems most frequently encountered in the design of open channels are those involving the resistance to steady, uniform flow.

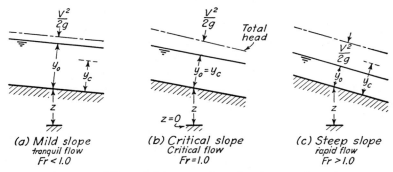

Fig. 8-1. Effect of slope on uniform flow in open channels.

When the flow is steady and uniform, the same basic principles of fluid resistance apply, regardless of whether the flow is tranquil or rapid. Therefore, in this section, no distinction will be made between tranquil and rapid flow.

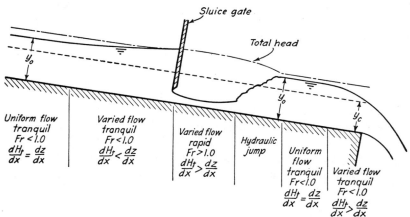

Fig. 8-2. Types of flow on a mild slope.

In an open channel, the boundary layer develops until it reaches the surface of the flow, beyond which point established flow exists. The pressure along any streamline within the fluid remains constant, but the elevation head z drops, which causes a gradient of piezometric head in the direction of motion.

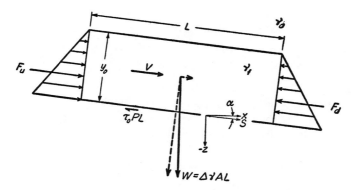

Fig. 8-3. Free-body diagram of segment of open channel flow.

• CHEZY DISCHARGE EQUATION

A free body of a segment of the full width of the channel may be selected for study, as shown in Fig. 8-3. Equating the forces in the s direction yields

$$F_u + W \sin \alpha = F_d + \tau_0 PL \qquad (8\text{-}1)$$

in which W is the weight of the entire segment of fluid;

> P is the wetted perimeter—that is, the length of the cross-sectional boundary which is in contact with the fluid flowing in the channel;
>
> F_u and F_d are the upstream and downstream hydrostatic forces acting on the free body (since the flow is uniform, $F_u = F_d$);
>
> τ_0 is the average boundary shear which is retarding the flow;
>
> A is the cross-sectional area of the flow;
>
> L is the length of the free-body segment;
>
> α is the angle which the channel slope makes with the horizontal;

The product $\Delta\gamma A L$ may be substituted for the weight W and the equation rearranged to solve for the boundary shear

$$\tau_0 = \Delta\gamma \frac{A}{P} \sin \alpha = \Delta\gamma RS \qquad (8\text{-}2)$$

in which $\Delta\gamma$ is the difference between the specific weight γ_f of the fluid flowing and the specific weight γ_a of the ambient fluid, normally the air.

R is called the hydraulic radius, defined as the area A divided by the wetted perimeter P;

S is $\sin \alpha = dz/ds$ which, for relatively flat slopes, may be considered more conveniently as $\tan \alpha = dz/dx$.

Equation 8-2, it may be noted, evaluates the boundary shear in terms of the static characteristics of the geometry and the fluid.

The shear τ_0 now remains to be evaluated in terms of the viscosity and velocity. From Eq 7-4 for pipe flow comes

$$\tau_0 = f\rho \frac{V^2}{8} \tag{8-3}$$

which may be applied equally well to flow in open channels, if the Darcy-Weisbach resistance coefficient f is evaluated properly for the different flow characteristics and the different shapes found in open channels.

Equating Eq. 8-2 to Eq. 8-3 and rearranging gives

$$V = \sqrt{\frac{8\Delta\gamma}{f\rho}} \, \sqrt{RS} \tag{8-4}$$

or
$$V = C\sqrt{RS} \tag{8-5}$$

in which C is the Chezy discharge coefficient defined as

$$C = \sqrt{\frac{8\Delta\gamma}{f\rho}} \tag{8-6}$$

Equation 8-6 shows that the Chezy discharge coefficient C varies inversely as the square root of the Darcy-Weisbach resistance coefficient f.

Equation 8-5 is the standard and widely-used Chezy equation for flow in open channels. In this form, the discharge coefficient contains the ratio $\Delta\gamma/\rho$, which makes the equation general for different combinations of fluids. When the fluid flowing is a liquid (generally water) and the ambient fluid is a gas (generally air), however,

$$\Delta\gamma = \gamma_f - \gamma_a \approx \gamma_f = \gamma \tag{8-7}$$

so that $\Delta\gamma/\rho$ may be expressed as

$$\frac{\gamma}{\rho} = g \tag{8-8}$$

in which g is the acceleration of gravity. The Chezy coefficient C can now be written in its more common form as

$$C = \sqrt{\frac{8g}{f}} \tag{8-9}$$

Strictly speaking, C is not a dimensionless coefficient, and the Chezy equation can be written more correctly as

$$V = \frac{C}{\sqrt{g}} \sqrt{gRS} = C' \sqrt{gRS} \qquad (8\text{-}10)$$

in which $C' = C/\sqrt{g}$. However, precedent and usage have fixed its form as Eq. 8-5 for the present.

In using Eq. 8-5, several matters to be remembered are:

1. The assumption that τ_0 is a constant along the wetted perimeter is generally acceptable for most problems. However, there are conditions, such as with marked secondary flow, wherein the variation of τ_0 around the wetted perimeter is appreciable, and hence considerable error may be involved in this assumption. The source of this error may be seen from Eq. 8-3, where the shear τ_0 varies with the square of the velocity.

2. The assumption has been made that the slope S equals sin α.

3. If $\Delta\gamma$ is to be used for flow of one gas under another gas of lesser specific weight, or one liquid under another liquid of lesser specific weight, then consideration must also be given to the shear and mixing which occur at the interface.

Now that the Chezy equation has been developed, it remains to develop a system for evaluating C. From the study of flow in pipes, it is known that both Reynolds number Re and the relative roughness e/R influence f, and hence C. Flow in pipes was restricted to pipe shapes, which do not correspond to the usual shape of open channels. Fortunately, it has been found that the hydraulic radius accounts automatically for the differences in shape—at least as a first approximation. Nevertheless, refinements need to be made, and no doubt will be made in the future, in evaluating the influence of cross-sectional shape upon the flow in an open channel.

Example Problem 8-1: A trapezoidal channel lined with concrete, $C = 130$, has a bottom width of 10 ft, side slopes of 1H:1V and depth of flow $y_0 = 5$ ft. If the channel slope, and slope of energy gradient S are equal to 0.0004, compute (a) the discharge, and (b) the average tractive force (shear stress) on the periphery of the channel.

Solution:

$$V = C \sqrt{RS}$$
$$Q = AC \sqrt{RS}$$
$$A = (10)(5) + 5^2 = 75 \text{ sq ft}$$

$$R = \frac{A}{P} = \frac{75}{10 + 10\sqrt{2}} = 3.12 \text{ ft}$$

$$Q = (75)(130)\sqrt{(3.12)(0.0004)} = \underline{\underline{344. \text{ cfs}}}$$

$$\tau_0 = \gamma R S$$

$$\tau_0 = (62.4)(3.12)(0.0004) = \underline{\underline{0.0777 \text{ psf}}}$$

• VELOCITY DISTRIBUTION

The evaluation of the Chezy coefficient C can be made in much the same manner as for pipes. For laminar flow in a **wide channel,** a parabolic distribution of velocity can be assumed, and an equation corresponding to Eq. 7-11 derived as

$$\frac{C}{\sqrt{g}} = \sqrt{\frac{\text{Re}}{8}} \tag{8-11}$$

in which $\text{Re} = 4VR/\nu$. From this equation, the Chezy C can be determined, and then the velocity computed from Eq. 8-5.

For turbulent flow in open channels, the velocity distribution may be assumed to be logarithmic, as it was for pipes, despite the fact that the boundary layer may be somewhat different from that for pipes.

For **wide channels** where the bank effects are negligible, the vertical distribution of the velocity v relative to the mean velocity V has been determined experimentally in terms of the depth y at which the velocity is measured relative to the total depth y_0 for uniform flow as

$$\frac{(v - V)C}{V\sqrt{8g}} = 2\log\frac{y}{y_0} + 0.88 \tag{8-12}$$

which does not apply near the bed or near the surface of the flow.

In **circular pipes,** the diameter D is equal to $4R$, which can be substituted into Eq. 7-18 and Eq. 7-19 to yield

$$\frac{C}{\sqrt{8g}} = 2\log\frac{\text{Re}}{C/\sqrt{8g}} - 0.8 \qquad (Smooth\ boundary) \tag{8-13}$$

and

$$\frac{C}{\sqrt{8g}} = 2\log\frac{4R}{e} + 1.14 \qquad (Rough\ boundary) \tag{8-14}$$

in which $\text{Re} = 4RV/\nu$

The actual velocity in an open channel is retarded near the banks as well as along the bed, see Fig. 8-4. This condition, together with other factors which are not well understood, causes a secondary spiraling motion which carries the low velocity flow from the bed upward along the bank and then out over the surface of the flow. Hence, the surface flow has a velocity less than that a short distance below the surface, where the velocity is a maximum. This can be seen in Fig. 8-4(a), which shows the lines of constant velocity called isovels.

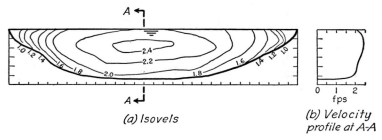

(a) Isovels   (b) Velocity profile at A-A

Fig. 8-4. Typical distribution of velocity in cross section.

Equation 8-14 is based upon the Nikuradse sand grain size e as a standard of roughness. However, Powell has proposed a simpler equation to take the place of both Eqs. 8-13 and 8-14. It is based upon ϵ as the roughness standard

$$C = -42 \log \left(\frac{C}{\mathrm{Re}} + \frac{\epsilon}{R} \right) \tag{8-15}$$

and is similar to Eq. 7-20 by Colebrook and White for the transition between smooth and rough boundaries. When the boundary is completely smooth, ϵ/R becomes very small, and Eq. 8-15 simplifies to

$$C = 42 \log \frac{\mathrm{Re}}{C} \tag{8-16}$$

and for a completely rough boundary ϵ/R becomes large, so that Eq. 8-15 simplifies to

$$C = 42 \log \frac{R}{\epsilon} \tag{8-17}$$

Equations 8-13, 8-14, 8-16, and 8-17 have been tried on Fig. 7-6 and found to have a difference that is insignificant. Hence, either these equations or Fig. 7-6 can be used to determine the Chezy C in Eq. 8-5. The magnitude of the Chezy C, however, depends upon the roughness of the boundary, as discussed in the following section.

• MANNING EQUATION

In an effort to correlate and systematize existing data from natural and artificial channels, Manning in 1889 proposed an equation which was developed into

$$V = \frac{1.5}{n} R^{2/3} S^{1/2} \qquad (8\text{-}18)$$

or

$$Q = AV = A \frac{1.5}{n} R^{2/3} S^{1/2} \qquad (8\text{-}19)$$

in which n is the Manning roughness coefficient, which has the dimensions of $L^{1/6}$. By comparing Eq. 8-5 with Eq. 8-18, the Chezy discharge coefficient C can be expressed as follows

$$C = 1.5 \frac{R^{1/6}}{n} \qquad (8\text{-}20)$$

and is related to the Manning coefficient n and the hydraulic radius R. The Manning n was developed empirically as a coefficient which remained a constant for a given boundary condition, regardless of slope of channel, size of channel, or depth of flow. As a matter of fact, however, each of these factors causes n to vary to some extent. In other words, the Reynolds number, the shape of the channel, and the relative roughness have an influence on the magnitude of Manning's n. Furthermore, for a given alluvial bed of an open channel, the size, pattern, and spacing of the dunes varies with slope, discharge, channel shape, and character of bed material, so that n also varies. Despite the shortcomings of the Manning roughness coefficient, however, it has an amazing utility and is used extensively in Europe, India, Egypt, and the United States.

• BOUNDARY ROUGHNESS

The type of roughness found in the rigid beds of open channels usually can be represented adequately by means of the Nikuradse roughness standard—the magnitude of which is independent of the flow. Under these conditions, the roughnesses are spaced closely together, so that the wake of one roughness interferes with the development of flow around the roughnesses immediately downstream, in much the same manner of interference as for the Nikuradse roughness. However, with an alluvial channel which has a moveable sand or gravel bed, the roughness elements are dependent upon the flow, and the bed material and usually consist of ripples, dunes, plane bed, symmetrical sand waves, or antidunes which have much greater spacing than the Nikuradse roughness, and hence one roughness interferes much less with those downstream. This condition results in a hypothetical e value which, under some conditions with dunes, would be equal to or even greater than the depth of flow—a situation that is ridiculous, since this would require that the Nikuradse roughness elements reach or penetrate the surface of the flow.

The forms of bed roughness which occur in alluvial channels, as observed and described by Simons and Richardson (1959), are illustrated in Fig. 8-5. Qualitatively, when the median size of bed material d is less than approximately 0.4 mm, the dunes wash out at a Froude number Fr $\approx$ 0.5

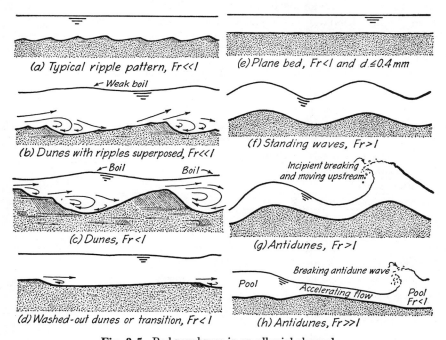

Fig. 8-5. Bed roughness in an alluvial channel.

and a plane bed and plane water surface persists until Fr $= 1$ at which time antidunes develop. If the median diameter is larger than about 0.4 mm, the Froude number Fr ≈ 1 when the dunes are completely washed out. With further increase in shear and Froude number, standing sand and water waves which are in phase develop and persist until Fr $= 1.2$ at which time antidunes develop. Considering coarser alluvial materials, ripples will not form when the median diameter $d \geqslant 2.0$ mm. That is, when sufficient shear is developed on the bed to cause bed material movement, conditions which yield ripples have been exceeded and dunes form. For similar reasons neither ripples nor dunes will develop when the median diameter of bed material $d \geqslant 6.0$ mm. In this case when motion of bed material begins $Fr \geqslant 1$ and the only forms of bed roughness which develop are standing sand and water waves which are in phase followed by antidunes at still larger values of Froude number.

The magnitude of Manning roughness is given in Table 8-1 for rigid channels and alluvial channels. Considering alluvial channels, note that as the form of bed roughness changes from dunes through transition to plane bed or standing waves the magnitude of Manning n decreases by approximately 50 per cent. Such a change in channel roughness will cause a discontinuity in the stage-discharge relation for an alluvial channel. That is, even though discharge is increasing, the change in roughness will cause a decrease in depth.

Table 8-1. MANNING ROUGHNESS COEFFICIENTS FOR VARIOUS BOUNDARIES

Boundary	Manning Roughness n in $(L)^{1/6}$
Very smooth surfaces such as glass, plastic, or brass	0.010
Very smooth concrete and planed timber	0.011
Smooth concrete	0.012
Ordinary concrete lining	0.013
Good wood	0.014
Vitrified clay	0.015
Shot concrete, untrowelled, and earth channels in best condition	0.017
Straight unlined earth canals in good condition	0.020
Rivers and earth canals in fair condition — some growth	0.025
Winding natural streams and canals in poor condition — considerable moss growth	0.035
Mountain streams with rocky beds and rivers with variable sections and some vegetation along banks	0.040–0.050
Alluvial channels, sand bed, no vegetation	
1. Tranquil flow, Fr < 1	
plane bed	0.014–0.02
ripples	0.018–0.028
dunes	0.018–0.035
washed out dunes or transition	0.014–0.024
plane bed	0.012–0.015
2. Rapid flow Fr > 1	
standing waves	0.011–0.015
antidunes	0.012–0.020

• OPTIMUM SHAPE OF CROSS SECTION

For a given area of cross section, there is a shape of cross section for which the hydraulic radius is a maximum. From Eq. 8-19, it is obvious that if the slope S and the roughness coefficient n are also fixed, the discharge is a maximum when the hydraulic radius is a maximum. Therefore, when other factors are held constant, the shape of cross section giving the maximum hydraulic radius is the optimum shape from the hydraulic viewpoint. For a given area, the maximum hydraulic radius is obtained with a minimum wetted perimeter. Therefore, the optimum cross section not only results in the maximum discharge, but also in a minimum bank and bed area of channel-lining material to build and maintain. Apparently then, the optimum cross section might be used for open channel design whenever possible.

The shape of cross section giving a minimum wetted perimeter for a given area is the circle or semicircle, see Fig. 8-6(a). A semicircle, however, has sides that are curved and are vertical near the surface, which makes both initial construction and maintenance difficult to accomplish, unless the channel is a flume of prefabricated metal. Therefore, trapezoidal shapes are generally used. The optimum cross section having a trapezoidal shape is a half hexagon, see Fig. 8-6(b), in which can be inscribed a semi-

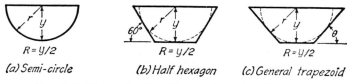

(a) Semi-circle   *(b) Half hexagon*   *(c) General trapezoid*

Fig. 8-6. Hydraulically optimum shape of cross section.

circle with its center at the free surface of the flow. Likewise, the optimum cross section of any trapezoidal channel where the side slope is fixed at an angle θ, see Fig. 8-6(c), is one in which a semicircle can be inscribed, with its center at the surface of the flow. This also means that $R = y/2$. Mathematical proof of these statements may be had by solving for the wetted perimeter, differentiating it with respect to depth, with the area held constant, and setting this equal to zero to obtain a minimum value.

Despite the fact that the smallest wetted perimeter and greatest discharge for a given cross-sectional area of an open channel are obtained when $R = y/2$, this particular shape is usually practical only when the canal is lined with a stabilizing material, so as to maintain the original shape. The width of canals constructed in natural earth must be several times as great as the depth in order to insure stability against bank erosion, and the bank slopes must be flatter than the angle of repose of the saturated material—which requires that the angle θ be considerably less than 60 degrees, as shown in Fig. 8-6(b) and (c), in order to insure stability against sliding. In other words, the shapes shown in Fig. 8-6 are optimum from the hydraulic viewpoint only. Usually, in practice other factors such as bank stability, excavation costs, and problems of maintenance require the shape to be wider and shallower than indicated in Fig. 8-6.

Example Problem 8-2: Considering a lined trapezoidal channel $Q = 500$ cfs, $S = 0.0004$, $\theta = 60°$, and Manning $n = 0.015$, determine (a) the hydraliucally optimum dimensions of the channel, (b) the equivalent dimensionless Chezy roughness coefficient, and (c) state briefly the limitations of the "hydraulically optimum" concept as it pertains to design.

Solution:

(a) $A = By_0 + \dfrac{y_0^2}{\sqrt{3}}$

$P = B + \dfrac{4}{\sqrt{3}} y_0$

$B = \left(\dfrac{2}{\sqrt{3}}\right) y_0$

$R = \dfrac{y_0}{2}$

$$\frac{500}{\dfrac{2}{\sqrt{3}} y_0^2 + \dfrac{y_0^2}{\sqrt{3}}} = \frac{1.50}{0.015} \left(\frac{y_0}{2}\right)^{2/3} (0.0004)^{1/2}$$

From which the normal depth

$$y_0 = \underline{\underline{7.48 \text{ ft}}}$$

and the bottom width

$$B = \frac{2}{\sqrt{3}} y_0 = \underline{\underline{8.64 \text{ ft}}}$$

(b) $C = \dfrac{1.50}{n} R^{1/6}$

$$\frac{C}{\sqrt{g}} = \frac{1.50}{0.015} \frac{(3.74)^{1/6}}{\sqrt{g}} = \underline{\underline{21.7}}$$

(c) The hydraulically optimum shape is usually practical only when the channel is lined with a stabilizing material such as concrete, and even then other factors such as cost and problems of maintenance may require some other shape.

• CLOSED CONDUITS FLOWING PARTLY FULL

When closed conduits are flowing partly full, the same principles govern the flow as for open channels, provided the pressure on the surface of the flow is maintained at atmospheric throughout. The shape of the flow section is accounted for through use of the hydraulic radius and the area. For circular conduits, such as sewers, drains, and tunnels, the relative velocity and discharge are plotted in Fig. 8-7, which is based upon the

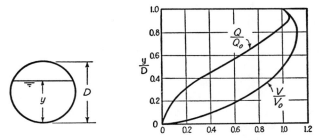

Fig. 8-7. Flow properties for circular conduit flowing partly full.

Manning equation (Eq. 8-19). It is assumed that n is a constant for all depths, and Q_0 and V_0 are the discharge and velocity respectively when the conduit is flowing full. From Fig. 8-7, it can be seen that the maximum discharge occurs at a depth of $y = 0.94D$. Strictly speaking, the coefficient n seldom remains a constant, but may vary as much as 25 percent with the depth of flow. Although operating a closed conduit at the partial depth which gives maximum discharge is an interesting possibility, it is not very practical since the slightest obstruction, increase in resistance, or underestimate of the boundary roughness would immediately cause the conduit to flow full.

Example Problem 8-3: Circular conduits which flow partly full are quite commonly encountered, such as sewer lines and drain pipe. Compute the effective area, wetted perimeter and hydraulic radius of a 6-in. circular conduit in which the depth of flow is 2 in.

Solution:

$$\cos \frac{\beta}{2} = \frac{r - y}{r} = \frac{1}{3}$$

$$\beta = 141°$$

$$A = \frac{\beta}{360} \frac{\pi D^2}{4} - \sqrt{2ry - y^2}\,(r - y)$$

$$A = \frac{141}{360} \pi \frac{0.5^2}{4} - \sqrt{2(0.25)(0.167) - 0.167^2}\,(0.25 - 0.167)$$

$$A = 0.057 \text{ sq ft}$$

$$P = \frac{\beta}{360}\pi D = \frac{141}{360}\pi(0.5) = 0.62 \text{ ft}$$

$$R = \frac{A}{P} = \frac{0.057}{0.62} = 0.092 \text{ ft}$$

• NATURAL CHANNELS

The foregoing discussions have been concerned principally with the design of artificial open channels. Although the same basic principles apply to natural channels, there are certain additional factors which must be considered.

The natural shape of an open channel may be markedly different from the shapes discussed thus far. However, it is usually possible to break down the complex shape of a natural open channel into simple elementary shapes for analysis. For example, consider Fig. 8-8 in which flow is occurring not only in the main channel, but also in the overbank or flood plain area. In this case, the hydraulic radius R which would be obtained by using the area

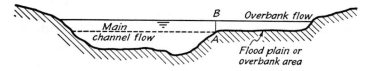

Fig. 8-8. Shape of natural stream channel.

and the wetted perimeter for the entire section would not be truly representative of the flow. Furthermore, the roughness coefficient in the overbank area is usually different from the coefficient in the main channel. Therefore, such a section should be divided along AB and treated as two separate sections. The plane AB, however, is not considered as a part of the wetted perimeter, since there is no appreciable shear in this plane.

Along a natural channel, there are frequently pools with a flatter slope, and riffles or rapids with a steeper slope, than the average slope of the channel taken over an appreciable distance. Therefore, care must be taken in studies of natural streams to consider the correct slope for the particular discharge and reach of stream in question.

• SIMPLE WAVES AND SURGES

Various types of waves and surges may occur in open channels and cause a locally unsteady flow. The simplest is the small surface wave which is observed to progress radially outward from the point of a small disturbance. The rate at which this wave progresses is called its celerity c, see Fig. 8-9(a). Like velocity, the dimensions of celerity are L/T. The unsteady flow of Fig. 8-9(a) can be transformed into steady flow simply by superposing a velocity on the system to the right, which is equal in magni-

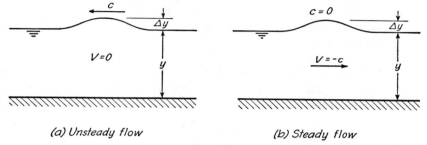

(a) Unsteady flow   (b) Steady flow

Fig. 8-9. Small gravity waves.

tude but opposite in direction to the celerity. In this way, the flow passes under the stationary wave at a velocity of $V = c$ in somewhat the same way that the flow passes through a stationary wave of compression in Chapter 4.

Assuming the energy loss in the wave to be negligible, the Bernoulli equation together with the continuity equation can be used to develop an analytical solution for rate of propagation of a solitary wave in terms of the depth of flow y and the height of the wave Δy. Under the wave, the velocity has been reduced slightly by the increased depth. Hence, the Bernoulli equation becomes

$$\frac{V^2}{2g} + y = \frac{V^2}{2g}\left(\frac{y}{y+\Delta y}\right)^2 + (y + \Delta y) \qquad (8\text{-}21)$$

which can be rearranged as

$$V = c = \sqrt{g}\left[\frac{2(y+\Delta y)}{1 + y/(y+\Delta y)}\right]^{1/2} \qquad (8\text{-}22)$$

When the wave height Δy is small compared with the depth y, this equation reduces to

$$c = \sqrt{gy}\left(1 + \frac{\Delta y}{y}\right)^{1/2} \qquad (8\text{-}23)$$

Furthermore, as the ratio of wave height Δy to depth y goes to zero, Eq. 8-23 becomes

$$c = \sqrt{gy} \qquad (8\text{-}24)$$

Derivation of the foregoing equations can be accomplished also by use of the continuity and momentum equations, as done in Chapter 4 for compressible flow.

From the foregoing analysis, it is obvious that the celerity c of a wave depends upon both the influence of gravity g and the depth of flow y. This type of wave is a shallow-water wave which is different from the deep-water wave (where the depth is half the wave length or greater), because the latter is independent of depth.

The entire discussion of waves in this chapter is based on shallow water waves. However, for the purpose of comparison, the equation for deep water waves is presented. For waves of small amplitude, the equation for celerity of deep-water waves is independent of depth y and is approximately

$$c = \sqrt{g\lambda/2\pi}, \qquad (8\text{-}25)$$

in which λ is the length of the wave and $\lambda < y/2$. As the amplitude or height of the deep-water wave increases, this equation must be modified until the limiting condition $2\Delta y/\lambda = 1/7$, where the wave becomes unstable and breaks, as commonly illustrated by ocean or lake waves generated by the wind.

In Eqs. 8-22, 8-23, and 8-24, it can be seen that if the velocity is equal to or greater than the celerity of the wave, the wave cannot move upstream. That is, when $V > c$ the wave moves downstream, and when $V = c$ the wave is stationary. For the very small stationary wave

$$V = \sqrt{gy} \qquad (8\text{-}26)$$

which may be rearranged as

$$\frac{V}{\sqrt{gy}} = 1.0 \qquad (8\text{-}27)$$

This equation is the principal definition of critical flow. That is, *critical flow* occurs when the velocity of flow is just equal to the celerity of a small wave in quiet water at the same depth. Equation 8-27 is the Froude number Fr for critical flow conditions. The general equation of the Froude number is

$$\text{Fr} = \frac{V}{\sqrt{gy}} \qquad (8\text{-}28)$$

which shows that the *Froude number is the ratio of the velocity of flow to the celerity of a very small gravity wave*. When $\text{Fr} < 1.0$ the wave can move upstream, and when $\text{Fr} > 1.0$ the wave is carried downstream. These facts are of considerable significance in the study of flow in open channels. Note the similarity of the Froude number to the Mach number in Chapter 4.

As the magnitude of the wave height Δy is increased, Eq. 8-23 (for a small wave) becomes less and less applicable and, at $\Delta y \approx y$, the wave becomes unstable and breaks. This may be understood by considering a large wave or surge as a series of small incremental surges of height Δy, see Fig. 8-10(a). Each of these small individual surges has a celerity corre-

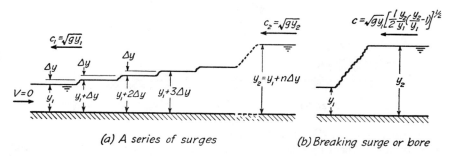

(a) A series of surges   (b) Breaking surge or bore

Fig. 8-10. Analysis of the breaking surge.

sponding to the depth of flow over which it is moving. That is, at the left side the depth is y_1 so that the celerity c is

$$c = \sqrt{gy_1} \tag{8-29}$$

while at the right side the depth is $y_2 = y_1 + n\Delta y$ and the celerity c is

$$c = \sqrt{gy_2} \tag{8-30}$$

Since $y_2 > y_1$, the incremental surges on the right have a celerity to the left which is greater than those surges on the left. Hence, the surges on the right will overtake those on the left, and a very steep surge will develop. As the height of this surge $n\Delta y$ increases, it becomes undulatory, and as it approaches y_1, the flow at the face of the surge becomes unstable and breaks, see Fig. 8-10(b). Consequently, considerable energy is lost due to the resulting turbulence. The net celerity of the breaking surge is less than $c = \sqrt{gy_2}$ but greater than $c = \sqrt{gy_1}$. This is developed further in the following paragraphs.

• HYDRAULIC JUMP

The breaking surge can be changed to steady flow by superposing a velocity of flow to the right which is equal in magnitude and opposite in direction to the net celerity of the breaking surge as shown in Fig. 8-11. Here, an analysis can be developed by use of the continuity and momentum

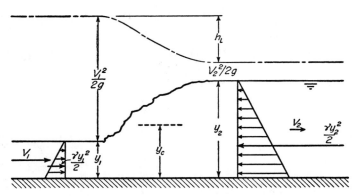

Fig. 8-11. Hydraulic jump or standing surge.

equations, in much the same manner as for the sudden expansion in closed conduit flow. According to the momentum equation, the sum of the total force and the momentum flux at section 1 must equal the sum of the total force and the momentum flux at section 2 (see the discussion of Fig. 3-16). That is,

$$\frac{\gamma y_1^2}{2} + q\rho V_1 = \frac{\gamma y_2^2}{2} + \rho q V_2 \tag{8-31}$$

in which q is the discharge per unit width. This equation can be combined with the continuity equation

$$q = V_1 y_1 = V_2 y_2 \tag{8-32}$$

to yield

$$V_1 = c = \sqrt{g y_1} \left[\frac{1}{2} \frac{y_2}{y_1} \left(\frac{y_2}{y_1} + 1 \right) \right]^{1/2} \tag{8-33}$$

which relates the upstream and downstream depths of the hydraulic jump to the velocity of approach V_1. By applying the Bernoulli equation to Fig. 8-11, it can be seen that the total head is reduced by the amount h_L. This fact explains why the hydraulic jump is very frequently used as an energy dissipator when supercritical or rapid flow (Fr > 1.0) is encountered in open channels.

For a more complete understanding of the hydraulic jump, Eq. 8-33 can be rearranged as

$$\frac{V_1}{\sqrt{g y_1}} = \mathrm{Fr_1} = \left[\frac{1}{2} \frac{y_2}{y_1} \left(\frac{y_2}{y_1} + 1 \right) \right]^{1/2} \tag{8-34}$$

to show that the Froude number Fr can be expressed in terms of the upstream and downstream depths only. Another significant arrangement of Eq. 8-33 expresses the depth ratio y_2/y_1 explicitly in terms of $\mathrm{Fr_1}$ alone

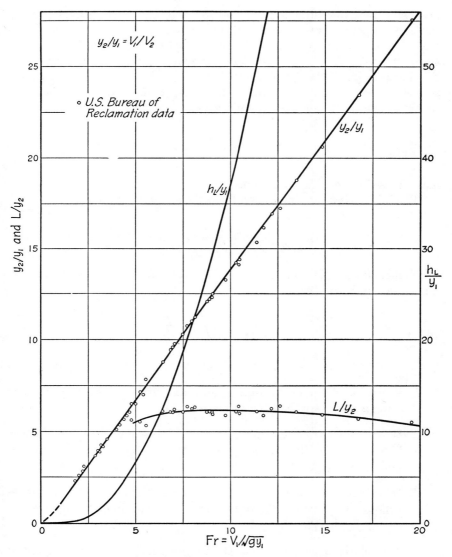

Fig. 8-12. Dimensionless characteristics of the hydraulic jump.

$$\frac{y_2}{y_1} = \frac{1}{2} \left[\sqrt{1 + 8\mathrm{Fr}_1^2} - 1 \right] \tag{8-35}$$

which shows that when $\mathrm{Fr}_1 = 1.0$ (critical depth) the depths are equal—that is, $y_2 = y_1$.

Figure 8-12 is a dimensionless representation of the characteristics of the hydraulic jump. The characteristics presented are the relative depths upstream and downstream from the jump y_2/y_1, the length of jump L relative to the downstream depth, and the energy loss h_L relative to the upstream depth, as functions of the Froude number of approach Fr_1. The data agree well with Eq. 8-35, except for large values of Froude number, where the downstream depth seems to be slightly smaller than indicated by the simple momentum flux principle. Because Eq. 8-35 is not simple to solve, an approximate empirical equation

$$\frac{y_2}{y_1} = 1.4 \ Fr_1 - 0.4 \tag{8-36}$$

may be used for ease in computation.

The head loss h_L is determined from the energy equation

$$\frac{V_1^2}{2g} + y_1 = \frac{V_2^2}{2g} + y_2 + h_L \tag{8-37}$$

which includes the head loss term and, when combined with the momentum flux equation, becomes

$$\frac{h_L}{y_1} = \frac{1}{2} Fr_1^2 \left[1 - \left(\frac{y_1}{y_2}\right)^2 \right] - \frac{y_2}{y_1} + 1 \tag{8-38}$$

Certain significant facts may be observed from a study of Fig. 8-12.

1. When $Fr_1 < 1.0$, no hydraulic jump can exist, because the celerity of a wave is greater than the velocity of flow.

2. When $1.0 < Fr_1 < 1.7$ and $y_2/y_1 < 2.0$ there is very little head loss, because the depth ratio y_2/y_1 is less than 2, and hence the surge is undulatory and not breaking.

3. When $Fr_1 > 2$ and $y_2/y_1 > 2.4$, the surge is breaking and h_L increases very rapidly—which shows that the hydraulic jump is an excellent energy dissipator.

4. The relative length of the hydraulic jump L/y_2 is nearly constant at about 5.5 to 6.0 beyond $Fr = 5$.

Example Problem 8-4: Five hundred cfs of water flows into a horizontal rectangular channel 10 ft wide at a depth of 2.5 ft. Compute (a) the Froude number to determine if a hydraulic jump is possible, (b) the depth of flow after the jump, and (c) the power corresponding to the loss of kinetic energy through the jump.

Solution:

(a) $V_1 = \dfrac{50}{2.5} = 20$ fps

$\text{Fr}_1 = \dfrac{20}{\sqrt{(32.2)(2.5)}} = \underline{\underline{2.23}}$

The hydraulic jump can occur.

(b) $\dfrac{y_2}{y_1} = \dfrac{1}{2}\left(\sqrt{1 + 8\,\text{Fr}_1^2} - 1\right)$ from Eq. 8-35

$y_2 = \dfrac{2.5}{2}\left(\sqrt{1 + (8)(2.23)^2} - 1\right)$

$y_2 = \underline{\underline{6.73 \text{ ft}}}$

(c) $V_2 = \dfrac{50}{6.73} = 7.43$ fps

Loss of kinetic energy $= \dfrac{V_1^2}{2g} - \dfrac{V_2^2}{2g} = \dfrac{20^2 - 7.43^2}{2g} = 5.35$ ft-lb/lb,

Power $= (5.35)(500)(62.4) = \underline{\underline{167,000 \text{ ft-lb/sec or 304 hp}}}$

The hydraulic jump may occur as a ***moving surge*** which is traveling either upstream if $V < c$, or downstream if $V > c$. To analyze this phenomenon, the flow system is first changed to steady flow by superposing a velocity on the entire system which is equal in magnitude and opposite in direction to the absolute velocity of the surge. Once the flow is steady, the problem may be analyzed by the usual procedure for the hydraulic jump.

A wave caused by a sudden release of additional water upstream in the channel may develop into a moving surge which is either stable or unstable (the moving hydraulic jump), in which case either Eq. 8-23 or Eq. 8-33 respectively is applicable. If the channel is dry, a sudden release of flow may be represented approximately by

$$c \approx 1.5 \sqrt[3]{gq} \approx 2\sqrt{g\Delta y} \qquad (8\text{-}39)$$

in which c is the velocity (celerity) of the wave front;

q is the steady discharge introduced into the channel;

Δy is the height of the wave.

● SUDDEN DECREASE IN FLOW

When the flow in an open channel is suddenly decreased, as shown in Fig. 8-13, a wave progresses downstream which has markedly different qualities than the waves discussed in the foregoing paragraphs. Upon applying the analysis, using incremental waves illustrated in Fig. 8-10(a), it can be seen that the upper portion of the wave is traveling more rapidly than the lower portion. Hence, the wave becomes flatter as it progresses downstream, as illustrated by the broken line in Fig. 8-13. This flattening

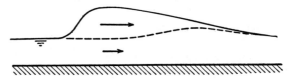

Fig. 8-13. Sudden decrease in flow.

of the wave is in contrast to the sudden increase in elevation which steepens and results in a breaking surge, as shown in Fig. 8-10(b) for a sudden increase in elevation.

• COMPARISON OF WAVE EQUATIONS

The various wave equations presented in this chapter may be seen to have certain important similarities. Each equation is of the form

$$c = K \sqrt{gy} \tag{8-40}$$

in which K serves as a coefficient which converts the celerity of a very small wave $\sqrt{gy}$ to the celerity of the wave in question. As a simple wave increases in size, the coefficient K increases from 1.0 to approximately $\sqrt{2}$, as may be seen by Eq. 8-23, until the wave becomes so large that it is unstable and breaks. For the breaking surge or hydraulic jump, the coefficient K varies from about $\sqrt{3}$ to 10 and greater. Finally, with a flood wave progressing down a dry channel, the coefficient K is approximately 2.0.

• INTERMITTENT SURGES

Under certain conditions on steep slopes, surges of an intermittent nature may occur which are called roll waves or slug flow, see Fig. 8-14. Such flow is not at all uncommon with harmless thin sheets of flow on slop-

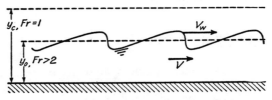

Fig. 8-14. Intermittent surges or roll waves.

ing side walks, for example. When these roll waves occur in large open channels, however, they may cause considerable damage, or force the operation of the channel at inefficient discharges in order to prevent damage.

Roll waves consist of waves superposed over the normal flow in an open channel. They travel at velocities greater than the normal flow and grow in size as they progress downstream.

There is no simple criterion for determining the size of roll waves, since their size depends upon the magnitude of the discharge, the type of flow (laminar or turbulent), the roughness and slope of the channel, the length of the channel, and the nature and frequency of the initial disturbances which cause the waves to form. However, the conditions of flow which are required to make the normal flow unstable so that roll waves can form is

$$\text{Fr} = \frac{V}{\sqrt{gy_0}} > 2 \tag{8-41}$$

which can be expressed in alternate form for a wide channel as

$$S \geq 4\frac{g}{C^2} \tag{8-42}$$

for turbulent flow with a rough boundary
in which y_0 is the normal depth;

S is the slope of the channel;

C is the Chezy discharge coefficient.

When the flow in a wide channel is turbulent with a *smooth boundary*, roll waves can form if

$$\text{Fr} \geq 1.5 \tag{8-43}$$

or

$$S \geq 2.25\frac{g}{C^2} \tag{8-44}$$

and when the flow in a wide channel is *laminar*, roll waves can form if

$$\text{Fr} \geq 0.5 \tag{8-45}$$

These empirical equations show that, for turbulent flow in a wide channel with a rough boundary, roll waves can occur when the velocity of flow is greater than twice the celerity of a wave (that is, the Froude number is greater than 2), or when the slope is four times as great as the slope required for critical depth. They can form also for turbulent flow in a wide channel with a **smooth** boundary if the velocity of flow is greater than 1.5 times the celerity of a wave, or the slope is 2.25 times the slope required for critical depth. By way of contrast, roll waves can form on laminar flow in a wide channel if the velocity is half the celerity of a gravity wave; in other words, the flow may never pass through critical flow (Fr = 1.0).

• INFLUENCE OF SPECIFIC WEIGHT

The foregoing wave and surge phenomena are developed for the specific condition of a liquid-gas interface. However, it is possible for the same phenomena to occur at a liquid-liquid interface, or a gas-gas interface, when there is relative motion between the fluids and the lower fluid has a greater specific weight than the upper fluid. Examples occur in the atmosphere where a cold air mass is flowing under a warm air mass, and in a large reservoir of water where a current of cold water, or sediment laden water, or salt water flows under the large body of water which has a lesser specific weight.

Under these conditions, the standing wave or hydraulic jump depends upon the Froude number written as

$$\text{Fr} = \frac{V}{\sqrt{\dfrac{\Delta\gamma\, L}{\rho}}} \tag{8-46}$$

in which $\Delta\gamma$ is the difference in specific weight across the interface;

ρ is the density of the heavier fluid;

L is a characteristic vertical length, such as the depth of flow.

It may be noted in Eq. 8-46 that, as $\Delta\gamma$ becomes small, the V also becomes small for a constant Froude number. Hence, on the one hand, the standing wave or surge or the hydraulic jump may occur under conditions which result in extremely small celerities. On the other hand, extremely large Froude numbers may occur in nature, such as in the atmosphere or in the estuary of a large river, where a liquid-liquid interface or a gas-gas interface exists with the correspondingly small $\Delta\gamma/\rho$ but an appreciable velocity.

• EFFECT OF RESISTANCE ON WAVES AND SURGES

When Scott Russell rode on horseback following waves in navigation canals, he found that the waves became progressively smaller with increasing distance until eventually they became so small they ceased to be discernible. Such a decrease in the size of waves must be due to forces of resistance which reduce their energy.

In the wave analyses presented in this chapter, the assumption has been made that the influence of boundary resistance and internal resistance is negligible. This assumption is justified for wave phenomena which occur over short distances, but the assumption may be considerably in error when relatively large distances are involved.

• TRANSITIONS IN OPEN CHANNELS

Flow through relatively short transitions in open channels may be described by the Bernoulli or energy equation, since the resistance forces over these short distances are relatively small. As shown in Fig. 8-15, an increase in velocity caused by a reduction in the area of cross section, either

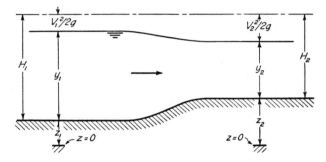

Fig. 8-15. Flow in transitions.

by raising the floor or by contracting the sides, causes an increase in velocity head, and hence a decrease in water surface elevation. The Bernoulli equation

$$\frac{V_1^2}{2g} + y_1 + z_1 = \frac{V_2^2}{2g} + y_2 + z_2 \tag{8-47}$$

is applicable to this flow pattern and may be used to determine the drop in water-surface elevation.

In connection with a detailed study of Fig. 8-15 and various other transitions, Eq. 8-47 can be modified as

$$H_1 + z_1 = H_2 + z_2 \tag{8-48}$$

in which

$$H = \frac{V^2}{2g} + y \tag{8-49}$$

The new term H is known as the ***specific head*** or the total head above the floor of the channel. As will be seen in the following pages, the specific head H is very useful in analyzing the flow in open channels. By use of the continuity equation, Eq. 8-49 can be expressed in terms of q, the discharge per unit width of rectangular channel, as

$$H = \frac{q^2}{2gy^2} + y \tag{8-50}$$

This equation can be analyzed from two viewpoints:

1. Variation of y with H when q is held constant, which yields the specific head diagram;
2. Variation of y with q when H is held constant, which yields the discharge diagram.

• SPECIFIC HEAD DIAGRAM

The specific head equation, Eq. 8-50, may be plotted as shown in Fig. 8-16(a) to show how the specific head H varies with the depth of flow y for progressively increasing values of discharge per unit width q_1, q_2, q_3, etc. This diagram shows that two different depths can exist with a given specific head H and discharge q—see for example point A in Fig. 8-16(a),

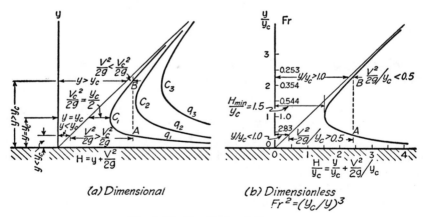

Fig. 8-16. Specific head diagrams.

where the depth is small and the velocity is great, and point B, where the depth is great and the velocity is small. These depths are termed **alternate depths,** because they can occur independently of each other and at the same specific head—depending only upon the conditions of the boundaries. Also of significance is the fact that there is a minimum value of specific head. This minimum means that, for a given discharge, there is a minimum specific head below which the discharge will decrease—say from C_2 to C_1 in Fig. 8-16(a)—until the specific head diagram for the correct discharge q is reached.

The depth of flow for the minimum value of the specific head H may be evaluated by differentiating Eq. 8-50 with respect to y and setting it equal to zero

$$\frac{dH}{dy} = -\frac{q^2}{gy^3} + 1 = 0 \tag{8-51}$$

which may be rearranged to yield

$$q = \sqrt{gy_c^3} \tag{8-52}$$

or

$$y_c = \sqrt[3]{\frac{q^2}{g}} = 2\frac{V_c^2}{2g} \tag{8-53}$$

in which y_c and V_c are the critical depth and critical velocity, respectively. Critical velocity, it will be recalled, is the velocity of a very small gravity wave in quiet water. The first and last terms of Eq. 8-53 may be rearranged as

$$\frac{V_c}{\sqrt{gy_c}} = 1 = \text{Fr} \tag{8-54}$$

which shows again that the Froude number equals 1.0 for critical flow conditions. From these equations and Fig. 8-16(a), it may be seen that under critical conditions the velocity head $V_c^2/2g$ is half as great as the depth of flow y_c, and that the critical depth is dependent only on the discharge per unit width of flow $q^{2/3}$, and the acceleration of gravity $g^{1/3}$. Furthermore, the critical depth does *not* depend upon the roughness, the width, or the slope of the channel, or upon the properties of the water.

Figure 8-16(a) and Eq. 8-51 also show that, for a given discharge with rapid flow, an increase in depth results in a decrease in specific head. That is, dH/dy is negative until the minimum value is reached (critical flow) where $dH/dy = 0$. If the depth is increased still further, tranquil flow occurs and the slope dH/dy reverses to become positive, so that the specific head must increase with increasing depth.

Flow at or near critical may be seen to require only a small change in specific head for a relatively large change in depth of flow. This fact results in the surface elevation being rather unstable, so that undulations form easily on the surface when the flow is at or near critical.

The dimensional specific head equation, Eq. 8-49, may be arranged in a significant dimensionless form by simply dividing each term by y_c which yields

$$\frac{H}{y_c} = \frac{y}{y_c} + \frac{V^2/2g}{y_c} \tag{8-55}$$

to express the variation of H with y entirely in terms of the critical depth y_c. This equation is plotted in Fig. 8-16(b), where it may be seen that

1. For rapid flow $y/y_c < 1.0 < \text{Fr}$ and $\dfrac{V_c^2/2g}{y_c} > 0.5 < \text{Fr}/2$

2. For critical flow $y/y_c = 1.0 = \text{Fr}$ and $\dfrac{V_c^2/2g}{y_c} = 0.5 = \text{Fr}/2$

3. For tranquil flow $y/y_c > 1.0 > \text{Fr}$ and $\dfrac{V_c^2/2g}{y_c} < 0.5 > \text{Fr}/2$

The fact that the minimum value of

$$H/y_c = y/y_c + \frac{V^2/2g}{y_c} = \frac{y}{y_c} + \frac{q^2}{2gy^2y_c} \tag{8-56}$$

occurs at 1.5 means that, if the specific head H is not sufficiently large to hold the ratio H/y_c at 1.5, then the discharge, and hence y_c, will decrease until 1.5 is reached as a minimum for the critical conditions of flow.

• BOUNDARY CONTRACTIONS

The foregoing principles may be illustrated by two types of simple transitions in an open channel:

1. A rise in floor level, and

2. A side wall contraction.

Although these are frequently used in combination in design practice, they are considered separately here. The combined case can be solved without difficulty by use of the same principles in combination.

Flow in an open channel having parallel walls, with a rise in the floor elevation which acts as a contraction, is illustrated in Fig. 8-17. In Fig. 8-17(a), the floor has a small rise so that the water surface drops over the rise, line AB, if the approaching flow is tranquil, $\text{Fr} < 1.0$, and the water surface rises even more than the floor, line DE, if the approaching flow is rapid, $\text{Fr} > 1.0$. The water surface profile downstream from the rise depends upon the downstream conditions. If the channel downstream is the same as that upstream, then the profiles will follow lines ABC and DEF

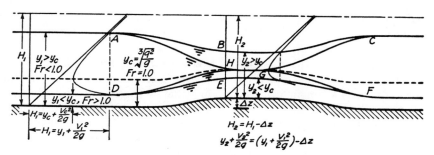

(a) Small rise in bed elevation

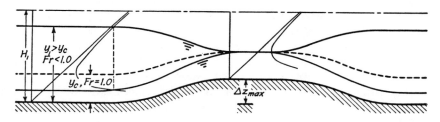

(b) Critical rise in bed elevation

Fig. 8-17. Flow with rise in bed elevation.

for Fr < 1.0 and Fr > 1.0, respectively, of the approaching flow. However, if the channel is such that upstream Fr < 1.0 and downstream Fr > 1.0, then the profile will follow line AHF and be at critical depth, where Fr = 1.0, over the crest of the rise. Likewise, if the channel is such that upstream Fr > 1.0 and downstream Fr < 1.0, then the profile will follow line $DHGC$, passing through critical depth, where Fr = 1.0, on the downstream slope of the rise in floor elevation. Although the latter profile does not occur frequently, it does occur with some conditions of rapid flow in alluvial channels.

Whereas the profiles ABC, AHF, and DEF are frequently encountered in natural channels and are not difficult to design in artificial channels, the profile $DHGC$ in pure form is seldom found under natural conditions and is usually difficult to incorporate into the design of a transition for a range of discharges. This is caused by the fact that the GC portion of the flow is replacing a standing wave, as discussed earlier in this chapter. Furthermore, as shown in the discussion of the breaking surge, see Fig. 8-10, the upper layers of flow tend to move upstream at a greater velocity (celerity) than the lower layers, and hence the wave is unstable and tends to break, so that a hydraulic jump is formed. The GC portion of the profile in Fig.

8-17(a), therefore, is inherently unstable, and hence the surface tends to be irregular, regardless of how gradual is the downstream expansion of the rise in floor elevation.

The depth over the rise in Fig. 8-17(b) corresponds to the minimum specific head at which the given discharge can be maintained. In other words, there is a limiting or maximum value of the rise in floor elevation Δz_{max} which can be made, and yet have the same approach conditions upstream. This is shown in the figure, in which the rise has been carried up to the point where all water surface profiles flow at critical depth over the rise. If the rise were made still higher, either one or both of two possible conditions would occur. Either the discharge would decrease, if the specific head were held constant; or the specific head would have to increase, if the discharge were held constant. Frequently, a combination of these occurs. The foregoing facts are readily seen from a study of Eqs. 8-50, 8-52, 8-53, and Fig. 8-16(a). When the specific head is held constant with a rise higher than the critical rise, the discharge is decreased, e.g., from the q_2 curve to the q_1 curve in Fig. 8-16(a), until the minimum value H_{min} of the specific-head curve equals the available specific head.

When the side walls contract for an open channel in which the bottom is on a uniform slope, the flow pattern has many similarities to the flow patterns discussed in the foregoing paragraphs—especially when the flow is tranquil throughout. The principal difference occurs with rapid flow, on which large waves are created by any change in alignment of the side walls. In the remainder of this discussion, these waves which occur under rapid flow conditions are not given further consideration, except to say that they can be highly undesirable, and may even cause a transition to fail to function as intended.

In Fig. 8-18, the total discharge Q must remain the same from section 1 to section 2, but the discharge per unit width q changes from q_1 to a greater value of discharge q_2 as the walls contract. Hence, the specific-head curve changes from one curve to another which is for a greater q value, see Fig. 8-16(a). Since $y_c = \sqrt[3]{q^2/g}$, the critical depth also increases when the sides contract.

Upstream from the contraction, two alternate depths may occur for a given specific head and discharge. As the sides contract more and more, however, specific-head curves for greater and greater q values come into use, and the two alternate depths approach each other until both are equal to the critical depth. If the sides are contracted still further, the discharge per unit width q increases and the specific head curve for the new q value has a minimum which is greater than the available specific head. Hence, either the discharge must decrease, if the specific head is to be held constant, or the specific head must increase, if the discharge is to be held constant.

The various profiles which may occur with contracted sides are illustrated by curves B_2 and B_3 in Fig. 8-18(b), depending on the relative up-

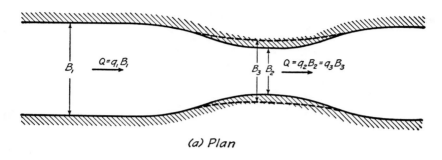

(a) Plan

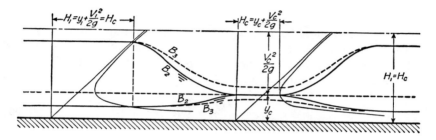

(b) Surface profiles

Fig. 8-18. Flow with change in width.

stream and downstream channel conditions, as discussed for the rise in bed elevation. In Fig. 8-18, the solid lines show the contraction which will cause critical depth in the throat. Lesser contractions, of course, are possible (see broken lines), in which case for an entering tranquil flow the depth in the throat is greater than critical, and for an entering rapid flow the depth in the throat is less than critical.

• DISCHARGE DIAGRAM

The discharge q in Eq. 8-50 may be plotted as a function of the depth of flow y for a constant value of the specific head H, see Fig. 8-19(a). Such a plot reveals that, for a given specific head, the same discharge may occur at two different depths called alternate depths, except at the maximum discharge possible, which occurs at the critical depth, where $y = y_c$. Figure 8-19(a) may be compared with Fig. 8-16(a) by taking for a constant specific head H, the value of y/y_c from the various different discharge curves. The maximum discharge q_{max} may be determined exactly by rearranging Eq. 8-50 as

$$q = y\sqrt{2g(H - y)} \tag{8-57}$$

and then differentiating and equating to zero

$$\frac{dq}{dy} = \sqrt{\frac{g}{2}}\, \frac{2H - 3y}{(H - y)^{1/2}} = 0 \tag{8-58}$$

which becomes

$$q_{max} = \sqrt{g\left(\frac{2}{3}H\right)^3} = \sqrt{gy_c^3} \tag{8-59}$$

or

$$y_c = \frac{2}{3}H = 2\frac{V_c^2}{2g} \tag{8-60}$$

These equations should be compared with Eqs 8-52 and 8-53 for the minimum specific head. The similarity of these equations shows that, for critical conditions, the discharge is a maximum for a given specific head, and the specific head is a minimum for a given discharge. Hence

$$y_c = \sqrt[3]{\frac{q^2}{g}} = \frac{2}{3}H \tag{8-61}$$

Equation 8-57 can be arranged in significant dimensionless form upon dividing by

$$\sqrt{gy_c^3} = \sqrt{gy^2\left(\frac{2}{3}H\right)} \tag{8-62}$$

to yield

$$\frac{q}{\sqrt{gy_c^3}} = \frac{q}{q_{max}} = \frac{y}{y_c}\left(3 - 2\frac{y}{y_c}\right)^{1/2} \tag{8-63}$$

which expresses the discharge q, relative to the maximum possible discharge q_{max} for a given specific head H as a function of the depth y relative to the critical depth y_c. Figure 8-19(b) is a plot of Eq. 8-63 which shows that $q/q_{max} = 1.0$ when $y/y_c = 1.0$ and Fr $= 1.0$. Furthermore, it shows that a zero discharge is reached when either the depth goes to zero ($V^2/2g = H$ and $y = 0$) or when the ratio $y/y_c = 1.5$ ($y = H$ and $V^2/2g = 0$).

As the discharge in Fig. 8-19 is decreased from $q/q_{max} = 1.0$, there will result an increase in depth for tranquil flow and a decrease in depth for rapid flow. These facts are readily seen when the sides of a rectangular open channel are gradually contracted to the configuration shown in Fig. 8-18, so that the discharge q per foot of width is gradually increased, for a given specific head, until it reaches the critical discharge in Eq. 8-59. In this case, the depth increases for rapid flow ($y < y_c$), and decreases for tranquil

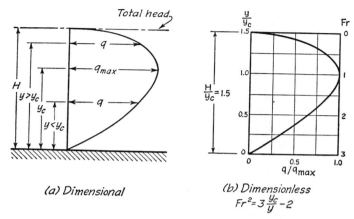

Fig. 8-19. Discharge diagrams.

flow $(y > y_c)$, until they meet at critical depth y_c. As the sides are contracted still further, the discharge per unit width q cannot increase any more since, for the given specific head, the maximum discharge is governed by Eq. 8-59

$$q_{max} = \sqrt{g \left(\frac{2}{3} H \right)^3} \tag{8-64}$$

and the specific head H must be increased before q_{max} can be increased.

By means of a gate on the horizontal crest of a spillway, Rouse has illustrated very graphically the foregoing arguments, see Fig. 8-20. Here, the specific head is held constant by the constant reservoir elevation upstream. When the gate is open, see Fig. 8-20(c), the discharge is a maximum and the depth of flow is y_c, as indicated by Eq. 8-53. Therefore, the gate can be closed until its lower edge is at the height y_c above the crest before it contacts the water and can have any effect on the flow. As the gate is closed still further, however, see Fig. 8-20(b), it acts as an obstruction to the flow and the discharge is decreased, with a corresponding increase in water surface elevation upstream and a decrease in elevation downstream. When the gate reaches complete closure, see Fig. 8-20(a), the discharge goes to zero as upstream $y \rightarrow H$ and $V \rightarrow 0$, and downstream $y \rightarrow 0$ as $V \rightarrow V_{max} = \sqrt{2gH}$ for flow through the orifice under the gate.

- ## ALTERNATE DEPTHS AND SEQUENT DEPTH

Flow under a sluice gate in Fig. 8-20 illustrates *alternate depths,* one greater than critical $y > y_c$ and the other less than critical $y < y_c$, which occur for each discharge at a constant specific head. This is illustrated also

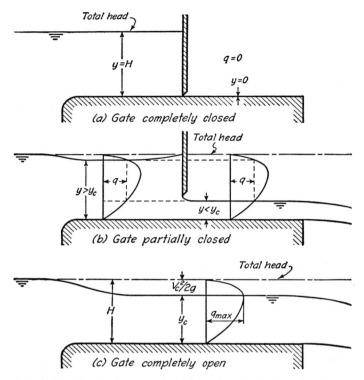

Fig. 8-20. Flow from reservoir with crest controlled by sluice gate.

in Fig. 8-21 (a), in which one alternate depth occurs upstream from the gate where $y_0 > y_c$, and the other occurs downstream from the gate where $y_1 < y_c$. Because the total energy or head is the same for each of the alternate depths, it is possible for one alternate depth to occur followed by the other, without loss of total head. This has been discussed in detail in connection with Figs. 8-16 through 8-20.

For the hydraulic jump, a head loss h_L occurs which results in the downstream depth y_2 being less than the alternate depth y_0 upstream from the gate. The depth y_2 is greater than y_c and is called the **sequent depth,** because it must always follow as a sequence from $y < y_c$ to $y > y_c$, with the head loss making the process irreversible. On the specific head diagram at the right of Fig. 8-21 (a), the **alternate depths** are directly above one another, since there is a constant specific head. The sequent depth, however, is to the left of the alternate depths, because the breaking surge causes a loss which results in a reduction of specific head through the hydraulic jump. The magnitude of this reduction may be obtained either from Eq. 8-38 or from Fig. 8-12. The other sloping broken lines in the specific head diagram indicate the sequent depth which results from different gate openings.

In the development of Eq. 8-31, it was pointed out that the hydrostatic force F_1 plus the momentum flux M_1 upstream is equal to the corresponding sum $F_2 + M_2$ downstream. On this basis, a sequent-depth diagram can be constructed as shown in Fig. 8-21(a). This diagram is especially useful

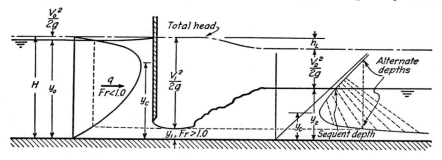

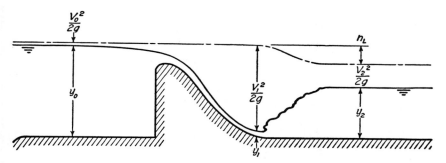

Fig. 8-21. Alternate depths and sequent depth.

for nonrectangular channels. When water flows over the crest of a dam, see Fig. 8-21(b), it passes from tranquil flow $y > y_c$ upstream from the crest, to rapid flow $y < y_c$ downstream from the crest. If the floor or bed elevation is the same upstream and downstream, then y_0 and y_1 are alternate depths, except for the head loss which occurs from the resistance to flow caused by the downstream face of the overflow spillway. For low dams where y_0 is relatively small, this resistance loss is negligible, while for high dams it may be very great. If the resistance loss is appreciable, its effect is to reduce the velocity V_1 and to increase the depth y_1, compared with these values which correspond to the alternate depth. Further downstream the hydraulic jump may occur as in Fig. 8-21(a). Both Fig. 8-21(a) and Fig. 8-21(b) are typical water-control systems using the hydraulic jump to decrease the total head, and at the same time to convert much of the remaining kinetic energy $V^2/2g$ to potential energy by increasing the depth y. This explains why the hydraulic jump is frequently called both an "energy dissipator" and a "head increaser"

• NONRECTANGULAR CROSS SECTIONS

The entire discussion in the foregoing pages is confined to flow in rectangular channels. For nonrectangular channels, see Fig. 8-22, Eq. 8-50 must be arranged as

$$H = K_e \frac{Q^2}{2gA^2} + y \tag{8-65}$$

in which Q is the total discharge through the cross-sectional area A and K_e is the kinetic energy coefficient. The area A is a function of the depth y. If this function is known and is simple (as it is for the rectangular channel), then a specific-head diagram may be made immediately. However, if the variation of A with y is unknown or too complex, the specific-head diagram is best determined by a trial solution or a graphical solution.

The minimum specific head and critical flow conditions may be determined by differentiating Eq. 8-65 with respect to y

$$\frac{dH}{dy} = 1 - K_e \frac{Q^2}{2g} \left(\frac{2}{A^3} \frac{dA}{dy} \right) \tag{8-66}$$

and setting it equal to zero to yield for critical conditions

$$\frac{K_e Q^2 B_c}{g A_c^3} = 1.0 \tag{8-67}$$

in which the simplification $dA = B\,dy$ is used as illustrated in Fig. 8-22. Like Eq. 8-65, however, Eqs. 8-66 and 8-67 depend upon a knowledge of the variation of A with y for its complete solution.

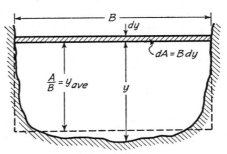

Fig. 8-22. Nonrectangular cross section.

Equation 8-67 for critical flow may be rearranged to solve for the critical velocity V_c

$$V_c = \sqrt{\frac{gA_c}{K_e B_c}} \tag{8-68}$$

The critical depth y_c must be determined as that depth which corresponds to the critical area A_c in a plot of A versus y. The ratio A_c/B_c is an average depth which is less than critical depth, since it is the average over the entire width B_c, see Fig. 8-22. The velocity head $V_c^2/2g$ for critical flow

$$K_e \frac{V_c^2}{2g} = \frac{1}{2} \frac{A_c}{B_c} \tag{8-69}$$

can be seen to be equal to half the average depth A_c/B_c.

In Eqs. 8-65 to 8-69, the kinetic energy coefficient may vary from 1.0 to as much as 2.0, depending upon the velocity distribution. Usually, K_e is 1.05 to 1.10 for turbulent flow, but under unusual circumstances of locally nonuniform flow it may be increased to 2.0 or even more. Because of the uncertainty normally associated with the computations of flow in channel transitions, and because K_e is most usually about 1.05, it may be assumed as 1.00 — at least as a first approximation.

Example Problem 8-5: Calculate the critical depth in a trapezoidal channel if the side slopes are 1H:1V, $Q = 200$ cfs, $K_e = 1$, and the bottom width is 10 ft.

Solution:

$$K_e \frac{V_c^2}{2g} = \frac{1}{2} \frac{A_c}{B_c} = \frac{y_{ave.}}{2}$$

Since $K_e = 1$

$$\frac{V_c^2}{2g} = \frac{1}{2} \frac{A_c}{B_c}$$

This problem will be solved using a tabulated trial and error procedure

Depth y (assume)	2	2.2	2.13	2.15
Area $A = 10y + y^2$	24	26.84	25.85	26.13
Surface width $B = 10 + 2y$	14	14.4	14.26	14.30
$\frac{1}{2}A/B = y_{ave}/2$	0.857	0.934	0.908	0.91
Average velocity, $V = Q/A$	8.35	7.44	7.73	7.65
Velocity head $V^2/2g$	1.08	0.86	.928	0.91

Based on the tabulated data since, $\frac{1}{2}A/B = \frac{V^2}{2g}$, $y_c = \underline{\underline{2.15\,\text{ft}}}$

• CHARACTERISTICS OF CRITICAL FLOW

In the foregoing discussions, the flow has continually been referred to critical flow conditions. In summary, these characteristics of critical flow may be stated as follows:

1. Critical flow occurs when the velocity of flow V is equal to the celerity c of a very small wave in quiet water at the same depth y

$$V = c = \sqrt{gy} \qquad (8\text{-}70)$$

2. Critical flow occurs when the Froude number Fr is unity

$$\text{Fr} = \frac{V}{\sqrt{gy}} = 1.0 \qquad (8\text{-}71)$$

3. Critical flow occurs when the specific head H is a minimum for a given discharge q

$$H_{\min} = \frac{3}{2} \sqrt[3]{\frac{q^2}{g}} \qquad (8\text{-}72)$$

4. Critical flow occurs when the discharge q is a maximum for a given specific head H

$$q_{\max} = \sqrt{g \left(\frac{2}{3} H\right)^3} \qquad (8\text{-}73)$$

5. Critical flow occurs when the velocity head $V^2/2g$ is half the depth y

$$\frac{V_c^2}{2g} = \frac{y_c}{2} \qquad (8\text{-}74)$$

and the depth of flow is two-thirds the specific head

$$y_c = \frac{2}{3} H \qquad (8\text{-}75)$$

For flow in nonrectangular channels, this depth is the average depth A_c/B_c.

6. When flow is at or near critical, the surface elevation is quite unstable and undulations form very readily. This is due to the very small change in specific head which is required for a relatively great change in depth, when the depth is at or near critical, see Fig. 8-16.

• GRADUALLY VARIED FLOW

The types of flow in open channels which have been considered thus far involve only steady-uniform flow where the depth is called normal depth, or nonuniform flow for which changes in cross section take place in a relatively short distance. For uniform flow, the slope of the water surface, the slope of the bed, and the slope of the total head line are all equal. When the nonuniform flow extends over only a short distance through a stream-lined transition, resistance is negligible.

Backwater curves and drawdown curves, which are usually included in the term "backwater curves," describe the water surface in an open channel where changes in cross section take place very gradually with distance along the channels. The assumption is made that the resulting changes in velocity take place so slowly that the accelerative effects are negligible. Because the changes take place gradually, the flow is commonly known as *gradually varied flow.*

Changes in the cross section of the flow may result either from a change in geometry of the channel—such as a change in slope or cross-sectional shape, or an obstruction—or from an unbalance between the forces of resistance tending to retard the flow and the forces of gravity tending to accelerate the flow.

In order to analyze the various types of backwater curves, the total head H_T can be expressed as

$$H_T = K_e \frac{V^2}{2g} + y + z \tag{8-76}$$

$$= K_e \frac{Q^2/A^2}{2g} + y + z \tag{8-77}$$

in which K_e is the kinetic energy correction coefficient;

y is the depth of flow;

z is the elevation of the channel bed above some arbitrary datum.

Since the variation of these terms with distance x along the channel is desired, Eq 8-77 can be differentiated with respect to x to obtain

$$\frac{dH_T}{dx} = -\frac{Q^2}{gA^3}\frac{dA}{dx} + \frac{dy}{dx} + \frac{dz}{dx} \tag{8-78}$$

$$= -\frac{Q^2 B}{gA^3}\frac{dy}{dx} + \frac{dy}{dx} + \frac{dz}{dx} \tag{8-79}$$

in which K_e is assumed again to be unity and dA is set equal to $B\,dy$. The gradient of total head dH_T/dx can be set equal to the negative of the slope obtained from the Chezy equation $S = (Q/A)^2/C^2R$, and the bed slope is $dz/dx = -(Q/A_0)^2/C_0^2R_0 = -S_0$ for normal uniform flow conditions. For simplicity, however, a wide rectangular channel can be assumed, so that $Q/B = q$ and $R = A/B = y$. Eq. 8-79 then becomes

$$-\frac{q^2}{C^2y^3} = \frac{dy}{dx}\left(1 - \frac{q^2}{gy^3}\right) - \frac{q^2}{C_0^2y_0^3} \tag{8-80}$$

Furthermore, $q^2/g = y_c^3$, so that Eq. 8-80 can be rearranged to solve explicitly for dy/dx, the rate of change of depth of flow y with respect to distance x along the channel,

$$\frac{dy}{dx} = \frac{(q^2/C_0^2y_0^3) - (q^2/C^2y^3)}{1 - (y_c/y)^3} \tag{8-81}$$

which simplifies to

$$\frac{dy}{dx} = S_0\left[\frac{1 - \left(\dfrac{C_0}{C}\right)^2\dfrac{y_0^3}{y^3}}{1 - (y_c/y)^3}\right] \tag{8-82}$$

If the change in the Chezy C is not great from one point to another along the channel, the ratio C_0/C can be considered equal to 1.0. Usually, however, the Manning n is more nearly constant from section to section. Hence, $C = 1.5y^{1/6}/n$ can be used in Eq. 8-82 to yield

$$\frac{dy}{dx} = S_0\left[\frac{1 - \left(\dfrac{n}{n_0}\right)^2\left(\dfrac{y_0}{y}\right)^{10/3}}{1 - \left(\dfrac{y_c}{y}\right)^3}\right] \tag{8-83}$$

With Eq. 8-83, it is possible to classify the various surface profiles of backwater curves which may occur in open channels.

• CLASSIFICATION OF SURFACE PROFILES

The analysis of surface profiles depends first upon the sign of dy/dx, the slope of the water surface relative to the bed. If dy/dx is positive, the depth is increasing downstream, and if it is negative, the depth is decreasing downstream. From Eq. 8-83, it can be seen that the slope dy/dx depends upon the slope of the bed S_0, the ratio n/n_0 of the Manning n for the actual depth to that for the normal depth which would occur if uniform flow existed, the ratio y_0/y of the normal depth to the actual depth, and the ratio y_c/y of the critical depth to the actual depth. In the following analysis, it is assumed that $n/n_0 = 1.0$. Although for some conditions this assumption is not justified (see the discussion of alluvial channels), it is assumed to be sufficiently exact for the purposes of this analysis. Hence

$$\frac{dy}{dx} = S_0 \left[\frac{1 - \left(\frac{y_0}{y}\right)^{10/3}}{1 - \left(\frac{y_c}{y}\right)^3} \right] \qquad (8\text{-}84)$$

The slope of the channel serves as the primary means of classification. If the bed slope S_0 is negative, the bed rises in the direction of flow. This is called an adverse slope and the curves of the water surface over it are known as A curves. If $S_0 = 0$, the bed slope is horizontal and the curves over it are H curves. When $S_0 > 0$, the bed slope is either mild, steep, or critical, and the corresponding curves of the water surface are either M curves, S curves, or C curves—depending upon the ratio y_0/y_c. When $y_0/y_c > 1.0$ an M curve exists, when $y_0/y_c = 1.0$ a C curve exists, and when $y_0/y_c < 1.0$ an S curve exists.

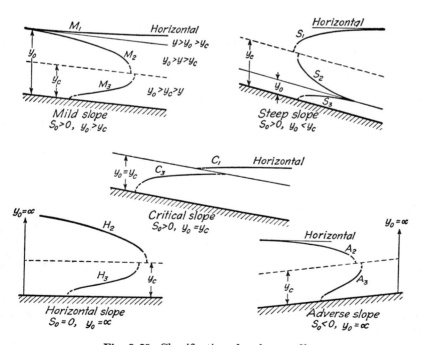

Fig. 8-23. Classification of surface profiles.

A further classification of surface curves depends upon the ratios y_c/y and y_0/y. If both y_c/y and y_0/y are less than 1.0, then the curve is designated as type 1—for example M_1, S_1, C_1 and H_1. If the depth y is between the normal depth y_0 and the critical depth y_c, then it is type 2, and if both y_c/y and y_0/y are greater than 1.0, then the curve is type 3.

Each of the foregoing curves are plotted in Fig. 8-23, where the longitudinal distance has been shortened and the slopes have been exaggerated for the sake of clarity. Actually, the longitudinal distances are so great under natural conditions that the naked eye usually cannot distinguish between flow that is entirely uniform and flow that is gradually varied. Furthermore, the slopes, even when steep, are actually so flat that it makes no difference whether the depth is taken vertically or normally to the bed of the channel. In fact, this stems from the original condition that the effects of acceleration are negligible, and in the regions where there is appreciable surface curvature and acceleration, the foregoing and subsequent analyses do not apply. It should be noted that, for both the adverse and the horizontal slopes, the normal depth y_0 goes to infinity, so that the A_1 and H_1 curves do not exist. Furthermore, with a critical slope, $y_0 = y_c$, so that the C_2 curve does not exist.

The general characteristics of surface profiles are summarized in the following table:

Table 8-2. CHARACTERISTICS OF SURFACE PROFILES

Class	Bed slope	$y : y_0 : y_c$	Type	Symbol
Mild	$S_0 > 0$	$y > y_0 > y_c$	1	M_1
Mild	$S_0 > 0$	$y_0 > y > y_c$	2	M_2
Mild	$S_0 > 0$	$y_0 > y_c > y$	3	M_3
Critical	$S_0 > 0$	$y > y_0 = y_c$	1	C_1
Critical	$S_0 > 0$	$y < y_0 = y_c$	3	C_3
Steep	$S_0 > 0$	$y > y_c > y_0$	1	S_1
Steep	$S_0 > 0$	$y_c > y > y_0$	2	S_2
Steep	$S_0 > 0$	$y_c > y_0 > y$	3	S_3
Horizontal	$S_0 = 0$	$y > y_c$	2	H_2
Horizontal	$S_0 = 0$	$y_c > y$	3	H_3
Adverse	$S_0 < 0$	$y > y_c$	2	A_2
Adverse	$S_0 < 0$	$y_c > y$	3	A_3

The limiting conditions for the water surface profiles may be studied by the conditions $y \to y_0$, $y \to y_c$, and $y \to 0$. When $y \to y_0$, it may be seen by Eq. 8-84 that the numerator goes to zero, which means that the curves always approach normal depth y_0 asymptotically. As $y \to y_c$ the denominator goes to zero, which makes dy/dx go to infinity and each of the curves cross the critical depth line perpendicularly. Such a condition requires rapid acceleration in the vicinity of the critical depth, which is contrary to the original assumption of negligible accelerative effects and resulting hydrostatic pressure distribution. Hence, Eq. 8-84 cannot be expected to do more than approximate the surface profile in the vicinity of $y = y_c$. Finally, as $y \to 0$ the surface curve approaches the bed of the channel perpendicularly, and again the original assumptions are no longer applicable.

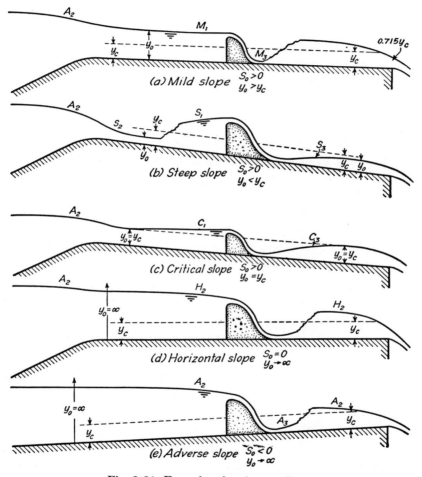

Fig. 8-24. Examples of surface profiles.

Figure 8-23 illustrates actual conditions where backwater curves of the various types occur. Again, the longitudinal distances are very much foreshortened. A sluice gate could be placed where the dam is located in each case, and the same curves would exist. It should be observed that a smooth, although relatively sharp, curve is created when the profile crosses the critical depth line from a greater to a lesser depth, see the A_2 curve and the S_2 curve in Fig. 8-24(b). As the profile crosses from a lesser to a greater depth, however, there is a hydraulic jump which causes a sudden change in elevation—see the downstream sections of Fig. 8-24(a), (d), and (e).

The location of the hydraulic jump can be determined by computing the pair of surface profile curves toward the jump, see Fig. 8-24, until the computed downstream depth reaches the point where it is the sequent depth for the computed upstream depth. In other words, the surge is propagated upstream or downstream until the approaching velocity of flow is just equal to the celerity of the surge in quiet water at the same depth.

In summary, the following statements can be made about the surface profile curves:

1. The type-1 curves are all above both y_c and y_0.

2. The type-2 curves are all between y_c and y_0.

3. The type-3 curves are all below both y_c and y_0.

4. The type-3 curves all approach the bottom perpendicularly to it.

5. All curves approaching the y_0 line approach it asymptotically, except for the C curves, where $y_0 = y_c$.

6. All curves approaching the y_c line approach it perpendicularly, except for the C curves, where $y_0 = y_c$.

7. All curves approaching a horizontal line approach it asymptotically, except for the C curves, which are horizontal throughout.

8. All curves where $y_0 < y_c$ are unaffected upstream from any disturbance, because the wave of the disturbance is swept downstream by the rapid flow. Compute in the downstream direction.

9. All curves where $y_0 > y_c$ are influenced by any downstream disturbance, because the celerity of the wave of the disturbance is so great it travels upstream against the oncoming flow. Compute in the upstream direction.

10. At a dropoff, the brink depth is $0.715y_c$, see Fig. 8-24(a). Critical depth occurs approximately $5y_c$ upstream from the brink, although this is often neglected and y_c is assumed to occur at the brink.

The foregoing discussion of the characteristics of surface profiles in open channels is applicable in general not only to wide uniform channels, but also to natural open channels, provided the changes in shape, roughness, and slope are taken into consideration properly. In the following section, the step method is explained for the actual computation of backwater curves.

• COMPUTATION OF BACKWATER CURVES

The foregoing material explains the various types of backwater curves which can occur in an open channel. The system of computing these curves is explained in this section.

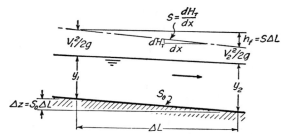

Fig. 8-25. Definition sketch for computation of backwater curves.

Figure 8-25 illustrates a reach of channel sufficiently short so that the water surface can be approximated by a straight line. Equation 8-78 may then be rearranged as

$$\frac{\Delta H}{\Delta L} = S_0 - S \quad \text{or} \quad \Delta L = \frac{\Delta H}{S_0 - S} \qquad (8\text{-}85)$$

in which H is the specific head, $S = -dH_T/dx$ the average of the total energy gradient, and $S_0 = -dz/dx$ the slope of the channel bed. By rearranging the Manning equation $S = Q^2 n^2 / (1.5 A R^{2/3})^2$ to compute the average of the total energy gradient S for the two ends of the reach, the backwater curve may be computed from Eq. 8-85 by a step procedure. The computation is carried upstream for tranquil flow and downstream for rapid flow, so that the effect of any error in the assumed initial conditions will decrease rather than increase as the computation proceeds.

The basic Eq. 8-85 can also be derived by writing the Bernoulli equation between points 1 and 2 in Fig. 8-25 and solving for ΔL.

That is, $\dfrac{V_1^2}{2g} + y_1 + \Delta z = \dfrac{V_2^2}{2g} + y_2 + h_f$

or $\dfrac{V_1^2}{2g} + y_1 + S_0 \, \Delta L = \dfrac{V_2^2}{2g} + y_2 + S \, \Delta L$

and $\Delta L = \dfrac{H_2 - H_1}{S_0 - S} = \dfrac{\Delta H}{S_0 - S}$ \qquad (8-85)

The procedure for computation is to assume a depth y and compute the length of reach ΔL required to reach that depth, as follows:

1. Determine or assume the initial conditions of depth, channel characteristics, and discharge. Insert into a table, as in Example Problem

8-6, the initial values for the depth y, the area A, the wetted perimeter P, the hydraulic radius R, the velocity V, the velocity head $V^2/2g$, the specific head $H = y + V^2/2g$, and the corresponding slope S for the assumed value of Manning's n.

2. Assume a new depth of flow y for the other end of the reach and insert the corresponding values of A, P, R, V, $V^2/2g$, H and S.

3. Compute the average S value and the change in specific head ΔH for the reach.

4. Insert the ΔH value and the average value of S in Eq. 8-85 and solve for ΔL.

5. Repeat the process for each reach, adding the resulting ΔL values to obtain the total required distance $\Sigma \Delta L$, until the depth or distance desired is reached.

The magnitude of the incremental increase or decrease in depth selected for each reach must be a result of a balance between speed of computation on the one hand, and accuracy on the other hand. That is, the smaller the increment, the greater the accuracy but the slower the speed of computation. Experience will soon guide the engineer in this matter.

The foregoing step procedure for computing backwater curves is very simple and is applicable to channels of variable cross section, roughness and slope, provided this information is known accurately for the entire length of channel under study. Other methods have been developed for computing backwater curves which are more rapid. In most cases, however, they involve assumptions which limit their application to uniform channels, or to first approximation computations.

Example Problem 8-6: A lined rectangular channel carries 200 cfs, its width is 10 ft, its slope is 0.001, and $n = 0.02$. If an overflow weir is installed across the channel which raises the water level at the weir to a depth of 6.0 ft, (a) compute the normal depth of flow, (b) compute the distance upstream to where the depth of water is 5.8 ft, and (c) classify the surface profile.

Solution:

(a) $Q = A \dfrac{1.5}{n} R^{2/3} S^{1/2}$

$$200 = (10)y_0 \frac{1.5}{0.02} \left(\frac{10y_0}{10 + 2y_0} \right)^{2/3} (0.001)^{1/2}$$

from which

$$y_0 = \underline{\underline{4.75 \text{ ft}}}$$

(b) The distance L is determined as follows:

y	A	V	$V^2/2g$	H	P	R	n	S^*	S^*_{ave}	ΔH	ΔL	ΣL
6.0	60	3.34	.174	6.17	22	2.73	.02	.000517	.000542	.18	386	
5.8	58	3.45	.185	5.99	21.6	2.68	.02	.000567				386

$$* \; S = \frac{n^2 Q^2}{2.25 A^2 R^{4/3}}, \quad S_{ave} = \left(\frac{n^2 Q^2}{2.25 A^2 R^{4/3}}\right)_{ave}$$

The distance is $L = 393$ ft

Other points on the backwater curve can be determined in the same manner.

(c) M_1 backwater curve

- ## SUMMARY

1. Flow in open channels involves the following *types of flow* and what they express:

    a. Steady $\left(\dfrac{\partial v}{\partial t} = 0\right)$ c. Laminar (Re < 500)

       Unsteady $\left(\dfrac{\partial v}{\partial t} \neq 0\right)$ Turbulent (Re > 500)

    b. Uniform $\left(\dfrac{\partial v}{\partial s} = 0 = \dfrac{\partial v}{\partial n}\right)$ d. Tranquil (Fr < 1.0)

       Nonuniform $\left(\dfrac{\partial v}{\partial s} \neq 0 \neq \dfrac{\partial v}{\partial n}\right)$ Rapid (Fr > 1.0)

2. The **Chezy** *equation* for *steady, uniform flow* is

$$V = C\sqrt{RS} \tag{8-5}$$

3. The **Manning** *equation* is a *variation of the Chezy equation*

$$V = \frac{1.5}{n} R^{2/3} S^{1/2} \tag{8-18}$$

4. For design values of the *Manning* **roughness** *coefficient* see Table 8-1.

5. When **closed conduits** are *flowing partly full*, the same principles govern the flow as for open channels, provided the pressure at the free surface is maintained constant throughout.

6. The *wave equations* for a free-surface are:

$$c = K\sqrt{gy} \qquad\qquad \text{(\textit{Basic equation})} \qquad (8\text{-}40)$$

$$c = \sqrt{gy}\left(1 + \frac{\Delta y}{y}\right)^{1/2} \qquad\qquad \begin{array}{l}\textit{(Wave height small}\\ \textit{compared with depth)}\end{array} \qquad (8\text{-}23)$$

$$c = \sqrt{gy} \qquad\qquad \text{(\textit{Very small wave})} \qquad (8\text{-}24)$$

$$c = \sqrt{gy_1}\left[\frac{1}{2}\frac{y_2}{y_1}\left(\frac{y_2}{y_1}+1\right)\right]^{1/2} \qquad \text{(\textit{Hydraulic jump})} \qquad (8\text{-}33)$$

$$c \approx 2\sqrt{gy} \qquad\qquad \text{(\textit{Flood wave in dry channel})}$$

$$c = \sqrt{g\lambda/2\pi} \qquad\qquad \text{(\textit{Deep-water waves})}$$

7. In *equation form, the Froude number* is

$$\text{Fr} = \frac{V}{\sqrt{\dfrac{\Delta\gamma}{\rho}y}} \qquad (8\text{-}46)$$

which becomes

$$\text{Fr} = \frac{V}{\sqrt{gy}} \qquad (8\text{-}28)$$

for a liquid-gas interface.

8. *Froude number can be defined* as:
    a. Ratio of the influence of inertia to the influence of gravity.
    b. Ratio of velocity of flow to celerity of a small wave in quiet, shallow water.

9. The *basic energy equation* relating specific head H, depth y, and discharge per unit width q is

$$H = \frac{q^2}{2gy^2} + y \qquad (8\text{-}50)$$

which can be analyzed from two viewpoints
    a. Variation of y with H when q is held constant. This yields the **specific head diagram,** for which the critical depth in terms of discharge is

$$y_c = \sqrt[3]{q^2/g} = 2\frac{V_c^2}{2g} \qquad (8\text{-}53)$$

    b. Variation of y with q when H is held constant. This yields the **discharge diagram,** for which the critical depth in terms of specific head is

$$y_c = \frac{2}{3}H = 2\frac{V_c^2}{2g} \qquad (8\text{-}60)$$

10. *Alternate depths* are the two depths at which flow can exist for a given discharge and specific head. One depth is $y < y_c$ and the other depth is $y > y_c$, and flow can change from one to the other without loss of energy.

11. The *sequent depth* is that depth downstream from a hydraulic jump. As the name implies, the process is irreversible, because of the turbulence and resulting loss of specific head.

12. The *characteristics of critical flow* are:

  a. $V = c = \sqrt{gy}$ (8-70)

  b. $\mathrm{Fr} = \dfrac{V}{\sqrt{gy}} = 1.0$ (8-71)

  c. $H_{min} = \dfrac{3}{2}\sqrt[3]{\dfrac{q^2}{g}}$ (8-72)

  d. $q_{max} = \sqrt{g\left(\dfrac{2}{3}H\right)^3}$ (8-73)

  e. $y_c = 2\dfrac{V_c^2}{2g}$ (8-74)

  f. $y_c = \dfrac{2}{3}H$ (8-75)

  g. $y_c = \sqrt[3]{q^2/g}$

  h. Waves form easily because a large change in water surface elevation results in only a small change in specific head.

13. The *basic equation for classification of water surface profiles* is

$$\frac{dy}{dx} = S_0 \left[\frac{1 - \left(\dfrac{y_0}{y}\right)^{10/3}}{1 - \left(\dfrac{y_c}{y}\right)^{3}} \right]$$ (8-84)

14. The *characteristics of water surface profiles* are tabulated in Table 8-2.

15. The basic equation for *computation of backwater curves* is

$$\Delta L = \frac{H_2 - H_1}{S_0 - S} = \frac{\Delta H}{S_0 - S}$$ (8-85)

References

1. Addison, Herbert, *Applied Hydraulics*. New York: John Wiley & Sons, Inc., 1945.

2. Bakhmeteff, B. A., *Hydraulics of Open Channels*. New York: McGraw-Hill Book Co., Inc., 1932.

3. Francis, J. R. D., *Fluid Mechanics*. London: Edward Arnold, Ltd., 1958.

4. Ippen, A. T., "Channel Transition and Controls," Chapter VIII of *Engineering Hydraulics*, ed. Hunter Rouse. New York: John Wiley & Sons, Inc., 1950.

5. Jaeger, Charles, *Engineering Fluid Mechanics*. London: Blackie & Son, Ltd., 1956.

6. Keulegan, G. H., "Wave Motion," Chapter XI of *Engineering Hydraulics*, ed. Hunter Rouse. New York: John Wiley & Sons, Inc., 1950.

7. King, Horace W. and Ernest F. Brater, *Handbook of Hydraulics*, 4th ed. New York: McGraw-Hill Book Co., Inc., 1954.

8. Posey, C. J., "Gradually Varied Open Channel Flow," Chapter IX of *Engineering Hydraulics*, ed. Hunter Rouse. New York: John Wiley & Sons, Inc., 1950.

9. Powell, Ralph W., *Hydraulics and Fluid Mechanics*. New York: The Macmillan Co., 1951.

10. Rouse, Hunter, *Fluid Mechanics for Hydraulic Engineers*. New York: McGraw-Hill Book Co., Inc., 1938.

11. Simons, D. B. and E. V. Richardson, *Forms of Bed Roughness in Alluvial Channels*. Ft. Collins, Colorado, USGS Paper, 1959.

12. Woodward, S. M. and C. J. Posey, *The Hydraulics of Steady Flow in Open Channels*. New York: John Wiley & Sons, Inc., 1941.

PROBLEMS

8-1. A rectangular channel 10 ft wide is carrying 300 cfs at a depth of 6 ft. Is the slope steep or mild?

8-2. A variable-slope flume 8 ft wide carries uniform flow at the following depths: $\frac{1}{2}$ ft, $\frac{3}{4}$ ft, 1 ft, 1.5 ft, and 2 ft. If the discharge is 30 cfs, determine whether the flow is tranquil or rapid for each depth.

8-3. If the Froude number for each flow in a rectangular channel 3 ft wide is 0.5, what is the discharge condition for each of the following depths: 0.5 ft, 1.0 ft, and 2.0 ft?

8-4. A trapezoidal channel with a bottom width of 5 ft and with side slopes of 2H:1V has a slope of 1 ft in 3000 ft. If the depth of water flowing in the channel is 8 ft, what is the average shearing stress at the boundary?

8-5. What is the average shear stress at the sides and bottom of a trapezoidal flume 8 ft wide at the bottom with side slopes of 2H:1V, flowing 3 ft deep with a gradient of 0.004? What is the Manning n of the channel? $Q = 100$ cfs.

8-6. Using Chezy's discharge coefficient C and Manning's formula, determine the relationship between the resistance coefficient f and the Manning n.

8-7. If the value of the friction coefficient f is 0.025, what is the corresponding value of the Chezy coefficient C?

8-8. A trapezoidal channel has a bottom width of 8 ft and side slopes of $1\frac{1}{2}$ H:1V. If the uniform flow is 4 ft deep, calculate the average shearing stress at the boundary for a slope of 1 ft in one mile. Repeat for a depth of 8 ft.

8-9. The discharge in a trapezoidal channel is 3000 cfs. The depth of flow is 10 ft and the bottom width is 35 ft. The side slopes are 2H:1V. Is the flow rapid or tranquil?

8-10. List six equations used for determining the average velocities in open channels. List the author and the approximate limitations for each formula. Consult the library.

8-11. A rectangular channel 8 ft wide of smooth concrete has a slope of 0.002. For a uniform flow of 300 cfs, what is the depth of flow?

8-12. A trapezoidal channel with a bottom width of 12 ft and side slopes of 2H:1V has an estimated n value of 0.017. For a discharge of 500 cfs, what is the maximum permissible slope if the maximum mean velocity to prevent scour is limited to 2.5 fps?

8-13. A concrete canal with a trapezoidal cross-section has a bottom width of 30 ft and side slopes of $1\frac{1}{2}$H:1V. The depth of uniform flow in the canal is 12 ft. If the slope of the canal is 3 ft per mile, what is the discharge?

8-14. A rectangular channel 4 ft wide is made of old concrete with an n value of 0.015. If the slope of the flume is 0.008, the discharge is 130 cfs, and the flow is uniform, (**a**) what is the depth of flow? (**b**) Is the type of slope mild or steep?

8-15. In Problem 8-14, if the channel is repaired and carefully trowelled so that it is now very smooth concrete, how will this affect (**a**) the depth of flow, (**b**) the type of slope? The discharge and slope are still 130 cfs and 0.008 respectively.

8-16. 300 cfs of water flow in an open rectangular channel ($n = 0.02, b = 10$ ft, $y = 6$ ft). What is the slope of the channel?

8-17. An earth canal in fair condition has a bottom width of 10 ft and side slopes of $1\frac{1}{2}$H:1V. What is the maximum discharge this canal will carry if the slope is 1 ft in 4000 ft and the maximum depth is 8 ft?

8-18. A trapezoidal channel of good wood has a 20 ft bottom width and a depth of 10 ft. The side slopes are $1\frac{1}{2}$H:1V. If the discharge is measured by current meter and found to be 1250 cfs, what is the slope of the channel?

8-19. A planed timber triangular flume is carrying water at a depth of 3 ft. The apex of the triangle is the bottom of the flume, and the angle at the apex is 90°. If the slope of the flume is 0.0003, what is the discharge for *uniform flow?*

8-20. An 18-in. circular pipe is carrying water at a uniform depth of 15 in. down a slope of 0.001. If the value of n is 0.019, what is the discharge in the pipe?

8-21. At a point in a wide channel where the depth is 10 ft, the discharge per unit width is 70 cfs. Plot the vertical velocity distribution curve, $f = 0.10$.

8-22. If a vertical velocity curve in a channel gives a surface velocity of 4.04 fps and a maximum velocity of 4.26 fps at 0.21 depth below the surface, and if the vertical velocity curve is assumed to be a parabola with axis horizontal, what is the mean velocity in the vertical and at what depth does it occur?

8-23. At what point below the water surface does the average velocity exist in a wide channel whose slope is 0.0002, if the discharge is 10 cfs/ft and the Manning coefficient is 0.018? How does this compare with the 6/10 depth below the surface commonly used as the point at which the average velocity occurs?

8-24. Plot the velocity distribution curve given by Eq. 8-12. Plot v/V (abscissa) against y/y_0 (ordinate).

8-25. Find the diameter for circular pipes such that four of them flowing half full will just carry the water delivered by an open channel of half-square section 12 ft wide and 6 ft deep. The slope of both channel and pipes is 0.0009 ft/ft. Assume a value of Chezy C of 124.

8-26. For wide rivers, the hydraulic radius may be taken equal to the depth of water. (**a**) Derive the relation between slope and depth for constant discharge, assuming that the Manning n remains constant. (**b**) If the depth is reduced to one-half, how much will the slope change?

8-27. For Fig. 8-6, prove that the optimum shape of an open channel of trapezoidal cross-section is $R = y_0/2$ where $y_0 = $ depth of flow.

8-28. Prove that for a rectangular channel, $R = y_0/2$ and $B = 2y_0$ for a hydraulically optimum shape.

8-29. A concrete rectangular channel is to be designed to carry 300 cfs down a slope of 0.00071. If the flow is uniform and the channel is to have the optimum hydraulic cross-section, design the size of the channel.

8-30. A rectangular flume carries a discharge of 306 cfs. The flow is uniform, the slope of the flume is 0.0016, and n equals 0.018. If at this discharge condition, the flume has the optimum cross-section for the given area, what is the width of the flume and the depth of flow? Is the flow rapid or tranquil, and is the slope steep or mild?

8-31. A semicircular canal of radius 6 ft discharges 250 cfs with Chezy's C equal to 114. Determine the slope of the canal.

8-32. A vitrified sewer pipe is laid on a slope of 0.0017 ft/ft and is to carry 71.8 cfs when the pipe flows 0.75 full. Determine the diameter of the pipe.

8-33. A trapezoidal channel in the form of a half hexagon when flowing full carries 250 cfs a distance of 5000 ft with a fall of 1.9 ft. What is the area of the cross section? What area is required with the same side slopes but with the bottom width three times the depth? In both cases $n = 0.018$.

8-34. For a constant cross-sectional area of 100 sq ft in a rectangular channel, plot the relationship between the width of the channel and the hydraulic radius. Briefly discuss the significance of this plot.

8-35. For a constant cross-sectional area of 3 sq ft in a triangular flume with the apex at the bottom, plot the relationship between the angle at the bottom and the hydraulic radius. Briefly discuss the significance of this plot.

8-36. A rectangular channel with an n value of 0.014 and a slope of 0.0002 carries 100 cfs for a channel width of 100 ft. Is the flow in this channel tranquil or rapid?

8-37. Find the radius of a smooth concrete pipe to carry 40 million gallons of water per day flowing $\frac{3}{4}$ full at a gradient of 1 in 4000.

8-38. A smooth concrete pipe 6 ft in diameter has a gradient of 1 in 4000. Find the discharge when flowing (**a**) $\frac{1}{4}$ full, and (**b**) $\frac{3}{4}$ full, using the curve of Fig. 8-7.

8-39. Determine the most efficient or optimum section of a trapezoidal channel, $n = 0.018$, to carry 600 cfs. To prevent scouring, the maximum velocity is to be 2.6 ft/sec and the side slopes of the trapezoidal channel are $1\frac{1}{2}$V:2H. What slope of the channel is required?

8-40. Find the rate of flow, by means of Manning's formula, of a river in fair condition with a surface slope of 0.00013, if the mean area of channel section is 4500 sq ft. The average width is 460 ft and the hydraulic radius may be taken as equal to the mean depth.

8-41. A trapezoidal canal having a base width of 30 ft with side slopes of 2H:1V has an earth bottom in good condition. The velocity of flow is 2.8 ft/sec and the measured discharge is 400 cfs. Determine the slope of the canal, using Manning's formula.

8-42. The area of flow in an ordinary concrete pipe was found to be 0.28 sq ft; the wetted perimeter 1.60 ft; the fall 1 ft in 38.7 ft. The mean velocity of flow was found to be 6.12 fps. Find the value of Chezy discharge coefficient C.

8-43. An aqueduct 10 miles long consists of 3 miles of siphon, and the remainder of smooth concrete culvert pipe 84 in. in diameter with a gradient of 1 ft per mile. The siphon consists of cast iron pipe 48 in. in diameter with a change of elevation of 10 ft in 3 miles. Determine the maximum discharge, if the concrete pipe cannot flow more than $\frac{1}{2}$ full.

8-44. An irrigation canal, with side slopes of $1\frac{1}{2}$:1, receives 1000 cfs. Design a suitable channel of smooth earth sides and bottom having a depth of 3.5 ft and determine its dimensions and slope. The mean velocity is not to exceed 2.25 ft/sec. Assume a freeboard of 1.5 ft.

8-45. A large river in flood stage has the cross section shown. What is the discharge if the slope is 1 ft per mile? The Manning n for the main channel is 0.018, and for the overflow section n is estimated to be 0.040.

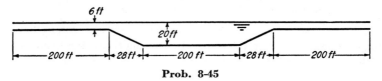

Prob. 8-45

8-46. A flood occurs in a main channel which is 40 ft wide with the flow in this section 12 ft deep. The flood spills out over an almost horizontal flood plain which is 200 ft wide, and the depth of overflow is 3 ft. If $n = 0.025$ for the main channel and is three times as large for the overflow section, estimate the discharge if the slope is 0.0003.

8-47. If the depth of water is 10 ft, calculate the celerity of a gravity wave, using Eq. 8-24. Using Eq. 8-22, determine the maximum wave height which will give a wave celerity within 5% of the value determined by Eq. 8-24.

8-48. Make a plot of wave celerity c (ordinate) against wave height y (abscissa) for a depth of flow of 10 ft.

8-49. A deep water wave has a measured celerity of 30 mph. What is the wave length?

8-50. The average distance between deep water wave crests is 30 ft. How fast is the wave travelling?

8-51. A rock lands in a pond 3 ft deep. If the wave height is very small, estimate the wave celerity.

8-52. Compare Eq. 8-22 with Eq. 8-23 by plotting celerity (ordinate) against y (abscissa), for a depth of water of 8 ft.

8-53. A deep water wave has a celerity of 15 fps. How high can the wave get before it breaks?

8-54. The velocity upstream from a hydraulic jump is 40 ft/sec and the depth is 4 ft. Find the depth y_2 downstream from the jump.

8-55. A hydraulic jump forms at the base of a dam and the depth downstream from the jump is 10 ft, while the depth just upstream from the jump is 2 ft. What is the discharge over the spillway, if the spillway is 100 ft wide?

8-56. Compare Eq. 8-35 with Eq. 8-36 by plotting Fr_1 (abscissa) against y_2/y_1 for each equation. Discuss the result.

8-57. Using the results of Problem 8-55, determine the head loss across the hydraulic jump.

8-58. For a value of $Fr_1 = 3$ and $V_1 = 20$ fps, determine the value of V_2. Consider the decrease in velocity in the light of the decreased capacity of the flow to cause scour in an earth channel.

8-59. The discharge per ft over an ogee spillway may be evaluated by the equation $q = Ch^{3/2}$. Assuming a flood over the spillway of 40 cfs/ft and a value of $C = 4.0$, determine the depth of flow downstream from the spillway which is required to cause a hydraulic jump to form (see Fig. 8-20), if the dam is 20 ft high. Neglect friction losses down the spillway.

8-60. Derive Eq. 8-35 from application of the momentum and continuity equations.

8-61. Derive an equation for y_2 in terms of y_1 and q for the hydraulic jump.

8-62. What is the approximate celerity of a wave 3 ft high travelling down a dry channel?

8-63. Find the minimum slope required for roll waves to form in a channel with a Chezy discharge coefficient of 140. Flow is turbulent.

8-64. What is the minimum slope required for roll waves to form in a channel with a Chezy C equal to 80? Flow is turbulent.

8-65. The flow in a 60-ft wide rectangular channel is 4000 cfs. The n value is 0.015, and the slope is 0.0003. Assuming uniform flow, what would happen to the water surface if the bottom of the flume were suddenly raised 6 in.? Neglect losses in the constriction.

8-66. Plot the specific heat diagram for a wide channel carrying 40 cfs/ft. What is the critical depth, the minimum specific energy? Find the alternate depth for a depth of 3 ft.

8-67. A rectangular channel 20 ft wide carries 1000 cfs. Determine the minimum specific energy required to produce flow in the channel, by plotting depth of flow against specific head. What type of flow exists when the depth is 1.5 ft and when it is 6 ft? For $n = 0.015$, what slopes are necessary to maintain depths of flow of 2.0 ft and 5.0 ft?

8-68. A concrete rectangular canal with a bottom width of 10 ft carries a discharge of 275 cfs. If the total head measured above the bottom of the channel is 8.0 ft, what are the alternate depths?

8-69. The cross section of a mountain stream with a rocky bed approximates a rectangular channel with a bottom width of 24 ft. The grade is $2\frac{1}{2}$ ft per 1000 ft. The flow is estimated to be 250 cfs at a depth of 3 ft. Is this depth greater or less than the normal depth? Is it greater or less than the critical depth?

8-70. A discharge of 400 cfs is flowing in a rectangular channel 20 ft wide. If there is a rise in floor elevation of 1.0 ft and the flow passes through the transition at minimum specific head, what are the depths upstream and downstream from the rise in the channel bottom? Neglect losses.

8-71. The floor elevation of a 12-ft rectangular channel rises 9 in. If the mean velocity upstream from the rise in the floor is 4 ft/sec and the depth is 4 ft, what is the change of water surface elevation through the transition? Neglect losses.

8-72. Rapid flow occurs in a rectangular channel 15 ft wide. If the Froude number is 3.0 and the depth is 3.0 ft, what is the critical rise in bed elevation as shown in Fig. 8-17(b)?

8-73. Rapid flow occurs upstream from a small rise of 6 in. in bed elevation. If the discharge is 600 cfs, the rectangular channel is 12 ft wide, and the upstream depth is 2.5 ft, what is the depth of flow over the small rise of bed elevation?

8-74. Plot the discharge diagram for a constant value of $H = 5.3$ ft. Find the critical depth and the maximum q for this condition.

8-75. A rectangular channel 10 ft wide carries 360 cfs at a depth of 12 ft. What is the maximum side-wall contraction which can occur in this channel without changing the specific head? What will be the depth of flow in the contraction? Neglect losses.

8-76. 1000 cfs are flowing in a rectangular channel 20 ft wide. If the depth of flow is 2 ft, what will happen to the depth if the channel contracts to a 19 ft width? Neglect losses.

8-77. A rectangular channel 30 ft wide carries 1200 cfs at a depth of 2 ft. What is the maximum side-wall contraction which will not change the specific head? Neglect losses.

8-78. A rectangular channel 18 ft wide has a side contraction of 2 ft, making the width 16 ft in the contraction. The depth upstream from the contraction is 8 ft for a flow of 360 cfs. In addition to the side contraction, what can be done to keep the water surface horizontal through the contraction?

8-79. A rectangular channel 20 ft wide expands to a width of 21.5 ft. If in addition to the side expansion the floor elevation raises 1 ft, what is the elevation of the water surface downstream from the change of boundary? The discharge in the 20-ft channel section is 400 cfs at a depth of 5 ft. Neglect losses.

8-80. A sluice gate as shown in Fig. 8-20(b) is 1 ft from the bottom of the flume. If the contraction coefficient for the gate is 0.61 and the discharge is 10 cfs/ft, what is the depth upstream from the gate? What is the specific head? Find the Froude number upstream and downstream from the gate.

8-81. If the discharge is 800 cfs in a rectangular channel 20 ft wide, find the alternate depth and the sequent depth for a depth of 1 ft. What is y_c for this discharge?

8-82. A channel has 1H:1V side slopes and a bottom width of 8 ft. Find the critical depth when $Q = 100$ cfs. What is the critical velocity?

8-83. A flow of 10 cfs occurs in a semicircular conduit with a diameter of 4 ft. Find the critical depth.

8-84. A semielliptical conduit, when flowing full has a depth of 4 ft and a width of 12 ft. Find critical depth for a discharge of 30 cfs.

8-85. A trapezoidal channel with side slopes of 2H:1V and a bottom width of 10 ft flows with a depth of 5 ft. If the discharge is 600 cfs, is the flow rapid or tranquil?

8-86. In Problem 8-85, for what discharge would the flow be critical?

8-87. A wide rectangular channel made of ordinary concrete has a discharge of 20 cfs/ft. The slope of the channel is 0.0005. Find the slope of the energy line if the depth is 4 ft. Is the flow rapid or tranquil?

8-88. Starting with Eq. 8-78, derive Eq. 8-84.

8-89. A wide river in fair condition is discharging 60 cfs/ft. The slope of the stream is 1 ft per mile. If the depth of flow is 7 ft, determine the water surface slope, the slope of the energy line, the type of backwater curve, and whether the flow is tranquil or rapid.

8-90. The depth of uniform flow in a wide canal in poor condition is 6 ft. Find the discharge per foot of width and state whether the flow is tranquil or rapid. The slope of the canal is 0.0002.

8-91. The discharge per foot in a wide channel is 50 cfs/ft. If the channel slope is 0.002, $n = 0.012$, and the depth of flow is 2 ft, determine dy/dx, the type of backwater curve, and whether the flow is tranquil or rapid.

8-92. Water flows over the crest of a dam into a long rectangular chute ($n = 0.012$) which has a width of 10 ft and carries a flow of 500 cfs. The depth at point A some distance downstream from the crest is 4.0 ft. If the slope of the flume is 0.01, how far from point A is the depth 3.5 ft? What is the type of backwater curve?

8-93. A smooth concrete-lined rectangular canal 6 ft wide has a slope of 1 in 1200, and the normal depth of flow is 2.3 ft. If the depth is increased to 3.5 ft by a weir, find how far back from the weir the depth will be 2.8 ft. What type of backwater curve is this?

8-94. A long straight wide channel of ordinary concrete ends in an abrupt drop-off. If the depth at the brink of the dropoff is 3 ft and the slope of the channel is 0.0001, find (a) the discharge per foot, (b) the point upstream from the brink where $y = y_c$, and (c) the depth of flow 500 ft upstream from the brink.

8-95. The chute spillway for Deercreek reservoir in Utah has the approximate dimensions shown in the figure. Using a design flood of 10,000 cfs for the 50-ft wide spillway and $n = 0.013$, plot the backwater curve from A to D. Classify the type of backwater curve in each section and give the required elevation of point E for the jump to occur. Assume a straight line from A to B.

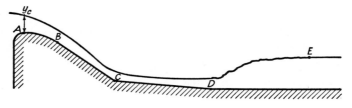

Prob. 8-95

Point	Station	Elevation
A	0 + 00	5270
B	5 + 00	5265
C	15 + 00	5235
D	35 + 00	5200

8-96. A river 100 ft wide is flowing at a rate of 10,000 cfs. Analyzing the problem as a wide stream, what is the backwater curve upstream from an irrigation diversion dam 12 ft high, if the head on the crest of the dam is 9 ft? If the river slope is 0.0003 and the Manning coefficient is 0.020, how far upstream from the dam does the M-1 backwater curve extend to $y = 0.99\ y_0$?

8-97. A horizontal flume of smooth concrete 20 ft wide drains a lake. If the flume is 1000 ft long and ends in a sudden drop-off, what is the discharge in the flume when the water surface in the lake is 8 ft above the bottom of the flume?

8-98. A long planed timber chute is 10 ft wide and carries 150 cfs. The slope of the flume is 0.02 and the depth of flow at point A is 1.6 ft. Find normal depth and the point B where the depth becomes or is within 1% of normal. How far is it from A to B, and what type of backwater curve exists?

8-99. In Fig. 8-24(d), the depth just upstream from the dam is 12 ft. If the channel is wide and $n = 0.025$, what is the depth of flow 1000 ft to the left of the dam, assuming $q = 72$ cfs/ft.

8-100. A slope of 0.0001 ends with a slope of 0.03. The n value for this wide straight channel is 0.014. The discharge is 45 cfs/ft. If the depth at the break in grade is critical depth, determine the depth 500 ft upstream and downstream from the break in grade. What type of backwater curves exist? Does normal depth occur within 500 ft of the break?

chapter 9

Flow Around
Submerged Objects

Submerged objects are frequently encountered in the study of fluid motion. Examples range from tiny objects such as water droplets in fog and fine sediment in air or water, to large objects such as aircraft, automobiles, and ships moving through air or water, and topographic features such as mountains and valleys, as well as buildings, bridges and other structures where all or part of the structure is submerged in a fluid such as air or water. The flow about the object may be due either to the movement of the fluid, or to the movement of the object, or a combination of both. However, in order to simplify the analysis of the flow patterns, only the steady flow of a fluid around the object is considered in this text. In other words, consideration is given only to those flow systems where the flow either is already steady, or can be made steady by application of the principle of relative motion.

It is obvious that the various submerged objects to be considered have innumerable shapes. However, for simplicity in understanding the basic phenomena involved, only simple shapes, such as bodies of revolution, plates and cylinders, are considered in detail.

Study of flow around submerged objects involves consideration of the distributions of pressure and shear around the object, the separation zones and wakes around and downstream from the object, and the resultant drag on the object. Each of these subjects, together with circulation and lift, is given detailed consideration in this chapter.

• TYPES OF DRAG

In Chapter 5 on Fluid Resistance, consideration was given to the two basic types of drag—*shear drag,* caused by tangential shear along the boundary, and *pressure drag,* caused by the pressure distribution which is applied normal to the surface of the boundary. The latter, for obvious reasons, is frequently called form drag. Pure shear drag is illustrated by the flow around a flat plate or a disk oriented parallel to the flow, see Fig. 6-1(b). Pure pressure drag is illustrated, see Fig. 6-2, by the flow around a flat plate or a disk oriented perpendicularly to the flow. Most problems of flow around submerged objects involve both shear drag and pressure drag, see Fig. 9-1(b). The *general drag equation,* see Eq. 6-4, is

$$F_D = \frac{C_D A \rho V_0^2}{2} \tag{9-1}$$

in which C_D is the dimensionless drag coefficient;

A is the projected cross-sectional area (L^2);

ρ is the density of the fluid (FT^2/L^4);

V_0 is the velocity of the ambient fluid (L/T).

The pressure distribution with irrotational flow serves as a basis of comparison with the pressure distribution and shear distribution when a viscous fluid is involved.

• IRROTATIONAL FLOW AROUND SUBMERGED OBJECTS

The pressure and velocity distribution for irrotational flow around a sphere is shown in Fig. 9-1(a). In this case, the flow pattern is symmetrical on the upstream side and the downstream side of the object. Hence, the corresponding velocity distribution and pressure distribution over the

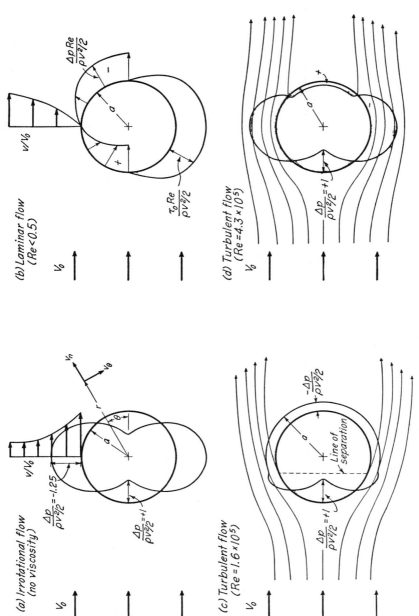

Fig. 9-1. Distribution of pressure and velocity for flow around a sphere.

object are also symmetrical. Furthermore, with irrotational flow the fluid is assumed to have no viscosity and, therefore, no shear can result. As a consequence, there is no drag on the object caused either by a pressure difference on the upstream and downstream sides of the object, or by shear on the boundary of the object.

Of particular importance is the fact that, for irrotational flow, the velocity at the outer edges of the boundary is a maximum which, by the energy equation, results in a minimum pressure. These facts will be used later to compare and to contrast irrotational flow with laminar flow and turbulent flow around submerged objects.

• LAMINAR FLOW AROUND SUBMERGED OBJECTS AT SMALL REYNOLDS NUMBERS

Because flow around submerged objects, like flow in a conduit such as a pipe or open channel, involves the forces of both inertia and viscosity, it is logical that the Reynolds number (which is a ratio of these two forces) would be involved. When the Reynolds number is very small, say Re < 0.5, the flow about a submerged object is laminar and the shape of the object is of secondary importance in regard to the drag, compared with the size of the object, the viscosity of the fluid, and the velocity of flow. Stokes derived a mathematical solution for the purely *laminar* condition of *flow past a sphere*—the effects of inertia being ignored completely—see Eq. 5-28.

$$F_D = 3\pi d\mu V_0 \qquad (9\text{-}2)$$

in which F_D is the drag on the sphere (F);

$\qquad d$ is the diameter of the sphere (L);

$\qquad V_0$ is the velocity of flow past the sphere (L/T);

$\qquad \mu$ is the dynamic viscosity (FT/L^2).

This equation is developed for the condition of viscous flow, and has been proved valid for small Reynolds numbers. The velocity, pressure, and shear distribution based on the Stokes theory are illustrated in Fig. 9-1(b). If Eq. 9-2 is combined with Eq. 9-1, the drag coefficient C_D is found to be

$$C_D = \frac{24}{\text{Re}} \qquad (\textit{Sphere } \text{Re} < 0.5) \qquad (9\text{-}3)$$

in which Re is the Reynolds number, $\text{Re} = V_0 d/\nu$;

$\qquad \nu$ is the kinematic viscosity of the fluid (L^2/T).

Equations similar to Eq. 9-3 have been derived for other objects at small Reynolds numbers. For a *disk perpendicular to the laminar flow*, the drag coefficient is

$$C_D = \frac{20.4}{\text{Re}} \qquad (\text{Re} < 0.5) \qquad (9\text{-}4)$$

and for a *disk parallel to the flow*, the drag coefficient is

$$C_D = \frac{13.6}{\text{Re}} \qquad (\text{Re} < 0.1) \qquad (9\text{-}5)$$

The drag coefficient for two-dimensional laminar flow past a *circular cylinder* is

$$C_D = \frac{8\pi}{\text{Re} \ (2.0 - \ln \text{Re})} \qquad (\text{Re} < 0.1) \qquad (9\text{-}6)$$

and for a *flat plate perpendicular to the flow* it is

$$C_D = \frac{8\pi}{\text{Re} \ (2.2 - \ln \text{Re})} \qquad (\text{Re} < 0.1) \qquad (9\text{-}7)$$

Two-dimensional laminar flow past a *flat plate* which is oriented *parallel to the flow* is somewhat different from that flow discussed in Chapter 6 for which the drag coefficient C_D varies inversely with $(\text{Re})^{1/2}$. For very small Reynolds numbers, the drag coefficient is

$$C_D = \frac{4.12}{\text{Re}} \qquad (\text{Re} < 0.01) \qquad (9\text{-}8)$$

The area A in Eq. 9-1 must include both sides of the plate or disc when C_D is computed by Eq. 9-8 or Eq. 9-5, respectively. This is in contrast to the area used in Eq. 9-1 when applied to other equations, Eqs. 9-3-4-6-7, where it is the projected area normal to the direction of flow.

The coefficients expressed in Eqs. 9-3 to 9-8 are applicable at *small Reynolds numbers*. However, in the case of a gas, the size of the particle must be large compared with the length of the mean free path of the gas molecules, or the gas can no longer be considered a continuous fluid medium. When the particles are smaller than the mean free path, Brownian movement results. Since the mean free paths of gas molecules are in themselves extremely small, the submerged objects (and likewise Reynolds number) can be quite small, as small as $d = 4 \times 10^{-6}$ in., before difficulty is encountered. The objects considered in this chapter are larger than this size.

In order to gain an understanding of the basic causes of the drag and its variation with Reynolds number and shape of the object, a detailed study of the distribution of velocity, pressure and shear is necessary.

Although laminar flow is simply that flow in which all momentum transfer and mixing takes place by molecular action only (i.e., no turbulent mixing), there are two extremes of laminar flow—one at small values of Reynolds number and the other at moderate values of Reynolds number.

On the one hand, when Reynolds number is small, there are no effects of inertia either near the submerged object or far away from it. Under these conditions, there is widespread deformation of the flow pattern extending to a great distance from the object. Furthermore, the theory of Stokes shows that the velocity of flow progresses from zero at the boundary to the ambient velocity V_0 at some great distance from the boundary, as shown in Fig. 9-1(b). The distance from the boundary must be four diameters before the velocity reaches even 92% of the ambient velocity V_0. As one progresses around the sphere, the *pressure changes* as follows:

1. A pressure at the upstream center of the sphere, which is 6/Re times stagnation pressure $\rho V_0^2 / 2$ greater than the ambient pressure,

2. Ambient pressure at the circumference perpendicular to the flow,

3. Negative pressure having a magnitude 6/Re times stagnation pressure at the downstream centerline.

The *shear,* on the contrary, increases from zero at the upstream and downstream centerline points to a maximum at the circumference equal in magnitude to that of the maximum pressure on the fore part of the sphere. The longitudinal components of these distributions of pressure and shear (based upon the theory of Stokes) combine to give the total drag on the sphere—those components of both shear and pressure in the transverse direction balancing each other. The contribution of the shear to the total drag is twice the contribution of the combined negative and positive pressures. These proportions of shear drag and pressure drag will be compared later with those at larger Reynolds numbers.

In the direction of flow, and at any point on the sphere, the longitudinal components of shear drag and pressure drag combine to equal the same value of unit drag over the entire surface of the sphere. Hence, the total drag is a product of the surface area πd^2 and the combined longitudinal components of pressure and shear $3\mu V_0 / d$ at all points on the sphere, which gives Stokes equation

$$F_D = 3\pi d\mu V_0$$

The distribution of velocity, pressure, and shear in Fig. 9-1(b) is in sharp contrast to that for irrotational flow past a sphere, see Fig. 9-1(a), where the velocity is a maximum at the boundary—decreasing to the ambient velocity a relatively short distance from the boundary. The pressure distribution for irrotational flow varies from a maximum pressure on the upstream centerline (somewhat like the case of laminar flow) and on the downstream centerline (just the opposite from the laminar flow case), to a minimum at the circumference which is even less than *negative stagnation.*

For laminar flow, however, the pressure is positive on the upstream side and negative on the downstream side of the sphere by the same amount. Furthermore, the magnitude of the pressure is dependent not only upon the density ρ and the velocity V_0 (as it is for irrotational flow), but also upon the viscosity μ. Hence, the pressure at any point relative to stagnation pressure is a function of Reynolds number Re. For a sphere, the longitudinal components of shear and pressure combine to give a unit drag f_D at any point of

$$\frac{f_D}{\rho V_0^2 / 2} = \frac{6}{\text{Re}} \tag{9-9}$$

In other words, for small Reynolds numbers (say Re < 1.0), the maximum pressure on the upstream part of the sphere is many times greater than stagnation pressure, which is the maximum possible for irrotational flow. In fact, when Re = 0.1, the maximum pressure rise is sixty times as great as stagnation pressure $\rho V_0^2 / 2$.

The striking contrast between laminar flow and irrotational flow becomes even greater when shear is considered—no shear whatsoever being involved in irrotational flow.

The variation of the drag coefficient with Reynolds number (Stokes law, Eq. 9-3) is plotted in Fig. 9-2. Plotted also is the theoretical equation, Eq. 9-4, for a circular disk perpendicular to the flow. It should be noted that the difference between the coefficient 24 for a sphere and 20.4 for a disk is rather small, despite the fact that the entire drag for the disk is pressure drag—compared with pressure drag being only one-third of the total for a sphere. All shear drag is radially outward on the face of the disk, and hence balances itself around the center point of the disk. Therefore, the pressure drag for a disk is nearly three times the pressure drag for a sphere of the same diameter in laminar flow. The rather close agreement between the total drag for a disk and a sphere is due to the fact that the widespread deformation of the flow is essentially the same for a given projected area, regardless of the shape of the object. Furthermore, experiment has shown that the drag for a disk and the drag for a square plate of the same area are almost identical, for a given Reynolds number.

In contrast to the disk perpendicular to the flow, which has purely pressure drag, is the disk parallel to the flow, which has purely shear drag. For the disk parallel to the flow, the coefficient in the drag equation is 13.6, which shows that the drag on a disk (considering the area on both sides) parallel to the laminar flow is only slightly more than the drag on a sphere of the same diameter.

Laminar flow about a cylinder is similar in many respects to the laminar flow around a sphere, except that, if the cylinder is long, the influence of flow around the ends is insignificant. The drag on a cylinder and flat plate perpendicular to the flow is apparently different in some respects, as evidenced by the nature of the equation for the drag coefficient. With the logarithmic nature of these equations, the curve approaches a -1 slope only in the limit, as Re goes to zero. No data exist to check these equations.

The drag on a flat plate parallel to the flow is purely shear drag, as it is for the disk. However, the drag per unit area for the flat plate, which has a coefficient of 4.12, is about one-third that for the disk, which has a coefficient of 13.6.

- ## TRANSITION FROM SMALL TO MODERATE REYNOLDS NUMBERS

The discussion thus far has been limited to small Reynolds numbers, where the influence of inertia is insignificant compared with the influence of viscosity. As the Reynolds number of the flow about a submerged object is increased, the influence of inertia becomes increasingly pronounced, until eventually, at large Reynolds numbers, the situation is completely reversed and the influence of viscosity becomes small compared with the influence of inertia.

The change from one condition to another usually takes place gradually, as shown for the flat plate in Fig. 9-3. For the range of Reynolds numbers where Re < 0.5, the drag coefficient C_D clearly varies inversely with the Reynolds number (Eq. 9-8). With increasing Reynolds numbers, however, a transition develops and the drag coefficient approaches the curve where it varies inversely with $\sqrt{\text{Re}}$, which is the Blasius curve, Eq. 6-14, presented in Chapter 6. This equation is applicable in the range $10^4 < \text{Re} < 4 \times 10^5$ —beyond which the flow is entirely turbulent. Kuo has derived a theoretical equation which expresses the variation of C_D with Re throughout the range of laminar flow

$$C_D = \frac{1.33}{\sqrt{\text{Re}}} + \frac{4.12}{\text{Re}} \qquad (\textit{Flat plate parallel to flow}) \qquad (9\text{-}10)$$

For each of the other shapes of submerged objects, there is a transition from $C_D \sim 1/\text{Re}$ to $C_D \sim 1/\text{Re}^n$, in which the exponent n varies gradually from 1.0 to zero. For the sphere and circular cylinder, however, this transition takes place over a larger range of Reynolds number than it does for the disk and the flat plate perpendicular to the flow. This range of transition is associated with the type of drag involved. With the flat plate parallel to the flow, the drag is entirely shear drag, and hence the transition is the

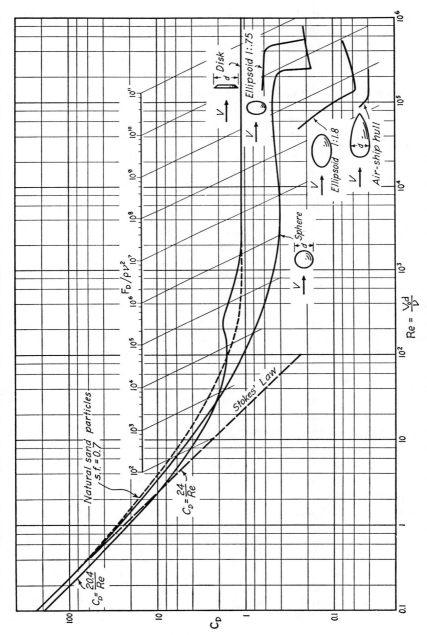

Fig. 9-2. Drag coefficients for spheres and other bodies of revolution.

most gradual—the influence of inertia becoming predominant only at relatively larger Reynolds numbers. A similar situation exists for the shear drag on a circular cylinder, see Fig. 9-3.

With the bluff objects, such as the disk and flat plate perpendicular to the flow, the drag is entirely pressure drag and the influence of inertia becomes predominant at considerably smaller Reynolds numbers.

In Fig. 9-3, there is a striking similarity between the variation of shear drag with Reynolds number for the circular cylinder and the flat plate. The best theoretical solution available for the drag on a circular cylinder is Eq. 9-6, for which the slope of the curve approaches -1.0 at very small values of Reynolds number, as it does for the flat plate. Since the shear drag contributes two-thirds of the total drag and the pressure drag contributes the other one-third for small Re (as in the case of the sphere), the curves for shear drag and pressure drag (which have been developed for large Reynolds numbers by the boundary layer theory and actual pressure measurements, respectively) can be extended approximately to small Reynolds numbers for the circular cylinder.

In the curves for the disk (where the drag is pressure drag exclusively), and in the pressure drag for the cylinder (which is the difference between the two curves for the total drag and the shear drag), there is at about Re = 80 an actual increase in the drag coefficient with increasing Reynolds number. The cause of this rise is not understood exactly, but it is no doubt associated in some way with the development of the separation zone and the initial shedding of the vortices downstream from the cylinder, as discussed in the following section.

As the Reynolds number increases from small values to large values for a circular cylinder, the contribution of the shear drag changes from two-thirds of the total to an insignificant amount at about Re = 10^4—the pressure drag gradually increasing until it is responsible almost exclusively for the total drag. This reversal in the contribution of shear drag and pressure drag occurs also for the sphere and other rounded objects.

Another feature of the transition from small to moderate Reynolds numbers is that of orientation and stability of objects moving unrestrained through a fluid—such as gravel, sand, dust, or other objects falling through air, water, oil or some other fluid, under the influence of gravity or some other outside force. For Re < 0.1, any object—such as a spheroid, a circular cylinder, and a rectangular bar—which is symmetrical about three mutually-perpendicular planes will fall at any initial orientation to the flow without changing orientation as it falls.

As an unsymmetrical object falls, however, it will rotate slowly until its greatest projected area is perpendicular to the flow, and the center of the drag force is in line with the center of gravity or other outside force. In this **way, the resultant** drag force and the gravitational force are collinear.

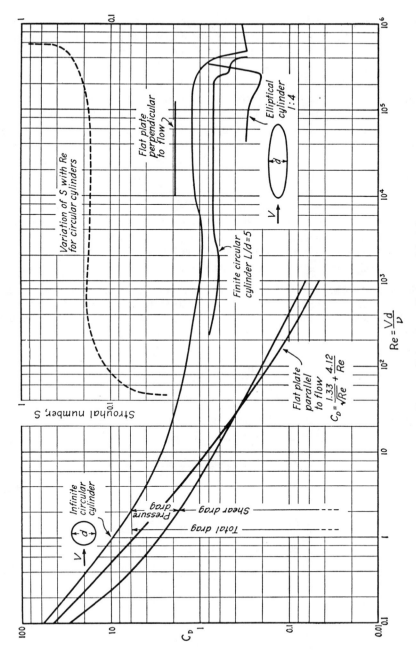

Fig. 9-3. Drag coefficient for cylinders and flat plates.

With increasing Reynolds numbers, the object will then continue to fall or move in this stable position until a critical value is reached at about Re = 80 to 500, when the object begins to oscillate from side to side, and in some cases even rotate and tumble. This oscillation seems to occur at about the same Re values as the strange and unexplained rise in pressure drag which can be seen for Re = 80 for the disk in Fig. 9-2. This same hump occurs for the pressure drag of the circular cylinder, which is the difference between the total drag and the shear drag in Fig. 9-3.

As already pointed out, there is widespread deformation of the flow at small Reynolds numbers, and the velocity varies, in an extremely gradual manner, from zero at the boundary, to the ambient velocity at a great distance from the boundary, see Fig. 9-1(b). As Reynolds number increases, the inertia effects also increase and the deformation of the flow gradually becomes confined to a region closer and closer to the object. Hence, the velocity profile reaches ambient velocity at a point closer and closer to the boundary. This confining of the flow to a region near the boundary is one of the essential features of the boundary-layer theory presented in Chapter 6.

Example Problem 9-1: Calculate the total drag force, the shear drag force, and the pressure drag force exerted on a 2.0 ft length of an infinite circular cylinder which has a diameter $d = 0.1$ ft, if $V_0 = 0.2$ fps and the fluid is standard air at 60°F.

Solution:

$$\text{Re} = \frac{(0.2)(0.1)(10)^4}{1.58} = 1.27 \times 10^2$$

From Fig. 9-3

Total drag coefficient, $C_D = 1.4$ (upper curve)

Shear drag coefficient $C_D = 0.185$ (lower curve)

$$\text{Total Drag Force } F_D = C_D A \rho \frac{V_0^2}{2} = \frac{(1.4)(2)(0.1)(0.00237)(0.2)^2}{2}$$

$$\underline{\underline{1.33 \times 10^{-5} \text{ lbs}}}$$

$$\text{Shear Drag Force } F_D = 1.33 \times 10^{-5} \frac{0.185}{1.40} = \underline{\underline{1.76 \times 10^{-6} \text{ lbs}}}$$

$$\text{Pressure Drag Force} = (1.33 - 0.176)10^{-5} = \underline{\underline{1.15 \times 10^{-5} \text{ lbs}}}$$

- SEPARATION DOWNSTREAM FROM A SUBMERGED
   OBJECT

The increasing influence of inertia with increasing Reynolds number has another important effect—that of separation. For small Reynolds numbers, say Re < 0.5, the flow pattern is essentially symmetrical about a vertical axis through the object, that is, on the upstream and downstream sides of the object. As the Reynolds number is increased, however, a small zone of separation forms downstream from the object and, at Re = 10 for a cylinder, a separation zone of appreciable size is clearly visible, with a pair of eddies or vortices rotating in the separation zone.

The flow in the separation zone and the flow downstream remains laminar, despite the fact that the influence of inertia is becoming increasingly greater and the separation zone together with the vortices in it are increasing in length, until Reynolds number reaches approximately the range of 50 to 100 for a cylinder. At about Re = 50, a waviness begins to develop downstream from the wake, and a periodic oscillation of the wake is clearly evident at Re = 60. This periodic motion is caused by an unstable balance between the vortices on each side of the centerline. Initially, the two vortices oscillate in place, but as Reynolds number is increased still farther, the vortices alternately separate from the cylinder and move downstream. With additional increase in Reynolds number, this entire flow pattern breaks down into turbulent flow, and the Karman vortex street is formed (see Fig. 9-4).

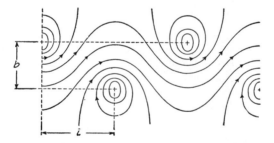

Fig. 9-4. Karman vortex street.

Downstream from a flat plate perpendicular to the flow, the same gradual formation of a separation zone, creation of the vortices, and alternate shedding of the vortices takes place—but at different Re values.

The separation zone downstream from a sphere has many similarities to the separation zone downstream from a cylinder. The principal difference is that a vortex ring forms with the sphere, instead of a pair of vortices. As a consequence, the pressure imbalance and resulting alternate shedding of vortices and the vortex street do not develop with the sphere.

The stationary vortex ring forms at Re ≈ 15, and its location moves downstream with increasing Reynolds numbers. At Reynolds numbers of 300 to 2000, the vortex ring becomes unstable and occasionally moves on downstream, with a new one forming immediately. However, there is no definite periodicity of this action as there is with the cylinder.

• VORTEX LAYERS AND THE KARMAN VORTEX STREET

On each side of a two-dimensional object, such as a cylinder or a flat plate, are formed planes of discontinuity immediately downstream from the points of separation. Across each of these planes of discontinuity is a large velocity gradient, and hence intense shear. If the Reynolds number of the flow is sufficiently small, the viscous forces are predominant, and the shear will transmit momentum without becoming unstable in the wake downstream—hence the entire flow remains laminar. *Downstream from the point of separation, the zone of discontinuity becomes wider.* With increasing Reynolds number, the forces of inertia become increasingly important. Eventually, the shear region becomes unstable and mixing by eddies begins at about Re = 60. With still further increases in Reynolds number, the point of instability moves upstream and eventually reaches the cylinder itself, at about Re = 10^4.

It is interesting to note, see Fig. 9-3 for a circular cylinder, that the contribution which pressure drag makes to the total drag gradually increases from one-third the total at small Reynolds numbers to about half the total as the vortices begin to form; it then jumps rather quickly to 75 percent at Re = 200, where the well-developed system of alternating vortices has been established. As Reynolds number increases still further, the vortex layer gradually becomes turbulent throughout, and the pressure drag continues to increase until about Re = 10^4 is reached. Another factor contributing to this variation in C_D is the point of separation which is gradually moving upstream with increasing Re. Consequently, the pressure at the point of separation is decreasing and, since essentially the same pressure is continued around the rear of the cylinder, the drag is increasing. As the vortex layer becomes turbulent throughout, however, there is increased momentum transfer within the vortex layer as it continues around the cylinder, and there is some recovery of pressure on the downstream side of the cylinder, so that the drag is not as great as it would have been without the turbulent vortex layer. From Re = 10^4 to Re = 10^5, the point of separation and the relative intensity of turbulence within the vortex layer remain stable, and hence the contribution of pressure drag (which at these large Reynolds numbers is almost entirely responsible for the total drag) to C_D remains essentially constant.

Once the creation of an alternating system of vortices has been well established downstream from a cylinder, it takes on certain characteristics which can be expressed mathematically. Karman derived a theoretical relationship between the longitudinal spacing l and the transverse spacing b, see Fig. 9-4, which showed that, when $l/b \approx 3.55$ for a circular cylinder and $l/b \approx 3.3$ for a flat plate, the vortex formation is a consistent pattern of flow. Experiments have verified this spacing.

The frequency f of the alternate shedding of the vortices may be determined by the spacing l and the velocity V_v at which the vortex street is moving relative to the cylinder to yield $f = V_v/l$.

In general, the frequency is a function of the viscosity of the fluid ν, the velocity of the flow V_0, and the size of the cylinder d, which may be combined into two dimensionless parameters to yield

$$\frac{fd}{V_0} = \phi \text{ (Re)} \qquad (9\text{-}11)$$

This function, in which the dependent variable is called the Strouhal number, is plotted in Fig. 9-3, from experimental data. The alternate vortex formation begins at about Re = 45. The Strouhal number increases with Reynolds number until it reaches approximately 0.18 at Re = 1000. Further increases in Reynolds number do not change appreciably the Strouhal number, until about Re = 10^5, beyond which it increases very rapidly with only a small increase in Re. Throughout the range of 45 < Re < 10^6 the alternate vortices are forming, but beyond Re $\approx$ 2500, the wake no longer has the appearance of a systematic vortex street. The speed of rotation and intensity of shear associated with the vortices cause them to break down nearer and nearer the cylinder, with increasing Re, and blend into the flow as random turbulence.

The alternate formation and shedding of vortices downstream from a circular cylinder creates corresponding fluctuations in pressure in the regions where the vortices form. As a consequence, there is an alternating thrust transversely against the cylinder. This alternating thrust sets up vibrations which may be harmful to the cylinder, or to the structure with which it is associated. This is true especially for flow past round bridge piers and pile bents, round smoke stacks, submarine periscopes, and electric power and telephone lines. The frequency of thrust of the wind or water combines with the natural frequency of the cylindrical object to create stresses which commonly have caused failure. The frequency of the tone of singing telephone wires in a cross wind has been related directly to the frequency of the vortex formation.

One method of eliminating the undesirable vibration of cylindrical objects is to attach a longitudinal fin or plate to the cylinder on the downstream side and extend it downstream between the vortices for a distance of about one diameter. Such a fin or plate stabilizes both vortices so that they remain fixed in position, and the alternate shedding and associated side thrust are stopped.

• CRITICAL REYNOLDS NUMBER

One may note that, at Re $\approx 2 \times 10^5$ for both the sphere and the cylinder, the drag coefficient C_D suddenly decreases to about one-third its original magnitude. This phenomenon is caused by a change in the nature of the boundary layer from a laminar boundary layer to a turbulent boundary layer, on the fore part of the object. As a consequence, the point of separation suddenly moves downstream, and the flow is permitted to expand, which reduces the local velocity at the point of separation. This reduction in local velocity is accompanied by an increase in pressure which extends over the downstream side of the sphere (see Fig. 9-1).

The Reynolds number at which this drop takes place depends upon both the amount of turbulence in the ambient flow and the nature of the surface of the cylinder upstream from the point of separation. If the ambient turbulence is sufficiently great, or if the surface of the cylinder is sufficiently rough, the boundary layer changes from laminar to turbulent at a smaller Reynolds number than for a smooth boundary—that is, Re $< 2 \times 10^5$. In fact, simply placing a small roughness such as a wire on the surface of the cylinder upstream from the point of separation will create sufficient turbulence to cause the boundary layer to change at a smaller Reynolds number.

The point of separation is fixed for bluff objects such as a disk and a flat plate, and for this reason the drag coefficient remains constant for Re $> 10^3$. Drag coefficients for these and other objects are given in Table 9-1.

Example Problem 9-2: The screen of an outdoor theater is 40 ft long and 20 ft high. What is the overturning moment about the base of the screen, assuming standard air at 60°F, and a wind velocity of 60 mph normal to the screen?

Solution:

Assume the characteristic length dimension is the screen height.

$$R_e = \frac{(88)(20)(10)^4}{(1.58)} = 1.112 \times 10^7$$

$C_D = 1.17$ based upon Table 9-1.

$$F_D = (1.17)(20)(40)(0.00237)\frac{(88)^2}{2} = 8600 \text{ lbs}$$

$$M = (8600)(10) = \underline{\underline{86,000 \text{ lb-ft}}}$$

Table 9-1. DRAG COEFFICIENTS FOR CYLINDERS AND FLAT PLATES

Object (Flow from L to R)	L/d	$\mathrm{Re} = Vd/\nu$	C_D
1. Circular Cylinder, Axis Perpendicular to the Flow	1 5 20 ∞	10^5	0.63 0.74 0.90 1.20
	5 ∞	$>5 \times 10^5$	0.35 0.33
2. Circular Cylinder, Axis Parallel to the Flow	0 1 2 4 7	$>10^3$	1.12 0.91 0.85 0.87 0.99
3. Elliptical Cylinder (2:1) (4:1) (8:1)		4×10^4 10^5 2.5×10^4 to 10^5 2.5×10^4 2×10^5	0.6 0.46 0.32 0.29 0.20
4. Airfoil (1:3)	∞	$>4 \times 10^4$	0.07

5. Rectangular Plate for which
 	
   L = length
   d = width
	1
	5
	20
	∞
		$>10^3$	1.16
			1.20
			1.50
			1.90

6. Square Cylinder		3.5×10^4 $10^4 \times 10^5$	2.0 1.6

7. Triangular Cylinder
    120°
    60°
    30°
		$>10^4$
		
		$>10^5$
			2.0
			1.72
			2.20
			1.39
			1.80
			1.0

8. Hemispherical Shell		$>10^3$ 10^3 to 10^5	1.33 0.4
9. Circular Disk, normal to the flow		$>10^3$	1.12
10. Tandem Disks, spacing is L	0 1 2 3	$>10^3$	1.12 0.93 1.04 1.54

• DRAG ON MISCELLANEOUS BODIES

In Fig. 9-2 are plotted the curves for other bodies of revolution, such as the ellipsoid and an airship. It may be noted that, for the more streamlined bodies, the decrease in C_D is more gradual than for a sphere, although the order of magnitude of the drop is the same. However, for the bluffer body, such as the 1: 0.75 ellipsoid, the decrease is just as sudden but not as great as for the sphere. The limit of the bluffness is the disk, for which there is no decrease in C_D. The same phenomena may be observed in Fig. 9-3 for two-dimensional objects.

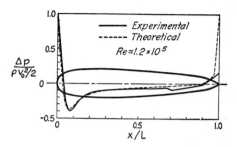

Fig. 9-5. Pressure distribution around an airship hull.

Figure 9-5 shows pressure distribution on an airship, based upon both actual measurement and theoretical irrotational flow. Because of the excellent streamlining of the body, the velocity head is converted to pressure head with very little loss in the expanding flow on the downstream side of the body. Because irrotational flow results in no drag whatsoever, this shape is nearly as efficient as possible from the viewpoint of eliminating pressure drag. Shear drag, however, will remain dependent upon the area and the roughness of the surface. Therefore, it is desirable to make the object as short as possible to reduce shear drag, but it must be sufficiently long and streamlined to eliminate pressure drag. The marked effect of such streamlining can be seen in Fig. 9-3, where the shear drag becomes negligible compared with the pressure drag for the circular cylinder, and in Fig. 9-2 where C_D for a disk is reduced from 1.1 to about 0.04 (a reduction of about 98 per cent) by streamlining it like an airship hull.

Another type of submerged object frequently encountered is natural material such as dust, sand, and gravel. The variation of the drag coefficient for these is shown in Fig. 9-2. Extremely small particles such as dust usually fall at Re $<$ 0.5, and hence they tend to follow the Stokes variation. As the particles become larger, however, their angularity and lack of symmetry increase the pressure drag, and the curve moves away from the curve for spheres toward the curve for disks. For this broken-line curve, the nominal diameter (diameter of a sphere of the same volume) is used for both C_D and Re, with an average value of the Corey shape factor $s.f.$

$$s.f. = \frac{c}{\sqrt{ab}} = 0.7 \qquad (9{:}12)$$

in which a, b, and c are respectively the major, intermediate, and minor axes which are mutually perpendicular. This shape factor expresses the relative thickness of the particle, e.g., the thickness c in the direction of motion relative to the size of the particle in the plane perpendicular to the flow.

• DIRECT DETERMINATION OF FALL VELOCITY AND DIAMETER

Frequently, in the analysis of sediment and other falling bodies, the size and weight are known and it is desirable to determine the velocity of fall. By using the plot of C_D and Re, the velocity must be determined by trial and error. However, a direct solution can be made by plotting, with either C_D or Re, a parameter which does not contain the velocity. This has been done in Fig. 9-2 by the scale $F_D/\rho\nu^2$ which can be obtained either by dimensional analysis, or in the following way as a product of the drag coefficient C_D and the square of the Reynolds number

$$C_D(\text{Re})^2 = \frac{F_D}{A_\rho V^2/2}\left(\frac{Vd}{v}\right)^2 \sim \frac{F_D}{\rho\nu^2} \qquad (9\text{–}13)$$

At any point on the diagonal lines shown for the constant values of $F_D/\rho\nu^2$, the corresponding values of C_D and Re are those required to satisfy the particular values of F_D, ρ, and ν. If the object is a disk, these values are satisfied at a smaller Re than for a sphere, but C_D is considerably greater.

The $F_D/\rho\nu^2$ scale can be employed to determine the fall velocity of an object by using the submerged weight of the object as F_D (the weight and drag must be equal in magnitude if the object is not accelerating) and following down the proper $F_D/\rho\nu^2$-line to the curve for the shape of object involved, from which either C_D or Re can be determined to solve for the fall velocity.

Example Problem 9-3: Calculate the fall velocity of a spherical droplet of water through standard air and the rate of rise of a bubble of air through water. Assume a diameter of 0.01 ft for the droplet and bubble, and a temperature of 40° F. Why is the difference in the two velocities so large when $\Delta\gamma$ is the same?

Solution:

The gravitational force causing the droplet to fall and the buoyant force causing the bubble to rise are equal.

$$F_D = \Delta\gamma \tilde{V} = (62.4)(\pi)(\tfrac{4}{3})(0.005)^3 = 3.27 \times 10^{-5}\,\text{lb}$$

For droplet in air $\dfrac{F_D}{\rho\nu^2} = \dfrac{(3.27)(10)^3}{(0.00247)(1.46)^2} = 6.2 \times 10^5$

From Fig. 9-2, $C_D = 0.42$

$$V = \left(\frac{2F_D}{C_D \rho A}\right)^{1/2} = \left[\frac{(2)(3.27)(10)^{-5}}{(0.42)(0.00247)(\pi)(0.005)^2}\right]^{1/2} = 28.3 \text{ fps down}$$

For air bubble in water $\dfrac{F_D}{\rho \nu^2} = \dfrac{(3.27)(10)^5}{(1.94)(1.664)^2} = 6.08 \times 10^4$

From Fig. 9-2, $C_D = 0.55$

$$V = \left[\frac{(2)(3.27)(10)^{-5}}{(0.55)(1.94)(\pi)(.005)^2}\right]^{1/2} = 0.88 \text{ fps upward}$$

The large difference between the two velocities is largely due to the difference in magnitude of the viscous forces.

• CIRCULATION

On the preceding pages, the submerged objects discussed have had a constant orientation to the flow. However, if the submerged object is rotated, the additional motion gives rise to a phenomenon called circulation. When a baseball is pitched with a certain spin, it curves one way or another depending upon the direction of the spin. Likewise, when a golf ball is sliced, and when a tennis ball or ping-pong ball is cut, the ball follows a curved path. By the momentum principle, there must be a side thrust which causes this change in direction of motion. This phenomenon was discovered by Magnus in 1852, and hence has been named the Magnus effect.

The side thrust which causes the Magnus effect is created by a combination of movement through the air and a rapid spin placed on the ball, either intentionally or unintentionally. The analysis of the Magnus effect from the viewpoint of fluid mechanics follows.

Consider steady **irrotational flow** *past a circular cylinder* as shown in Fig. 9-6(a). The velocity distribution indicated by the streamline pattern

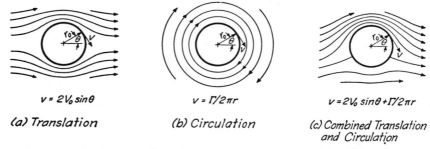

$v = 2V_0 \sin\theta$ $v = \Gamma/2\pi r$ $v = 2V_0 \sin\theta + \Gamma/2\pi r$

(a) Translation   *(b) Circulation*   *(c) Combined Translation and Circulation*

Fig. 9-6. The Magnus effect for irrotational flow around a circular cylinder.

produces a pressure distribution similar to that in **Fig. 9-1**(a). Along the streamline constituting the boundary of the cylinder, the velocity is

$$v = 2V_0 \sin \theta \tag{9-14}$$

in which v is the local velocity at a point on the cylinder (L/T);

V_0 is the ambient velocity (L/T);

θ is the angle included between the horizontal and the point in question.

Next consider a **vortex with a cylindrical core** having a radius r_0, see **Fig. 9-6**(b). The velocity distribution within this vortex varies inversely with the radial distance outward. Hence, the product vr of the velocity v and the radius r remains a constant

$$vr = c = \omega r^2 \tag{9-15}$$

in which ω is the rate of rotation in radians per unit time $(1/T)$;

c is the strength of the vortex (L^2/T).

This type of flow has a direct relationship to circulation Γ, which is now considered.

Mathematically, **circulation** is defined as

$$\Gamma = \oint v_L \, dL \tag{9-16}$$

in which Γ is the circulation (L^2/T);

L is the length along the circuit taken **(L)**;

v_L is the component of velocity along the path of L (L/T);

and the integral sign indicates a line integral which begins at a given point, follows a circuitous path through the fluid, and returns to the starting point, see **Fig. 9-7**. This integral is very similar to the integral of the work done by a force over a distance L

$$W = \oint_0^L F_L \, dl \tag{9-17}$$

in which W is the work done (LF);

L is the length of the line of motion considered (L);

F_L is the component of force in the direction of $L(F)$.

The circulation Γ within irrotational flow is zero, provided there is no rotational zone within the area considered. With the vortex of **Fig. 9-6**(b), however, there is a rotational core of radius r_0, and hence the circulation is not zero. If the circulation is taken along one of the streamlines, the velocity v_L is the tangential velocity v and the integral of length is simply $2\pi r$. Hence

$$\Gamma = v \oint dL = v(2\pi r) \tag{9-18}$$

Since for the vortex $v = c/r$

$$\Gamma = \frac{c}{r}(2\pi r) = 2\pi c \tag{9-19}$$

which shows that, for the vortex, the circulation is everywhere the same (that is, independent of the radius), provided the rotational core is always included inside the line integral. Equation 9-18 can be rearranged as

$$v = \frac{\Gamma}{2\pi r} \tag{9-20}$$

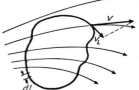

The flow pattern and velocity in Fig. 9-6(a) can be combined mathematically with those in Fig. 9-6(b) to yield Fig. 9-6(c), which clearly is unsymmetrical. For the velocity v at the surface of the cylinder in Fig. 9-6(c), Eqs. 9-14 and 9-20 can be combined to give

Fig. 9-7. Circulation around a closed curve in a field of flow.

$$v = 2V_0 \sin \theta + \frac{\Gamma}{2\pi r_0} \tag{9-21}$$

or

$$\frac{v}{V_0} = 2 \sin \theta + \frac{\Gamma}{\pi d V_0} \tag{9-22}$$

$$\frac{v}{V_0} = 2 \sin \theta + \frac{v_p}{V_0} \tag{9-23}$$

in which

v_p is the peripheral speed of the cylinder (L/T);

V_0 is the ambient velocity (L/T);

d is the diameter of the cylinder or core of the vortex (L);

v is the surface velocity at the point where the angle is $\theta(L/T)$.

From the spacing of the streamlines in Fig. 9-6(c), it is obvious that a greater pressure exists against the bottom of the cylinder than against the top. The distribution of the pressure around the cylinder can be determined by combining Eq. 9-23 with the energy equation

$$\frac{\Delta p}{\rho V_0^2/2} = 1 - \left(\frac{v}{V_0}\right)^2 \tag{9-24}$$

in which Δp is the difference between the pressure at the point in question and the ambient pressure. This combination yields

$$\frac{\Delta p}{\rho V_0^2/2} = 1 - \left(2\sin\theta + \frac{v_p}{V_0}\right)^2 \tag{9-25}$$

Equation 9-25 shows that a stagnation point exists when the sum in parentheses is equal to zero. Evidently, as the peripheral speed increases relative to the ambient velocity v_p/V_0, the stagnation points on the centerline of Fig. 9-6(a) move downward until they meet at the bottom of the cylinder, when $v_p/V_0 = 2$. If the ratio is increased beyond 2, the stagnation point moves away from the cylinder.

The foregoing analysis has direct application to circulation associated with a rotating solid circular cylinder having a radius r_0, as discussed in the following paragraphs.

• LIFT

From the pressure distribution expressed in Eq. 9-25 and from Fig. 9-6(c), it is evident that a side thrust exists on the cylinder. This thrust is known as lift F_L and it can be evaluated for irrotational flow by integrating the pressure of Eq. 9-25 over the entire surface. Thus

$$F_L = L \int_0^{2\pi} \sin\theta (\Delta p \, r_0 \, d\theta) \tag{9-26}$$

from which

$$F_L = \frac{r_0 L \rho V_0^2}{2} \int_0^{2\pi} \sin\theta \left[1 - \left(2\sin\theta - \frac{v_p}{V_0}\right)^2\right] d\theta \tag{9-27}$$

and

$$F_L = 2\pi \frac{v_p}{V_0}(2r_0 L)\frac{\rho V_0^2}{2} = \rho L V_0 \Gamma \tag{9-28}$$

or

$$F_L = \frac{C_L A \rho V_0^2}{2} \tag{9-29}$$

in which

A is the projected area Ld of the cylinder (L^2);

C_L is the lift coefficient, defined as

$$C_L = \frac{2\pi v_p}{V_0} = \frac{2\Gamma}{dV_0} \qquad (9\text{-}30)$$

Equation 9-29 is of the same form as Eq. 9-1 for the drag coefficient. In Eq. 9-30, the lift coefficient is seen to vary directly with the circulation and with the velocity ratio v_p/V_0 for irrotational flow. For the case of $v_p/V_0 = 2$, the two stagnation points of Fig. 9-6(c) meet at the bottom, and $C_L = 4\pi = 12.6$, which is a theoretical maximum for the lift coefficient. Eq. 9-28 is an important one which shows that the lift F_L is directly proportional to the density of the fluid, the ambient velocity V_0, the strength of the circulation Γ, and the length of the cylinder L.

Thus far, circulation and lift have been considered for the flow of an ideal fluid having no viscosity—that is, irrotational flow. Real fluids, however, have viscosity which results in certain differences from the irrotational theory developed in the foregoing paragraphs. Among these differences are:

1. The actual flow pattern around a cylinder (without circulation) is not like Fig. 9-6(a), because separation occurs as in Fig. 9-1.
2. The circulation established by the boundary shear on a rotating cylinder is not the same as that of an irrotational vortex with a rotational core, as assumed in the derivation.
3. The theoretical derivation of lift for a cylinder of infinite length is markedly different from that for a relatively short cylinder of finite length, where flow occurs along the cylinder and around the ends.

Despite these limitations to the theory of circulation and lift, a circulation is actually developed by the viscous shear and a lift force is created. This has been given special study by Flettner with respect to his rotor on a ship which actually made crossings of the Atlantic Ocean. It was abandoned only because it was uneconomical compared with other modes of propulsion.

Figure 9-8 shows variation of the lift coefficient C_L with the velocity ratio v_p/V_0 for a circular cylinder. The broken line is Eq. 9-30 for irrotational flow, and the solid line represents actual experimental data when $\mathrm{Re} \approx 10^5$ and the boundary is smooth. From this plot, it can be seen that an actual speed is required which is greater than twice that required theoretically to produce a given lift coefficient. Furthermore, as the peripheral speed increases to about four times the ambient velocity, the lift coefficient appears to approach a maximum of about 9.0, as compared with the theoretical maximum of about 12.6. In addition, Fig. 9-8 shows that the drag coefficient C_D also varies with the velocity ratio from about 1.0 (see also Fig. 9-3) for small velocity ratios, to about 5 at a ratio of 4. There is a drop in the drag coefficient for $v_p/V_0 \approx 1.5$, beyond which it rises steeply.

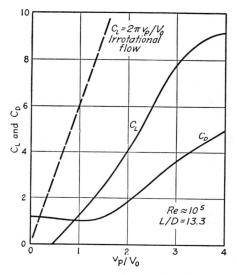

Fig. 9-8. Coefficients of lift and drag for a rotating cylinder.

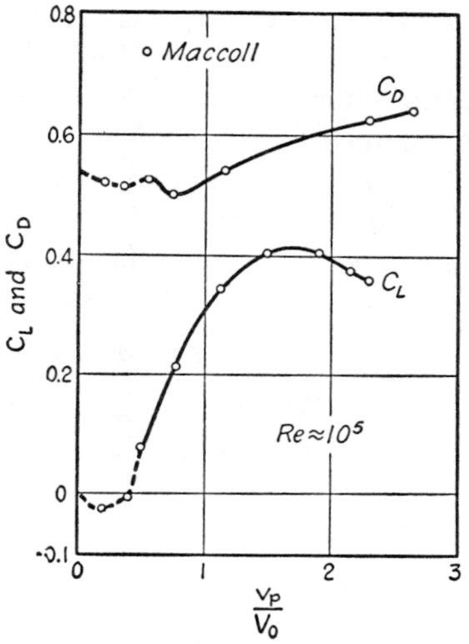

Fig. 9-9. Coefficients of lift and drag for a rotating sphere.

If the cylinder is relatively short, say $L/d < 10$, there is considerable effect of flow about the ends, which reduces appreciably the coefficient of lift. In fact, when the aspect ratio $L/d = 5$, the lift coefficient is about half that for the longer cylinder. Flow around the end also causes an induced drag, in addition to the normal drag. This longitudinal flow may be essentially eliminated by placing disks with a diameter of $2d$ on the ends of the cylinder.

Figure 9-9 shows lift and drag coefficients for a smooth rotating sphere. The lift coefficient drops to a negative value for $v_p/V_0 < 0.4$, and then increases to a maximum at about $v_p/V_0 = 1.8$. Meanwhile, the drag coefficient is relatively constant at $C_D = 0.5$ until the maximum lift coefficient is approached—beyond which C_D gradually increases to about 0.65.

• THE AIRFOIL

Although the rotating circular cylinder has virtually no use today as a means of supplying a thrust to lift or drive a vehicle, the study of a cylinder provides an understanding of the basic principles of circulation and lift which are important for grasping the theory of the airfoil. Furthermore, Eq. 9-28

$$F_L = \rho L V_0 \Gamma \tag{9-28}$$

which is known as the Kutta-Joukowsky lift equation, has direct application to shapes other than a circular cylinder. This equation was derived first by Kutta in 1902, and independently by Joukowsky in 1906. By means of a mathematical manipulation known as a conformal transformation, this equation can be applied not only to cylinders, but also to an infinite variety of shapes of airfoil. The mathematics of this operation, however, are beyond the scope of this book.

The application of circulation to an airfoil is illustrated in Fig. 9-10, where both the shape of the airfoil and the irrotational flow pattern about

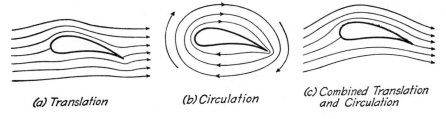

(a) Translation   *(b) Circulation*   *(c) Combined Translation and Circulation*

Fig. 9-10. Irrotational flow around an airfoil.

it are determined by the Kutta-Joukowsky transformation. As in the case of the cylinder, the translation of uniform flow around the airfoil, Fig. 9-10(a), is combined with circulation, Fig. 9-10(b), to yield the final flow pattern in Fig. 9-10(c). From the theoretical analysis, it has been found that the circulation required to move the center streamline, which is trailing in Fig. 9-10(a), to the tip of the airfoil as shown in Fig. 9-10(c) is

$$\Gamma = \pi c V_0 \sin \alpha_0$$

which combines with Eq. 9-30 to yield

$$C_L = 2\pi \sin \alpha_0 \tag{9-31}$$

In these equations

c is the chord length or width of the airfoil (L);

α_0 is the angle of attack, in radians, see Fig. 9-11.

The cambered shape (lack of symmetry about a longitudinal center-line) is selected so that the separation on the upper side of the airfoil is reduced. The angle of attack is α_0, and obviously this has considerable effect upon the lift. Therefore, α_0 is an important variable in determining

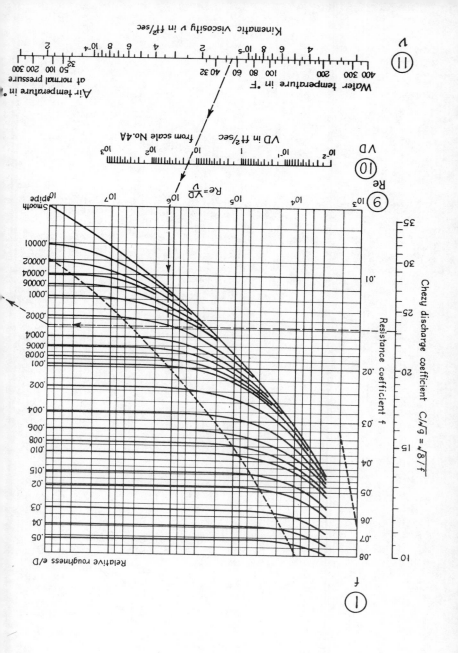

PIPE RESISTANCE DIAGRAM
after the Manual of British Water Supply Practice

the characteristics of an airfoil. Figure 9-11 shows actual variation of both C_L and C_D with the angle of attack for a Joukowsky airfoil. Plotted as a broken line is Eq. 9-31, the theoretical variation of lift with angle of attack. Note that the experimental curve is not greatly different from the theoretical one for small angles of attack (including negative angles). When the angle reaches about 8 degrees, however, a marked deviation develops. The experimental C_L values even reach a maximum, and then decrease in magnitude for $\alpha > 10$ degrees. This sudden change in the curve occurs where stalling takes place because separation develops on the upper side of the airfoil and the entire flow system (including the circulation) is radically modified.

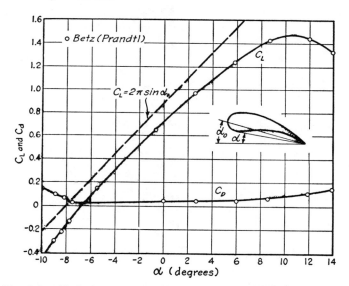

Fig. 9-11. Variations of lift and drag coefficients with angle of attack for infinite length of span.

The lift and drag equations for an airfoil correspond to Eq. 9-29 and Eq. 6-4, except that the area is now represented not by the projected area in the direction of flow, but by the product of the length or span L and the chord c, as it is for the flat plate parallel to the flow in Fig. 9-3. Thus

$$F_L = C_L \, cL \, \rho \, \frac{V_0^2}{2} \tag{9-32}$$

and

$$F_D = C_D \, cL \, \rho \, \frac{V_0^2}{2} \tag{9-33}$$

The coefficients C_L and C_D are determined experimentally, and for a wing of finite length they differ somewhat from the infinite wing length, because of induced drag.

Example Problem 9-4: A jet plane which weighs 6000 lb and has a wing area of 200 ft², flies at a velocity of 600 mph when the engine delivers 10,000 horse-power. Assuming $p_a = 10$ psia, $T = 30°$F, and that 65 per cent of the power is used to overcome the drag resistance of the wing, calculate the lift coefficient C_L and the drag coefficient C_D of the wing.

Solution:

$$V_0 = \frac{(600)(44)}{30} = 880 \text{ fps}$$

$$\rho = \frac{(10)(144)}{(32.2)(53.3)(490)} = 0.00171 \text{ lb-sec}^2/\text{ft}^4$$

$$C_L = 2\frac{F_L}{A\rho V_0^2} = \frac{(2)(6000)}{(200)(0.00171)(880)^2} = 0.0454$$

$$F_D = 0.65\frac{P}{V_0} = \frac{0.65\,(10,000)(550)}{880} = 4070 \text{ lb}$$

$$C_D = C_L\frac{F_D}{F_L} = \frac{(0.0454)(4070)}{6000} = 0.0307$$

• INDUCED DRAG

When there is no motion around an airfoil, there obviously cannot be any circulation. Furthermore, as motion begins, the flow pattern tends to be the irrotational flow shown in Fig. 9-10(a). However, the velocity of the fluid leaving the under side of the trailing edge is greater than the velocity of fluid leaving the upper side, and hence a plane of discontinuity and shear is established downstream from the airfoil. This shear causes a vortex which has circulation, see Fig. 9-12. Once circulation is created in one part of a flow system, the starting vortex, there must be circulation in the opposite direction in another part, so that the net circulation is zero. According to Prandtl,

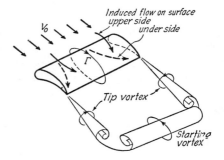

Fig. 9-12. Vortex system associated with a wing of finite length.

this is the means by which circulation around an airfoil becomes possible in a real fluid.

When a wing or airfoil of finite length is considered in the foregoing manner, the starting vortex and the vortex around the wing appear to terminate in space at the ends of the wing. This condition is not possible, and hence a secondary flow is established along the length of the wing— toward the tips on the under side (see broken-line arrows in Fig. 9-12), and toward the center on the upper side (see solid arrows in Fig. 9-12)— so that there is a continuous system. This effect is most pronounced at the ends of the wing, and hence a tip vortex is created at each end, as shown in Fig. 9-12, which completes the vortex system so that it is continuous, in a somewhat regular rectangular pattern. In actual flight, the tip vortices cause such a reduced pressure (and temperature) that condensation or vapor trails may sometimes be seen in the sky for miles after the passage of an aircraft. Actually, the secondary flow causes vortices to be shed all along the wing, with the strength of the vortices varying from zero at the center to a maximum at the wing tips.

Because of the drop in pressure along the wing to zero pressure at the wing tips (and the resulting secondary flow along the wing), the lift is reduced and there is created an induced drag which increases the total drag on the wing. The distribution of the lift has a shape which is approximately that of half an ellipse, varying from zero at the wing tips to a maximum at the center of the wing—the maximum being greater than the theoretical value for an infinitely long airfoil. Hence, the net lift is not reduced as much as it might at first appear.

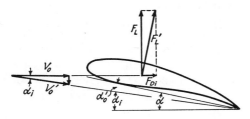

Fig. 9-13. Diagram of induced drag.

The induced drag F_{Di} is a consequence of the downward flow of air at the wing tips caused by the secondary flow along the wing, and the resulting vortices. This induced velocity or downwash causes the effective velocity of approach to have a downward component which decreases the effective angle of attack by an amount α_i, as shown in Fig. 9-13.

From the foregoing analysis, it can be seen that the actual drag F_D on a wing is

$$F_D = F_{D0} + F_{Di} \tag{9-34}$$

in which

F_{D0} is the drag on an airfoil of infinite length due to $V_0(F)$;

F_{Di} is the induced drag (F);

and

$$\alpha = \alpha_i + \alpha_0' \tag{9-35}$$

Upon dividing the terms in Eq. 9-34 by $\rho c L V_0^2 / 2$, the drag coefficients can be related as

$$C_D = C_{D0} + C_{Di} \tag{9-36}$$

Since the angle α_i caused by the induced drag is usually small,

$$F_L' \approx F_L, \quad V_0' \approx V_0, \quad \text{and} \quad F_{Di} \approx F_L \alpha_i \tag{9-37}$$

and on the basis of an elliptic distribution of lift it can be found that

$$C_{Di} \approx \frac{C_L^2}{\pi L / c} \tag{9-38}$$

and

$$\alpha_i \approx \frac{C_L}{\pi L / c} \tag{9-39}$$

in which

L is the length or span of the wing (L);

c is the width or chord of the wing (L);

L/c is the aspect ratio;

α_i is the induced angle of attack, in radians.

Equation 9-38 shows that, as the span of wing L becomes relatively great or the chord becomes relatively small, the coefficient of induced drag goes to zero. This explains why long-range airliners have wings of such great length and a chord as small as structural design will permit.

• THE POLAR DIAGRAM FOR LIFT AND DRAG

On the basis of the foregoing developments, the important variables are C_{Di}, C_D, C_{D0}, C_L, and α. These variables can be presented on a single plot called the polar diagram, which was developed by Prandtl, see Fig. 9-14. The broken line at the left is the parabolic equation, Eq. 9-38, for $L/c = 6$. When $L/c \to \infty$, this line becomes vertical (coinciding with the ordinate axis at $C_D = 0$), and as L/c becomes smaller, it moves farther to the right.

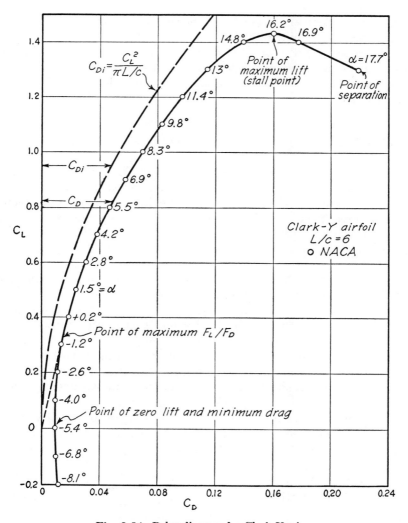

Fig. 9-14. Polar diagram for Clark-Y wing.

The data for the solid curved line are for the Clark-Y wing having a span of 48 ft and a chord of 8 ft. They lie to the right of the broken line representing Eq. 9-38. At $C_L = 0$, this curve is vertical, indicating that it is the point of minimum drag and zero lift. Note that this occurs at an angle of attack α of -5.3 degrees. In other words, the angle of attack can be downward or negative as much as 5 degrees before the lift goes to zero. As C_L becomes either negative or positive, C_D increases because of the increased pressure drag. The point of maximum C_L/C_D is reached at about $\alpha = -1$ degree, the most efficient angle of attack, beyond which C_D increases more rapidly than C_L. Finally, C_L reaches a maximum at about $\alpha = 16$ degrees, beyond which it decreases. This is the point of stalling, where maximum lift occurs. With only a small additional increase of α, widespread separation takes place and the circulation and lift decrease rather suddenly.

• DECREASING THE DRAG COEFFICIENT

The blunt shape of an airfoil such as that in Fig. 9-10 supplies a large lift at relatively slow velocities, and at these velocities the drag is not excessive. At greater velocities, however, the drag on the aircraft becomes of major importance and the necessary lift can be attained with some sacrifice in the lift coefficient itself. Therefore, the trend in recent years has been to try to decrease the drag coefficient, even at the expense of the lift coefficient.

As pointed out in Chapter 6 and in the early part of this chapter, there are two types of drag—shear drag and pressure drag. By means of streamlining—particularly in the downstream part of the airfoil—the pressure drag has been reduced to a very small value. Hence, little can be accomplished by attempted reduction in pressure drag.

As can be seen in Fig. 6-7, the velocity gradient near the boundary is considerably less in a laminar boundary layer than in a turbulent boundary layer. Therefore, it is advantageous to maintain a laminar boundary layer over as much of the airfoil as possible.

The turbulent boundary layer is created at especially small values of Reynolds number in regions of expanding flow, such as downstream from the thickest portion of an airfoil. Therefore, it is logical to move the thickest portion of the airfoil as far to the rear as possible. Compare shape (a), which is an older one, with shape (b), which is more recent, in Fig. 9-15. With the older shape, the theoretical pressure becomes a minimum (maximum negative pressure) on the forepart of the airfoil, and then rises gradually until it is again positive at the trailing edge. It is in this region of expanding flow and increasing pressure that instability increases and that the boundary layer is apt to become turbulent.

$C_D = 0.0063$ at $Re = 3 \times 10^6$

(a) Older shape (Conventional aerofoil)

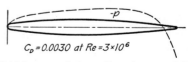

$C_D = 0.0030$ at $Re = 3 \times 10^6$

(b) More recent shape (Low-drag aerofoil)

Fig. 9-15. Change in shape of airfoil for minimum drag.

In Fig. 9-15(b), however, the minimum pressure is reached more gradually and occurs near the midpoint of the airfoil. Furthermore, the pressure does not begin to increase appreciably until about the $\frac{3}{4}$-point, and hence the boundary layer can be maintained more easily as laminar over three-fourths of the surface, and the shear drag is thereby held to a minimum. In fact, for zero angle of attack and Re = 3×10^6, the drag coefficient for the more recent shape is cut to less than half that for the older shape.

The foregoing comments are intended to apply to subsonic flow. Supersonic flow is considered in a later section.

• INCREASING THE LIFT COEFFICIENT

Although the lift coefficient for high-speed aircraft may be relatively small, the necessary lift is attained by the speed itself (lift varies directly with C_L and also directly with V_0^2, see Eq. 9-29). However, it is desirable to have landing and take-off speeds as slow as possible. Therefore, it is necessary to have some means of temporarily increasing the lift coefficient, even though the drag coefficient may also be increased.

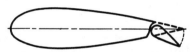

Fig. 9-16. Flaps used to increase lift coefficient.

The lift coefficient is usually increased by the use of flaps, as shown in Fig 9-16. By this means, the rear portion of the wing can be turned down to increase the curvature of the wing, and hence the lift. With split flaps, there is a low-pressure flow through the space between the flap and the wing, which reduces the pressure markedly on the upper side of the flap. By the proper use of flaps, the lift coefficient can be approximately doubled.

• COMPRESSIBILITY EFFECTS

Thus far, consideration has been given only to the influence of Reynolds number and shape upon the drag on a submerged object. Reynolds number involves the relative influence of two fluid properties, density and viscosity. The fluid property of compressibility is now given consideration.

To do this, the principles of compressible flow presented in Chapter 4 are extended to the problem of drag on submerged objects.

The types of flow most frequently involved with compressibility are transonic flow and supersonic flow—flows in which a part or all of the fluid is moving past the object faster than the speed of sound. Typical of such an object are certain aircraft, rockets, and projectiles. The types of flow in which compressibility effects are important are:

a. Subsonic

b. Transonic

c. Supersonic

d. Hypersonic

These are illustrated in Fig. 9-17. With subsonic flow, the velocities are everywhere less than the speed of sound, Ma < 1.0, and the compressibility effects are confined to such relations as those developed in Chapter 4 for stagnation pressure. Transonic flow involves subsonic flow everywhere except in certain local regions where the velocity is increased or the density, which decreases the local speed of sound, is decreased, or both, so that

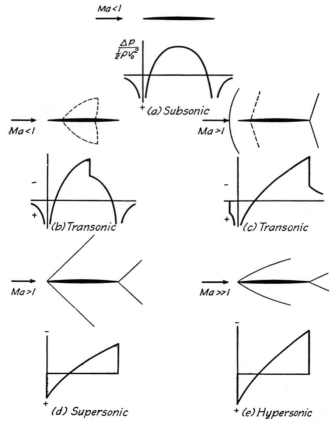

Fig. 9-17. Types of flow of compressible fluid about a submerged object.

Ma > 1.0 locally. This occurs, as shown in Fig. 9-17(b), at the center of the object, where the velocity is increased locally and, by the Bernoulli equation, where the pressure is reduced, which in turn reduces the density and the speed of sound. On the upstream side of the local supersonic region, there is a gradual increase from subsonic to supersonic velocities, while on the downstream side there is a shock wave through which the velocity suddenly decreases from supersonic to subsonic.

As the Mach number of approach increases to Ma > 1.0, the region of supersonic flow spreads fore and aft, and a shock wave develops upstream from the object, as shown in Fig. 9-17(c). Across the shock wave, the velocity decreases suddenly from supersonic to subsonic, and a region of subsonic flow exists, because of the increased density.

Once the flow is supersonic throughout, Ma > 1.0 throughout, see Fig. 9-17(d), and the local subsonic region is eliminated, as the shock wave moves down to the leading edge of the object, the wave pattern becomes relatively independent of Mach number. In supersonic flow, the variations in Mach number are due to changes in both the velocity of flow and the speed of sound.

At very large Mach numbers, the flow becomes hypersonic, the flow is much more confined to the immediate vicinity of the object, see Fig. 9-17(e), irrotational flow no longer exists, and certain of the theories developed for supersonic flow are no longer applicable. The variations in Mach number are due almost entirely to local changes in the speed of sound rather than local changes in velocity of flow.

Supersonic flow has certain similarities to, and certain differences from, subsonic flow about a submerged object. Figure 9-18 illustrates the flow around a flat plate, a diamond section, and a lens section. With the plate section, the streamlines are deflected across the leading shock wave to parallel the plate itself. This is in distinct contrast to subsonic flow, for which a rather complex pattern develops. The streamlines above the plate pass through rarefication waves and are turned gradually, until they are parallel to the boundary. The streamlines have a wider spacing, since the air is expanded and has less density. Consequently, the pressure is reduced.

On the underside of the plate, the streamlines pass through a compression shock wave and are turned suddenly, to be parallel with the boundary. They are spaced closer together—which indicates the air is compressed and the pressure increased.

At the downstream end of the plate, a compression shock wave is set up on the top side and rarefaction waves develop at the underside, which nearly restores the streamlines and pressure distributions to their normal upstream condition.

The waves of rarefaction coming from the upper side of the nose and the lower side of the tail spread outward to meet the shock wave coming upward from the tail and downward from the nose, respectively. As the waves of rarefaction meet the shock wave, they interact and tend to nullify or attenuate each other so that, further beyond, the flow is relatively unaffected by the object.

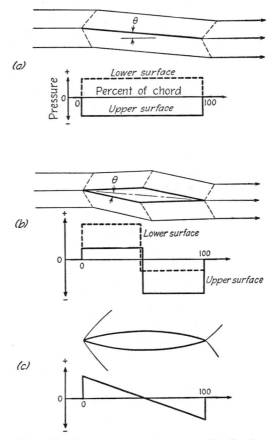

Fig. 9-18. Flow patterns and pressure distribution around thin sections at supersonic speeds.

• DRAG IN COMPRESSIBLE FLOW

Drag on a submerged object in supersonic flow is caused by both shear and pressure differences on the object—as already discussed for subsonic flow. The pressure and shear distributions, however, are modified appreciably. Figure 9-17 and Fig. 9-18 illustrate schematically the pressure distributions for various types of flow and various shapes of objects. The pressure is negative and uniform on the upper side of the plate and positive and uniform on the under side. Note how this differs from the pressure distribution at subsonic speeds. The increased pressure on the under side develops as the flow passes through the shock wave which is associated with a sudden increase in pressure, as already seen in Chapter 4. This increase in pressure increases the density and decreases the local Mach number. On the upper side, the nose of the plate causes a wave outward at the same initial angle as on the underside, but the pressure and density are being decreased, so that a rarefaction wave is initiated.

This wave forms continuously as it pivots about the nose, until it has turned the flow through the angle so that it is parallel with the plate.

The downstream edge of the rarefaction wave moves outward at the angle corresponding to the new Mach number (for the new velocity of flow and speed of sound) in the region above the plate. A similar process occurs downstream from the tail of the plate, so that essentially the same pressure and velocity are established as those which exist upstream from the plate. The pressure distributions on the upper side and under side of the plate cause a thrust, which can be divided into components of lift and drag.

Shear is developed on each side of the plate under a very thin boundary layer. When the plate is parallel to the flow, the resulting drag is entirely shear drag—pressure being the same on each side of the plate—as determined by Fig. 6-10. If the boundary is smooth, the drag per unit length transverse to the flow is a function of the Reynolds number, the Mach number, and the width (in the direction of flow) of the plate. As the plate is rotated at an angle θ to the flow, the pressure difference and resulting pressure drag across the plate rapidly become of such great magnitude that the shear drag is small and even insignificant by comparison.

For supersonic flow, the center of pressure is at the center of the plate. With subsonic flow, however, the center of pressure is nearer the leading edge.

Figure 9-18(b) illustrates a double wedge or diamond section, which is geometrically a simple combination of four flat plates. Furthermore, the wave patterns and the distributions of velocity and pressure are strikingly similar to a combination of those for the plate. On the upstream faces of the diamond section, shock waves are formed, and increased pressure is developed. At the midpoints, rarefaction waves are formed with a reduced pressure. The longitudinal components of the increased pressure on the upstream faces make up the *drag* on the body, and the corresponding upward components make up the *lift*. Again, the center of pressure is approximately through the center of the object.

A rounded section is illustrated in Fig. 9-18(c) which is a biconvex or lens section. This section can be considered as a series of infinitely short sections of flat plates. The largest angle is at the leading edge, with the angle becoming progressively smaller downstream. Each of these minute reductions in angle causes a rarefaction wave to develop, so that the pressure along the entire surface is being reduced after the initial sudden increase in pressure across the shock wave from the leading edge. Upstream from the midpoint the pressure is positive, and downstream from the midpoint the pressure is negative. At the trailing edge, a shock wave again is created, through which the pressure is increased suddenly to ambient pressure.

Projection of shock waves and rarefaction waves in Fig. 9-18(c) shows that they will meet if carried out some distance. As already mentioned, this meeting of a shock wave and a series of rarefaction wavelets attenuates all of the waves involved. In fact, if carried a sufficient distance, the waves are entirely dissipated. When a rarefaction wavelet meets a shock wave, the attenuation causes the shock wave to bend slightly downstream, which results in a curved shock wave in Fig. 9-18(c). The same curvature of the shock waves in Fig. 9-18(a) and 9-18(b) would develop if carried out farther than is done in the drawings.

The total drag on an object must be accompanied by a resistance to the flow system in general. This resistance results in a general reduction in the momentum of the flow which is equal in magnitude and opposite in direction to the drag.

In supersonic flows, separation usually occurs as a result of the inter-action of a shock wave with the boundary layer. Although the boundary layer develops along a gradually curving surface, such as that in Fig. 9-18(c), it can not separate until a shock wave forms along the central section, under transonic conditions. The shock causes the boundary to thicken, and generally to become turbulent and then separate. The separation which develops changes the pressure distribution so that the lift is reduced and the drag is increased, until the section is stalled.

• STREAMLINING FOR COMPRESSIBLE FLOW

Streamlining the shape of an object for supersonic flow is radically different from streamlining for subsonic flow. In subsonic flow, the tail portion of the object is brought gradually to a sharp edge downstream from a rounded but blunt nose, whereas in supersonic flow the nose must be a sharp edge and the tail portion can be rounded or abrupt without having more than secondary influence on drag and lift. This is illustrated by a comparison of Fig. 9-18(b) with Fig. 9-19. If the nose angle 2θ is greater than a certain maximum value for a given Mach number greater than 1.0, the oblique shock wave in Fig. 9-18(b) can no longer remain attached. Consequently, the wave becomes detached and rounded, as shown in Fig. 9-19(a). Such detachment can occur also if the Mach number is reduced to a value closer to 1.0 for a given nose angle.

When the shock wave is detached, as shown in Figs. 9-19 and 9-17(c), there is a region of subsonic flow immediately downstream from the wave. As already discussed, the Mach number gradually increases to $Ma \geq 1.0$ as it progresses downstream. Upstream from a rounded nose, the shock is always detached, because the angle 2θ is 180°.

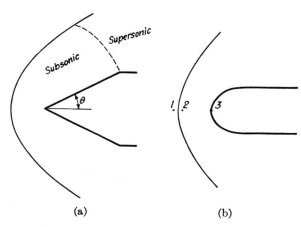

(a)   (b)

Fig. 9-19. Influence of the nose shape on shock wave.

• NOSE PRESSURES AND TEMPERATURES

The stagnation pressure on the upstream centerline of the objects in Fig. 9-19 is complicated by the fact that two pressure increases must be determined—the increase across the shock wave from point 1 to point 2 in Fig. 9-19(b), and the increase across the region from the wave to the object from point 2 to point 3 in Fig. 9-19(b). The ratio of pressures on each side of the shock wave can be determined from Eq. 4-56 as

$$\frac{p_2}{p_1} = \frac{2k}{k+1}\left[Ma_1^2 - \frac{k-1}{2k}\right] \tag{9-40}$$

in which

Ma_1 is the Mach number upstream of the wave;

p_1 is the pressure upstream of the wave;

p_2 is the pressure downstream of the wave.

In the region of $Ma < 1.0$ between the shock wave and the object Eq. 4-24 is applicable as

$$\frac{p_3}{p_2} = \left[1 + \frac{k-1}{2}Ma_2^2\right]^{k/(k-1)} \tag{9-41}$$

in which

       Ma_2 is the Mach number downstream from the shock wave;

       p_3 is the stagnation pressure at the upstream centerline of the object.

In order to eliminate Ma_2, the following relationship between Ma_1 and Ma_2 must be used

$$Ma_2^2 = \frac{Ma_1^2(k-1)+2}{2kMa_1^2 - k + 1} \tag{9-42}$$

Equations 9-40, 9-41 and 9-42 can be combined to yield

$$\frac{p_3}{p_1} = \frac{\left(\dfrac{k+1}{2}Ma_1^2\right)^{k/(k-1)}}{\left(\dfrac{2k}{k+1}Ma_1^2 - \dfrac{k-1}{k+1}\right)^{1/(k-1)}} \tag{9-43}$$

which relates the Mach number and static pressure of approach to the stagnation pressure at the nose of an object—such as a Pitot tube—in supersonic flow. Equations 9-43 and 9-41 are plotted in Fig. 9-20. These curves show that a transition is involved between subsonic and supersonic flow where the shock wave is first formed. This has particular application to the Pitot tube discussed in Chapter 10 on Flow and Fluid Measurements.

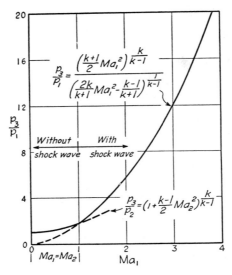

Fig. 9-20. Variation of stagnation pressure with Mach number.

The temperature change across a shock wave can be determined by combining Eq. 9-40, the ratio of densities

$$\frac{\rho_1}{\rho_2} = \frac{\text{Ma}_1^2 (k - 1) + 2}{\text{Ma}_1^2 (k + 1)} \tag{9-44}$$

with the temperature equation

$$\frac{T_2}{T_1} = \frac{p_2}{p_1} \frac{\rho_1}{\rho_2} \tag{9-45}$$

to yield

$$\frac{T_2}{T_1} = \frac{[k (2 \text{ Ma}_1^2 - 1) + 1][\text{Ma}_1^2 (k - 1) + 2]}{\text{Ma}_1^2 (k + 1)^2} \tag{9-46}$$

Equation 9-46 can then be combined with the temperature change within the region of Ma < 1.0 upstream from the object as expressed by Eq. 9-47.

$$\frac{T_3}{T_2} = 1 + \frac{k - 1}{2} \text{ Ma}_2^2 \tag{9-47}$$

Equations 9-47 and 9-42 may then be combined to give the temperature ratio of T_3/T_1.

The extreme increase in temperature resulting from very large Mach numbers explains in part why missiles and meteorites become so extremely hot as they enter the earth's atmosphere.

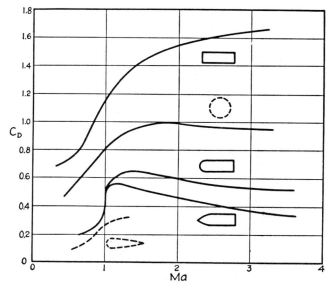

Fig. 9-21. Variation of drag coefficient with Mach number.

• DRAG COEFFICIENT AND MACH NUMBER

The drag coefficient C_D as a function of the Mach number for flow around submerged objects is illustrated in Fig. 9-21. At small Ma values, there is a slight decrease in C_D as Ma increases, because Re is also increasing. When Ma increases (Ma $\approx$ 0.5 for spheres) so that transonic flow is initiated, then C_D increases rapidly with Ma—due to a combination of increasing shock wave effects and the expansion of the separation zone. The magnitude of C_D continues to increase with increasing Ma for blunt objects. But for objects with a rounded or pointed nose, C_D reaches a maximum, decreases somewhat, and then tends to a constant with further increase in Ma_1. This is due to such factors as the change in form of the shock wave, the stabilization of the separation zone, and the limiting minimum pressure of an absolute vacuum on the downstream side of the object. The continual increase in drag for a bluff object bears a striking similarity to the variation of stagnation pressure with Mach numbers, as shown in Fig. 9-20.

• SUMMARY

1. The ***general drag equation*** for flow around submerged objects is

$$F_D = \frac{C_D A \, \rho V_0^2}{2} \tag{9-1}$$

2. The *variation of **drag coefficient** with Reynolds number for laminar flow* is

$$C_D = \frac{24}{\text{Re}} \qquad (Sphere,\ Re < 0.5) \tag{9-3}$$

$$C_D = \frac{20.4}{\text{Re}} \qquad (Disk\ perpendicular\ to\ flow,\ Re < 0.5) \tag{9-4}$$

$$C_D = \frac{13.6}{\text{Re}} \qquad (Disk\ parallel\ to\ flow,\ Re < 0.1) \tag{9-5}$$

$$C_D = \frac{8\pi}{\text{Re}(2.0 - \ln \text{Re})} \qquad (Circular\ cylinder,\ Re < 0.1) \tag{9-6}$$

$$C_D = \frac{8\pi}{\text{Re}(2.2 - \ln \text{Re})} \qquad \begin{array}{l}(Flat\ plate\ perpendicular\ to\ flow,\\ Re < 0.1)\end{array} \tag{9-7}$$

$$C_D = \frac{4.12}{\text{Re}} \qquad Flat\ plate\ parallel\ to\ flow,\ Re < 0.01) \tag{9-8}$$

$$C_D = \frac{1.33}{\sqrt{\text{Re}}} + \frac{4.12}{\text{Re}} \qquad \begin{array}{l}(Flat\ plate\ parallel\ to\ flow,\\ Re < 6 \times 10^5)\end{array} \qquad \begin{array}{l}(6\text{-}15)\\ (6\text{-}10)\end{array}$$

3. From the theory of **Stokes** *for flow around a* **sphere**

   a. The shear drag is $\frac{2}{3}$ and the pressure drag is $\frac{1}{3}$ of the total drag.

   b. At all points on a sphere the longitudinal components of shear drag and pressure drag combine to equal the same value of unit total drag over the entire surface of the sphere.

   c. The Stokes equation is the product of the surface area of the sphere and the unit total drag

$$F_D = (\pi d^2)(3\mu V_0/d) = 3\pi d\mu V_0 \qquad (9\text{-}2)$$

4. The **Karman vortex street** *downstream from cylinders* is a result of separation, and the alternate formation of vortices, which causes alternate side thrusts on the cylinder. This is described by the Strouhal number and the Reynolds number as

$$\frac{fd}{V_0} = \phi(\text{Re}) \qquad (9\text{-}11)$$

5. The **sudden decrease** in C_D at $\text{Re} \approx 2 \times 10^5$ for rounded objects is caused by a change from a laminar boundary layer to a turbulent boundary layer, and the resulting change in location of the point of separation of flow.

6. The combination of **circulation** and *translation flow* around an airfoil creates a lift (the Magnus effect)

$$F_L = \rho L V_0 \Gamma = \frac{C_L A \rho V_0^2}{2} \qquad \begin{matrix} (9\text{-}28) \\ (9\text{-}29) \end{matrix}$$

The same phenomenon causes a spinning tennis ball, baseball, or golf ball to curve.

7. **Stalling** of a wing occurs because of *separation* on the upper side of the airfoil.

8. **Induced drag** is caused by a secondary flow along the length of a wing.

9. The **polar diagram** (Fig. 9-14) shows the relation between the coefficients of lift and drag and the points of zero lift, minimum drag, maximum lift-drag ratio, and maximum lift.

10. The *types of flow* in which **compressibility** *effects* are important are

a. Subsonic	c. Supersonic
b. Transonic	d. Hypersonic

11. **Streamlining** *for compressible flow* is radically different from streamlining for incompressible flow. For compressible flow there must be a pointed and tapered upstream section. For incompressible flow there must be a rounded and blunt upstream section and a tapered down-stream section.

12. See Eqs. 9-40 to 9-47 for the **pressures** *and* **temperatures** associated with compressible flow around the nose of an object.

13. The **drag coefficient** *for compressible flow* increases continuously with Mach number for bluff objects. For streamlined objects it increases abruptly at about Ma = 1.0, then decreases gradually with increasing Mach number for larger Ma values.

References

1. Binder, R. C., *Fluid Mechanics*, 3rd ed. Englewood Cliffs, N. J.: Prentice-Hall, Inc., 1955.

2. Dodge, Russell A., and Milton J. Thompson, *Fluid Mechanics*. New York: McGraw-Hill Book Co., Inc., 1937.

3. Durand, W. F., *Aerodynamic Theory*. Pasadena: California Institute of Technology, 1934.

4. Goldstein, S., *Modern Developments in Fluid Dynamics*. New York: Oxford University Press, 1938.

5. Leipmann, H. W., and A. Roshko, *Elements of Gasdynamics*. New York: John Wiley and Sons Inc., 1957.

6. Prandtl, Ludwig, *Fluid Dynamics*. New York: Hafner Publishing Co., 1952.

7. Rouse, Hunter, *Elementary Mechanics of Fluids*. New York: John Wiley & Sons, Inc., 1946.

8. Shapiro, Ascher H., *The Dynamics and Thermodynamics of Compressible Fluid Flow*, Vol. I. New York: The Ronald Press, 1953.

9. Vennard, John K., *Elementary Fluid Mechanics*, 3rd ed. New York: John Wiley & Sons, Inc., 1954.

Problems

9-1. An object with a projected cross-sectional area of 0.5 sq ft is held in a stream of water which has a velocity of 20 ft/sec. If the drag coefficient is 2.0, what is the force on the submerged object?

9-2. An object falls at a constant velocity in a barrel of oil (sp. gr. = 0.85). The projected cross-sectional area is 0.1 sq ft and the force on the object is 0.2 lb. If the drag coefficient is 10, what is the velocity?

9-3. A very small dew drop is 10^{-4} ft in diameter and spherical in shape. If the temperature is 40°F and the air is absolutely still, how many seconds will it take the dew drop to fall 5 ft?

9-4. An air bubble, $d = 0.001$ ft, with same diameter as the water droplet in Problem 9-3, rises through still water whose temperature is 70°F. How long does it take the air bubble to rise 5 ft through the water? Since water and air are both involved in Problems 9-3 and 9-4, explain why there is a difference between the two answers.

9-5. A flat disk of mica (sp. gr. = 2.8) 0.001 ft in diameter and 0.00001 ft thick is dropped in a bucket of SAE 10 oil whose temperature is 50°F. If the particle falls so that the disk is normal to the direction of motion, what is the velocity of the disk?

9-6. A balloon 10 ft in diameter rises at a rate of 15 ft/sec. Estimate the drag force if the air temperature is 80°F.

9-7. Estimate the maximum load which may be carried by a 30-ft parachute when the maximum permissible velocity of descent is 18 ft/sec. Use an air temperature of 70°F.

9-8. Determine the drag force acting on the following bodies of revolution when placed in an airstream having a velocity of 100 ft/sec: (**a**) circular disk, (**b**) sphere, (**c**) circular cylinder, and (**d**) airship hull. Use an air temperature of 60°F and a diameter of one foot.

9-9. The viscosity of a heavy bearing oil was tested by dropping a steel sphere 0.25 in. in diameter in the oil through a full distance of 6 ft, the oil being contained in a vertical glass cylinder 10 in. in diameter. The time taken to fall through the middle 2 ft was 10 seconds. The specific gravity of the ball was 7.73, that of the oil 0.80. Find the absolute viscosity of the oil.

9-10. A hailstone of 0.72 in. in diameter falls at a constant velocity of 67 ft/sec. Assuming that the hailstone is a sphere, what is its specific gravity if the specific weight of the air is 0.075 lb/cu ft and the air viscosity is 1.5×10^{-4} sq ft/sec?

9-11. Does the drag on a cylinder (flow normal to the axis) increase with a decrease in the Reynolds number? Explain.

9-12. A cylinder 8 in. in diameter moves through the water at 40°F in a direction perpendicular to its axis. If its L/d ratio is very large, determine the drag force per foot of length at velocities between 2 ft/sec and 25 ft/sec, taking enough values to determine definitely the drag in the critical range. Plot a curve of F_D versus Re.

9-13. In Problem 9-9, assuming that the law of resistance during acceleration is $F = 3\pi d\mu V$, find the terminal velocity and the time taken to attain it. The ball is released from rest just below the oil surface.

9-14. A smoke stack 100 ft high has a mean diameter of 22 ft. Estimate the force on the stack in a wind which is 100 mph at an air temperature of 0°F.

9-15. A long signboard is perpendicular to a 60 mph wind. If the signboard is 10 ft high, what is the drag per foot. Assume an air temperature of 40°F.

9-16. A long power cable $\frac{1}{2}$-in. in diameter is 300 ft long between poles. If the wind is 90 mph normal to the direction of the cable, what is the force on the cable between poles? Use an air temperature of 0°F.

9-17. For an infinite circular cylinder, what percentage of the total drag is a result of pressure drag when (**a**) Re = 10^3, (**b**) Re = 10, (**c**) Re = 0.1.

9-18. A flat plate is parallel to the flow of an oil whose velocity is 6 ft/sec. The plate is 3 ft long, the specific gravity of the oil is 0.8 and the kinematic viscosity is 10^{-4} sq ft/sec. What is the force on the plate if flow is on both sides of the plate?

9-19. A long circular cylinder with a diameter of 2 in. is submerged in water at 60°F. At what velocity will the drag coefficient suddenly decrease, if the flow is normal to the axis of the cylinder?

9-20. What is the drag on a model airship hull at a Reynolds number of 2×10^5 if $d = 8$ inches, and the air in the wind tunnel is at 50°F and standard atmospheric pressure?

9-21. An auto with a projected area of 30 sq ft has a drag coefficient of 0.3. What is the horsepower required to overcome air friction at 70 mph in still air at 60°F?

9-22. What horsepower is required in Problem 7-21 if the car is travelling into a wind of 40 mph?

9-23. A man weighing 130 lb jumps from an airplane with a 20-ft diameter parachute. With what velocity will he strike the ground if the air temperature is (**a**) 0°F, (**b**) 90°F? Assume still air and standard atmospheric conditions.

9-24. In Problem 7-23, calculate the height from which the man would have to jump to attain the same velocity without a parachute. Neglect air resistance.

9-25. The air velocity normal to a $\frac{1}{2}$-in. suspended cable is 60 mph. Assuming standard air at sea level, what is the frequency of the alternating system of vortices?

9-26. A submarine periscope is $\frac{1}{2}$ ft in diameter and is travelling at 20 mph. What is the frequency of the alternating shedding of vortices if the water temperature is 60°F? Assume that $\gamma = 64$ lb cu ft and that the viscosity is the same as that of fresh water. What is the force per ft on the periscope?

9-27. A quartz sphere (sp. gr. = 2.65) 1 in. in diameter falls in a large container of SAE 30 oil at 70°F. What is the terminal velocity of the sphere?

9-28. If the sphere in Problem 9-27 were a plastic sphere (sp. gr. = 1.10), what would be the rate of fall in the same oil?

9-29. A rubber ball 3 in. in diameter weighs 1 oz. If the ball is dropped from a high building on a still day, when the temperature is 70°F, how long will it take to fall the last 200 ft if it descends at a constant velocity for the entire 200 ft?

9-30. Steady irrotational flow approaches a horizontal cylinder at a uniform velocity of 10 ft/sec normal to the axis of the cylinder. What are the maximum and the minimum velocities along the boundary of the cylinder, and where do these velocities occur? What is the velocity at a point on the front of the cylinder at an angle of 30° above the horizontal?

9-31. The velocity at a point on the boundary of a cylinder is measured as 32 ft/sec. If uniform irrotational flow with a velocity of 20 fps is approaching the cylinder normal to its axis, at what point on the cylinder was the velocity measured?

9-32. Derive Eq. 9-28 and carefully explain each step in the solution.

9-33. If the cylinder in Problem 9-31 is 3 in. in diameter and is rotated at 100 rpm, what is the velocity at the ends of the diameter which is perpendicular to the flow? Assuming air at standard conditions, what are the pressures at these points? Where are the stagnation points?

9-34. What rotation in rpm will cause a single stagnation point for the cylinder in Problem 9-33? Assume irrotational flow, as before.

9-35. A counterclockwise free vortex at point A has a circulation of 200 sq ft/ sec. What is the velocity at a radius of 10 ft? If a clockwise vortex located 15 ft from the first has a strength of 150 sq ft/sec, what is the velocity 5, 7.5 and 10 ft from the center of the first vortex?

9-36. A horizontal cylinder 52 in. long and 4 in. in diameter is rotated at 600 rpm in the counterclockwise direction. Water at 70°F with a uniform flow of 3 ft/sec is flowing normal to the axis of the cylinder. What are the drag and the lift on the cylinder?

9-37. A cylinder 1 ft in diameter and 13 ft long rotates at 60 rpm in a wind which has a velocity of 20 mph normal to the axis of the cylinder. Use air at 90°F and sea level. Assuming irrotational flow or no slip, calculate: **(a)** the circulation, **(b)** the lift force, and **(c)** the location of the stagnation points.

9-38. In Problem 9-37, if the actual circulation is just half the theoretical or no-slip value, calculate the lift and the drag, and give the magnitude and direction of the resultant force on the cylinder. Also locate the stagnation points for this case.

9-39. A ball 12-in. in diameter is thrown through still air at 68°F at sea level so that the Reynolds number is 10^5. If the ball is rotating 600 rpm counterclockwise about a vertical axis, find the transverse lift and the drag on this ball.

9-40. In Problem 9-39, if the weight is neglected, what is the approximate radius of curvature of the path of the ball?

9-41. A baseball is thrown toward home plate with an initial velocity of 75 mph. The ball is 2.9 in. in diameter and weighs 5.25 oz. The ball is released with a horizontal velocity headed directly toward the inside edge to a right-handed batter and 7 ft above home plate, and is spinning counterclockwise about a vertical axis at 5200 rpm. The ball must travel 60 ft to home plate which is 17 in. wide.

The still air is at 80°F and at sea level. If the drag is neglected, what is the elevation and the distance of the ball from the inside edge of the plate when it arrives at home plate?

9-42. In Problem 9-41, assume that the average drag is 0.5 the initial drag and the average lift is 0.5 the initial lift. If the air resistance in the vertical direction is neglected, estimate the location and the speed of the ball when it arrives at home plate.

9-43. An airfoil of infinite length has a chord length of 9 ft. If the air speed past the foil is 300 mph at a temperature of −30°F at a barometric pressure of 18 in. of mercury, what is the lift force and the drag per foot of air foil for angles of attach of $\alpha = 6°$, $0°$, and $-6°$?

9-44. For a Clark-Y wing of aspect ratio 6, find the maximum lift force when the air velocity is 200 mph and the air temperature is 60°F at sea level. What is the horsepower required to overcome the drag at this condition? Use $c = 1.0$ ft.

9-45. For Problem 9-44, find the lift and the drag forces when the ratio of F_L/F_D is a maximum. At what angle of attack does this occur?

9-46. For a Clark-Y airfoil with an angle of attack of 8°, what percentage of the drag is induced drag?

9-47. Assuming that the boundary layer for a Clark-Y wing is turbulent, estimate what part of the total drag is a result of skin friction.

9-48. Derive Eq. 9-46 by filling in all missing steps.

9-49. Fill in the missing steps for the derivation of Eq. 9-47.

9-50. A Pitot tube is placed on a body moving through still air at 12 psia and 20°F. The stagnation pressure at the end of the Pitot tube is 5 psi. Assuming a frictionless adiabatic process, how fast is the body going?

9-51. If the air temperature is −40°F upstream from a pressure wave and the pressure is 8 psia, what is the approximate temperature at the nose of a missile with a Mach number of 4.0?

9-52. Each of the objects shown in Fig. 9-21 travels through still air at 80°F and 14.7 psia. If each has a diameter of 6 in., what is the drag on each object when its velocity is 900 mph?

9-53. If the atmospheric pressure upstream from a shock wave of a missile with a Mach number of 2.0, is 13 psia, what is the stagnation pressure at the nose of the missile?

chapter 10

Flow and
Fluid Measurements

Measurements are a necessary part of the study of a fluid system. Such a study arises from the need for information regarding fluid properties or flow characteristics associated with some physical process or geophysical occurrence. If it is desired either to understand or to control the physical process or geophysical occurrence, measurements are an indispensable adjunct to such understanding or control.

Geophysical processes involve many different types of fluids—principal among these being water and air. In connection with such a study, questions arise regarding the magnitude of the discharge at various times in rivers and streams of different sizes. The sediment content of the water which is flowing may be the deciding factor in the design of some structure or works associated with the river. The U. S. Geological Survey (U.S.G.S.) is charged with the responsibility of making measurements of the various rivers and streams, and publishing these records for the future use of designers. The behavior of the atmosphere and understanding and forecasting the weather have an effect upon nearly every human being. In order to aid those individuals and groups most directly associated with the weather, the U. S. Weather Bureau operates thousands of instruments which perform the measurements necessary for them to discharge their responsibility.

The operation of an industrial machine or a vehicle, such as an automobile or an airplane, requires instruments which make the measurements that the operator needs to accomplish his job safely and successfully. Certain industrial processes require a careful control of the flow of various types of fluids. This control cannot be accomplished without measurements which guide the operator in controlling the process. In the research laboratory, equipment and instruments for making measurements are an indispensable part of studying the properties of fluids and characteristics of the flow associated with various phenomena. In the home, in the office, and in the school, many different instruments are utilized to measure and to control flow systems associated with the gaseous and liquid environments. In short, the twentieth-century world is almost completely dependent for its operation upon measurements of fluid properties and flow characteristics.

This chapter considers the techniques of measuring certain properties of fluids and characteristics of flow systems. Proper design and control of measuring equipment, and intelligent use of the results, depend very largely upon a good understanding of the *basic principles* of fluid mechanics which have been presented in previous chapters of this book. Consequently, these principles are used as the basis for the treatment of instrumentation in the remainder of this chapter.

The *fundamental dimensions* of length (L), time (T), and force (F), or mass (M), are an important part of a presentation on flow measurement. Any fundamental or complex quantity that is to be measured involves one or more of these fundamental dimensions. In this chapter, treatment is given to the measurement of each of these fundamental dimensions singly, as well as in various combinations, in the consideration of fluid properties and various characteristics of the flow systems.

In the analysis of a problem and the selection of instruments, consideration must be given to making measurements in various *types of flow systems*—such as steady and unsteady, uniform and nonuniform, compressible and incompressible, laminar and turbulent, and confined and unconfined systems.

The methods of measurement can usually be divided into one or a combination of observational, mechanical, electrical, and chemical techniques. In all cases, visual or acoustical observations are necessary in the ultimate determination. Observational techniques are even older than man, since animals have always observed various fluid phenomena by means of their senses. Mechanical measuring devices are frequently involved in measuring techniques. Such mechanical systems employ to advantage various basic principles of mechanics.

Perhaps the most rapidly expanding field of measurements is that of electronic systems, primarily because of the needs in the aircraft and flow-process industries. Almost unbelievable strides have been made in the development of instruments for measuring various characteristics of flow systems electronically. This type of equipment is especially useful where phenomena are encountered that involve rapid changes, small magnitudes, sensing elements at remote location, or a fluid property or flow characteristic outside the range of human or mechanical sensitivity.

General selection of a method of measurement, and specific selection of the system or instruments to be used, depend to a large extent upon such factors as the necessary accuracy, the permissible energy loss, the desired simplicity, the dependability of operation, and the allowable cost.

Fluid Properties

As pointed out in previous chapters, the fluid properties most frequently encountered in fluid mechanics are density, specific weight, elasticity, viscosity, vapor pressure, and surface energy. These various fluid properties have been discussed in Chapter 1.

• DENSITY AND SPECIFIC WEIGHT

Density and specific weight are related simply by the acceleration of gravity, as pointed out in Eq. 1-1. Consequently, if one is measured the other can be determined. A balance is one of the most accurate instruments, as well as the one most frequently used to determine either density or specific weight. If the **balance** uses standard calibrated weights, the counterbalance is the object in question. Therefore, the mass, and hence the density, of the object are determined in terms of the calibrated weights. A balance depending upon standard weights would give the same magnitude of mass or density at any location, regardless of the gravitational system. If the counterbalancing force is supplied by a spring, however, the specific weight is measured, since the reading on a spring balance depends upon the gravitational system as well as the mass (volume and density) of the object or material being weighed.

For liquids, a **pycnometer** is frequently used in the determination of density or specific weight. A pycnometer is usually a glass container which has its volume and weight, and variations of these factors with temperature, determined very accurately. The net weight of the liquid in the pycnometer, divided by the volume of the pycnometer, then gives the specific weight of the liquid.

Another method of measuring the density of a liquid is to weigh an object first in the air, and then while it is suspended in the liquid. If the volume of this object has been determined accurately, the difference in the two weights, $W_a - W_e$, divided by its volume $\tilde{V}$, will give the specific weight of the liquid in question, that is,

$$\gamma = \frac{W_a - W_e}{\tilde{V}} \qquad (10\text{-}1)$$

A special type of device for this measurement is the **Westphal balance** which may include a thermometer in the suspended object to determine the temperature of the liquid.

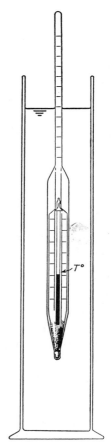

An object which is suspended within the liquid without connection to a balance is the **hydrometer.** Special types of hydrometers are commonly used in connection with specific weight determinations for anti-freeze in the radiator, and water in the battery of a vehicle. The hydrometer consists of a large hollow glass tube with a slender neck protruding from it, as shown in Fig. 10-1. The bottom of the tube is weighted so that the specific weight of the hydrometer as a whole is less than that of the liquid to be studied. If the liquid is particularly dense, the hydrometer will float high in the liquid and the calibrated graduations on the stem will so indicate. The determination of density and specific weight by using submerged or floating objects is based upon the principles of buoyancy discussed in Chapter 2. Density and specific weight vary with temperature for various pure liquids, as shown in Table III.

Fig. 10-1.
Hydrometer.

• ELASTICITY

The elasticity and compressibility of a gas can be determined rather simply, for either adiabatic or isothermal conditions, by compressing a known volume in a cylinder by means of a piston, and carefully determining the changes in pressure and temperature; or by increasing the temperature of a known volume of gas in a container and measuring the change in pressure for the known change in temperature. If the process is isothermal Eq. 1-8, $E = p$, is applicable, and if the process is adiabatic Eq. 1-9, $E = kp$, is applicable.

• VISCOSITY

The measurement of viscosity is usually associated either directly or indirectly with Eq. 5-2, Newton's law of viscosity, or Eq. 5-28, Stokes' law for spheres. One of the simplest methods of measuring viscosity is the flow of a fluid in a small **capillary tube**, so that viscosity can be determined by Eq. 5-26, the Hagen-Poiseuille equation,

$$\mu = \frac{h_f \gamma D^2}{32VL} \tag{5-26}$$

This method is illustrated schematically in Fig. 10-2, and has the advantage of being accurate, simple, inexpensive and reliable, provided the

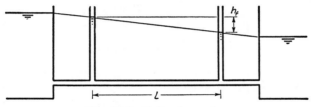

Fig. 10-2. Tube viscometer.

diameter of the tube is accurately determined. It has the disadvantage of being cumbersome and inconvenient to move from one place to another. The **Saybolt viscometer** is a special adaptation of the capillary tube, as illustrated in Fig. 10-3. This is an instrument which is commonly used in the petroleum industry, and it has been standardized by the American Society for Testing Materials. The liquid to be tested is placed in the cylindrical chamber, and the Saybolt reading is the time in seconds required for 60 milliliters of the fluid to pass through the capillary tube at the bottom of the chamber. The following equation, in which t is in seconds, permits evaluation of the kinematic viscosity ν with close approximation.

$$\nu = (2.37 \times 10^{-6}t) - \left(1.94 \times \frac{10^{-3}}{t}\right) \tag{10-2}$$

The Saybolt viscometer has the distinct advantage of being standardized, accurate, simple, dependable, and convenient. It has the disadvantage of being more expensive than some of the simpler viscometers assembled from standard equipment in the laboratory.

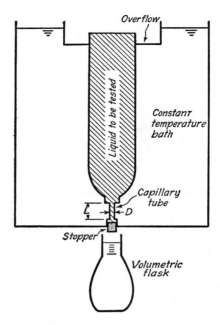

Fig. 10-3. Saybolt viscometer.

The **rotational** type of **viscometer,** see Fig. 10-4, also utilizes fundamental laws of viscous flow, Eq. 5-2. The torque is measured for a given
velocity of rotation and temperature of liquid. This type of viscometer is
frequently used in industrial processes to check the viscosity of the product
or some liquid used in the process. The rotational viscometer has essentially
the same advantages and disadvantages as the Saybolt viscometer.

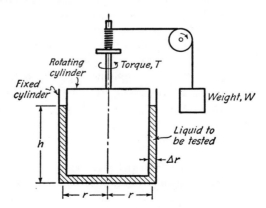

Fig. 10-4. Cylinder viscometer.

The *falling-sphere viscometer* is illustrated in Fig. 10-5. In order to determine the viscosity, it is necessary to determine the time required for a standard sphere to fall over a given distance. The viscosity is then determined by the Stokes equation, Eq. 5-29,

$$\mu = \frac{\Delta \gamma d^2}{18 V_0} \tag{5-29}$$

provided Re < 0.1. Certain precautions must be taken in using such a viscometer—the size and weight of the sphere must be determined accurately, the sphere must not be in a tube which is too small (or special corrections must be applied), and temperature must be controlled carefully so

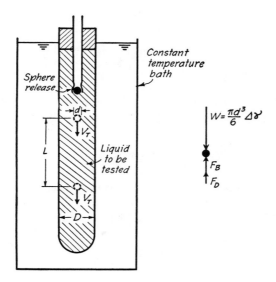

Fig. 10-5. Falling-sphere viscometer.

that there is no secondary flow due to density variations within the tube. The principal disadvantages of this type of viscometer are the need for carefully calibrated spheres and careful control of temperature.

If Re < 0.1, the correction in fall velocity is made by the McNown and Lee equation

$$\frac{V_0}{V_r} = 1 + \frac{9d}{4D_t} + \left(\frac{9d}{4D_t}\right)^2 \tag{10-3}$$

in which V_r is the reduced velocity, as observed in the container of diameter D_t. This equation shows that, if the diameter of the cylinder in which the sphere is falling is a hundred times the diameter of the sphere, the fall velocity will be 2.5 per cent less than the velocity predicted by the Stokes equation. Figure 1-2 shows a variation of viscosity with temperature for a variety of fluids.

Example Problem 10-1: Oil flows through a small tube as illustrated in Fig. 10-2. If 50 lb of oil, which has a specific weight of 50 lb/ft³, flows from the ½ in. tube in 3 minutes and the head loss per ft of length is 1.0 ft, calculate the dynamic viscosity.

Solution:

$$V = \frac{Q}{A} = \frac{(50)}{(3)(60)(50)(0.00136)} = 4.08 \text{ fps}$$

$$\mu = \frac{h_f \gamma D^2}{L\, 32V} = \frac{(1.0)(50)(1/24)^2}{32(4.08)} = 0.000664 \text{ lb-sec/ft}^2$$

• SURFACE ENERGY

The *capillary rise* or depression in a tube, see Figs. 1-3 and 1-4, is a common method of determining the surface energy by use of Eq. 1-14.

$$\sigma = r_0 \gamma \frac{h}{2} \tag{1-14}$$

Great care must be taken, however, to be certain that all surfaces are clean and that the liquid is pure. Another method is illustrated in Fig. 1-3(b), in which a *bubble* is created at the end of a tube submerged in the liquid.

$$\sigma = r_0 \frac{\Delta p}{2} \tag{1-12}$$

Equation 1-12 can be used to determine the surface energy σ by using the pressure difference across the bubble, when the radius of the bubble is the same as the radius of the tube.

• VAPOR PRESSURE

The vapor pressure of a liquid can be measured by use of a tube or chamber in which is placed the pure liquid, with no entrained air or other foreign gases. The pressure in the container is reduced until the liquid separates from the wall or bubbles begin to form. At this point, the pressure is equal to the vapor pressure. It can be measured for various temperatures of liquid.

Flow Characteristics

Measurement of the characteristics of flow systems involve various measurements of length, time, pressure, velocity, acceleration, and discharge. Each of these is considered separately.

• LENGTH MEASUREMENTS

Measurement of length is accomplished with such equipment as *tapes*, *scales*, and *calipers*, in accordance with the accuracy required for the particular problem. In fluid mechanics, however, special attention must be given to accurate measurement of surface elevation and depth of flow. For example, slope of a water surface is frequently needed for a given length of a flow system. This requires determination of both the length of the system and the drop in water surface elevation over a given distance. Measurement of the horizontal distance does not require the precision that the vertical measurement requires. The important part of precision in the length measurement is the percentage error involved. If a horizontal distance is ten feet, an error of $\frac{1}{8}$ in. is only 0.1 per cent, which would be no more significant than 0.001 in. in an elevation or piezometric head variation of 1 in.

The *staff gage* is a common method of measuring the surface elevation of a liquid. This is a graduated gage consisting of a scale usually mounted vertically. The surface elevation is measured where the liquid surface intersects the gage. A common staff gage is the dip stick used to determine the amount of oil in an automobile engine. The staff gage is used as a primary method of measurement or as a reference gage for stream-gaging stations of the U.S.G.S. The elevation of the water surface of a stream is usually referred to as the stage of the stream. The staff gage can be calibrated to whatever degree of refinement is desired for the particular measurement to be made. If it is placed on a slope, the accuracy of reading can sometimes be increased.

Float gages are also common in the measurement of surface elevation. They are frequently used in water storage tanks, with a wire or chain connection which goes over a pulley at the top of the tank and down the outside, where an indicator on a scale permits a reading of the inside liquid level. Float gages are also used with instruments which record the variation in water surface elevation in a stream, as associated with the U.S.G.S. stream-gaging station. Figure 10-6 shows a sketch of a stream-gaging station which includes a float gage attached to a recorder. The float also has considerable use as an indicator of liquid surface elevation in an automobile gasoline tank, and in tanks and containers associated with processing industries.

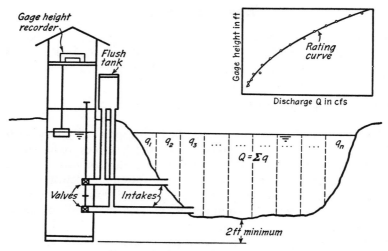

Fig. 10-6. Stream-gaging station.

Water surface elevations are also determined by **differential leveling** with the surveyor's precise level. Although very accurate measurements can be made with this system, the accuracy depends also upon how accurately the scale, which is observed in the telescope, is set on the water surface.

The most common instruments for measuring the elevation of a liquid surface in the laboratory are the **point gage** and **hook gage.** These are illustrated in Fig. 10-7. The point gage is simply a sharp-pointed rod which projects down to the surface. The rod is mounted on a shaft which is controlled by a rack and pinion mechanism. Attached to the sliding shaft is a scale and vernier, which permits accurate determination of the location of the point. By slowly lowering the point to the liquid surface, it is possible to determine quite accurately the elevation of a quiet liquid surface. Waves or surging of the surface, of course, make accurate measurement more difficult. In some cases, a hook

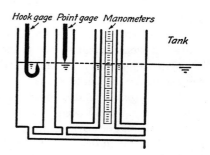

Fig. 10-7. Measurement of liquid surface elevation.

gage, which is a point gage brought to the water surface from below, provides more accurate measurements. In each case, it is necessary to have a light placed in such a manner that reflection shows immediately the disturbance to the water surface caused by the point of the gage.

A *differential bubble gage* has recently been developed by the U.S.G.S. A simplified version of this gage, which illustrates the basic principle of operation, is given in Fig. 10-8. This instrument bubbles an inert gas against two liquid heads. The gas pressure in each line is related to the head which the gas must bubble against. These pressures are transmitted to a sensing element and an automatic recorder which records this

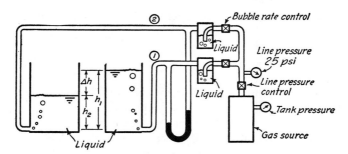

Fig. 10-8. Differential bubble gage.

differential head Δh. The magnitude of h_1 or h_2 can also be recorded by opening line 2 or line 1, respectively, to the atmosphere. This instrument is extremely useful in measuring the drop in water surface between two fixed points, such as in a natural stream or a laboratory flume.

The U.S.G.S. has also developed a *gage height follower,* which is similar in principle to the bubble gage, but simpler in design. As its name implies, the gage height follower records the piezometric head of liquid with respect to a fixed datum. It is becoming more common for field installations, in place of the float gage shown in Fig. 10-6.

Depth measurements are usually a measurement from the water surface to the bottom of the stream or container. A common method of measuring depth is with a *sounding weight* fastened to a tape or cable of some known length. Of more recent origin are *sonic probes.* By varying the frequency of the sound wave emitted, and by other aspects of the instrumentation, it is possible to determine electronically the time required for the *sound waves* to be transmitted to the bottom and reflected back to a receiver. In this way, depths of tens of feet in rivers, hundreds of feet in lakes, and thousands of feet in oceans can be determined.

Recently, a sonic depth sounder has been developed at Colorado State University by the U.S.G.S. for depths of less than five feet, with an accuracy of a few hundredths of a foot. If a sonic probe is mounted on a traveling carriage or boat, it is possible to record depth with respect to distance, which permits measurement of the shape of sand waves. It is also possible to record the depth with respect to time, which permits measurement of the rate of movement of sand waves, by having the sonic probe stationary and connected to a recorder. Two sonic probes can be inverted and used to map water surface waves, and bed configuration simultaneously.

Manometers can be used to measure the elevation of a water surface by means of a tube or hook gage well, as illustrated in Fig. 10-7. The well for the point gage or hook gage can be considered an enlarged manometer. The manometer, however, must have a scale connected to it, so that the height of the water surface in the manometer can be measured directly.

• TIME MEASUREMENTS

Time is most commonly recorded by a standard *stop watch* or *clock.* Usually, the error involved in standard equipment is very small compared with the human error involved in operating the timepiece. Recently, *electronic timing* equipment has been developed which eliminates or greatly decreases the human error, and greatly increases the accuracy of timing. A mechanical recorder can have one marker which indicates the time, so that it can be correlated with the other items being recorded regarding the characteristics of the flow system. The *stroboscope* is a convenient method of synchronizing a repeating flow phenomenon with the frequency of light flashes. In this way, it is possible to measure the frequency of occurrence of the phenomenon. The variation of a phenomenon with time can also be recorded by *high-speed motion pictures,* if it is a rapidly-occurring phenomenon. If it is a slowly-occurring phenomenon, it can be recorded by *time-lapse motion pictures.*

• PRESSURE MEASUREMENTS

As pointed out in Chapter 1, in fluid mechanics, pressure is the force per unit area exerted by a fluid at a point. A common method of measuring pressure is provided by piezometer openings in a boundary which lead to *manometers* for determination of the height of a column of balancing fluid, as illustrated in Fig. 2-9. If the fluid is moving past the boundary, it is of paramount importance that the piezometer openings have certain characteristics, in order that the true ambient pressure be registered. The ambient pressure is sometimes called the static pressure. As pointed out by Howe, however, this is a misnomer, since the pressure is that at the particular point within the flow, regardless of whether the pressure distribution is hydrostatic. Figure 10-9 illustrates some of the various types of openings which might occur. If the opening is not made properly it will not register accurately the ambient pressure at the point of the opening. The opening must be normal to the boundary and should be cylindrical in shape, with at least two diameters of length. The opening must be flush at the boundary surface. If the opening is somewhat rounded, or if there is a projection or burr, it will cause an erroneous measurement of pressure. There must be no net flow through the piezometer opening, or else the manometer reading will reflect a decrease in pressure in the direction of flow in the connecting tube between the opening and the manometer. If there is a surging through the opening with no net flow, then both ends of the tube must have the same entrance loss coefficient.

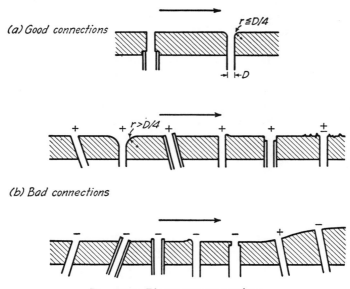

Fig. 10-9. Piezometer connections.

If the piezometer opening projects on the downstream side, a stagnation pressure tends to develop upstream, so that the manometer will read too high. If the burr projects on the upstream side, a separation zone tends to develop over the piezometer opening, which gives a lower reading than the actual ambient pressure. If the tube itself projects inward beyond the boundary, a low-pressure zone develops above the opening and a smaller pressure is registered.

The manometer systems used to measure pressure with both gases and liquids, under conditions of both negative and positive pressure, are illustrated in Chapter 2.

Electronic **pressure cells,** such as piezoelectric cells, are manufactured commercially. The quartz crystal, for example, will respond to pressures as great as 5000 psi and to fluctuations of pressure from 1 to 10,000 cps (cycles per second). Such cells can be mounted flush with the surface of a boundary, and thereby measure not only the average pressure, but also fluctuations of pressure with time. If these gages are not flush with the surface, the same difficulties are encountered as with the piezometer connections already discussed. Great care must be taken to make either the piezometer connection or the pressure cell flush with the surface. Even the slightest irregularity will cause significant errors in measurement. These difficulties cannot be overemphasized.

Because of the irregularities and variations which are bound to occur in a closed conduit, a series of piezometer connections are sometimes connected to a *closed annular ring* around a pipe, which is in turn connected into a single manometer or other type of pressure gage, so that the reading from the ring is an average of all of the piezometer openings. This type of system is quite effective if there are not large differences in pressure from one side of the pipe to the other (caused by faulty piezometer openings or a secondary flow system). If large differences in pressure exist, however, errors can develop because of flow from one opening to another and, consequently, the reading becomes a function of rate of flow and the flow pattern through the piezometer system.

The pressure within a fluid can be measured by an *ambient pressure tube,* as illustrated in Fig. 10-10. The nose of the tube is streamlined so

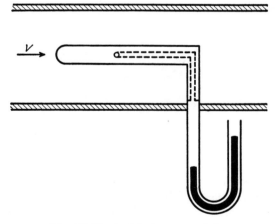

Fig. 10-10. Ambient-pressure tube.

that it creates as little disturbance as possible to the flow. When the nose of the tube is hemispherical, the piezometer opening should be placed between three and four diameters downstream, to allow sufficient distance for the streamlines to return to the same spacing they had upstream from the tube. The arm downstream which holds the tube in the flow may influence the flow pattern, and hence the pressure reading upstream from it, for a considerable distance. Therefore, this arm should be at least nine diameters downstream from the piezometer openings. The ambient pressure tube is sensitive to direction and, consequently, must be aligned so as to be parallel with the streamlines. Usually, the misalignment can be as much as plus or minus 5 degrees without serious error. If a boundary is near the tube, however, the tube creates a local acceleration or retardation of the flow, which causes erroneous readings.

If the flow around the tube is *supersonic,* the piezometer holes should be approximately ten diameters downstream from the nose.

Mechanical gages for measuring pressure usually involve a special bellows, diaphragm, or piston, which is activated as the pressure changes. The movement of any one of these elements is connected to an indicator which, when calibrated, gives the reading of the pressure. Bellows- and tube-type gages are illustrated in Chapter 2. A common tube-type gage is the Bourdon gage of Fig. 2-7, and a common bellows gage is the barograph illustrated in Plate 2-1.

When a liquid is involved, difficulties are encountered sometimes if a piezometer opening is placed where an air pocket accumulates, or where there is considerable surging of the flow. These phenomena usually occur in regions of separation zones or low pressure zones. The air pocket can accumulate as a result of air entrained in the water, which separates from the water and is collected in pockets, or it can be an accumulation of air which comes out of solution in upstream regions of low pressure. Pulsating or surging flow can, as already discussed, cause inaccurate readings because of the greater resistance to flow in one direction as compared to the other direction in the piezometer tube.

• SHEAR MEASUREMENTS

The measurement of shear can be accomplished by both direct and indirect methods. Shear can be measured directly by means of a very sensitive *dynamometer* attached to a small plate which replaces a part of the boundary. This plate must not touch the boundary at any point, but there must also be only a very small gap between the plate and the boundary. The plate must be flush with the boundary throughout.

Indirect methods can also be used to determine shear. For laminar flow, the boundary shear τ_0 can be calculated, if the velocity distribution above the boundary is known, from the Newton law of shear

$$\tau_0 = \mu \left(\frac{dv}{dy}\right)_{y=0} \tag{5-2}$$

Considering steady flow in a pipe, a concentric cylinder of fluid of radius r and length dx can be selected as a free body, and the forces on the free body summed up and equated to zero, because the fluid moves without acceleration, to obtain

$$\tau = -\frac{r}{2}\frac{dp}{dx} \tag{5-22}$$

in which τ is the shear at distance r from the center of the pipe;

r is the radial distance from the center of the pipe;

— dp/dx is the pressure gradient with respect to distance along the pipe.

Shear can be calculated from Eq. 5-22 for turbulent flow and laminar flow, since its derivation is independent of the nature of the flow.

The shear at a point on the boundary of an open channel where the depth is y can be derived in the same manner as Eq. 8-2,

$$\tau_0 = \gamma\, y S \tag{10-4}$$

The shear at a point on the boundary can also be determined for turbulent flow by

$$\tau_0 = \rho \left[\frac{v_2 - v_1}{\dfrac{2.3}{\kappa} \log \dfrac{y_2}{y_1}} \right]^2 \tag{10-5}$$

in which v_1 and v_2 are the velocities parallel to the channel boundary, at distances y_1 and y_2 measured from and normal to the boundary. Equation 10-5 can be derived by solving for the difference between the velocities v_1 and v_2, as described by the von Karman logarithmic velocity distribution law, assuming $\kappa = 0.4$, and then solving for τ. Best results are obtained from Eq. 10-5 when y_1 and y_2 are relatively close to the boundary. The accuracy of Eq. 10-5 is limited by variation in κ and the accuracy with which the velocities at y_1 and y_2 can be measured.

• VELOCITY MEASUREMENTS

Since velocity is a vector quantity, both ***direction and magnitude*** must be measured. The measurement of velocity usually involves visual observation of the flow, the use of the Bernoulli equation to find the velocity from the difference between stagnation pressure and the ambient pressure (Pitot tube, etc.), the use of rotating vanes, the trajectory of a jet, and special sensing probes which permit electrical measurements of the velocity.

Floats: Floats on the surface of a liquid and zero-lift balloons in the air permit determination of flow patterns which will give velocity magnitude and direction, and variations of each of these quantities with respect to time. In the atmosphere, the motion is sufficiently slow for time-lapse motion pictures to aid appreciably in visualizing and measuring the flow pattern. When flow patterns are needed on the surface of a liquid, photographs taken over a definite period of time will indicate both magnitude and direction by the streaks which are created. A flow pattern below the surface of the liquid can be measured by special zero-lift submerged objects. In the field, the flow pattern at the surface can frequently be observed by floating debris. In the laboratory, confetti can be placed on the water-surface and, by means of time exposures, photographs will indicate both magnitude and direction of the velocity. If a clock is included in the field of exposure, it is not necessary to depend upon an estimate of shutter speed to determine the magnitude of the velocity. Velocities beneath the surface of the flow can be determined by means of droplets of mixtures of liquids which are immiscible with water, but are adjusted to have the same specific weight. One such mixture is benzine and carbon tetrachloride, with powdered anthracene added to create a milky appearance. This liquid can be introduced into the flow, as droplets a fraction of a millimeter in diameter, by means of a tiny hypodermic needle. These droplets, like confetti, can be photographed to determine both magnitude and direction of velocity.

Pitot tubes, Pitot cylinders, and Pitot spheres: Figure 10-11 illustrates the Pitot tube, Pitot cylinder, and Pitot sphere—which are simply submerged objects as discussed in Chapter 9. As shown in Fig. 10-11(a), the Pitot tube is an adaptation of the ambient pressure tube of Fig. 10-10. The ambient pressure is measured through the holes in the side of the tube, as illustrated in Fig. 10-10, but the stagnation or total pressure is measured through the hole in the tip of the tube.

The principle of the Pitot tube, for determining the **magnitude of the velocity,** is illustrated in the special form of the Bernoulli equation

$$\frac{V^2}{2g} = \frac{\Delta p}{\gamma} = \Delta h \qquad (10\text{-}6)$$

or

$$V = \sqrt{2\,g\Delta h} = \sqrt{\frac{2}{\rho}\,(\Delta p)} \qquad (10\text{-}7)$$

in which Δp is the difference in pressure between the tip of the tube and the side holes. In this way, the magnitude of velocity is determined in a simple and straightforward fashion. At the tip of the tube, the stagnation pressure represents the total head which includes the pressures due both to the velocity and to the ambient pressure, whereas the side pressure is only the ambient pressure—the difference then being directly proportional to the square of the velocity.

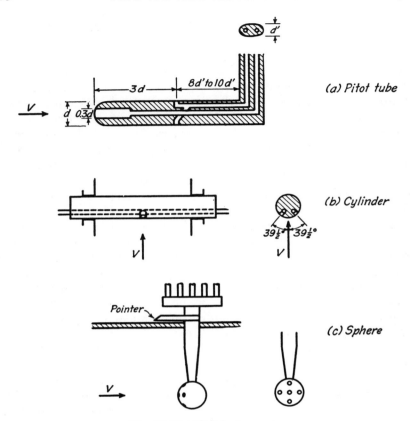

Fig. 10-11. Pitot instruments.

The Pitot cylinder and the Pitot sphere depend upon the same basic principles, but the ambient pressure must be determined by the particular location of the ambient pressure piezometer hole. Experiment has shown that the point where ambient pressure exists on the surface of a cylinder is at an angle to the flow of $39\frac{1}{2}$ degrees. This location of the ambient pressure openings is applicable up to velocities of approximately one-fourth the speed of sound. When velocities are increased beyond this, the point of ambient pressure moves further around the cylinder. The location of the ambient pressure openings for the sphere should be at an angle of 45 degrees for Re $< 10^5$ and 41 degrees for Re $> 10^5$.

The foregoing discussion has been applicable to the determination of the magnitude of velocity. These same instruments can also be used to determine the ***direction of the velocity.*** In some cases, it is desired to have the instrument very sensitive to direction, whereas in other cases it is desired to have the instrument very insensitive to misalignment with the

flow. The Pitot tube is a particularly good instrument for measuring velocities under conditions where misalignments are apt to occur. The Prandtl-
type tube illustrated in Fig. 10-11(a) has approximately a 1 per cent error
at an angle of 20 degrees to the direction of flow.

The Pitot cylinder and Pitot sphere are particularly adaptable to determination of direction. It is evident from Figs. 10-11(b) and 10-11(c)
that a slight rotation of the cylinder or the sphere will create a magnified
difference in pressure between the two ambient pressure openings. In order
to determine the exact direction of flow, the cylinder or sphere can be adjusted until the difference in pressure between these two openings is zero.
The Pitot cylinder is particularly adaptable for determination of both
magnitude and direction of velocity in two-dimensional flow, whereas the
Pitot sphere makes the same determination possible in three-dimensional
flow.

At *supersonic velocities,* the static openings must be approximately
ten diameters downstream from the tip of the Pitot tube. Determination of
the velocity from the difference in head for a Pitot tube in supersonic flow
must take into consideration the energy loss and increased pressure through
the shock wave, as well as the increase of pressure which takes place between
the shock wave and the tip of the Pitot tube. This results in the following
equation, which was developed from Eq. 9-43 in Chapter 9.

$$p_s = p_1 \frac{\left(\dfrac{k+1}{2}\,\mathrm{Ma}_1^2\right)^{k/(k-1)}}{\left(\dfrac{2k}{k+1}\,\mathrm{Ma}_1^2 - \dfrac{k-1}{k+1}\right)^{1/(k-1)}} \tag{10-8}$$

in which

p_1 is the ambient pressure;

p_s is the stagnation pressure;

Ma_1 is the Mach number of the approaching flow;

k is the adiabatic exponent.

For velocities between the limiting conditions of one-fourth the speed
of sound and the speed of sound itself, the velocity can be determined by
use of a Prandtl tube and Eq. 10-9, which is based upon Eq. 4-24, in which
the two pressures must be measured separately:

$$p_s = p_1 \left[1 + \frac{k-1}{2}\,\mathrm{Ma}_1^2\right]^{k/(k-1)} \tag{10-9}$$

The Pitot tube must be small compared with the flow system in which it
is placed. If the Pitot tube is relatively large or if it is near a boundary,
the ambient flow is accelerated around the tube and the reading is erroneous.

The Pitot tube also gives incorrect readings in a flow system where the velocity gradient is appreciable, so that the velocity on one side of the tube is much greater than on the other side.

Rotating wheel meters: There are several mechanical devices which measure the velocity of the flow by rotating wheels. The rotating wheel can be mounted on a vertical shaft or on a horizontal shaft. ***Anemometers*** are devices which measure the speed of air movement. ***Current meters*** are devices which measure the speed of water movement. The cup type of wheel is best for relatively slow speeds, whereas the vane type measures larger speeds more accurately. Each of these meters must be calibrated, because they do not automatically give the magnitude of the velocity itself. Meters of the vane or propeller type offer the least resistance to flow. The standard procedure for determining the magnitude of velocity by use of current meters is to count the number of revolutions during a certain period of time, as determined by a stop watch or a clock. For an *anemometer*, the number of revolutions is read from a dial for a given period of time.

Jet trajectory: The trajectory of a jet issuing from an orifice or a nozzle follows a path which depends upon the initial efflux velocity. This principle can be used to advantage to determine the velocity of a jet, Fig. 3-19, from a pipe in the field or the laboratory. On the basis of the Bernoulli and Newton's equations, the following equation can be derived for the *velocity of efflux V* of a horizontal jet issuing from a pipe or an orifice

$$V = x \sqrt{\frac{g}{2z}} \qquad (10\text{-}10)$$

in which

x is the horizontal coordinate;

z is the vertical coordinate measured downward from the center of the orifice

The discharge is determined by the continuity equation $Q = AV$.

Hot-wire anemometer: Measurement of the velocity of a gas is frequently determined in the laboratory by means of the hot-wire anemometer. As explained in Chapter 5, this device consists of a fine wire of approximately 0.0003 in. diameter. It is usually less than $\frac{1}{4}$ of an inch in length and mounted on the ends of two pointed prongs. A small electrical current is passed through the wire, in order to heat it above the temperature of the surrounding air or gas. As the gas flows around the wire, the wire is cooled and, since the resistance in the wire is a function of its temperature, the resistance of the wire can be calibrated against the velocity of gas movement. This device requires frequent calibration and is subject to breakage. Despite these limitations, it has widespread use in gas flow studies under carefully-controlled conditions.

Because of its very small size, the wire of the *hot-wire anemometer* contains so little heat that the temperature of the wire fluctuates almost immediately with any fluctuation in velocity. Consequently, it serves as an excellent device for measuring the intensity of the turbulence in a gas stream, see Chapter 5. By electronic equipment which measures the rms value of the velocity fluctuations, it is possible to read the intensity of turbulence directly. The scale of turbulence can be determined by using two hot wires and varying the distance between them. A correlation of the readings of these two anemometers provides a measure of the scale of the turbulence.

The hot wire anemometer has not been developed adequately for widespread use in liquids at the present time. Consequently, the turbulence in a liquid is usually measured by means of a special Pitot device, or by photographing and measuring the movement of tiny zero-lift droplets within the liquid. No method available is entirely satisfactory, however. The zero-lift droplets, for example, give accurate results but their use is very time consuming.

Radioactive tracers: Radioactive tracers have been used to measure the velocity distribution in a closed conduit. It is quite probable that tracer techniques will be used increasingly in the future for measurement of both magnitude and direction of velocity, and the fluctuation of magnitude and direction with time.

• ACCELERATION MEASUREMENTS

In certain experiments with submerged or floating objects it is desired to measure the acceleration of the object. Accelerometers usually note the apparent weight of a known mass attached to a pendulum or cantilever spring. The piezoelectric crystal and heated elements in a convection tube are used as sensing elements.

• DISCHARGE MEASUREMENTS

Measurement of discharge or flow of a fluid can be accomplished from several basically different points of view. Perhaps the simplest method is a volumetric determination or a weight determination. The variation in piezometric head associated with localized changes in velocity, as reflected by the Bernoulli equation, is the principle used with the Venturi meter and the Parshall flume. Nozzles, orifices, and gates utilize a combination of the Bernoulli equation and geometric considerations discussed in Chapter 3. Bridges and culverts also serve as contractions for approximate measurement of discharge. Weirs, spillways, and Venturi meters under free-flow conditions depend upon the Bernoulli equation and the critical depth consideration for open channel flow. The deflected vane and rotometer depend upon drag of a submerged object in confined flow. The momentum principle is utilized in the bend or elbow meter.

There are a number of instruments which determine the mass rate of flow. It is also possible to determine the discharge by integrating the velocity distribution across the flow prism. If certain information is available regarding the basic velocity distribution, it is possible to measure the discharge by the use of floats.

The salt-velocity and the dilution methods involve the addition of dielectric chemicals and measurements of their concentration and time of arrival downstream.

Discharge measurements by volume and weight: To measure the discharge by volume, it is necessary to have a calibrated container in which the fluid can be discharged for a certain period of time. The container has been calibrated so that the volume at various elevations in it is known. The container can be a reinforced concrete tank, in which case hydrostatic pressure may have very little influence on the volume of the tank. With tanks that are not cylindrical, however, it is necessary to remember that, as they are filled with a liquid, the hydrostatic pressure may deflect them to some extent. With some steel rectangular tanks, this deflection is appreciable, so that the calibration of the tank must be made under loaded conditions.

As a measurement of discharge is being made by the volumetric method, care must be taken to have the same conditions existing at the beginning and at the end of the measurement. In other words, if the measurement is made "on the run" the same conditions of inflow to the tank must be present in each case. This sometimes requires that special precautions be taken to eliminate waves which are accompanied by a jet impinging on the surface. If the discharge is turned on at the beginning of the run and then turned off at the end of the run, the rate of deflecting the incoming flow into and out of the tank must be the same.

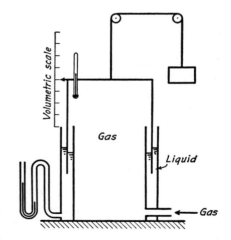

Fig. 10-12. Gas meter.

Weighing tanks can be employed for the measurement of discharge in much the same way that volumetric tanks are employed. The same precautions must be taken with respect to deflecting the flow into and out of the tank. Such a tank is either mounted on a platform scale or, in the case of large tanks, a special supporting system and balance must be designed for this purpose.

Volumetric determination of the discharge of a gas is frequently necessary in connection with calibration of gas meters which are used for distribution systems by a public utility. One of the most common methods is the use of a gas-meter shown in Fig. 10-12. This is composed simply of three concentric cylinders with a liquid seal and counter balancing system. The method of operation can be determined from Fig. 10-12.

Mechanical meters: Various types of mechanical meters are almost universally employed to measure the water delivered to homes and small businesses and industries such as a public utility. These meters do not measure the rate of flow but simply the total volume over a period of time. Meters of this type are based on various principles, but in each case there is some internal moving device which activates a mechanism attached to an indicator wheel or dial. The disk meter is commonly used in water distribution systems. It consists of a wobbling disk which activates a system of gears (see Fig. 10-13). The disk wobbles on the central ball bearing in much the same way as a coin which is dropped on a flat surface. It does not rotate about its own axis.

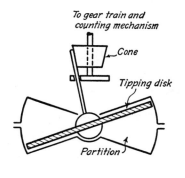

Other mechanical meters involve various types of oscillating systems or rotating wheels with rings or blades on them. One type of rotary meter involves two lobes which are similar to the gears of certain types of gear pumps (see Fig. 12-11).

Venturi meters: The rate of discharge can be determined quite accurately with very little loss of head by means of the Venturi meter. Figure 10-14 illustrates the standard Venturi meter which is frequently used in main lines of flow systems in closed conduits. For simplicity of construction, the changes in contour are angular, which increases the resistance to flow and the formation of separation zones. At extremely high velocities with liquids, there is also the danger of cavitation in the throat of the Venturi meter. Nevertheless, the Venturi meter is an effective, useful, and common instrument for measuring

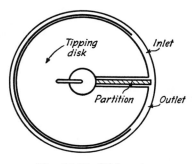

Fig. 10-13. Disk meter.

discharge. It is based upon the Bernoulli equation which can be re-arranged as

$$\frac{V_2^2}{2g} - \frac{V_1^2}{2g} = h_1 - h_2 = \Delta h \qquad (10\text{-}11)$$

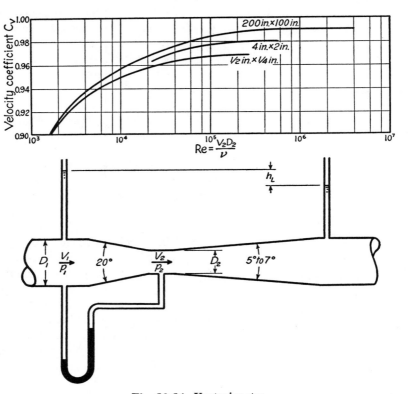

Fig. 10-14. Venturi meter.

wherein

h_1 is the upstream piezometric head;

h_2 is the throat piezometric head;

V_1 is the velocity upstream;

V_2 is the average theoretical velocity in the throat.

At large velocities, separation occurs in the entrance of the throat because of the angular shape. There is also a loss of energy between the entrance to the meter and its throat. Consequently, the theoretical velocity V_2 in the throat is larger than the actual velocity. To obtain the actual average velocity in the throat, V_2 is multiplied by the velocity coefficient C_v.

The continuity equation therefore is

$$Q = V_2 C_v A_2 = V_1 A_1 \tag{10-12}$$

which can be combined with Eq. 10-10 and rearranged to yield

$$Q = \frac{C_v A_2}{\left[1 - \left(\frac{D_2}{D_1}\right)^4\right]^{1/2}} \sqrt{2g\,\Delta h} \tag{10-13}$$

or

$$Q = C_d A_2 \sqrt{2g\,\Delta h} \tag{10-14}$$

in which

C_d is a dimensionless discharge coefficient;

D_2 is the throat diameter;

D_1 is the upstream diameter.

Figure 10-14 shows a typical variation of the velocity coefficient C_v with the Reynolds number $\mathrm{Re} = V_2 D_2 / \nu$.

The discharge coefficient C_d varies not only with the Reynolds number, due to the resulting influence of viscosity on the nature of the approaching velocity and turbulence distribution, but also upon other variables such as the diameter ratio D_2/D_1; the upstream bends, valves, and other geometric factors; and the actual geometric shape of the meter. Consequently, it is necessary to calibrate a Venturi meter in place to obtain very accurate results.

Because of the gradually-expanding transition downstream from the throat of the Venturi, there is considerable recovery of the kinetic energy. In the equation for head loss

$$h_L = C_L \frac{V_2^2}{2g} \tag{10-15}$$

the loss coefficient C_L depends on the diameter ratio and the Reynolds number but it is approximately 0.1. Hence, the Venturi meter is quite efficient from the viewpoint of energy loss in a pipe line and has received widespread use, despite its being greater in cost than other instruments of similar accuracy and simplicity.

Use of the Venturi meter for measuring discharge of compressible fluids is based upon the principles presented in Chapter 4, under converging flow and diverging flow. The equation for weight rate of flow for compressible fluids is:

$$W = \frac{C_v Y A_2 \gamma_1 \sqrt{\dfrac{2g(p_1 - p_2)}{\gamma_1}}}{\sqrt{1 - \left(\dfrac{D_2}{D_1}\right)^4}} \qquad (10\text{-}16)$$

in which

$$Y = \sqrt{\frac{1 - \left(\dfrac{D_2}{D_1}\right)^4}{1 - \left(\dfrac{D_2}{D_1}\right)^4 \left(\dfrac{p_2}{p_1}\right)^{2/k}} \left[\frac{\dfrac{k}{k-1}\left(\dfrac{p_2}{p_1}\right)^{2/k}\left[1 - \left(\dfrac{p_2}{p_1}\right)^{(k-1)/k}\right]}{1 - \dfrac{p_2}{p_1}}\right]} \qquad (10\text{-}17)$$

Values of Y to be used in Eq. 10-16 for Venturi meters, flow nozzles, and pipe orifices are given in Fig. 1 at the back of the book.

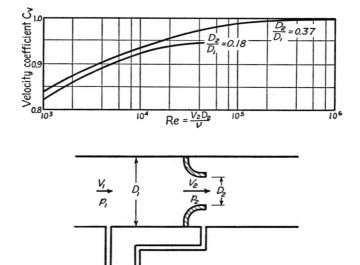

Fig. 10-15. Nozzle meter.

Nozzles in pipes: A truncated and somewhat contracted form of the Venturi meter is the nozzle. It is simply a rounded construction placed in the conduit, as shown in Fig. 10-15. The discharge equations for the nozzle are the same as for the Venturi meter and can be derived in the same manner. See Eqs. 10-11 and 10-14.

The principal variation of the coefficient of velocity C_v for the nozzle is with the Reynolds number, see Fig. 10-15. The coefficient of velocity is larger for the nozzle than for the Venturi meter, principally because the coefficient of contraction $C_c = 1$ for all discharges for the nozzle. There is very little or no contraction in the nozzle, because of its streamlined geometry.

To obtain accurate results, the flow nozzle (like the Venturi meter) must be calibrated in place.

Example Problem 10-2: A Venturi meter with a 4-in. throat is installed in an 8-in. pipe which is inclined upward at an angle of 45° to the horizontal. If the distance between pressure taps along the pipe is 5 ft, the differential pressure is 10 psi, and the water temperature is 60°F, estimate the discharge of water in cfs.

Solution:

$$\frac{V_2^2}{2g} - \frac{V_1^2}{2g} = \frac{p_1 - p_2}{\gamma} + z_1 - z_2 = \Delta h = \frac{(10)(144)}{62.4} - (5)(0.707) = 19.56$$

From Fig. 10-14 assume $C_v = 0.985$

$$Q = \frac{C_v A_2}{[1 - (D_2/D_1)^4]^{1/2}} \sqrt{2g\Delta h} = \frac{(0.985)(0.087)\sqrt{2g(19.56)}}{[1 - (4/8)^4]^{1/2}} = 3.15 \text{ cfs}$$

$$V_2 = \frac{3.15}{.0872} = 36.1 \text{ fps} \qquad \text{Re} = \frac{(36.1)(4/12)}{1.217 \times 10^{-5}} = 9.9 \times 10^5$$

From Fig. 10-14 $C_v = 0.985$ as assumed and

$$\underline{\underline{Q = 3.15 \text{ cfs}}}$$

Orifices in pipes: Orifices in pipes consist of a thin plate with a central orifice which is clamped between pipe flanges. An orifice causes the flow to accelerate like that in a Venturi meter or nozzle, so that it can be used as a flow meter. An orifice in a pipe line is illustrated in Fig. 10-16. For the same diameter opening, the orifice contracts the flow much more than the nozzle. Orifice contraction coefficients can be as small as 0.61. Selection of C_c for the orifice can be made from the curve of $\alpha = 90°$ in Fig. 3-24.

Utilizing the Bernoulli equation, as for the Venturi meter, the discharge equation for the pipe orifice is

$$Q = \frac{C_v C_c A_2}{\sqrt{1 - C_c^2 \left(\frac{D_2}{D_1}\right)^4}} \sqrt{2g\,\Delta h} \qquad (10\text{-}18)$$

The foregoing discharge equation can be expressed as

$$Q = C_d A_2 \sqrt{2g \, \Delta h} \qquad (10\text{-}19)$$

in which

C_d is the discharge coefficient;

A_2 is the area of the orifice;

Δh is the change in piezometric head.

The variation of discharge coefficient C_d with Reynolds number is given for a 2-in. sharp-edged pipe orifice with flange taps in Fig. 10-16. This coefficient varies with C_v, C_c, and the area ratio of the orifice D_2/D_1.

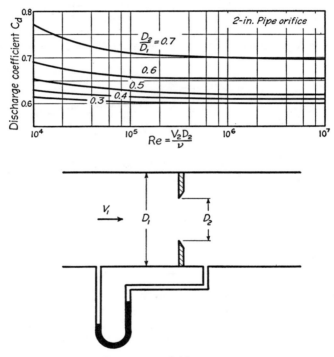

Fig. 10-16. Orifice meter.

Supersonic flow from a nozzle or an orifice is greatly simplified, because the ambient or back pressure has no influence upon the rate of efflux. As shown in Eq. 4-46, the discharge W depends only upon the upstream pressure p_1 and specific weight.

Example Problem 10-3: Calculate the weight rate of flow of air W through a 2 in. orifice in a horizontal 4-in. pipe line. The pressures upstream and immediately downstream of the orifice are 70 psia and 49 psia respectively, and the temperature in the line is 60°F.

Solution:

Assume $C_d = 0.62$

$$\gamma_1 = \frac{(70)(144)}{(53.3)(520)} = 0.364 \text{ lb/ft}^3$$

From Fig. I using $p_2/p_1 = \dfrac{49}{70} = 0.7$ and $D_2/D_1 = \dfrac{2}{4} = 0.5$

$Y = 0.907$

$$W = \frac{C_v C_c Y A_2 \gamma_1 \sqrt{2g \dfrac{p_1 - p_2}{\gamma_1}}}{\sqrt{1 - C_c^2 \left(\dfrac{D_2}{D_1}\right)^4}} = C_d Y A_2 \gamma_1 \sqrt{2g \dfrac{\Delta p}{\gamma_1}}$$

$$W = (0.62)(0.907)(0.0218)(0.364)\sqrt{\frac{(2)(g)(21)(144)}{0.364}} = \underline{\underline{3.26 \text{ lbs/sec}}}$$

Check C_d

$$\gamma_2 = \frac{(49)(144)}{(53.3)(520)} = 0.255 \text{ lb/ft}^3$$

$$V_2 = \frac{W}{A_2 \gamma_2} = \frac{3.26}{(0.0218)(0.255)} = 586 \text{ fps}$$

$$\text{Re} = \frac{V_2 D_2}{\nu_2} = \frac{(586)(\frac{1}{6})(10)^5}{4.72} = 2.07 \times 10^6$$

From Fig. 10-16

$$C_d = 0.62 \text{ as assumed}$$

$$W = \underline{\underline{3.26 \text{ lbs/sec}}}$$

Parshall flume: An instrument which uses the Venturi principle for free-surface flow in open channels is the Parshall flume illustrated in Fig. 10-17. Under free-flowing, unsubmerged conditions, the Parshall flume is a critical-depth meter, somewhat like the weir or free overfall. This meter is used extensively throughout the world for measurement of discharge in open channels. The Parshall flume has the following special advantages:

1. The discharge is measured with only a small loss in energy.

2. Sediment and debris enter and pass through the flume without difficulty.

3. It is simple to install.

4. It can be operated and maintained by personnel with limited technical training.

The difficulties which are sometimes encountered with the Parshall flume are:

1. Setting the structure at the proper elevation to have incoming flow conditions without creating either drawdown or backwater.

2. Setting the structure at the proper elevation to insure a maximum recovery of kinetic energy and a minimum of erosion downstream.

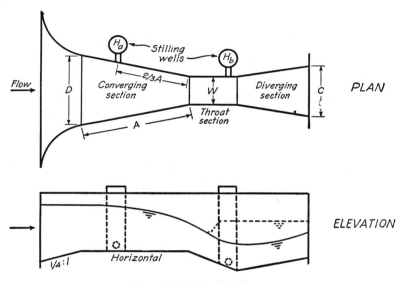

Fig. 10-17. Parshall flume.

Under free-flowing (unsubmerged) conditions, the Parshall flume has been calibrated for different widths of throat W, and the corresponding equations for determining the discharge are as follows:
For W from 1 ft to 8 ft

$$Q = 4WH_a^{1.522W^{0.026}} \qquad (10\text{-}20)$$

and for W from 8 ft to 50 ft

$$Q = (3.688W + 2.5)H_a^{1.6} \qquad (10\text{-}21)$$

in which

Q is the discharge in cfs;

W is the size of flume or throat width;

H_a is the head in feet, see Fig. 10-17

When there is free flow over the crest of the Parshall flume, only one measurement of head H_a is needed. The discharge is considered to be free flow for values of $H_b/H_a < 0.75$. If the flume is operating under submerged conditions, however, it is necessary to measure the head downstream as well as upstream.

Gates: By means of the Bernoulli principle, nearly any kind of a contraction to the flow that is involved in a pipe line or open channel can be employed as a metering device. Gates are especially adaptable to this use. A *sluice gate* makes an excellent metering device, as shown in Fig. 3-26. In effect, it is simply a special type of orifice. Equation 3-57 is applicable for the discharge under the sluice gate, and can be simplified and generalized to

$$Q = C_d A \sqrt{2g \, \Delta y} \qquad (10\text{-}22)$$

in which

    C_d is the discharge coefficient and is a function of the size of the opening b, the upstream depth y_1, and the coefficient of contraction and coefficient of velocity;

    Δy is the difference between the upstream depth y_1 and the downstream depth y_2;

    A is the area of the opening under the gate.

The coefficient of contraction can be determined from the curve $\alpha = 90°$ in Fig. 3-24. The velocity coefficient is assumed to be 1.0. Under submerged conditions, y_2 is simply the downstream depth of flow which may vary in magnitude from the depth of the jet, as illustrated in Fig. 3-26, to the depth y_1 at which Δy and the discharge go to zero in Eq. 10-22.

Various other types of gates can serve effectively as discharge measuring devices. These include a radial gate or roller gate on the crest of a spillway or in a flume or open channel; and a slide gate in a pipe line such as the Armco 101 meter-gate. Measurement of discharge by each of these gates depends upon the same basic equation, Eq. 10-22,

in which

    Δy becomes Δh, the difference in piezometric head at arbitrarily-chosen points upstream and downstream from the gate;

    C_d is the discharge coefficient, as determined by a calibration either in place or as a model study in the laboratory;

    A is the area of the opening which is related to the distance the gate is open.

Care should be taken not to extrapolate Eq. 10-22 beyond the range of Δh for which the gate is calibrated—particularly with respect to the upstream elevation of the water surface. As the water surface upstream becomes very low and approaches the lip of the gate, the entrance flow pattern is modified and the discharge coefficient changes appreciably.

It is not expected that calibrations of this nature will give precise discharge measurements. The accuracy, however, is usually between 1 and 5 per cent.

The ***contracted-opening method*** of measuring discharge through a sluice gate in open channels is applicable when laboratory or field measurements of A, V and Δy can be made. Application of the basic Bernoulli equation across the sluice gate yields the equation for discharge

$$Q = A \sqrt{2g \left(\frac{V^2}{2g} + \Delta y \right)} \qquad (10\text{-}23)$$

in which

A is the area of the water section at the *vena contracta* downstream from the contracted opening;

V is the average velocity upstream from the sluice gate;

Δy is the difference in water surface elevation upstream and downstream from the sluice gate.

This method can be modified to apply to other types of contracted openings. Although a head loss is usually considered to take place from upstream to the section where the area is measured, this loss can usually be ignored, because it is very small compared with the errors which are bound to occur in even the best of the rough field measurements which are possible. The best measurements are obtained if they are made while the water is flowing through the contracted opening. In this case, care should be taken to discount any water which is in a separation zone immediately adjacent to the boundary of the contracted opening. In other words, it is necessary to measure the cross-sectional opening of the water which is flowing at a high velocity through the contracted opening. If it is necessary to make the measurements after the flow has passed, each of the items in the equation is extremely difficult to obtain accurately.

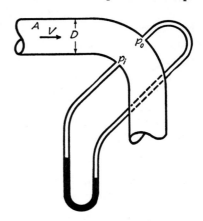

Fig. 10-18. Bend meter.

Bend meters: As pointed out in Chapters 3 and 7, the bend involves a greater pressure at the outside of the bend than at the inside of the bend, because of the momentum principle. Lansford has utilized this property of flow in the bend to measure discharge by correlating the discharge with the difference in pressure from the inside to the outside of the bend (see Fig. 7-9).

The relationship describing the discharge through the bend is the same as Eq. 10-14, in which

A is the cross-sectional area of the pipe;

Δh is the difference in head, in terms of fluid flowing as shown in Fig. 10-18.

Rather accurate results can be obtained from the bend meter if it is calibrated in place. If the bend meter cannot be calibrated in place, however, the discharge for large Reynolds numbers can be measured within an accuracy of approximately 10 per cent by determining the discharge coefficient as follows, provided there is at least thirty diameters of straight pipe upstream from the bend

$$C_d = \sqrt{\frac{r}{2D}} \qquad (10\text{-}24)$$

in which

r is the centerline radius of the bend;

D is the diameter of the pipe.

The discharge coefficient C_d of Eq. 10-24 is used in Eq. 10-14, in which Δh is the difference in piezometric head between the inside and outside of the center of the bend. Since many piping systems have elbows in them which are suitable for adaptation as a measuring device, the bend meter is particularly useful—the accuracy approaching that of the Venturi meter or nozzle, if the bend meter is calibrated in place, while the sensitivity is somewhat less.

Weirs: The weir is probably the most extensively used measuring device for flow in open channels. Although many types of weirs exist, the weir is

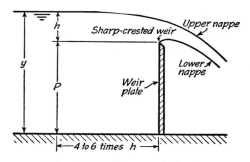

Fig. 10-19. Sharp-crested weir.

essentially an overflow structure extending across the channel normal to the direction of flow. Weirs are generally classified according to shape, and in this chapter the standard uncontracted weir (known as the suppressed weir), the contracted weir, the V-notch weir, the trapezoidal weir, and the broad-crested weir are considered. The first four are sharp crested, as illustrated in Fig. 10-19 while the fifth is the broad-crested weir shown in Fig. 10-21.

Many formulas have been suggested by various experimenters, but a few of the generally-accepted formulas will be tabulated for each of the types listed above. All of these equations were developed for water flowing over the weir. In using these equations, it must be remembered that the same conditions must prevail as those for which the equation was developed. Since this is usually very difficult to insure, weirs should be calibrated in place.

The **uncontracted or suppressed weir** is a standard weir which should meet the following requirements in order for the equations to apply:

1. The weir plate is vertical and the upstream face essentially smooth.

2. The crest is horizontal and normal to the direction of flow. The crest must be sharp so that the water springs free from the edge.

3. The pressure along the upper and the lower nappe is atmospheric.

4. The approach channel is uniform in cross section and the water surface is free of surface waves.

5. The sides of the channel are vertical and smooth, and they extend downstream from the crest of the weir.

In 1823, Francis suggested a simple equation for this type of weir and, although this is limited in its usefulness, it is still one of the pioneers of weir equations.

$$Q = 3.33Lh^{3/2} \qquad (10\text{-}25)$$

in which

> L is the length of the crest in ft;
> h is the head on the weir in ft.

The basic equation suggested for standard suppressed weirs by Kindsvater and Carter in 1957 is

$$Q = C_d L_e h_e^{3/2} \qquad (10\text{-}26)$$

From past experimenters, they summarized the data for applying Eq. 10-26 in which

$$C_d = 3.22 + 0.40\,\frac{h}{P} \qquad (10\text{-}27)$$

$$L_e = L - 0.003 \tag{10-28}$$

$$h_e = h + 0.003 \tag{10-29}$$

in which

P is the height of the weir in ft.

The crest of a weir with end contractions is shorter than the width of the channel, so that the water must contract horizontally as well as vertically to flow over the crest. Francis found that, if the head h on the weir does not exceed one third of the length L, the equation for a contracted weir becomes:

$$Q = 3.33 \left(L - \frac{nh}{10} \right) h^{3/2} \tag{10-30}$$

in which n is the number of horizontal end contractions.

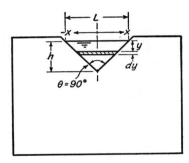

Fig. 10-20. Triangular weir.

The side contractions on a **trapezoidal weir** are not vertical. When the sides of the weir slope at four vertical to one horizontal, the equations for the suppressed weir can be applied to the trapezoidal weir.

The **triangular weir** is particularly useful where there is a large range in discharge and the same accuracy is desired for both small and large discharges. The triangular weir is illustrated in Fig. 10-20. The variation of flow through the triangular weir can be determined theoretically by assuming that the discharge through the small section dy is issuing at a rate which is proportional to the area of the small section times the head y on this section. The differential equation is then

$$dQ = C_d x \sqrt{2gy} \, dy \tag{10-31}$$

since

$$x = L\left(\frac{h - y}{h} \right) = 2(h - y) \tan \frac{\theta}{2} \tag{10-32}$$

Equation 10-32 can be substituted into Eq. 10-31 and then integrated from 0 to h to yield

$$Q = \frac{8}{15} C_d \sqrt{2g} \tan \frac{\theta}{2} h^{5/2} \tag{10-33}$$

Determination of the coefficient C_d is through experiment. Experiments for $\theta = 90°$ have shown that

$$Q = 2.5h^{2.5} \tag{10-34}$$

is an accurate representation of the flow.

The **broad-crested weir** employs the Bernoulli equation in somewhat the same manner as the Parshall flume—except that the broad-crested weir contracts the flow from the bottom and the Parshall flume contracts the flow from the sides. In each case, however, the contraction of the flow decreases the elevation of the water surface, which provides an indirect measure of the discharge. Figure 8-17 in Chapter 8 for a rise in the bottom of an open channel is a special case of the broad-crested weir.

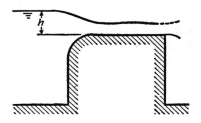

Fig. 10-21. Broad-crested weir.

As the height of the broad-crested weir is increased, the water surface elevation over the weir decreases until the depth of flow over the weir is critical. If the height of the broad-crested weir is increased still further, the depth of flow upstream must be increased in order for the discharge to remain the same. This is explained in greater detail in Chapter 8. The equation for the **broad-crested weir** is approximately

$$q = 3h^{3/2} \tag{10-35}$$

in which q is the discharge per unit width,

h is the upstream height of the water surface above the weir, as indicated in Fig. 10-21.

Note the similarity between this equation and Eq. 8-64. The broad-crested weir is particularly useful because it does not cause appreciable upstream backwater. It is simple to construct and it controls the flow for measurement purposes. Some degree of submergence is permissible.

Example Problem 10-4: Calculate the discharge over a rectangular suppressed weir in a 20 ft wide rectangular channel if the crest of the weir is 4 ft above the channel floor and the head on the weir is 2 ft. (a) Using the Francis equation, (b) using the Kindsvater-Carter equation. (c) What per cent error is introduced when the Francis equation is used?

Solution:

(a) $Q = 3.33Lh^{3/2} = (3.33)(20)(2)^{3/2} = \underline{\underline{188.30 \text{ cfs}}}$

(b) $C_d = 3.22 + 0.40 \dfrac{h}{P} = 3.22 + \dfrac{(0.40)(2)}{(4)} = 3.42$

$L_e = L - 0.003 = 19.997$

$h_e = h + 0.003 = 2.003$

$Q = C_d L_e h_e^{3/2} = (3.42)(19.997)(2.003)^{3/2} = \underline{\underline{193.5 \text{ cfs}}}$

(c) Per cent error $= \dfrac{(188.3 - 193.5)\,(100)}{(193.5)} = \underline{\underline{2.69\%}}$

Spillways: Like gates in closed and open conduits, spillways in open channels serve as an excellent means of measuring discharge. If the spillway is not submerged the depth of flow over the crest of the spillway is critical depth. Usually the spillway must be calibrated either by field measurements or by measurements on a model of the spillway in the laboratory. An approximate equation for the determination of discharge over an ogee spillway (which has a rounded crest shape similar to the under side of the nappe springing from a sharp-crested weir) is

$$q = 4.0h^{3/2} \qquad\qquad (10\text{-}36)$$

in which h is the upstream height of the water surface above the crest of the dam and the velocity of the approaching flow is negligible.

Deflected vane and rotometer: Several different meters act as a sub-merged object, on which the flow causes a drag that depends upon the velocity of flow or the discharge. If the area of a submerged portion of the object does not vary with discharge or velocity, the drag in general varies with the velocity squared. Since the drag on the submerged object is a function of the velocity of flow, it is possible to calibrate the meter in a given section of flow for the discharge. If the submerged object such as a vane is attached to an arm, the deflection of the meter or the bending moment in the arm can be used for this calibration. Springs or balances can also be attached to the arm in order to measure the drag on the vane.

Parshall has recently developed a **vane meter,** for measurement of discharge in open channels, which is independent of the depth of flow. Because of the unusual curvature of the meter, the drag moment on the meter is a constant for any depth of flow, provided the discharge is also constant. For shallow flows the velocity is large and, consequently, the drag per unit area is great and the area exposed to the flow must be increasingly small. As the depth for a given discharge is increased, however, the velocity is correspondingly decreased and the area exposed to the flow must be increased. This meter has an accuracy of approximately 5 per cent. Once it has been calibrated for a given flow section, it has an unusual flexibility for measurement of discharge in many types of open channels—including sewer pipes flowing partly full.

The **rotometer** consists of a graduated glass tube which is tapered so that it is larger at the top than at the bottom. This meter is illustrated in Figure 10-22. Flow through the tube is upward and causes a drag on the submerged object so that it is suspended at the point where the drag is just equal to the submerged weight of the object.

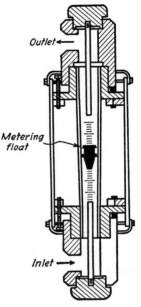

For small discharges, the clearance between the tube and the object must be small, so that the drag is equal to the submerged weight of the object. As the discharge is increased, the clearance is increased as the object moves upward through the tapered tube. In this way, the object seeks that particular level where the clearance is just sufficient for the drag to balance the submerged weight. The object has grooves in it, so that it rotates in position, which helps to center the object and to prevent dragging on the walls of the tube.

Mass flow: A unique method originated by Kolin determines the discharge in a closed conduit by use of the principle of electromagnetic induction created by the mass of fluid passing through a magnetic field. A magnetic field must be created around the closed conduit, and a liquid or gas that is slightly ionized must pass

Fig. 10-22. Rotometer.

through the magnetic field. Electric connections made across the conduit measure the induced emf (electromotive force). This meter must be calibrated in place, but it has the obvious advantage of no energy loss to the flow. Furthermore, it has the advantage of measuring the discharge of a variety of unusual liquids and gases, provided they are ionized at least slightly. The section of pipe in which the electrodes are located and the measurement is made must be non-conducting.

Velocity integration: Discharge in a closed conduit or open channel can be determined quite accurately by integration of the velocity distribution across the flow section. This is standard procedure for the U.S.G.S., in connection with the determination of discharge in an open channel. It is also used frequently in closed conduits by means of velocity traverses across the flow. Figure 10-23 illustrates the method of integrating the velocity distribution across the flow of the closed conduit. From this figure, it can be seen that the discharge can be divided into concentric cylinders for which the velocity is a constant. The plot of $2\pi r v$ against the radius gives the discharge variation with radius, and the area under this curve represents the discharge flowing through the pipe.

For discharge in an open channel, the Geological Survey utilizes the fact that the average velocity in a vertical section is located at approximately 0.6 depth below the surface, and that it can also be represented by an average of the 0.2 and 0.8 depth. Actually, according to the log distribution of velocity, the average velocity is located more nearly at 0.62 times the depth below the water surface, that is $0.62y$, but the 0.6 average is sufficiently accurate for most purposes in shallow flow. In deeper flow, the velocity at the 0.2 and 0.8 depth is measured, and the discharge determined as illustrated in the following example problem.

Example Problem 10-5: The computation of stream discharge based on current meter measurements is illustrated in the following table, see Fig. 10-6.

Distance from Bank	Depth	Observation Depth	Velocity At Point	Velocity Mean in Vertical	Mean in Section	Area	Mean Depth	Width	Discharge q
0	0	0	0	0					
					0.78	1.70	0.85	2	1.33
2	1.70	0.35	1.52	1.56					
		1.35	1.60						
					1.73	4.40	2.20	2	7.61
4	2.70	0.54	1.91	1.89					
		2.16	1.88						
					2.08	13.80	3.45	4	28.7
8	4.20	0.84	2.35	2.27					
		3.36	2.19						
					2.33	8.60	4.30	2	20.1
10	4.40	0.88	2.41	2.38					
		3.52	2.34						
					2.37	8.40	4.20	2	19.94
12	4.00	0.80	2.31	2.36					
		3.20	2.40						
					2.05	13.60	3.40	4	27.9
16	2.80	0.56	1.92	1.74					
		2.24	1.57						
					1.49	4.30	2.15	2	6.41
18	1.50	0.30	1.35	1.24					
		1.20	1.13						
					0.62	1.50	0.75	2	0.93
20	0	0	0	0					

$$Q = \Sigma q = \underline{\underline{112.9}} \text{ cfs.}$$

Float measurements: Although floats are particularly useful in connection with the measurement of velocity, various types of floats have been used in an attempt to make an approximate measurement of discharge. The submerged part of the float can either be connected to a weight which is suspended at a lower level, or by a continuous shaft which extends to a lower level within the flow. By this process, an approximate averaging of the drag upon the floating object can produce an approximately average velocity for that section. Obviously, this type of measurement is subject

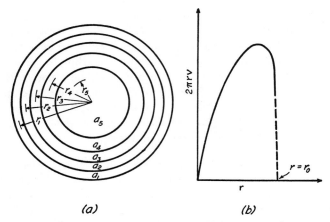

(a) *(b)*

Fig. 10-23. Velocity integration.

to many errors, including changes in section and wind blowing on the upper part of the float.

Salt solution determinations: A method of determination of velocity within a closed conduit has been developed which depends upon the greater electrical conductivity of salt water as compared with the water normally flowing in the pipe. The salt water is suddenly introduced into the flow and, at some measured location downstream, the variation of conductivity of the mixture is determined with respect to time. The center of gravity of the time-concentration curve determines the time T it has taken the fluid to flow that distance L, which in turn gives the average velocity $V = L/T$. The discharge is a product of the average velocity and the cross-sectional area, $Q = AV$. This method can be quite accurate, but requires special equipment for accurate determination of time.

Another method which is somewhat simpler than the salt-velocity method is the salt-dilution method. In this case, the salt solution is introduced at a given point upstream under steady state conditions, and the mixing along the way results in a diluted solution at the point of measurement. The conductivity of the mixture is measured, so that the rate of flow can be determined in terms of the known rate of flow of the salt water introduced upstream. The concentration of the salt solution and the mixture downstream can be determined either by chemical titration or by electrical conductivity methods.

Slope-area method: For flow in open channels—particularly natural streams under flood flow conditions—it is sometimes impossible to measure the discharge during the maximum rate of flow. Consequently, the discharge must be determined indirectly by a survey after the passage of the flow. This is frequently done by means of the slope-area method. In this method, the slope and the cross-sectional area are determined by the high water marks along each of the banks. A straight reach of channel of considerable length must be selected, if the slope and area measurements are to be truly representative. This length should be a minimum of 50 depths of flow for average slopes. For very flat slopes, such as those in canals, however, the length of reach sometimes needs to be as much as 200 depths. In larger streams, this must be at least a thousand feet. Once the slope and cross-sectional areas are determined in as straight a reach as it is possible to find, an estimate must be made of the Manning resistance coefficient n. This, of course, is the source of considerable error in such measurements. However, with experience some engineers are able to make this estimate with considerable accuracy for channels with rigid boundaries. Alluvial beds in open channels cause considerable variation of the n value with discharge in a given channel, due to the successive formation of a plane bed, ripples, dunes, a flat bed, standing waves, and antidunes. These bed forms have caused variations from $n = 0.011$ for the plane bed, to $n = 0.035$ for the dune bed. With such a variation of more than 300 percent in a single channel, accurate estimation of the n value for a given discharge is obviously very difficult. Once n, S, R, and A have been determined, the discharge is found by Eq. 8-19

$$ Q = \frac{1.5}{n} A R^{2/3} S^{1/2} \tag{8-19} $$

The methods described in this chapter of measuring flow characteristics and fluid properties have variable accuracy, expense, and simplicity and ease of measurement. One or more of the methods can be selected for use, depending upon what is desired in this regard. New methods are continually being developed in this rapidly-expanding field of measurements.

• SUMMARY

1. Measurements are necessary to obtain information on
   a. Fluid properties
   b. Flow characteristics

2. Measurements need to be made in the following types of flow systems
   a. Steady or unsteady   d. Laminar or turbulent
   b. Uniform or nonuniform   e. Confined or unconfined
   c. Compressible or incompressible

3. The fundamental principles of fluid mechanics involved in making measurements are those presented in the earlier chapters of this book.

4. **Density** and **specific weight** are usually measured by a balance or a pycnometer which compare the fluid with standard weights.

5. **Elasticity** can be determined by compressing a known. volume in a cylinder or by heating a known volume in a container.

6. **Viscosity** is measured by applying the basic law of Newton (Eq. 5-2) to flow along boundaries such as: inside tubes, between flat plates, and around spheres.

7. **Surface energy** is a property exhibited by a liquid when it is in contact with a solid, a gas, or another liquid, see Chapter 1. It can be measured by the rise of the liquid in a capillary tube, see Figs. 1-3 and 1-4, or by the pressure required to force a bubble out the end of a tube immersed in the liquid, see Fig. 1-3(b).

8. **Vapor pressure** of a liquid can be measured as the pressure at which bubbles of the liquid vapor (gas) form for a given temperature.

9. **Length** is measured with such equipment as tapes, scales, and calipers. **Depth** and liquid surface **elevation** are measured with: a staff gage, a float gage, a level, a point gage, a hook gage, a differential bubble gage, a gage height follower, a sounding weight, a sonic probe, or a manometer.

10. **Time** is measured with: a stop watch, a clock, an electronic timer, a stroboscope, high-speed motion pictures, or time-lapse motion pictures.

11. **Pressure** is measured directly or indirectly with: a single manometer, a differential manometer, a pressure cell, or a mechanical gage.

12. **Shear** can be measured directly by a dynamometer, or indirectly by the Newton shear equation (Eq. 5-2), the equations involving the energy gradient (Eqs. 7-3 and 8-2), or the equation for velocity distribution (Eq. 10-5).

13. **Velocity** is a vector quantity. Therefore, both **magnitude** and **direction** must be measured. Measurement of velocity usually involves visual observation of the flow (floats or zero-lift balloons or droplets), the Bernoulli equation (Pitot tubes, cylinders, or spheres), rotating vanes (anemometers or current meters), trajectory of a jet (issuing from an orifice, a nozzle, or a pipe), or electrical sensing probes (hot-wire anemometer or radioactive tracers).

14. **Acceleration** can be measured as the rate of change of velocity with respect to time or by accelerometers which determine the apparent weight of a known mass attached to a pendulum or cantilever spring.

15. *Discharge* can be determined: volumetrically, by the Bernoulli equation together with the momentum and continuity equations (Venturi meters, Parshall flumes, nozzles, orifices, gates, weirs, spillways, other contracted openings, or bend meters), by drag on an object in the flow (vane meters, rotometers, rotating vanes or wheels, or wobbling disks), by mass flux measurements (electromagnetic induction), by velocity integration, by use of dielectric chemicals, and by the slope-area method.

References

1. Addison, Herbert, *Applied Hydraulics.* New York: John Wiley & Sons, Inc., 1945.

2. Addison, Herbert, *Hydraulic Measurements*, 2nd ed. New York: John Wiley & Sons, Inc., 1946.

3. Binder, R. C., *Fluid Mechanics*, 3rd ed. Englewood Cliffs, N. J.: Prentice-Hall, Inc., 1955.

4. Daugherty, R. L., and A. C. Ingersoll, *Fluid Mechanics.* New York: McGraw-Hill Book Co., Inc., 1954.

5. Howe, J. W., "Flow Measurements," Chapter III of *Engineering Hydraulics*, ed. Hunter Rouse. New York: John Wiley & Sons, Inc., 1950.

6. King, Horace W. and Ernest F. Brater, *Handbook of Hydraulics*, 4th ed. New York: McGraw-Hill Book Co., Inc., 1954.

7. Roberts, Howard C., *Mechanical Measurements by Electrical Methods.* Pittsburgh, Pa.: The Instruments Publishing Co., Inc., 1951.

8. Shoder, Ernest W. and Francis M. Dawson, *Hydraulics.* New York: McGraw-Hill Book Co., Inc., 1934.

9. Vennard, John K., *Elementary Fluid Mechanics*, 3d ed. New York: John Wiley & Sons, Inc., 1954.

10. *Flow Measurement*, New York: ASME, 1940.

11. *Fluid Meters, Their Theory and Application*, 4th ed. New York: ASME, 1937.

12. *Water Measurement Manual*, 1st ed. Washington, D. C.: U. S. Bureau of Reclamation, 1953.

Problems

10-1. A pycnometer with a volume of 1000 cc weighs 413 grams when empty. When filled with a liquid, the pycnometer weighs 1673 grams. What is the specific weight of the liquid in lb/cu ft, and what is the probable name of the liquid involved?

10-2. A pycnometer with a volume of 500 cc weighs 200 grams when empty. When filled with water it weighs 700 grams. A rock is dropped into the pcynometer, and 100 cc of water are forced out. The pycnometer is now weighed again and weighs 865 grams. What is the specific gravity of the rock?

10-3. A hydrometer which weighs 200 grams is placed in a can of liquid. The hydrometer displaces 20 cc per inch of submergence, and it floats in the liquid with 8 in. submerged. What is the specific weight of the fluid in lb/cu ft?

10-4. Oil ($S = 0.85$) is flowing in a small tube whose diameter is 3 mm. The horizontal tube is 12 ft long and is attached to a large tank in which oil stands at an elevation of 10 ft above the opening to the tube. The tube discharges freely into a container at the end and oil accumulates in the container at the rate of 12 gallons per hour. What is the absolute viscosity of the oil?

10-5. Oil ($S = 0.80$) flows through a small horizontal tube 5 mm in diameter. If the mean velocity is 0.4 ft/sec and the change of pressure in 10 ft of tube is 5 psi, what is the viscosity of the oil?

10-6. A liquid is placed in a Saybolt viscometer and the time required for 60 milliliters to pass through the meter is 150 sec. What is the absolute viscosity of the fluid if the specific gravity is 0.85?

10-7. A small glass sphere ($S = 2.65$) has a diameter of 0.01 ft. If the sphere falls 2 ft in 30 seconds through a liquid, what is the absolute viscosity of the liquid?

10-8. In Problem 10-7, find the corrected viscosity if the diameter of the liquid container is (**a**) 1 in., (**b**) 1 ft.

10-9. A torque of 4 ft-lb is required to rotate a cylinder viscometer at 40 rpm (see Fig. 10-4). If the viscometer dimensions are as follows: $h = 12$ in., $r = 6$ in., $\Delta r = 0.01$ ft, what is the absolute viscosity of the liquid?

10-10. The capillary rise in a small glass tube 1 mm in diameter is 1.1. in. What is the surface energy of the liquid whose specific gravity is 1.00?

10-11. Mercury is placed in a small capillary tube 1 mm in diameter, and the depression is $\frac{1}{2}$ in. What is the surface energy of the mercury?

10-12. A small bubble is trapped in a liquid at the end of a capillary tube 2 mm in diameter (see Fig. 1-3). If the diameter of the bubble is the same as the tube and the pressure change across the bubble is 0.30 lb/sq ft, what is the surface energy of the liquid?

10-13. With a good piezometer connection on a tank, the water in an open manometer rises 10 ft above point A. What is the pressure in the tank at point A?

10-14. A smooth pipe carries oil whose viscosity is 4×10^{-4} lb-sec/sq ft. If the diameter of the pipe is 1 in. and the average velocity is 3 ft/sec, what is the shear on the boundary of the pipe?

10-15. The pressure gradient along a horizontal 2-in. pipe is 1 psi per 100 ft. What is the shear on the boundary of the pipe?

10-16. The depth of water in a rectangular channel is 6 ft. What is the shear on the bottom, if the flow is uniform and the channel slopes 2 ft per mile?

10-17. The velocity is measured at two points near the boundary, with the following results:

$$y = 0.1 \text{ ft} \qquad v = 0.81 \text{ fps}$$
$$y = 0.2 \text{ ft} \qquad v = 1.21 \text{ fps}$$

What is the boundary shear?

10-18. The difference in pressure between the tip and the side holes of a Pitot tube is 3 psi. What is the velocity?

10-19. A standard Pitot cylinder records a pressure difference of 5 psi. What is the velocity?

10-20. A Pitot cylinder is built with the side holes at 90° to the flow. Assuming no separation, what is the theoretical velocity for a pressure difference of 6 psi?

10-21. The velocity in a wind tunnel is 800 mph. If the air temperature is 70°F and the ambient pressure in the tunnel is 5 psi, what is the stagnation pressure at the tip of a Pitot tube?

10-22. A Pitot tube in a wind tunnel gives a stagnation pressure of 3 psi. The ambient pressure is standard atmospheric and the temperature is 68°F. What is the velocity in the tunnel?

10-23. A horizontal jet issues from a sharp-edged orifice. At a horizontal distance of 6 ft from the orifice, the jet has dropped 3 ft below the centerline of the orifice. What is the velocity of efflux from the orifice?

10-24. A 6-in. drainage pipe flowing full discharges horizontally into a gully. If the pipe carries a maximum discharge of 2 cfs, where must riprap protection be placed if the floor of the gully is 10 ft below the centerline of the pipe? Plot the trajectory of the jet.

10-25. A 4-in. pipe discharges water into a swimming pool which is 30 ft long by 20 ft wide. If the water surface in the pool rises at the rate of 1 ft per hour, what is the flow in gpm in the 4-in. pipe?

10-26. A 4-in. pipe and a 2-in. pipe are both used to fill a water tank. Assuming the tank is circular with a 10 ft radius, what is the flow in each pipe in gpm, if the 4-in. pipe discharges twice as much as the 2-in. pipe? The water tank fills at the rate of 6 ft of depth in one hour.

10-27. In calibrating a Venturi meter, the water flowing through the meter is dumped into a weighing tank. In 30 seconds, 3000 lb are added to the tank. What is the flow through the meter in gpm?

10-28. A Venturi meter in a 4-in. pipe has a 2-in. throat. A differential mercury manometer connected to the meter as in Fig. 3-23 reads a difference of 10 in. of mercury. What is the flow if the water temperature is 60°F.?

10-29. If the Venturi meter in Problem 10-28 is carrying 300 gpm of water at 60°F, what is the reading on the differential manometer in inches of mercury?

10-30. Derive Eq. 10-10.

10-31. The head H_a on a Parshall flume reads 1.64 ft. If the flume throat width is 36 ft, what is the discharge through the flume?

10-32. The head H_a on a 6 ft wide Parshall flume is 2.6 ft. What is the discharge?

10-33. Derive Eq. 10-18.

10-34. A Venturi meter, 4 in. by 2 in., carries air. The temperature at point 1 is 60°F and the pressure is 17 psia. At point 2 the pressure is 14 psia. Find the flow rate if $C_v = 0.97$.

10-35. Derive Eq. 10-16.

10-36. The pressure drop across a 2-in. nozzle in a 4-in. pipe is 10 psi (see Fig. 10-16). What is the discharge if water is flowing at 60°F?

10-37. If the nozzle in Problem 10-36 were an orifice ($\alpha = 90°$), what would be the discharge?

10-38. The depth of flow upstream from a sluice gate is 10 ft. If the gate opening is 3 ft and the width of the gate is 8 ft, what is the discharge under the gate? There is no submergence downstream.

10-39. What would be the discharge through the sluice gate in Problem 10-38 if the downstream depth y_2 were 6 ft?

10-40. A standard elbow meter in a 3-in. pipe is connected to a differential mercury manometer. If oil of specific gravity 0.8 is flowing in the pipe line, what is the discharge through the meter when the manometer difference is 6 in. of mercury?

10-41. A standard elbow meter is to be placed in a 4-in. pipe which is expected to carry a maximum discharge of 0.8 cfs of water. If it is desired to connect the meter to a differential water-air manometer, what length of manometer is necessary to insure a measurement of the maximum discharge.

10-42. The head of water on a 3 ft suppressed weir is 8 in. What is the discharge over the weir whose length is 6 ft?

10-43. What would be the discharge over the weir in Problem 10-42 if the weir were a contracted weir with two end contractions?

10-44. If 30 cfs flow over a standard trapezoidal weir which is 10 ft wide and 5 ft high, what is the head on the weir?

10-45. A weir with two end contractions is to be installed to measure the flow out of a small storage pond. If the maximum discharge over the weir is 60 cfs of water and the maximum allowable backwater from the weir is 1 ft, what is the necessary minimum length of the new weir?

10-46. Water is 1 ft deep over the crest of a 90° V-notch weir. What is the discharge?

10-47. Thirty gallons per minute are expected to flow down an irrigation furrow. A V-notch weir is to be installed to measure the flow. Design the weir if a minimum head of 4 in. is desired for the flow of 30 gpm. Assume $C_d = 0.61$.

10-48. State four types of weirs and list the advantages and disadvantages of each type.

10-49. Using Eqs. 10-25, 10-26, and 10-30, compare the discharge obtained for a 1.5 ft head and a crest width of 3 ft.

10-50. Through research in the library, find at least six additional equations for flow over suppressed weirs. Give the source for each equation. Compare these equations by calculating the discharge per foot for a head of 1 ft and a height of weir of 4 ft.

10-51. The head on a broad-crested weir 30 ft wide is 2.5 ft. What is the discharge over the weir?

10-52. Salt water solution of 5000 parts per million (ppm) is added to the flow of fresh water in a penstock at the uniform rate of one liter per minute. Downstream from the turbines, the salt concentration is 10 ppm. Estimate the discharge in cfs.

10-53. Pitot tube readings resulted in the following velocities in a 2-in. pipe. Measurements were made every 0.2 in., starting at one edge and working across the pipe to the other.

Distance from pipe wall in in.	Velocity in fps
0.2	3.1
0.4	4.2
0.6	4.8
0.8	5.2
1.0	5.4
1.2	5.1
1.4	4.7
1.6	4.2
1.8	3.0

Plot the curve and calculate the discharge.

10-54. Briefly describe how a discharge measurement could be made in a rectangular open channel.

10-55. A flood survey is made to determine the approximate discharge in a recent flood. Cross sections indicate that the average cross-sectional area is 300 sq ft and the average hydraulic radius is 5.7 ft. The slope of the high water mark along the sides of the channel is about 0.007. If the value of n is estimated as 0.025, what was the flood discharge?

10-56. Briefly describe how you would use the slope-area method, if you were assigned the job of determining the magnitude of a recent but unmeasured flood. In your description, include the measurements you would have to make.

chapter 11

Flow Similitude
and Models

Models have a widespread use in many aspects of life. Children play
with models of such things as ships, railroad trains, trucks, and various
types of structures. These models are similar to the prototype which they
represent primarily in their appearance and, to a lesser extent, in their
behavior. More complicated models are self-propelled, such as model airplanes and model trains. In these cases, the similarity to the prototype
may be more complete in both appearance and behavior. Completeness of
similarity in fluid mechanics however, involves additional factors which go
far beyond simple geometrical appearance. For example, a toy airplane
appears to be similar to a prototype and may fly under its own power, but
there may be certain internal geometric features which are not completely
similar and, even more important, there are certain kinematic and dynamic
aspects which are not exactly similar. These aspects of similarity—geometric, kinematic, and dynamic—are considered in this chapter. Although

they are not discussed specifically in this chapter, mathematical models also are frequently employed in fluid mechanics. Furthermore, in some cases mathematical models (equations) are sufficient to answer the question more simply and at much less expense than a physical model.

The prototype is the original after which the model is patterned for some special purpose, and the model is a reproduction, imitation, or copy of the prototype which it represents. Hickox has defined a model as "a system by whose operation the characteristics of other similar systems can be predicted." Examples of models, either as sections or in total, which are associated with fluid mechanics include naval vessels, aircraft, hydraulic machinery, pipes, rivers, harbors, and structures such as dams, spillways and stilling basins. The model can be larger, smaller, or the same size as the prototype—depending upon the need and purpose.

The *purposes of a model* are:

1. To study the general appearance of the geometry and the relative proportions of the various features.

2. To study the flow patterns—magnitude and direction of velocity and acceleration—over, through, or around the object.

3. To study the pressure distribution and resulting forces on the object or its parts.

4. To determine flow capacities and calibrate the various flow passages.

5. To determine hydraulic efficiencies of hydraulic machinery.

6. To determine mechanical energy loss due to shear drag and pressure drag.

7. To determine whether the prototype will operate as intended.

8. To determine ways of improving the design and/or operation of the prototype.

9. To find ways of reducing the cost of the prototype.

The *advantages of a model* stem from the purposes and objectives of a model, which can be attained

1. Less expensively   4. More conveniently
2. More quickly   5. More completely
3. More simply

than if the same information were obtained from the prototype. Sometimes, other fluids can be used in a model more simply and conveniently than the fluid in the prototype. Furthermore, changes can be made quickly and economically in the model to determine the most desirable design or system of operation.

The principal *disadvantage of model studies* is the fact that complete similarity in the model is sometimes difficult or impossible to obtain. In these cases, it is frequently possible to obtain only a partial (qualitative) or approximate answer by means of a model.

• TYPES OF SIMILARITY

The types of similarity involved in mechanics of fluids are *geometric* similarity, *kinematic* similarity, and *dynamic* similarity. **Geometric similarity** exists if all aspects of the model are geometrically similar to those in the prototype. This is illustrated in Fig. 11-1, in which the only geo-

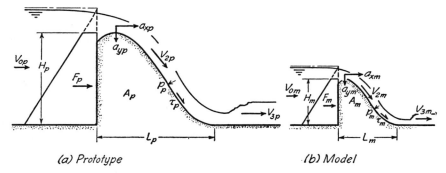

(a) Prototype   (b) Model

Fig. 11-1. Similarity between model and prototype.

metric difference between the model and prototype of the dam is the scale or size. Usually, it is necessary to reproduce only that part of a model directly associated with the flow. Consequently, a section of a dam or a section of an airplane wing is tested without involving the entire structure. Furthermore, that part of an object reproduced and tested for one purpose may not be the same as that part for another purpose.

Geometric similarity requires that all **length ratios** L_R from model to prototype be the same, so that

$$L_R = \frac{L_p}{L_m} = \frac{H_p}{H_m} \tag{11-1}$$

and that the **area ratios** A_R also be equal, as follows

$$A_R = \frac{A_p}{A_m} = \left(\frac{L_p}{L_m}\right)^2 = \left(\frac{H_p}{H_m}\right)^2 = L_R^2 \tag{11-2}$$

The geometry and scale of roughness of a boundary must also be described in terms of length.

Kinematic similarity requires that the streamline pattern in the model be the same as that in the prototype, so that the *velocity ratio* V_R and the *acceleration ratio* a_R are the same from point to point in the flow (see Fig. 11-1.) Thus

$$V_R = \frac{V_{0p}}{V_{0m}} = \frac{V_{1p}}{V_{1m}} = \frac{V_{2p}}{V_{2m}} = \frac{V_{3p}}{V_{3m}} \tag{11-3}$$

and

$$a_R = \frac{a_{np}}{a_{nm}} = \frac{a_{2p}}{a_{2m}} \tag{11-4}$$

By combining Eq. 11-2 with Eq. 11-3, the *discharge ratio* Q_R can be determined as

$$Q_R = A_R V_R = L_R^2 V_R \tag{11-5}$$

Equations 11-1, 11-3, and 11-4 can be related by the *time ratio* T_R, which is

$$T_R = \frac{L_R}{V_R} \tag{11-6}$$

to yield

$$a_R = \frac{L_R}{T_R^2} = \frac{V_R^2}{L_R} \tag{11-7}$$

Dynamic similarity requires that the *shear ratio* τ_R, the *pressure ratio* p_R, and the *force ratio* F_R be the same from point to point within the flow system. Consequently,

$$\tau_R = \frac{\tau_p}{\tau_m} \tag{11-8}$$

and

$$p_R = \frac{p_p}{p_m} \tag{11-9}$$

and if the force is due to the pressure distribution over an area—such as the hydrostatic pressure or pressure drag—

$$F_R = \frac{F_p}{F_m} = p_R A_R = p_R L_R^2 \tag{11-10}$$

since the product of pressure and area is force. If the force on the object is due to shear drag—that is, tangential shear along the boundary—then

$$F_R = \tau_R A_R = \tau_R L_R^2 \tag{11-11}$$

By Newton's equation, $F = Ma$, the acceleration can be related to the force.

• TYPES OF FORCES

The basic fluid properties have a considerable influence on the dependent variables, such as pressure and shear, which combine to make up all the forces acting on an object. These properties are density ρ, viscosity μ, specific weight γ, surface energy σ, and elasticity E. The inertia force, based on Newton's equation $F = Ma$, can be expressed dimensionally as

$$F = \rho L^3 \frac{L}{T^2} \tag{11-12}$$

This equation can be divided by an area term L^2 to yield the **inertia force** per unit area

$$f_i = \rho \frac{L^2}{T^2} = \rho V^2 \tag{11-13}$$

By a similar procedure, each of the following equations gives a force per unit area related to each of the fluid properties.

The **viscous force** per unit area,

determined from $\qquad F = A\mu \frac{dv}{dy},$

is $\qquad\qquad\qquad f_v = \tau = \mu \frac{V}{L} \tag{11-14}$

The **gravity force** per unit area, as determined from $F = Mg$, is

$$f_g = \rho g L = \gamma L \tag{11-15}$$

The unit **surface energy force,** based on the relation $F = \sigma L$, is

$$f_s = \frac{\sigma}{L} \tag{11-16}$$

Finally, the **elastic force** per unit area, from the expression $F = EA$, is

$$f_e = E \tag{11-17}$$

If complete dynamic similarity is to be established in the model, each **force ratio** from model to prototype must be the same at all points in the flow. As will be seen, however, this is not possible for all the forces simultaneously.

In most problems of flow, the force of inertia is involved and, consequently, certain standard ratios of forces have been developed—each of which involves the inertia force—as follows:

$$\frac{\text{Pressure}}{\text{Inertia force per unit area}/2} = \frac{p}{\rho V^2/2} = \text{Pressure number } P_n \tag{11-18}$$

$$\frac{\text{Shear}}{\text{Inertia force per unit area}/2} = \frac{\tau}{\rho V^2/2} = \text{Shear number } S_n$$

$$(11\text{-}19)$$

$$\frac{\text{Inertia force per unit area}}{\text{Viscous force per unit area}} = \frac{\rho V L}{\mu} = \text{Reynolds number Re}$$

$$(11\text{-}20)$$

$$\left(\frac{\text{Inertia force per unit area}}{\text{Gravity force per unit area}}\right)^{1/2} = \frac{V}{\sqrt{L\gamma/\rho}} = \text{Froude number Fr}$$

$$(11\text{-}21)$$

$$\left(\frac{\text{Inertia force per unit area}}{\text{Elastic force per unit area}}\right)^{1/2} = \frac{V}{\sqrt{E/\rho}} = \text{Mach number Ma}$$

$$(11\text{-}22)$$

$$\frac{\text{Inertia force per unit area}}{\text{Surface energy force per unit area}} = \frac{\rho V^2 L}{\sigma} = \text{Weber number We}$$

$$(11\text{-}23)$$

Some of these numbers (parameters) have been obtained by dimensional analysis in Chapter 1. The Froude number and the Mach number involve the square root of the forces as a result of common usage—since they are also ratios of velocities. Although each of these parameters and numbers involves inertia forces, there are conditions where either ρ or V is so small that inertia never requires consideration. Another fluid property then may be used in the ratio. For example, in two-phase flow (gas and liquid) through porous media, both viscosity and surface energy become important and a new dimensionless number involving these forces can be used which does not include the inertia force.

As pointed out in previous chapters of this text, the foregoing parameters are of primary importance in the correlation of experimental data. The shear parameter and the pressure parameter—and various modifications of these, such as the drag coefficient C_D, the lift coefficient C_L, and the resistance coefficient f—are usually dependent variables which depend upon one or more of the four dimensionless numbers. In each of the dimensionless numbers, if the inertia force is relatively great (either density or velocity or both are large), then the number is large and the forces caused by the other fluid properties are relatively insignificant. Under this condition, the dependent variable varies little or not at all with variation in the dimensionless number. Examples of this are the relatively small variation of f in Fig. 7-6 and C_D in Figs. 6-9, 9-2, and 9-3 with large Reynolds numbers; and of C_D in Fig. 6-10 with increasing Mach number.

On the other hand, if the inertia force is relatively small, the influence of the other fluid properties becomes relatively large and the dependent variable is very sensitive to variation of the dimensionless number. Examples are in the same plots of C_D and f, but for small Re- and Ma-values. Between these two extremes of small and large values of the dimensionless numbers, lies a transition which can be either abrupt (see Fig. 7-6) or gradual.

Application of the dimensionless parameters to model studies involves a combination of Eqs. 11-1 through 11-11 with the velocity ratio V_R, the pressure ratio p_R, and the shear ratio τ_R determined by Eqs. 11-18 through 11-23. For example, Eq. 11-10 can be combined with Eq. 11-18 to express a pressure drag force in terms of length and velocity.

$$F_R = p_R L_R^2 = \rho_R V_R^2 L_R^2 \tag{11-24}$$

in which ρ_R is the density ratio.

However, Eqs. 11-20 to 11-23 must be used to express the velocity in terms of length and the various fluid properties. The force ratios then become:

$$F_{R1} = \rho_R \left(\frac{\mu}{\rho L}\right)_R^2 L_R^2 = \left(\frac{\mu^2}{\rho}\right)_R \text{ for Reynolds number}$$

$$\tag{11-25}$$

$$F_{R2} = \rho_R \left(\frac{L\gamma}{\rho}\right)_R L_R^2 = (\gamma L^3)_R \text{ for Froude number}$$

$$\tag{11-26}$$

$$F_{R3} = \rho_R \left(\frac{E}{\rho}\right)_R L_R^2 = (EL^2)_R \text{ for Mach number}$$

$$\tag{11-27}$$

$$F_{R4} = \rho_R \left(\frac{\sigma}{\rho L}\right)_R L_R^2 = (\sigma L)_R \text{ for Weber number}$$

$$\tag{11-28}$$

Similar computations can be made for other characteristics of the flow to determine ratios given in Table 11-1.

Equation 11-25 shows that, for the Reynolds criterion, the force ratio is independent of the scale. Consequently, the force per unit area and the force ratio are of the same magnitude in the model and the prototype, if the ratio of viscosity squared to density is the same. The Froude number, however, shows in Eq. 11-26 that the force ratio depends upon the cube of the scale ratio, as well as upon the specific weight. Theoretically, it appears possible to establish similitude on the basis of both Reynolds number and

Table 11-1. SCALE RATIOS FOR MODELS

Characteristic	Dimension	Scale Ratios			
		$(Re)_R$	$(Fr)_R$	$(We)_R$	$(Ma)_R$
Length	L	L	L	L	L
Area	L^2	L^2	L^2	L^2	L^2
Volume	L^3	L^3	L^3	L^3	L^3
Time	T	$\rho L^2/\mu$	$(L\rho/\gamma)^{1/2}$	$(L^3\rho/\sigma)^{1/2}$	$L(\rho/E)^{1/2}$
Velocity	L/T	$\mu/L\rho$	$(L\gamma/\rho)^{1/2}$	$(\sigma/L\rho)^{1/2}$	$(E/\rho)^{1/2}$
Acceleration	L/T^2	$\mu^2/\rho^2 L^3$	γ/ρ	$\sigma/L^2\rho$	$E/L\rho$
Discharge	L^3/T	$L\mu/\rho$	$L^{5/2}\left(\dfrac{\gamma}{\rho}\right)^{1/2}$	$L^{3/2}(\sigma/\rho)^{1/2}$	$L^2(E/\rho)^{1/2}$
Mass	M	$L^3\rho$	$L^3\rho$	$L^3\rho$	$L^3\rho$
Force	$ML/T^2 = F$	μ^2/ρ	$L^3\gamma$	$L\sigma$	L^2E
Density	M/L^3	ρ	ρ	ρ	ρ
Specific weight	M/L^2T^2 $= F/L^3$	$\mu^2/L^3\rho$	γ	σ/L^2	E/L
Pressure	M/LT^2 $= F/L^2$	$\mu^2/L^2\rho$	$L\gamma$	σ/L	E
Impulse and momentum	ML/T	$L^2\mu$	$L^{7/2}(\rho\gamma)^{1/2}$	$L^{5/2}(\rho\sigma)^{1/2}$	$L^3(E\rho)^{1/2}$
Energy and work	ML^2/T^2 $= FL$	$L\mu^2/\rho$	$L^4\gamma$	$L^2\sigma$	L^3E
Power	ML^2/T^3 $= FL/T$	$\mu^3/L\rho^2$	$\dfrac{L^{7/2}\gamma^{3/2}}{\rho^{1/2}}$	$\sigma^{3/2}(L/\rho)^{1/2}$	$L^2E^{3/2}/\rho^{1/2}$

Froude number by selecting a fluid with the proper combination of density and viscosity in the proper gravitational system. Actually, however, with rare exceptions, such an attempt is both impractical and unnecessary, as discussed later in this chapter.

• GENERALIZED SIMILARITY

Generalized similarity expresses a flow phenomenon over a wide range of all variables found to have a significant influence. It therefore implies a thorough and essentially complete understanding of the fluid mechanics involved. Generalized similarity is illustrated in Figs. 6-9, 7-6, 9-2, and 9-3. In each of these cases, the coffiecient C_D or f is known over a wide range of Reynolds numbers for a given geometry of the boundary. Furthermore, in Fig. 7-6 the coefficient f is also known over a wide range of relative roughness e/D.

The generalized approach to similarity is valuable not only in connection with a broad and significant understanding of the phenomena involved, but also in connection with specific model studies. The model study is intended to determine some factor such as Q or C_D for a limited range of conditions of flow—sometimes only one condition—whereas the generalized function expresses the relationship for an infinite variety of conditions. From a knowledge of the generalized variation of similitude, such as Figs. 6-9, 7-6, or 9-2, it is usually possible to extrapolate the model data with considerable accuracy beyond the narrow limits of the model study itself. This generalized approach is the very heart of a thorough understanding of modern fluid mechanics.

Unfortunately, the practice has developed of putting considerable effort and sums of money into model studies to determine the flow pattern and pressure distribution for a particular object and a particular condition of flow, without attempting to develop a general understanding of the phenomena involved. The general understanding comes only from a generalized approach to fluid mechanics, which can greatly decrease the necessity for extensive model testing.

Generalized similarity can best be understood by considering the various dimensionless numbers and the conditions under which each is important. The pressure parameter and/or the shear parameter are involved as dependent variables, if the dynamic similarity of drag, lift, or pressure on some structure or object is to be studied. These parameters must then be related to independent variables, which are the various other dimensionless numbers based upon fluid properties, as stated in Eqs. 11-20 to 11-23. Kinematic similarity involves the flow pattern, as expressed by magnitude and direction of velocity, for which there is a specific flow pattern for each single combination of the independent dimensionless numbers.

*The **Reynolds number***

$$\text{Re} = \frac{\rho V L}{\mu} = \frac{V L}{\nu} \tag{11-29}$$

is frequently of primary importance in problems of fluid mechanics—partticularly with respect to flow in closed conduits, flow in open channels, and flow around submerged objects. Its importance is directly associated with the boundary past which the fluid is flowing. If purely laminar flow exists, see Chapter 5, Reynolds number is relatively small, the forces of viscosity control the entire phenomenon, and inertia is of no significance. Under these conditions, Reynolds number is of paramount importance. If turbulent flow prevails, however, the viscous effects are less important—the extent of their importance depending upon the nature of the laminar sublayer surrounding the boundary, see Chapter 6. If the flow is sufficiently turbulent, the laminar sublayer is destroyed, and the viscous forces and Reynolds number are no longer significant. Between this extreme and the other extreme of laminar flow lies a region of transition, in which Reynolds number is important throughout—but becoming less and less important with increasing Re-values, as the laminar sublayer is gradually destroyed.

Destruction of the laminar sublayer is caused by the turbulence forcing it to become thinner with increasing Reynolds number, so that the roughness projects through the layer and renders it ineffective. The drag then changes from shear drag, for which the laminar sublayer is in full control and the boundary is hydrodynamically smooth, to pressure drag (on individual roughness elements), for which the roughness controls the flow pattern and the boundary is hydrodynamically rough. This is illustrated by the curves for smooth pipes and rough pipes in Fig. 7-6. The laminar sublayer is also destroyed (or prevented from forming), and the influence of inertia due to nonuniform flow is increased, by relatively large boundary irregularities other than the usual type of roughness. An extreme example of this is a disk perpendicular to the flow or a parachute, see Chapter 10. Under these conditions, the laminar sublayer has little or no opportunity to develop upstream from the point of separation, and consequently it is ineffective. Furthermore, the extreme normal and tangential accelerations increase the relative influence of the inertia forces. The same is true for a sharp-edged orifice in a tank. At very small Reynolds numbers, the flow around a disk or parachute or through an orifice is laminar throughout, and consequently viscosity is again all-important (see Fig. 9-2).

The length term in the Reynolds number is selected as a characteristic length over which the shear drag and/or pressure drag is effective. The Reynolds number of the approaching flow, as it affects the velocity and turbulence distribution, may be significant. Furthermore, the Froude number and the Mach number are also important under certain conditions. Each of these is discussed in the following paragraphs.

Example Problem 11-1: Derive the Reynolds number scale ratios required for similarity for (a) time (b) velocity, (c) acceleration, (d) discharge, (e) roughness, (f) what factors are apt to cause the most difficulty when attempting to model according to Reynolds law.

Solution:

(a) $(Re)_R = 1$ so $\dfrac{V_R D_R \rho_R}{\mu_R} = 1$ or $\dfrac{L_R}{T_R} L_R \dfrac{\rho_R}{\mu_R} = 1$.

Solving for T_R, $T_R = \left(\dfrac{\rho}{\mu} L^2\right)_R$

(b) $V_R L_R \dfrac{\rho_R}{\mu_R} = 1$ or $V_R = \left(\dfrac{\mu}{\rho L}\right)_R$

(c) $\left(\dfrac{VD\rho}{\mu}\right)_R = 1$ or $\dfrac{L_R}{T_R} L_R \dfrac{\rho_R}{\mu_R} = 1$

Multiplying numerator and denominator by T_R

$$\frac{L_R}{T_R^2} L_R T_R \frac{\rho_R}{\mu_R} = 1 \quad \text{and} \quad \frac{L_R}{T_R^2} = a_R$$

Hence $a_R = \dfrac{\mu_R}{\rho_R L_R T_R}$. Substituting for T_R.

$$a_R = \frac{\mu_R \, \mu_R}{\rho_R L_R \rho_R L_R^2} = \underline{\left(\frac{\mu^2}{\rho^2 L^3}\right)_R}$$

(d) $\left(\dfrac{VD\rho}{\mu}\right)_R = 1 \quad \text{or} \quad \dfrac{V_R D_R \rho_R}{\mu_R} = 1$

Multiplying numerator and denominator by L_R

$$V_R \frac{L_R^2 \rho_R}{L_R \mu_R} = 1$$

and since $V_R L_R^2 = Q_R, \quad Q_R \dfrac{\rho_R}{L_R \mu_R} = 1$

from which $Q_R = \underline{\left(\dfrac{\mu}{\rho} L\right)_R}$

(e) $(h_f)_R = \left(f \dfrac{L}{D} \dfrac{V^2}{2g}\right)_R \quad \text{or} \quad L_R = \dfrac{f_R}{g_R} \dfrac{L_R}{L_R} \dfrac{L_R^2}{T_R^2} \quad \text{and} \quad f_R = g_R \dfrac{T_R^2}{L_R}$

Substituting for T_R

$$f_R = g_R \frac{\rho_R^2}{\mu_R^2} \frac{L_R^4}{L_R} = \underline{\left(g \frac{\rho^2}{\mu^2} L^3\right)_R}$$

(f) A relatively large model must be used because

$$V_R = \frac{1}{L_R} \quad \text{and} \quad a_R = \frac{1}{L_R^3}$$

when using the same fluid with both model and prototype.

Also

$f_R = L_R^3$ when using the same fluid which shows that the model must be large if roughness requirements are to be satisfied. For example, considering a scale ratio of 1:20:

$$f_m = \frac{1}{20^3} f_p = \frac{f_p}{8,000}$$

The **Froude number** Fr is of importance only when there is a sloping interface of two fluids of different density. The difference in density results in a difference in specific weight, due to the effect of gravity. This difference in gravitational pull on the two fluids does not affect the flow, however, unless there is a gradient or slope of the interface. Basically, the Froude number

$$\text{Fr} = \frac{V}{\sqrt{\dfrac{\Delta \gamma L}{\rho}}} = \frac{V}{c} \tag{11-30}$$

is a ratio of the velocity of flow V to the celerity c of a *small* wave in *quiet fluid,* see Chapter 8. Consequently, if $\text{Fr} < 1.0$ a wave can move upstream against the flow, and if $\text{Fr} > 1.0$ a wave is swept downstream.

Although most problems involving an interface are free surface flow of a liquid under a gas, there are numerous other problems, such as density currents in reservoirs, free convection in bodies of liquid or gas, and flow of air masses in the atmosphere, which also involve an interface of two liquids or two gases. In the case of a liquid-gas interface, the specific weight of the gas is insignificant, and hence $\Delta \gamma \approx \gamma$, so that

$$\text{Fr} = \frac{V}{\sqrt{gL}} \tag{11-31}$$

The movement and form of waves and other irregularities at the interface are a function of the Froude number. Consequently, in a study of flow phenomena involving irregularities at an interface, the Froude law must be taken into consideration. If the Reynolds number is also important, then apparently both criteria should be used. This is theoretically possible by selecting a fluid which has a certain density and viscosity. Practically, however, such fluids either do not exist or they are too difficult or expensive to use. Therefore, special techniques are employed to eliminate, or at least minimize, the viscous effects. Usually, this is accomplished by means of roughness to destroy the laminar sublayer, as discussed in the next section of this chapter.

The Froude criterion is most frequently used in studies of hydraulic structures involving free-surface flow. Examples are spillways, stilling basins, gates, and open channel flow. The length selected is usually a characteristic depth, since this is directly related to wave celerity.

Example Problem 11-2: An ogee spillway of a gravity dam is to be modeled using water. The spillway section is 40 ft high, its crest length is 60 ft, and the maximum discharge is 3,000 cfs when the head on the crest of the spillway is 5 ft. Using a model scale of 1:10 calculate (a) the height of the model, (b) the head on the crest of the model at maximum discharge, (c) the crest length of the model, (d) the capacity of the pump required to supply maximum required discharge to the model and (e) the length of time the model must be operated to check the equivalent of 36 hours of prototype operation.

Solution:

Referring to Table 11-1 and satisfying the Froude number scale ratios

(a) $H_m = \dfrac{H_p}{10} = \dfrac{40}{10} = \underline{\underline{4.0 \text{ ft}}}$

(b) $h_m = \dfrac{h_p}{10} = \dfrac{5}{10} = \underline{\underline{0.5 \text{ ft}}}$

(c) $L_m = \dfrac{L_p}{10} = \dfrac{60}{10} = \underline{\underline{6.0 \text{ ft}}}$

(d) $Q_R = \left(\dfrac{\gamma^{1/2}}{\rho^{1/2}} L^{5/2} \right)_R$

from which

$$Q_m = \left(\dfrac{1}{10} \right)^{5/2} Q_p = \dfrac{3,000}{(10)^{5/2}} = \underline{\underline{9.47 \text{ cfs}}}$$

(e) $T_R = \left(\dfrac{\rho}{\gamma} L \right)_R^{1/2}$ or $T_m = \left(\dfrac{36}{10} \right)^{1/2} = \underline{\underline{1.9 \text{ hr}}}$

The **Mach number** Ma is associated with high speed flow of a gas. It is the ratio of the ambient velocity V to the velocity of sound—the velocity of an elastic wave c—and assumes significance as the velocity V approaches or exceeds the sonic velocity c, say Ma > 0.8. Usually, when the Mach number is relatively small, say Ma < 0.5, the effect of the Reynolds number is important. As the Mach number approaches unity, however, the flow pattern and force system become essentially independent of the Reynolds number, and the Mach number becomes the parameter of principle importance. There is a striking similarity between the pattern of elastic waves as related to the Mach number, and the pattern of gravity waves as related to the Froude number.

Example Problem 11-3: The flow pattern around a high speed rocket is to be modeled in a wind tunnel. If the rocket travels at a speed of Ma $= 2$ through standard air at 40°F, calculate the corresponding model rocket velocity required for similarity in air which has a temperature of 80°F.

Solution:

The acoustic velocity for prototype conditions is:

$$c = \sqrt{kg\,R\overline{T}}$$

$$c = \sqrt{(1.4)(32.2)(53.3)(500)} = 1099 \text{ fps}$$

$$V_p = (2)(1099) = 2198 \text{ fps}$$

Then

$$V_m = \left(\frac{E}{\rho}\right)_R^{1/2} V_p = \left(\frac{\rho_p}{\rho_m} \frac{E_m}{E_p}\right)^{1/2} V_p$$

$$V_m = \left(\frac{0.00228}{0.00247}\right)^{1/2} \left(\frac{1.4(14.7)(144)}{1.4(14.7)(144)}\right)^{1/2} 2198 = \underline{\underline{2110 \text{ fps}}}$$

The **Weber number** We becomes significant when the forces of surface energy become relatively large—that is, when the Weber number is relatively small. As the forces of inertia increase, however, they become predominant, and hence the Weber number becomes more and more insignificant with increasing We-values. Therefore, the Weber number, like the Reynolds number, is important for relatively small velocities of flow or small hydrostatic systems. Unlike the Reynolds number and somewhat like the Froude number, however, the Weber number involves an interface of two fluids—one of which must be a liquid. It also may involve a third item—a boundary. The Weber number is particularly important in connection with the formation of small drops of liquid, small (capillary) waves, the shape of small jets issuing from an orifice, flow of thin sheets or jets over weirs, the breakup of a jet into liquid droplets, and capillary movement and hydrostatic equilibrium of liquids in porous media and other capillary tubes. Attempts to minimize effects of surface energy usually involve increasing the scale of the flow system or using a detergent, which can have only limited benefit.

Although the dimensionless parameters already presented and discussed in this chapter are those most frequently involved in problems of hydraulic similitude, certain other factors and parameters are also significant.

The vapor pressure p_v of a liquid is a fluid property which enters some problems of fluid mechanics. From this has been derived the **cavitation number**

$$K = \frac{(p - p_v)}{\rho V^2/2} \tag{11-32}$$

in which $p - p_v$ is the difference between the local absolute pressure p and the vapor pressure p_v. This number is similar to the parameters derived by dimensional analysis in Chapter 1. It is proportional to the ratio of the pressure difference $(p - p_v)$ to the stagnation pressure $\rho V^2/2$. The cavitation number must be considered only with liquids in which the local or ambient pressure becomes so small that it approaches or becomes less than the vapor pressure of the liquid. Such a reduced local pressure occurs, according to the Bernoulli equation,

1. If the local velocity is increased so that the local pressure must decrease,

2. If the elevation z is high, so that the local pressure decreases, or

3. If the general ambient pressure of the system is reduced to a very small magnitude.

Cavitation frequently occurs on the suction side of a pump, in a turbine, at gate slots and other irregularities in a boundary where high-velocity flow occurs, around ship propellers, and around fast-moving submerged objects such as under-water projectiles. Methods of preventing cavitation are given in Chapter 3. Similarity with respect to cavitation depends upon maintaining the cavitation number a constant from model to prototype. The limiting condition of a *siphon* is dependent also on the vapor pressure of the liquid. Frequently, however, gas pockets form at pressures greater than vapor pressure, because of gas which comes out of solution as the pressure is reduced. These gas pockets sometimes accumulate at a high point in a pipe line. This accumulation of gas can result in surging flow which gives rise to impact pressures against the pipe wall. These pressures are sometimes of sufficient magnitude to cause damage to the pipe line. The accumulation of gas can also result in a complete stoppage of the flow.

A factor of primary importance in similarity problems is that of *velocity* and *turbulence distribution.* The mixing and momentum transfer within the flow is sometimes a very important aspect of similarity considerations. The variation of velocity distribution is illustrated in Eq. 3-24 and Fig. 3-11. Consequently, if true dynamic similarity exists, the velocity distribution as determined by v/V and the turbulence as determined by the scale l and relative intensity v'/V at a point must be the same in the model as in the prototype. The local velocity is v, the average or ambient velocity is V, and the intensity of turbulence is v' (see Chapter 5). In atmospheric turbulence, the spectrum of the turbulent energy distribution is also considered.

Directly associated with the velocity and turbulence distribution is the formation of *separation zones.* This is discussed in Chapter 3. Separation zones greatly influence flow patterns—and consequently the drag is influenced through the distributions of pressure and shear. The shape and location of separation zones, therefore, should be maintained constant from model to prototype.

• SPECIFIC MODELS

Application of the principles discussed in this chapter can be understood better by considering specific examples of models. These are considered from the viewpoint of closed conduits, submerged objects, free surface flow, and transport and diffusion. Models of closed conduits for which there is no free surface flow involve only

$$f = \phi\left(\text{Re}, \frac{e}{D}\right) \tag{11-33}$$

as shown in Eq. 7-8. In this case, it is simply a matter of reproducing in the model a Reynolds number and a relative roughness equivalent in effect to that of the prototype. Since the velocity of the flow and characteristic length of the conduit vary inversely with each other, the velocity must increase as the size of the model decreases. Therefore, if the velocity in the prototype is not too great and if the size of the model is not too small, it may be possible to accomplish this model-prototype similarity. The relative roughness must be reproduced not only with respect to its magnitude but also with respect to the geometry and general form.

Special types of studies involving flow through closed conduits are those of **gates, valves,** and **metering devices.** In each of these cases, the distribution of the approach velocity and turbulence is of primary importance in the flow pattern through the device and the calibration of the device for discharge. The velocity and turbulence profile approaching the device exerts considerable influence on the separation zones and, consequently, also upon the flow pattern and pressure distribution. A strict similarity, therefore, depends upon reproducing in the model not only the Reynolds number and relative roughness, but also the approaching flow pattern. The approaching flow pattern is influenced in turn by upstream bends and other objects in the flow, as well as by the shape and roughness of the conduit itself. In the case of pressure distributions for gates, valves, and meters, which also involve discharge calibrations, the following equation is applicable:

$$\frac{p}{\rho V^2/2} = \phi\left(\text{Re}, \frac{e}{D}\right) \tag{11-34}$$

This pressure equation has application not only in determining drag and forces on boundaries, but also in studying cavitation, for which the following modification of Eq. 11-34 is applicable

$$\frac{p - p_v}{\rho V^2/2} = \phi\left(\text{Re}, \frac{e}{D}\right) \tag{11-35}$$

in which $(p - p_v)/\rho V^2/2$ is called the cavitation parameter, as discussed for Eq. 11-32. For a given dimensionless pressure distribution curve on the submerged object, the magnitude of the cavitation parameter must be greater than zero for any point on the boundary, or else cavitation is likely to occur in the region of that point. The critical value of the cavitation parameter approaches zero but must be determined by experiment.

Many different kinds of **submerged objects** are involved in model studies. These include studies on the different parts of aircraft and the aerodynamic aspects of suspension bridges, the motion of underwater craft and other objects, and the flow around sampling devices placed in air or water. The pressure distribution and the resulting drag on a submerged object are a function of the Reynolds number Re, the relative roughness e/d, and the geometry of the object as expressed in Eq. 11-34. For a given object, this relationship can be expressed also in terms of the cavitation parameter, see Eq. 11-35. If the increased velocity necessary to maintain the Reynolds number a constant, for a model which is smaller than the prototype, is too great for the laboratory setup, it is sometimes possible to reproduce the influence of the Reynolds number by artificially adjusting the roughness to a minor extent. In this way, the laminar sublayer in the model can be controlled so that it behaves approximately as it does in the prototype. Adjustments of this nature, however, are best done only after the experimenter has had considerable experience with this technique.

The wings and fuselage of an **aircraft** can be studied either in total or in parts. For example, a two-dimensional segment of the wing can be placed in a wind tunnel to determine the lift and the drag, either by the pressure and shear distribution, or by the use of a dynamometer. If the pressure and shear distribution are measured, they must be integrated to determine lift and drag. Lift and drag are determined directly if a dynamometer is used. The study of a fuselage usually includes the wings and tail, because of the interaction and interference of the various parts of the aircraft. Aircraft models are also studied in free flight, either in a wind tunnel or in the atmosphere, in order to determine take-off, landing, and flight characteristics.

Oscillating motions are studied in water tunnels and open channels in connection with vibration of submarine **periscopes** and **bridge piles,** and in the wind tunnel in connection with wing flutter and movement of **suspension bridges.** Since the failure of the Tacoma-Narrows Bridge in Washington, considerable experience has been gained regarding the behavior of such bridges in various types of winds.

The flow pattern around **buildings** *and various structures* on the ground can be studied in wind tunnels. In this way, the pressure and velocity distribution can be forecast with considerable accuracy. Fortunately, in the prototype, the Reynolds number is sufficiently large and the nature of the structures is sufficiently angular so that viscous effects are negligible. In the model, the viscous effects can be eliminated by means of special roughness or angular corners on the models. In connection with such model studies, however, it is important to recognize that in the prototype the velocity distribution depends upon the ratio of the local velocity to the mean or ambient velocity of flow.

Model studies associated with *free surface flow* are perhaps the most common. Since a free surface or interface is involved, both Froude number and Weber number must be considered. Furthermore, since there is a possibility that viscous effects will be important, Reynolds number should be given at least preliminary consideration. It is generally out of the question to have all of these dimensionless numbers the same magnitude in the model as in the prototype. Consequently, steps must be taken to compensate for this situation. Generally, the forces of gravity are predominant in free surface flow, and Froude number governs. In the prototype, the surface energy is usually insignificant, because of the large size of flow system. Furthermore, the Reynolds number is frequently insignificant in the prototype, because the laminar sublayer is destroyed as a result of the large Reynolds number and/or the extreme roughness. In a model study, the model must be sufficiently large so that the influence of surface energy is not significant. Eliminating the influence of the Reynolds number in the model, however, is not as easy as eliminating the Weber number. Furthermore, reproducing the same relative roughness in the model as in the prototype is not always convenient. These various factors are now considered with respect to specific model studies.

A *ship* experiences both shear drag and pressure drag. The shear drag follows very closely with that which is predicted from the boundary layer theory. The pressure drag, however, arises from two sources—the shape of the ship and the influence of the waves. The pressure drag readily follows the Froude criterion, but the shear drag depends upon the Reynolds number. Since these cannot both be made the same in the model as in the prototype, the drag on the model is determined for the combination of the shear and pressure drag, and the shear drag is then calculated and subtracted from the total drag to determine the pressure drag only. This pressure drag can then be changed to the pressure drag for the prototype, and a new shear drag, computed for the prototype, can be added to the prototype pressure drag to determine the prototype total drag. This technique has been developed so that total drag is quite accurately determined. Another aspect of the study of models of ships involves the behavior of ships in various wave systems. Ships are towed through a model basin to determine the accelerations which are experienced, as well as the other aspects of behavior such as roll, pitch, heave, and yaw. In making these studies, the ship is given various degrees of freedom in order to effect a complete analysis.

Wave action and surges, as well as inflow and outflow, are studied as they affect flow patterns and movement of sediment in connection with *coastal engineering.* Normally, there are littoral currents flowing along a shoreline. Furthermore, the tide flows in and out of estuaries. Finally, wave action is continually operating on beaches and against structures placed on the shore. Therefore, model studies are necessary to determine:

1. The cause of some undesirable situation, together with a method of correction; or
2. The proper design of some structure or proposed modification of a structure, shoreline, or channel.

Such studies involve determination of the flow pattern, both before and after a change is made. Obviously, such free surface flow involves the Froude number. It also involves Reynolds number, however, and special steps must be taken to eliminate the viscous effects. This is usually accomplished by increasing the roughness to destroy the laminar sublayer, or by distorting the model in the vertical direction, or by a combination of both. Breakwaters, jetties, and other beach and shore protection works are studied in this manner.

Various types of *hydraulic structures* are studied by models, either as total structures or as parts of the structure. The various parts of hydraulic structures include the following items. *Spillways* are frequently studied in order to determine whether the flow pattern and efficiency of passing the flow are as intended by the designer. Measurements are made of the pressure distribution on a spillway crest and the profile of the flow over the spillway.

The spillway must pass a certain quantity of water for a given head or reservoir elevation. Consequently, it is calibrated in the model. In such studies, the Froude number is predominant and the viscous effects must be eliminated. Fortunately, with contracting flow the viscous influence is relatively small and, consequently, no special steps need be taken in most cases to eliminate the effect of Reynolds number. *Chute spillways* involve long open channels and, consequently, the resistance to the flow must be the same in the model as in the prototype. If the viscous effects are negligible, only the roughness need be reproduced in the model. This, however, is seldom easy to do. With *siphon spillways*, the Reynolds number can be significant in the model. Consequently, special roughness elements may be required in order to destroy the laminar sublayer.

Outlet works through dams or walls of reservoirs involve three parts— the entrance, the conduit, and the exit. Because the flow is contracting in the entrance, the viscous effects are usually negligible and can be ignored. In the conduit, however, viscous effects can seldom be ignored and, consequently, special steps must be taken to compensate for Reynolds number effects. This is usually done by adjusting the length and size of the conduit. Such adjustments are determined from computations based on known information regarding flow in closed conduits, as discussed in Chapter 7. The exit for outlet works usually involves considerable transition over a short distance. Consequently, Reynolds number is relatively insignificant. Froude number is usually important at any point in an outlet works where there is free surface flow.

Stilling basins usually involve such extreme turbulence that Reynolds number is ineffective. Consequently, Froude number is assumed to control the flow entirely. In the stilling basin, it is desired to know the flow pattern, the height of the water surface, and in some cases the pressure distribution along certain boundaries. Each of these is a function of Froude number. The *spray* associated with a stilling basin (and also long spillways) is a function of the turbulence in the flow and the surface energy of the fluid. Consequently, since it is not possible to keep the Weber number a constant from model to prototype, the spray cannot be modeled exactly.

River models are frequently used to determine the effect of cutoffs, the nature of flood waves, the peak flow of sediment movement, the effect of levees and other flood protection works, and the flow in floodways. Sections of a river are frequently modeled downstream from the spillway of a dam, in order to determine the flow pattern downstream from the stilling basin. The phenomena associated with such models are governed primarily by the Froude number and the relative roughness. Unfortunately, however, the model sometimes must be built so small that Reynolds number also is important. Consequently, distorted models, which are exaggerated in the vertical scale (frequently ten times the horizontal scale), are sometimes used. This vertical distortion helps to reduce the influence of the laminar sublayer. The laminar sublayer is also destroyed in some cases by the use of extremely large roughness elements. These roughness elements are first adjusted for a known discharge and flow condition, until the model behaves as the prototype is known to behave for this given discharge. It is then assumed that the model will behave like the prototype for other discharges. By this means, the flow patterns and relative water surface elevations can be determined for larger discharges, both before and after special changes have been made along the river. Measurements usually made in a river model are water surface elevations and velocity patterns, both at the surface and near the bottom of the channel.

Bridges and *culverts* are also studied in larger river models to determine the flow pattern, the backwater, and the scour which is caused by the bridge. Again, Froude number controls and special precautions are taken to eliminate the effect of Reynolds number.

Model studies are also made on *diversion works* and *intake structures,* where water is diverted out of a river into a canal or special conduit. Because of the accelerations involved in such a model, the Reynolds effects are usually negligible.

All *flow in open channels* must be considered from the viewpoint of tranquil flow and rapid flow. If the flow is tranquil, the water surface profile is relatively controllable and predictable. If the flow is rapid, obstructions to the flow and bends will create undesirable wave patterns which may cause the structure to fail. Consequently, the Froude criterion is doubly important in both the model and the prototype.

Alluvial channels which have a movable sand bed are particularly difficult to study, because they involve not only the Froude number, but also other parameters which have not been clearly determined. Consequently, alluvial bed models must be interpreted with considerable care, and then usually only qualitatively, when applying the results to the prototype. Qualitative answers, however, are frequently very significant.

Models of *diffusion phenomena* in the *atmosphere* and in open channels have been studied to some extent. The principal difficulty involved in such studies is the duplication in the model of the velocity and turbulence characteristics at all points of the flow. If these are reproduced so that they are dynamically similar, the diffusion pattern, say from a smokestack, will be accurately produced in the model. It must be remembered, however, that in the prototype the velocity profile is not uniform. Instead, both the velocity profile and the turbulence characteristics are products of the upstream conditions, together with any special influence of overhead flow.

It can be seen, from the discussions presented in this chapter, that similitude should be considered from the over-all generalized viewpoint. In this way, considerable saving of time and effort can be had, because of the generalized information being available for a wide range of design conditions. For special cases, model studies will always be necessary to determine the influence of the special geometric conditions upon the flow pattern, pressure distribution, and other factors needing investigation. Such model studies, however, should be considered on the basis of fundamental fluid mechanics.

- ## SUMMARY

1. A *model* is a system by whose operation the characteristics of other similar systems can be predicted.

2. The *types of similarity* are *geometric similarity, kinematic similarity,* and *dynamic similarity.*

3. *Geometric similarity* requires that all *length ratios* from model to prototype be the same.

4. *Kinematic similarity* requires that the *streamline pattern* be the same in the model and the prototype so that the *velocity ratios* and the *acceleration ratios* are constant from model to prototype.

5. **Dynamic similarity** requires that the *shear ratios*, the *pressure ratios*, and the *force ratios* be the same from model to prototype.

6. The **types of forces** involved in similarity problems are *inertia forces* (due to density), *viscous forces* (due to viscosity), *gravity forces* (due to gravitational attraction), *surface energy forces* (due to surface energy), and *elastic forces* (due to elasticity).

7. For dynamic similarity between the model and the prototype:
   a. Use the **Reynolds law** Re (in model) = Re (in prototype) when viscous forces are predominant.
   b. Use the **Froude law** Fr (in model) = Fr (in prototype) when gravitational forces are predominant.
   c. Use the **Mach law** Ma (in model) = Ma (in prototype) when forces due to compressibility predominate.
   d. Use the **Weber law** We (in model) = We (in prototype) when surface energy forces predominate.
   e. Use the **cavitation number** K (in model) = K (in prototype) when cavitation may be involved.

8. The **scale ratios** based on the various laws for model-prototype relations are given in Table 11-1.

9. **Generalized similarity** expresses flow phenomena over a wide range of the significant variables. Examples of generalized similarity are Figs. 6-9, 7-6, 9-2, and 9-3.

10. Theoretically, **complete similarity** is obtained only if each of the ratios in item 7 is the same in the model and the prototype. From the practical viewpoint, however, this is not possible. Consequently, certain compromises must be made, based upon a *sound knowledge of the fundamentals of fluid mechanics*, see the section on *Specific Models*.

References

1. Allen, J., *Scale Models in Hydraulic Engineering*. New York: Long-mans, Green & Co., 1947.

2. Binder, R. C., *Fluid Mechanics*, 3rd ed. Englewood Cliffs, N. J.: Prentice-Hall, Inc., 1955.

3. Dodge, Russell A. and Milton J. Thompson, *Fluid Mechanics*. New York: McGraw-Hill Book Co., Inc., 1937.

4. Freeman, John R., *Hydraulic Laboratory Practice*. New York: ASME, 1929.

5. Hickox, George H., "Hydraulic Models," Chapter 24 of *Handbook of Applied Hydraulics*, ed. C. V. Davis. New York: McGraw-Hill Book Co., Inc., 1952.

6. Langhaar, H. L., *Dimensional Analysis and Theory of Models*. New York: John Wiley & Sons, Inc., 1951.

7. Murphy, Glenn, *Similitude in Engineering*. New York: The Ronald Press Co., 1950.

8. Rouse, Hunter, *Elementary Mechanics of Fluids*. New York: John Wiley & Sons, Inc., 1946.

9. Rouse, Hunter, *Fluid Mechanics for Hydraulic Engineers*. New York: McGraw-Hill Book Co., Inc., 1938.

10. Streeter, Victor L., *Fluid Mechanics*, 2nd ed. New York: McGraw-Hill Book Co., Inc., 1958.

11. Tiffany, J. B., *Laboratory Research Applied to the Hydraulic Design of Large Dams*. Vicksburg: U. S. Waterways Experiment Station Bulletin 32, 1948.

12. Warnock, J. E., "Hydraulic Similitude," Chapter II of *Engineering Hydraulics*, ed. Hunter Rouse. New York: John Wiley & Sons, Inc., 1950.

13. "Hydraulic Models," *ASCE Manual of Engineering Practice*, No. 25. New York: 1942.

Problems

11-1. If the length ratio from model to prototype is 1/40, what is the area ratio? What is the height of the model dam if the prototype is 200 ft high?

11-2. If in a model study $L_R = 1/40$ and $V_R = 1/6.34$, find the discharge in the model required for a design discharge of 40,000 cfs in the prototype. What length of time in the model represents one day in the prototype?

11-3. If the pressure ratios along the face of a model dam must be the same from point to point, what will a force of one pound in the model represent in the prototype if the length ratio $L_R = 1/50$?

11-4. A model of a 12-in. pipe line which carries oil (sp. gr. = 0.85) $\mu = 10^{-4}$ lb-sec/sq ft is to be modeled in the laboratory, using a 2-in. pipe carrying water at 70°F. If the discharge in the prototype is 1000 gpm, what is the required discharge in the model, assuming the Reynolds number in model and prototype are to be the same? What is the time ratio between model and prototype?

11-5. For a model study of a spillway of a dam, the Froude number is generally the criterion used for similarity. If the length ratio is 1/60, what is the flow in the model for a discharge of 200,000 cfs in the prototype?

11-6. A model of Bocono Dam in Venezuela had a scale ratio of 1/49.2. For a discharge of 10,000 cu m/sec over an overflow spillway, what was the discharge in the model in cfs?

11-7. Machio Dam in Japan was modeled with a model scale of 1/60. The prototype is an ogee spillway designed to carry a flood of 3200 cu m/sec. What was the required model discharge in cfs for the design flood? What time in the model represented one day in the prototype?

11-8. A model of an overflow spillway is to be constructed in a laboratory. The design flood for the prototype spillway is 20,000 cfs for a spillway 300 ft wide. If the maximum discharge available for the model is 5 cfs, what is the maximum model scale ratio which can be used for a model of the full width of the spillway?

11-9. A plane travels 500 mph through air at 0°F and 10 psia. A model with a scale ratio of 1 to 30 is to be tested in a pressure wind tunnel. If the tunnel operates at an air temperature of 70°F and a pressure of 50.0 psia, what is the required wind velocity in the tunnel? Assume an adiabatic condition with $k = 1.4$.

11-10. A missile travels at a speed of 1700 fps in air at standard atmospheric conditions and a temperature of −30°F. A model is to be tested in a wind tunnel at standard atmospheric conditions and a temperature of 80°F. What is the required air speed in the wind tunnel for dynamic similarity?

11-11. If water is rising in a capillary tube 0.2 mm in diameter at 1.0 ft/sec, at what rate is the water rising at a corresponding point in a tube which is 2 mm in diameter?

11-12. Assuming that the same fluid at the same temperature is to be used in the model as in the prototype, what is the force ratio F_R for the various similarity criteria; (**a**) Reynolds, (**b**) Froude, (**c**) Mach, (**d**) Weber?

11-13. For each of the model criteria, that is Reynolds, Froude, Weber, and Mach, what is the relationship between the time in the model and the time in the prototype. Prove the ratios given in Table 11-1.

11-14. A ship is to be tested by a small model which is 1/100 the length of the prototype. Assuming that sea water (sp. gr. = 1.03) is in the model tests, what is the model speed for a prototype speed of 30 knots, if surface wave resistance is to be studied. If the force on the model is 1 lb, what is the force on the prototype?

11-15. A torpedo is designed to travel through sea water at 50°F at 50 knots. Assume that the kinematic viscosity is the same as fresh water. A 1/20 model of the torpedo is to be studied in a water tunnel which circulates water at 70°F. What is the required velocity in ft/sec of the water past the stationary model of the torpedo?

11-16. A discharge of 400 gpm of SAE 10 oil at 70°F flows in a 6-in. pipe. If a model of this pipe is 1-in. in diameter, what is the velocity of air at 70°F required to give dynamic similarity?

11-17. If an airplane flies at 500 mph at an elevation of 20,000 ft, what speed must the airplane fly at sea level in standard air for dynamically similar conditions?

11-18. Water flows at 6 fps in a rectangular canal at a depth of 5 ft. In order to be dynamically similar, at what velocity must the water flow in a laboratory model of the canal where the depth is 4 in.?

11-19. A model airplane is to be tested in a wind tunnel which operates at a temperature of 50°F and a pressure of 50 psia. The prototype is to be tested for a speed of 500 mph at a pressure of 7 psia and a temperature of −10°F. What is the velocity in the wind tunnel?

11-20. Hirakud Dam in India was designed to carry a design flood of 1,535,000 cfs through 84 sluices and through 86 bays of an overflow spillway. Each bay of the overflow spillway discharges 5200 cfs. If a model of four bays is to be constructed using a scale ratio of 1/40, what would be the required model discharge in cfs?

11-21. An overflow spillway model is to be constructed with a scale ratio of 1/49. Assuming that water at 60°F flows in the prototype, what is the required kinematic viscosity of the fluid in the model for dynamic similarity, if both the Froude law and Reynolds law are to be satisfied?

11-22. A rigid-boundary river model 600 ft long is built to satisfy the Froude law. If the model scale is 1/196, what is the flow in the model for a prototype flood of 100,000 cfs. Assuming the flood peak requires 90 minutes to travel through the 600 ft of the model, how long would the prototype flood peak require to travel the corresponding distance in the actual river?

11-23. A large Venturi meter is to be placed in a water main to measure the flow. Briefly describe the type of model study which could be used to get the discharge coefficient of the prototype.

11-24. A 24-in. pipe carrying water at 60°F is to be modeled by a pipe carrying air at 60°F. Assume very low pressures and incompressible flow in the case of the air. Briefly outline the model study. If the maximum flow in the 24-in. pipe is 50 cfs and the maximum air velocity is 100 ft/sec, what size pipe should be used in the model study?

11-25. A spherical hailstone (sp. gr. = 0.8) diameter of $\frac{1}{4}$ in. falls at a constant velocity through still air whose temperature is 40°F. The falling of the hailstone is to be modeled in a tank of water at 70°F. If the model of the hailstone is 2 in. in diameter, what is the specific gravity of the model?

11-26. If in Problem 11-25 it is desirable to have a model hailstone which falls at a maximum velocity of $\frac{1}{2}$ ft/sec, briefly outline the experimental procedure you would use. Suggest a fluid and the size and weight of the model hailstone which you would use.

chapter 12

Hydraulic

Machinery

Hydraulic machines play an important role in practically all walks of life. For example, they find wide application in all fields of industry, agriculture, and municipal water supply and sewage treatment. Hydraulic machines are devices which can control the flow and/or convert fluid energy (kinetic and/or potential) to mechanical energy, or mechanical energy to fluid energy. All forms of hydraulic machines can be classified as either

1. Turbomachinery, or

2. Positive-displacement machinery.

The types of hydraulic machinery considered in this chapter include: centrifugal pumps, turbines, propellers, windmills, fluid couplings, torque convertors, and gear pumps—all of which can be classified as turbomachines with the exception of the gear pump, which is a positive-displacement machine. The analysis of turbomachines is a problem of fluid dynamics, whereas the positive-displacement machines must be analyzed in terms of purely mechanical concepts, since the fluid is displaced by some moving mechanical element such as a precisely machined gear system rotating in a closed housing, or a piston moving in a cylinder.

Centrifugal pumps, turbines, and related turbomachinery consist of two main parts:

1. The rotating element, which is usually called the impeller if it is a part of a pump, and the runner if it is a part of a turbine, and

2. The casing or housing which encloses the rotating element.

The power is supplied to or taken from a turbomachine by means of a shaft driven by a motor (for a pump) or by the machine (for a turbine). The impeller or runner is rigidly mounted on the shaft.

In the **pump,** the impeller gives the fluid rotary kinetic energy. This rotary motion creates a centrifugal force which enables the fluid to move from the vicinity of the shaft, where it enters the center of the casing at low pressure through an opening called an eye. The fluid then enters the region of rotation and higher pressure in the casing surrounding the impeller. Therefore, at the outlet opening of the casing, the rotary motion has imparted pressure and some kinetic energy to the fluid, as it flows into the discharge line for delivery. A simple example of a centrifugal pump invented by Demour in 1730, which illustrates the fundamental principle involved, is shown in Fig. 12-1. This simple pump consists of a rotating

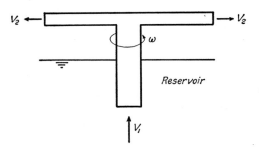

Fig. 12-1. Demour's centrifugal pump.

pipe in the form of a tee. If the lower end of the tee is submerged in a liquid and the complete unit is primed and rotated as indicated, the centrifugal force overcomes the gravitational force acting on the liquid, which then flows upward from the reservoir through the vertical shaft and out through the horizontal arms. A typical modern centrifugal pump is illustrated in Plate 12-1.

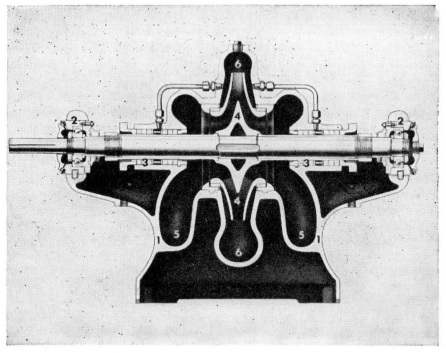

Plate 12-1. Single stage split case centrifugal pump.
Courtesy of Fairbanks Morse.

1. Pump casing
2. Bearings and housing
3. Stuffing box
4. Impeller
5. Intake chamber
6. Discharge chamber

The ***turbine*** performs a function opposite to that of the pump—the fluid flows in the opposite direction and energy is subtracted from it. The fluid with kinetic energy and under pressure enters the casing and flows inward through the turbine runner. This fluid kinetic energy and pressure is converted to mechanical energy as the fluid flows through the runner passages toward the outlet. In turn, the runner shaft can be connected to an electrical generator, which converts the mechanical energy to electrical energy. An example of a turbine is illustrated in Plate 12-2.

The ***similarity*** between centrifugal pumps and turbines is best illustrated by pointing out that if the reaction turbine runner were placed below the liquid level and rotated in reverse, it would perform as a pump. Turbomachines can be designed to function as either a pump or a turbine, the major disadvantage of the dual-purpose machine being a slight reduction in efficiency as compared with single-purpose machines.

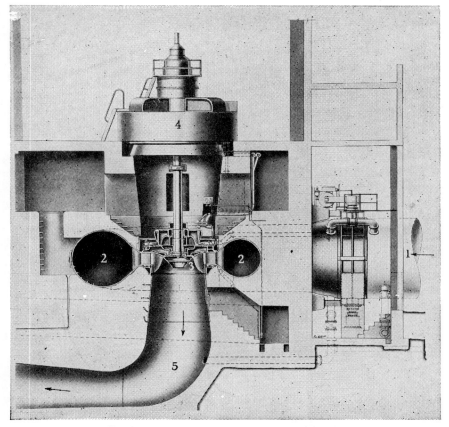

Plate 12-2. Reaction turbine unit for Pensacola Powerhouse, Oklahoma, 20,000 HP, 115 ft. head. Courtesy of Allis Chalmers.

1. Penstock	4. Generator
2. Scroll case	5. Draft tube
3. Turbine runner	

• TURBOMACHINERY THEORY

The theory of turbomachines is based on the momentum principle which states that, if a body in steady motion undergoes a change in angular momentum, the resultant of the external forces acting on the body is a *torque* which is *equal in magnitude to the time rate of change of angular momentum.* In equation form, this is

$$\text{Torque} = \left(\frac{\text{Mass}}{\text{Time}}\right)(\text{Radius})(\text{Velocity})$$

The relation between the flow pattern and the shape of the centrifugal pump impeller is shown in the velocity diagram of Fig. 12-2, for which

β is the vane angle;

v is the relative velocity of the liquid with respect to the impeller;

V is the absolute velocity of the liquid;

u is the speed of the impeller ($u = r\omega$);

r is the radius which becomes r_1 at the entrance and r_2 at the exit of the impeller.

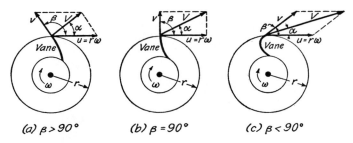

(a) $\beta > 90°$ (b) $\beta = 90°$ (c) $\beta < 90°$

Fig. 12-2. Variation of velocity diagram with shape of vane.

Useful trigonometric relations between V, v, and u are:

$$V^2 = u^2 + v^2 - 2uv \cos \beta$$

and $$V \cos \alpha = u + v \cos \beta$$

The moment of momentum or angular momentum can be computed at the entrance to and exit from the impeller by evaluating the integral

$$\rho \int_Q rV \cos \alpha \, dQ \tag{12-1}$$

in which

$\rho \, dQ$ is the mass of fluid flowing per unit time;

r is the radius from the center of the impeller shaft to any point on the impeller;

$V \cos \alpha$ is the tangential component of the absolute velocity.

The *torque* applied to the *pump* impeller is the difference between the moment of momentum at the inlet to and outlet from the impeller. In equation form, this is

$$T = \rho \int_Q r_2 V_2 \cos \alpha_2 \, dQ - \rho \int_Q r_1 V_1 \cos \alpha_1 \, dQ \tag{12-2}$$

Assuming the flow can be represented by an ***irrotational vortex***

$$r_1 V_1 \cos \alpha_1 = \text{constant}$$
$$r_2 V_2 \cos \alpha_2 = \text{constant}$$

and

$$T = \rho Q (r_2 V_2 \cos \alpha_2 - r_1 V_1 \cos \alpha_1) \tag{12-3}$$

The basic equations for ***power,*** LF/T (measured in ft-lb per sec) are

$$P = T\omega \tag{12-4}$$

in which

T is the torque applied to the pump;

ω is the angular velocity of the pump.

and

$$P = \gamma Q H \tag{12-5}$$

in which

γ is the unit weight of the liquid

Q is the discharge

H is the theoretical head LF/F (in ft-lb/lb of fluid flowing) added to the fluid by the runner or, in the case of a turbine, the head added to the runner by the fluid.

The magnitude of H can be computed from the energy equation

$$\frac{V_1^2}{2g} + \frac{p_1}{\gamma} + z_1 + H = \frac{V_2^2}{2g} + \frac{p_2}{\gamma} + z_2 + h_L$$

The effective head H_p imparted to the fluid by the pump is

$$H_p = e_p H = H - h_L \tag{12-6}$$

in which

e_p is the efficiency of the pump;

h_L is the sum of the hydraulic and mechanical losses in the pump.

Substituting T from Eq. 12-3 into Eq. 12-4 yields

$$P = \rho Q (r_2 V_2 \cos \alpha_2 - r_1 V_1 \cos \alpha_1)\omega$$

and since $u = r\omega$

$$P = \rho Q (u_2 V_2 \cos \alpha_2 - u_1 V_1 \cos \alpha_1) \tag{12-7}$$

If the liquid enters the impeller radially ($\alpha_1 = 90°$), then $\cos \alpha_1 = 0$ and the liquid has no moment of whirl at this point. Hence

$$u_1 V_1 \cos \alpha_1 = 0$$

and

$$P = \rho Q u_2 V_2 \cos \alpha_2 \tag{12-8}$$

The theoretical head H added to the fluid by the impeller can be determined for the centrifugal pump by equating Eq. 12-5 to Eq. 12-7 and solving for H. That is,

$$\gamma Q H = \rho Q (u_2 V_2 \cos \alpha_2 - u_1 V_1 \cos \alpha_1)$$

and

$$H = \frac{(u_2 V_2 \cos \alpha_2 - u_1 V_1 \cos \alpha_1)}{g} \tag{12-9}$$

For radial inflow to the impeller, $\cos \alpha_1 = 0$ and

$$H = \frac{u_2 V_2 \cos \alpha_2}{g} \tag{12-10}$$

The theoretical *water horsepower* of a pump is

$$\text{WHP} = \frac{\gamma Q H}{550} \tag{12-11}$$

The *brake horsepower* required to drive the pump is

$$\text{BHP} = \frac{\gamma Q H}{550 e_p} \tag{12-12}$$

in which e_p is the pump efficiency.

The *horsepower input* to the pump motor is

$$\text{HP} = \frac{\gamma Q H}{550 e_p e_m} = \frac{\gamma Q H}{550 e} \tag{12-13}$$

where e_m is the motor or mechanical efficiency, and the overall efficiency e is

$$e = e_p e_m \tag{12-14}$$

Example Problem 12-1: A 12 in. centrifugal pump which turns at 1200 rpm discharges 1.0 cfs of water. If $\beta_2 = 160°$, $v_2 = 40$ fps and $\alpha_1 = 90°$, calculate

(a) the angular velocity of the impeller in radians per second

(b) the linear speed u of the impeller

(c) the theoretical head H added to the water, and

(d) the theoretical horsepower input to the pump.

Solution:

(a) $\quad \omega = \dfrac{2\pi N}{60} = \dfrac{(2\pi)(1200)}{60} = \underline{\underline{126 \text{ radians per sec}}}$

(b) $\quad u_2 = r\omega = (0.5)(126) = \underline{\underline{63 \text{ fps}}}$

(c) $\quad V_2^2 = u_2^2 + v_2^2 - 2u_2v_2 \cos \beta_2$

$\quad\quad V_2^2 = \overline{63}^2 + \overline{40}^2 + (2)(63)(40)(0.938)$

$\quad\quad V_2 = 101 \text{ fps}$

$\quad\quad \dfrac{v_2}{\sin \alpha_2} = \dfrac{V_2}{\sin (180 - \beta_2)}$

$\quad\quad \sin \alpha_2 = \dfrac{(40)(.342)}{101} = 0.135$

$\quad\quad \alpha_2 = 7.75°$

$\quad\quad H = \dfrac{V_2 u_2 \cos \alpha_2}{32.2} = \dfrac{(101)(63)(.99)}{32.2} = \underline{\underline{195 \text{ ft}}}$

(d) $\quad HP = \dfrac{Q\gamma H}{550} = \dfrac{(1)(62.4)(195)}{550} = \underline{\underline{22.1 \text{ horsepower}}}$

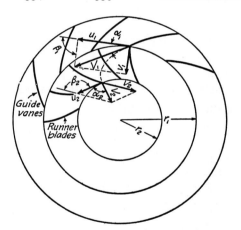

Fig. 12-3. Velocity diagram for a turbine runner.

The power equation derived for centrifugal pumps is applicable to reaction turbines if the subscripts are exchanged. That is, for *turbines*

$$P = T\omega = \rho Q(u_1 V_1 \cos \alpha_1 - u_2 V_2 \cos \alpha_2) \qquad (12\text{-}15)$$

or Eq. 12-15 can be derived, as the corresponding equation was for pumps, by utilizing the turbine velocity diagram of Fig. 12-3, in which

$$V_1^2 = u_1^2 + v_1^2 - 2u_1v_1 \cos \beta_1$$

and $\quad V_2^2 = u_2^2 + v_2^2 - 2u_2v_2 \cos \beta_2$

That is, $\gamma QH = \rho Q(u_1 V_1 \cos \alpha_1 - u_2 V_2 \cos \alpha_2)$

from which

$$H = \frac{1}{g}(u_1 V_1 \cos \alpha_1 - u_2 V_2 \cos \alpha_2) \qquad (12\text{-}16)$$

The angle α_2 is most commonly close to 90°, hence $u_2 V_2 \cos \alpha_2$ is small and, for purely radial outflow from the runner, $\alpha_2 = 90°$.

Then

$$H = \frac{u_1 V_1 \cos \alpha_1}{g} \qquad (12\text{-}17)$$

The *horsepower delivered* by the turbine is

$$\mathrm{HP} = \frac{e_t \gamma QH}{550} \qquad (12\text{-}18)$$

in which

e_t is the turbine efficiency.

The *horsepower output* from the generator is

$$\mathrm{HP} = \frac{e_t e_m \gamma QH}{550} \qquad (12\text{-}19)$$

in which

e_m is the mechanical efficiency of the generator

and the overall efficiency e of the turbine is

$$e = e_t e_m \qquad (12\text{-}20)$$

Example Problem 12-2: Calculate (a) the torque and power applied to the shaft and (b) the turbine efficiency e_t of a reaction turbine if $D = 50$ in. $H = 500$ ft, $Q = 200$ cfs, $N = 400$ rpm, $\alpha_1 = 20°$, $\alpha_2 = 90°$, and the area A_1 measured normal to V_1 is 1.20 ft².

Solution:

(a) $V_1 = \dfrac{Q}{A_1} = \dfrac{200}{1.20} = 166.8$ fps

$\omega = \dfrac{2\pi N}{60} = \dfrac{(2\pi)(400)}{60} = 41.8$ radians per sec

$r_1 = \dfrac{D}{2} = 25$ in.

$u_1 = r_1 \omega = \dfrac{(25)(41.8)}{12} = 87.2$ fps

$$P = \rho Q (u_1 V_1 \cos \alpha_1 - u_2 V_2 \cos \alpha_2)$$

$$P = (1.94)(200)(87.2)(166.8)(0.94) = \underline{\underline{5,300,000 \text{ ft-lb/sec}}}$$

$$P = \underline{\underline{9,630 \text{ horsepower}}}$$

$$T = P/\omega = \frac{5,300,000}{41.8} = \underline{\underline{126,800 \text{ ft-lb}}}$$

(b) $$e_t = \frac{550P}{\gamma Q H} = \frac{(550)(9,630)}{(62.4)(200)(500)} = \underline{\underline{0.85}}$$

$$e_t = \underline{\underline{83.7 \text{ per cent}}}$$

• DRAFT TUBE

The reaction turbine receives its supply of liquid through a pipe or penstock and converts the major portion of the fluid energy to mechanical energy as it is discharged through the turbine runner. As the fluid leaves the turbine runner it is carried to the tail race by a draft tube. A common type of draft tube is illustrated in Plate 12-2. The draft tube is a conduit which is usually slightly flared to reduce the velocity head at point of discharge to a minimum, and hence increase the efficiency of the turbine. Angle of flare measured with respect to the axis of the draft tube must be smaller than 9 degrees to avoid separation within the tube which would defeat its purpose. Flow conditions within the draft tube can be evaluated by applying the Bernoulli equation between the water surface in the tail race and any other point in the draft tube. In estimating friction and minor losses in the tube, the procedure used to evaluate losses in pipes is utilized.

The tube should have sufficient length to guarantee that the turbine wheel will not be submerged when maximum discharge occurs. However, the turbine must not be set so high above tail water level that small or negative pressures develop where the water discharges from the runner to the draft tube, because, if this condition exists, cavitation occurs.

• THE AFFINITY LAWS AND SPECIFIC SPEED

The principle of dynamic similarity can be utilized to derive the affinity laws. That is, to determine how Q, H, and P are affected by changing the speed or size of a particular geometric design. Turbomachines are geometrically similar, or of the same geometric design, if velocity diagrams at corresponding points are the same, see Fig. 11-1.

Considering the hydrodynamics of turbomachines and utilizing dimensional analysis

$$\phi_1 = (Q, H, P, N, D, \rho, \mu, E) = 0 \tag{12-21}$$

in which

> Q is the discharge in cfs;
>
> H is the theoretical "head," which in this case must be expressed as energy per unit of mass $(ML^2/T^2)(1/M) = L^2/T^2$
>
> P is the power;
>
> N is the angular velocity in rpm;
>
> D is the diameter of the impeller or runner;
>
> ρ is the mass density of the fluid;
>
> μ is the kinematic viscosity of the fluid;
>
> E is the modulus of elasticity of the fluid.

Selecting N, D, and ρ as repeating variables, the resulting dimensionless pi-terms are:

$$\pi_1 = \frac{Q}{ND^3} \qquad\qquad \pi_4 = \frac{\mu}{\rho ND^2}$$

$$\pi_2 = \frac{H}{N^2D^4} \qquad\qquad \pi_5 = \frac{E}{\rho N^2D^2}$$

$$\pi_3 = \frac{P}{\rho N^3D^5}$$

If the influence of ρ and μ are negligible, which is a close approximation to reality when dealing with a narrow range of operating conditions, and if a single fluid and only one machine are considered (i.e., D = constant), the first three pi-terms reduce to

$$\pi_1 = \phi_1\left(\frac{Q}{N}\right) \qquad \pi_2 = \phi_2\left(\frac{H}{N^2}\right) \qquad \pi_3 = \phi_3\left(\frac{P}{N^3}\right)$$

which show that

$$Q \propto N \tag{12-22}$$

$$H \propto N^2 \tag{12-23}$$

$$P \propto N^3 \tag{12-24}$$

Next consider a turbomachine in which N and ρ are constant. With these conditions the first three pi-terms show that

$$Q \propto D \tag{12-25}$$

$$H \propto D^2 \tag{12-26}$$

$$P \propto D^3 \tag{12-27}$$

provided an area term $A \propto D^2$ is eliminated from π_1 and π_3. Turbine tests have demonstrated that D^2 represents an area in π_1 and π_3.

Summarizing, for geometrically similar or **homologous turbines,** the foregoing equations state that:

<table>
<tr><td align="center">For D and ρ constant</td><td align="center">For N and ρ constant</td></tr>
<tr><td align="center">$\dfrac{Q_1}{Q_2} = \dfrac{N_1}{N_2}$</td><td align="center">$\dfrac{Q_1}{Q_2} = \dfrac{D_1}{D_2}$</td></tr>
<tr><td align="center">$\dfrac{H_1}{H_2} = \left(\dfrac{N_1}{N_2}\right)^2$</td><td align="center">$\dfrac{H_1}{H_2} = \left(\dfrac{D_1}{D_2}\right)^2$</td></tr>
<tr><td align="center">$\dfrac{P_1}{P_2} = \left(\dfrac{N_1}{N_2}\right)^3$</td><td align="center">$\dfrac{P_1}{P_2} = \left(\dfrac{D_1}{D_2}\right)^3$</td></tr>
</table>

$$(12\text{-}28)$$

These equations, which are called **affinity laws,** and the relationship for specific speed which is introduced later, can also be developed algebraically. They can be used to illustrate what can be accomplished by altering speed and/or diameter of a pump or turbine, or to predict the performance of a similar machine of different size.

In many cases, pumps originally selected to serve a particular duty can be modified to suit a different set of operating conditions. The alteration of one dimension of a turbomachine destroys true geometric similarity, but the effect of small changes can be predicted from the foregoing laws for similar turbines. However, care is necessary when attempting to predict the new performance characteristics of a machine, if the dimensional change is large.

Example Problem 12-3: The power input is 34 horsepower to a centrifugal pump that is discharging 900 gpm and which operates at 1800 rpm against a head $H = 112$ ft. If this pump is modified to operate at 1200 rpm, assuming its efficiency remains constant, estimate (a) its discharge in gpm (b) the theoretical head H it imparts to the liquid, and (c) the horsepower input to the pump.

Solution:

(a) $Q_2 = \dfrac{N_2}{N_1} Q_1 = \dfrac{1200}{1800}\, 900 = \underline{\underline{600 \text{ gpm}}}$

(b) $H_2 = \left(\dfrac{N_2}{N_1}\right)^2 (H_1) = \left(\dfrac{1200}{1800}\right)^2 112 = \underline{\underline{49.7 \text{ ft}}}$

(c) $P_2 = \left(\dfrac{N_2}{N_1}\right)^3 P_1 = \left(\dfrac{1200}{1800}\right)^3 34 = \underline{\underline{10.10 \text{ horsepower}}}$

An even more important parameter termed **specific speed** can be derived algebraically or by the manipulation of pi terms. That is, for **pumps**

$$N_s = \frac{(\pi_1)^{1/2}}{(\pi_2)^{3/4}} = \frac{NQ^{1/2}}{H^{3/4}}$$ (12-29)

in which

N is the angular velocity, in rpm;

Q is the discharge, in gpm;

H is the theoretical head in ft.

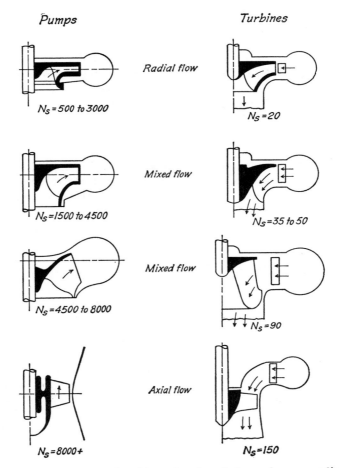

Fig. 12-4. Pumps and turbines of various design and corresponding specific speed, N_s.

Since N_s can vary from zero to infinity, N_s is computed for a particular machine where its efficiency is a maximum. Defining N_s in this way, it becomes an extremely useful parameter which is utilized to classify turbomachines with respect to their type or proportions, and as a means of predicting other important characteristics, such as the suction limitation of centrifugal pumps.

For *turbines*, the *specific speed* N_s is usually expressed in terms of H, P, and N. This form of N_s is obtained from π_2 and π_3, assuming $\rho =$ constant, as follows

$$N_s = \frac{(\pi_3)^{1/2}}{(\pi_2)^{5/4}} = \frac{NP^{1/2}}{H^{5/4}} \tag{12-30}$$

in which

H is the theoretical head utilized by the turbine, in ft;

N is the angular velocity of the turbine, in rpm;

P is the horsepower developed by the turbine.

Variation of specific speed N_s with turbomachine proportions is illustrated in Fig. 12-4.

The significance of N_s is further illustrated by considering specific cases. If a turbomachine has a small restricted inlet, Q must be kept small to avoid excessive loss of head associated with high velocities, and, if the diameter of

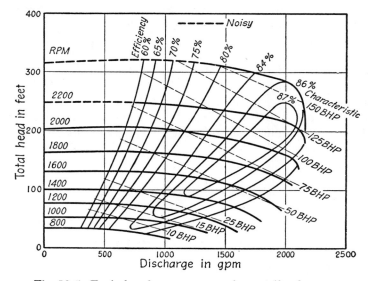

Fig. 12-5. Typical performance curve of a centrifugal pump.

the impeller is large, H is large because $H \propto N^2QD^2$. The equation for specific speed

$$N_s = \frac{NQ^{1/2}}{H^{3/4}} \qquad (12\text{-}29)$$

shows that if Q is small and H is large, the specific speed N_s is small. Thus, turbomachines with small specific speeds are **high-head, small-capacity** machines. By applying the same reasoning, the propeller-type turbomachine has a large value of N_s, since H is relatively small and Q is large. Thus, propeller-type machines are designed for **small head** and **large capacity.**

When selecting a turbomachine, it can be designed and built to meet the specific requirements, or it may be possible to select it from standard machines which are available and have been tested. To determine the suitability of a machine for a specific purpose, refer to its **performance curve,** known also as its **characteristic curve,** which has been established by test by its manufacturer. In Fig. 12-5, a typical performance curve of a centrifugal pump is given. This performance curve shows the interrelationship between head, discharge, impeller diameter, brake horsepower, and efficiency. Similar performance curves can be referred to in selecting a turbine for specific site conditions.

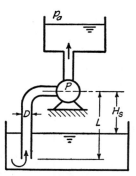

Fig. 12-6. Computation of positive net suction head.

In selecting a pump, it is also necessary to consider the **net positive suction head** NPSH, which is the head required to make the fluid flow through the suction piping to the pump impeller. The required NPSH varies with pump design, and with speed and capacity of any individual pump; its magnitude must be provided by the manufacturer. For any centrifugal pump to operate, the available NPSH must exceed the required NPSH. The available NPSH can be determined by referring to Fig. 12-6 and using Eq. 12-31. When the liquid source is below the pump, the actual NPSH can be computed as follows:

$$\text{NPSH} = \frac{p_a}{\gamma} - H_s - h_L - h_v \qquad (12\text{-}31)$$

in which

p_a/γ is the absolute atmospheric pressure head, in ft of liquid;

H_s is the static suction lift, in ft;

h_L is the head loss due to friction and bends in the suction piping, in ft;

h_v is the vapor pressure head of the liquid, in ft.

When the liquid source is above the pump, Eq. 12-31 is still applicable, but
the H_s term is positive.

• CENTRIFUGAL PUMP OPERATION

The centrifugal pump is, when properly selected for the service it is to
perform, a very simple reliable machine. Consequently, if it fails to start,
the source of difficulty is usually easily determined. Some of the most
common causes of failure are:

1. The pump is not primed.
2. The static head exceeds that for which the pump is designed.
3. The suction head is too large.
4. The strainer which is on the entrance to the suction pipe is clogged.
5. The impeller is clogged.

If the pump is not primed, it is probably the result of a leaky foot valve
or partly blocked foot valve caused by some solid object which allowed the
liquid to drain from the impeller casing and the suction line. If the foot
valve is working satisfactorily, prime the pump by filling the suction line
and impeller casing with the liquid to be pumped and start the pump.

The magnitude of the static head can be determined by measuring the
difference in elevation between the sump level and point of discharge, and
adding to this the friction and minor losses in the pump line. If this appreciably exceeds the static head for which the pump is designed, it will not
operate. The magnitude of the suction head can be checked in a similar
manner to see if it is excessive.

If necessary, the strainer and impeller can be checked visually. However, this will require draining the sump or disconnecting the suction line to
get at the strainer; and the impeller housing must be opened to check
the impeller.

• THE IMPULSE TURBINE

Although the impulse turbine is a turbomachine, its construction, and
hence its theory, are sufficiently different from those of centrifugal pumps
and reaction turbines to warrant special treatment.

The impulse turbine consists principally of a circular disk with vanes, more commonly designated as **buckets,** and a **nozzle.** Fluid is discharged in a direction tangential to the disk or runner, along the centerline of the buckets. The disk and buckets constituting the runner can be cast integrally, or the buckets can be cast separately and bolted onto the disk. The latter method has the advantage of easy bucket replacement. The buckets are usually cast of steel or bronze. The shape of the bucket is designed for effective operation and large efficiency, which means that the bucket in contact with the jet must deflect the jet so that it does not hinder or interfere with the free travel of adjacent buckets, and yet it must turn the fluid nearly 180°.

The diameter of the runner D is measured across the disk to the point of tangency of the jet to the disk. The size of the runner ranges from $D = 6d$ upward, where d is the diameter of the jet discharged by the nozzle against the buckets. The upper limit of D is a function of structural design, ease of transportation from the manufacturing site, and windage loss (air resistance to motion). The latter increases slightly with size of runner.

The nozzle, which is a fundamental part of the turbine, is a specially-designed needle valve which controls, by means of a governor, the discharge to the runner in accordance with power demand.

Impulse turbines can be mounted so their axes of rotation are either vertical or horizontal, and with one or several nozzles, depending upon the requirements, as dictated by the site and economic considerations. Figure 12-7 illustrates schematically typical single-jet and multi-jet turbines.

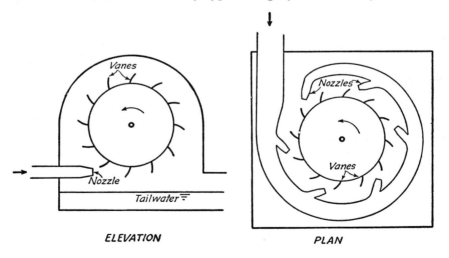

(a) *Horizontal axis, single-jet*   (b) *Vertical axis, multi-jet*

Fig. 12-7. Single-jet and multiple-jet impulse turbines.

To assist with the ***theoretical analysis*** of an impulse turbine, refer to Fig. 12-8, which is a definition sketch illustrating the physical relationship between liquid, nozzle, turbine, and tail race.

From Fig. 12-8, the total head available is the difference in elevation between reservoir surface and tail water level. The total head cannot be fully utilized, however, because of resistance loss and because of the fact

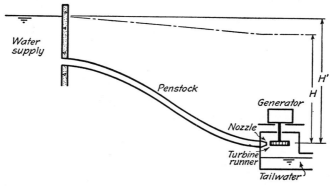

Fig. 12-8. Schematic diagram for an installation of an impulse turbine.

that the runner must be placed sufficiently high so that even at maximum tail water level it does not become even partially submerged. The ***effective head*** measured at the base of the nozzle is

$$H = \frac{V_1^2}{2g} + \frac{p_1}{\gamma} \qquad (12\text{-}32)$$

in which

p_1 is the pressure at the base of the nozzle, in psf;

V_1 is the velocity at the base of the nozzle, in fps.

The *loss* in the nozzle is

$$h_L = H - \frac{V^2}{2g} \qquad (12\text{-}33)$$

in which

V is the velocity of the jet in fps,

and the jet velocity is

$$V = C_v \sqrt{2gH} \qquad (12\text{-}34)$$

from which

$$\frac{V^2}{2g} = C_v^2 H \qquad (12\text{-}35)$$

and

$$h_L = H(1 - C_v^2) \qquad (12\text{-}36)$$

As with an orifice, the coefficient of velocity C_v is approximately 0.98.

The equations expressing the magnitude of torque and power can be derived in a manner similar to that used to derive equations of torque and power for pumps and reaction turbines, or they can be derived by beginning

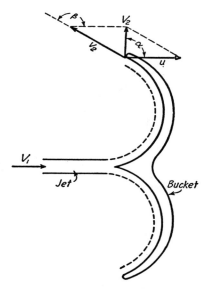

[**Fig. 12-9.** Typical impulse turbine bucket and velocity diagram.

with the impulse-momentum equation. Referring to the velocity diagram for an impulse turbine, Fig. 12-9, and utilizing the former method of analysis, the equation for *torque* is

$$T = \rho Q r (V_1 - V_2 \cos \alpha_2) \qquad (12\text{-}37)$$

The relative velocity of the jet with respect to the runner is

$$v_1 = V_1 - u$$

in which

v_1 is the relative velocity of the jet, in fps;

V_1 is the absolute velocity of the jet, in fps;

u is the linear velocity of the runner.

Based upon a trigonometric analysis of the velocity diagram of Fig. 12-9

$$V_2 \cos \alpha_2 = v_2 \cos \beta_2$$

Utilizing this relation, Eq. 12-37 reduces to

$$T = \rho Q r (V_1 - u - v_2 \cos \beta_2) \qquad (12\text{-}38)$$

The *power* is

$$P = T\omega \qquad (12\text{-}4)$$

and by combining Eq. 12-37 with Eq. 12-4 and substituting $u = \omega r$, the equation for power becomes

$$P = \rho Q u (V_1 - V_2 \cos \alpha_2) \qquad (12\text{-}39)$$

or, since

$$V_2 \cos \alpha_2 = u + v_2 \cos \beta_2$$

from the velocity diagram

$$P = \rho Q u (V_1 - u - v_2 \cos \beta_2) \qquad (12\text{-}40)$$

If the losses on the bucket are negligible

$$v_2 = v_1 = V_1 - u \qquad (12\text{-}41)$$

However, if losses are considered

$$v_2 = k(V_1 - u) \qquad (12\text{-}42)$$

Then by substitution in Eq. 12-38

$$T = \rho Q r (V_1 - u)(1 - k \cos \beta_2)$$

and

$$P = \frac{\gamma Q}{g} (V_1 - u)(1 - k \cos \beta_2)u \qquad (12\text{-}43)$$

The coefficient k has an approximate magnitude of 0.9.

From Eq. 12-43, note that if $\gamma Q /g$ and $k \cos \beta_2$ are constant

$$P \propto (V_1 - u)u \qquad (12\text{-}44)$$

The expression $(V_1 - u)u$ can be maximized to determine the speed relationship for maximum efficiency. That is, by setting the first derivative of Eq. 12-44 equal to zero

$$V_1 \, du - 2u \, du = 0$$

and

$$u = \frac{V_1}{2} \qquad (12\text{-}45)$$

This states that the best *runner speed for maximum efficiency* is equal to one-half the jet velocity. However, this is not achieved in practice, because of practical considerations such as a jet deflection angle which is less than 180°. That is, the linear speed $u < V_1/2$ and varies with the design between the limits which are indicated

$$u = 0.44 V_1 \quad \text{to} \quad 0.48 V_1$$

or

$$u = \phi V_1 = \phi \sqrt{2gH} \tag{12-46}$$

in which ϕ is defined as the relative speed.

The magnitude of ϕ ranges from 0.44 to 0.48, as indicated above, and once it is fixed by the design, the best speed u is also fixed.

The *overall efficiency* of the turbine is

$$e = \frac{\text{power delivered by the turbine shaft}}{\text{power being supplied by the water to the runner}} \tag{12-47}$$

as with reaction turbines

$$e = e_t e_m \tag{12-14}$$

In equation form, the impulse turbine efficiency is

$$e_t = \frac{(V_1 - u)(1 - k \cos \beta_2)_u}{gH} \tag{12-48}$$

and the mechanical efficiency is

$$e_m = \frac{\text{power obtained from the shaft}}{\text{power input to the shaft}} \tag{12-49}$$

Using the equations

$$V_1 = C_v \sqrt{2gH} \tag{12-34}$$

and

$$u = \phi \sqrt{2gH} \tag{12-46}$$

the *hydraulic efficiency* can be expressed as

$$e_t = 2(C_v \phi - \phi^2)(1 - k \cos \beta_2) \tag{12-50}$$

which illustrates that e_t is independent of H and D.

The *specific speed* is defined and applied as for reaction turbines and pumps. That is,

$$N_s = \frac{N P^{1/2}}{H^{5/4}} \tag{12-30}$$

Noting that

(1) $u = \phi \sqrt{2gH} = \pi \dfrac{DN}{60}$

(2) $V_1 = C_v \sqrt{2gH}$

(3) the ratio of u/V_1 necessary for peak efficiency approximately fixes the pitch diameter of the runner for a given head;

(4) the jet diameter is fixed to a certain degree by the discharge; the foregoing value of N_s can be expressed as

$$N_s = 129.3 \sqrt{eC_v}\, \frac{\phi}{D/d} \qquad (12\text{-}51)$$

From this expression, it can be concluded that D/d is the controlling factor in fixing N_s, since ϕ, C_v and e do not vary over a wide range of values.

A typical characteristic curve, comparable to that presented for the centrifugal pump, is shown in Fig. 12-10, for an impulse turbine.

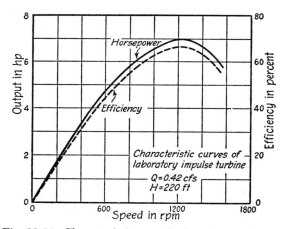

Fig. 12-10. Characteristic curve for impulse turbine.

Example Problem 12-4: Compute the power output of an impulse turbine which has a diameter $D = 60$ in., a speed $N = 350$ rpm, a bucket angle $\beta = 160°$, a coefficient of velocity $C_v = 0.98$, a relative speed $\phi = 0.45$, a generator efficiency $e_m = 0.90$, $k = 0.90$ and a jet diameter from the nozzle of 6 in.

Solution:

$$u = \frac{(350)(60)(\pi)}{(60)(12)} = 91.5 \text{ fps}$$

$$H = \frac{u^2}{2g\phi^2} = \frac{(91.5)^2}{(2g)(.45)^2} = 642 \text{ ft}$$

$$V_1 = C_v \sqrt{2gH} = 0.98 \sqrt{(2g)(642)} = 199 \text{ fps}$$

$$Q = V_1 A_1 = (199)(.196) = 39 \text{ cfs}$$

$$P = \frac{\rho Q}{550} (V_1 - u)(1 - k \cos \beta) u$$

$$P = \frac{(1.94)(39)}{(550)} (107.5)(1.85)(91.5) = 2{,}510 \text{ horsepower input}$$

$$e_t = 2(C_v \phi - \phi^2)(1 - k \cos \beta_2)$$

$$e_t = 2[(0.98)(0.45) - (0.45)^2](1.85) = 0.884$$

$$e = e_t e_m = (0.884)(0.90) = 0.795$$

Power output $= (0.795)(2510) = \underline{\underline{2000 \text{ horsepower}}}$

• GEAR PUMPS

The history of the gear pump, or at least of positive-displacement machinery of this type, dates back to 1588. A semi-rotary pump was described by Ramulli at that time. The gear pump has not been widely used until recently, however, because industry was not formerly able to manufacture economically this type of machine to the precise dimensions required for satisfactory operation.

Gear pumps and other similar positive-displacement machines are widely used to pump such fluids as light and heavy oils, gasoline, tars, pitch, asphalt, many industrial chemical solutions, and other viscous fluids,

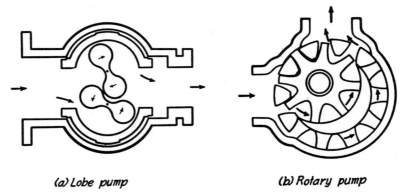

(a) Lobe pump   *(b) Rotary pump*

Fig. 12-11. Typical gear pumps.

especially when extreme pressures must be maintained. The only limitation is the grit content of the liquid. Anything gritty will cause excessive wear and allow back leakage, which reduces the volumetric efficiency of the pump.

There are many varieties of gear pumps. For example: the spur gear, the herringbone gear, the spherical gear and various types of lobe rotary pumps. Two typical gear pumps are illustrated in Fig. 12-11. The practical clearance between the teeth and the housing is of the order of 0.001 to 0.002 in. This clearance must be small, and yet large enough to allow for expansion due to the heating of the rotors.

These pumps have several characteristics which are advantageous under certain circumstances. They are noted for their ability to discharge a continuous flow of fluid which is practically free from pulsations. They work extremely well under high vacuum. They handle a wide range of viscosities and temperature without difficulty, and they are self priming. This type of pump can also be dismantled quickly for cleaning, which is an advantage in certain industries where periodic cleaning is essential.

Pumps of the positive-displacement type consist of three essential components, namely: the two geared rotors and a close-fitting housing. One of the geared rotors is connected to a drive shaft and motor. The other geared rotor is an idler mounted on a trunnion. Positive displacement from the housing is accomplished as the liquid is trapped between the gear teeth and the housing, and is transported from the inlet to the outlet chamber around the periphery of the wheel. This method of operation is illustrated in Fig. 12-11.

Gear pumps can be converted to motors by reversing the flow through them—that is, by allowing a fluid to flow from the high-pressure side to the low-pressure side of the pump through the gear train. The operation of servo-mechanisms is based upon this principle.

• THEORY OF GEAR PUMPS

The design of gear pumps does not involve the theory of fluid flow. Their principle of operation is purely mechanical.

The *overall efficiency e* of gear pumps, which have an average maximum value of about 75 percent, is defined as

$$e = \frac{60Q\,\Delta p}{2\pi N T} \tag{12-52}$$

or

$$e = \frac{0.746\Delta p}{EIe_m} \tag{12-53}$$

In these equations

Q is the actual discharge, in cfs;

Δp is the pressure difference across the pump, in psf;

N is the speed of the pump, in rpm;

T is the torque supplied through the shaft, in lb-ft;

I is the electrical current supplied to the motor;

E is the input voltage to the motor;

e_m is the efficiency of the motor.

The volumetric efficiency e_v of the pump is

$$e_v = \frac{Q}{Q_i} \tag{12-54}$$

in which

Q is the actual discharge, in cfs;

Q_i is the theoretical discharge, in cfs;

and

$$Q_i = \frac{\tilde{V}N}{60} \tag{12-55}$$

in which

$\tilde{V}$ is the volume displacement of the pump, in cubic ft per revolution;

N is the speed of the pump, in rpm.

The difference between actual and theoretical discharge is caused by slip and entrained gases or vapors. The magnitude of slip is directly proportional to the pressure difference across the pump, and inversely proportional to the viscosity. The entrainment of gases or vapors occurs with expansion in the low-pressure suction chamber, and is called a cavitation loss.

The equation for actual discharge, as given by Wilson, is

$$Q = \frac{\tilde{V}N}{60} - \frac{C_s \tilde{V}\, \Delta p}{2\pi\mu} Q_c \tag{12-56}$$

in which

C_s is the slip coefficient;

Q_c is the cavitation loss;

μ is the dynamic viscosity in lb-sec per sq. ft.

and other terms are as defined previously.

When high-pressure differences across the pump are involved, regular Bourdon-type gages are not suitable for measuring them. In situations where it is necessary to measure Δp, such as in pump testing, an electrical strain gage can be used. The wire resistors in the gage are connected externally to a Wheatstone bridge. The unbalance in the bridge is calibrated in terms of Δp in psi.

In a similar manner, the temperature can also be measured. That is, a temperature-sensitive resistor placed in the fluid is attached to a Wheatstone bridge and the measured unbalance is calibrated with fluid temperature.

• PROPELLERS AND WINDMILLS

A mathematical analysis sufficiently complete to serve as a means of designing propellers and windmills based on the impulse-momentum principle is not available. However, the application of the impulse-momentum principle provides a simple means of deriving basic laws which describe their operation.

The propeller and slipstream are illustrated in Fig. 12-12. The fluid can be moving to the right at velocity V_1 past a stationary propeller, or the pro-

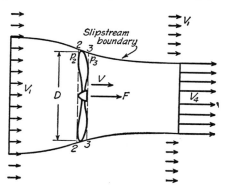

Fig. 12-12. Propeller and slipstream.

peller can be moving to the left through a stationary fluid at velocity V_1. The relative flow picture is the same in either case.

Considering the acceleration within the slipstream boundary, from the Bernoulli equation, $p_1 > p_2$ and, because of the action of the propeller, $p_3 > p_2$, which causes further acceleration of the fluid. The velocity V_2 is equal to V_3, and p_1 and p_4 are the pressures in the undisturbed fluid at 1 and 4, for the condition illustrated by Fig. 12-12.

Writing the momentum equation

$$F = \frac{M\Delta V}{\Delta t} \tag{12-57}$$

for the free-body confined by the slipstream boundary and sections 1 and 2,

$$p_3 A - p_2 A = \rho Q (V_4 - V_1) \tag{12-58}$$

or

$$p_3 - p_2 = \rho V (V_4 - V_1) \tag{12-59}$$

Application of the Bernoulli equation between sections 1 and 2 and 3 and 4 yields

$$p_1 + \frac{1}{2} \rho V_1^2 = p_2 + \frac{1}{2} \rho V^2 \tag{12-60}$$

and

$$p_3 + \frac{1}{2} \rho V^2 = p_4 + \frac{1}{2} \rho V_4^2 \tag{12-61}$$

Equating p_1 and p_4, as defined by Eqs. 12-60 and 12-61

$$p_3 - p_2 = \frac{1}{2} \rho (V_4^2 - V_1^2) \tag{12-62}$$

Eliminating $p_3 - p_2$ from Eqs. 12-59 and 12-62

$$\rho V (V_4 - V_1) = \frac{1}{2} \rho (V_4^2 - V_1^2) \tag{12-63}$$

and

$$V = \frac{V_1 + V_4}{2} \tag{12-64}$$

Hence, the velocity of the fluid normal to the plane of the propeller at the propeller is the average of the velocities at sections 1 and 4.

The ***power output*** of the system (force times distance the propeller is moving per unit of time), from Eq. 12-58 is

$$\text{Power output} = \rho Q (V_4 - V_1) V_1 \tag{12-65}$$

The ***power input*** which is required to accelerate the velocity V_1 at section 1 to V_2 at section 2 is

$$\text{Power input} = \rho \frac{Q}{2} (V_4^2 - V_1^2) \tag{12-66}$$

The ***efficiency of the propeller*** e is the ratio of power output to power input. That is,

$$e = \frac{2\rho Q (V_4 - V_1) V_1}{\rho Q (V_4^2 - V_1^2)} = \frac{2V_1}{V_4 + V_1} = \frac{V_1}{V} \tag{12-67}$$

The usual airplane propeller, when operating under ideal conditions and at a velocity less than 400 mph, can have an efficiency as large as 85 per cent. When the speed exceeds 400 mph, there is a rapid drop in efficiency due to compressibility effects.

The *windmill* is similar to the propeller, except that it extracts energy from the fluid, whereas the propeller adds energy to the fluid. Using the same procedure as was utilized in the analysis of propellers,

$$V = \frac{V_1 + V_4}{2} \tag{12-68}$$

and ignoring friction losses the power output is

$$\text{Power output} = \rho \frac{Q}{2} (V_1^2 - V_4^2) \tag{12-69}$$

Assuming the fluid is incompressible, the power input is

$$\text{Power input} = \rho Q \frac{V_1^2}{2} \tag{12-70}$$

and the *theoretical efficiency* is

$$e = \frac{V_1^2 - V_4^2}{V_1^2} \tag{12-71}$$

Considering friction losses, the **actual efficiency** of windmills is about 50 percent.

A more thorough treatment of the design of pumps, turbines, fans and propellers can be presented in terms of the airfoil theory, discussed in Chapter 9.

Example Problem 12-5: If an airplane which delivers 250 horsepower to its 6 ft diameter propeller flies at 150 mph through standard air at an elevation of 10,000 ft calculate:

(a) the velocity of the air in the plane of the propeller

(b) the discharge of air Q in cfs

(c) the horsepower output

(d) the propeller efficiency and

(e) the thrust of the propeller.

Solution:

(a) $250 = \dfrac{\pi (6)^2 (0.0565)}{(550)(4)(32.2)} \dfrac{(V_4 + 220)}{2} \dfrac{V_4^2 - (220)^2}{2}$

$V_4 = 266.5$ fps

$V = \left(\dfrac{266.5 + 220}{2}\right) = \underline{\underline{243.25 \text{ fps}}} \quad \rightarrow$

(b) $Q = \dfrac{\pi 6^2}{4} (243.25) = \underline{\underline{6870 \text{ cfs}}}$

(c) Power output $= \dfrac{(0.0565)}{(550)(32.2)} (6870)(266.5 - 220)(220) = \underline{\underline{225.8 \text{ HP}}}$

(d) $e = \dfrac{V_1}{V} = \dfrac{220}{243.25} = 0.905$

$e = \underline{\underline{90.5 \text{ per cent}}}$

(e) $F = \dfrac{(6870)(.0565)}{32.2} (266.5 - 220) = \underline{\underline{565 \text{ lb}}} \quad \rightarrow$

• FLUID COUPLINGS AND TORQUE CONVERTERS

A fluid coupling consists of a pump impeller which is attached to a driving shaft and a turbine runner attached to a driven shaft. These two units are enclosed in a single housing which contains a liquid, usually oil, because of its lubricating power, availability, and stability. No solid contact exists between the driving parts and the driven parts. The oil in the housing transmits the moment of momentum or torque from the pump impeller to the turbine runner. Binder suggests that the principle of operation can be simply and effectively illustrated by setting two electric fans face to face

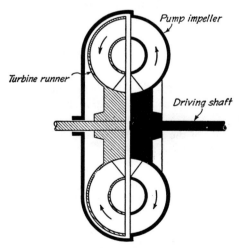

Fig. 12-13. Typical fluid coupling.

and starting one of them. As with a fluid coupling, when the fluid from the driving fan exerts sufficient torque on the opposite fan to overcome its inertia, the second, or turbine fan, will begin to turn.

Fluid couplings are widely used with both electrical and internal combustion engines to transmit torque. They were first used to connect marine engines with propeller shafts. Currently, they are employed in most industries, for example, in automobiles, in marine engines, and in railroad and oil field equipment.

The fluid coupling is particularly useful where smooth shock-free operation is required and where large initial loads are involved. They are widely used with both electric and internal combustion prime movers ranging in size from one to 36,000 HP.

A typical fluid coupling is illustrated in Fig. 12-13. As the driving shaft is started, the pump impeller causes the fluid to flow from the eye of the pump impeller to the outer periphery of the pump impeller, where it is discharged inwardly through the turbine runner to the pump. As the speed of the driver is increased, the fluid torque on the runner increases until it overcomɴs the inertia of the driven unit, and the turbine runner and shaft begin to rotate.

The **power input** to the coupling can be defined as

$$P = T\omega_p \tag{12-72}$$

in which

> T is the torque;
>
> ω_p is the angular speed of the pump or drive shaft.

The **power output** from the coupling is

$$P = T\omega_t \tag{12-73}$$

in which

> ω_t is the angular speed of the turbine shaft.

Defining the *efficiency* of the coupling as the ratio of power output to power input,

$$e = \frac{T\omega_t}{T\omega_p} = \frac{\omega_t}{\omega_p} \tag{12-74}$$

If *slip s* is defined as

$$s = \frac{\omega_p - \omega_t}{\omega_p} \tag{12-75}$$

then

$$e = 1 - s \tag{12-76}$$

The efficiency of fluid couplings is large, usually in excess of 94 percent. A typical efficiency versus speed-ratio curve for a fluid coupling is illustrated in Fig. 12-14.

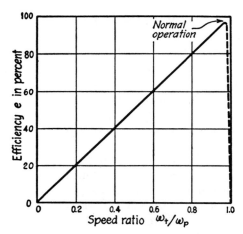

Fig. 12-14. Efficiency—speed ratio curve for a fluid coupling.

The efficiency of the fluid coupling starts at zero and increases uniformly with the speed ratio until $e > 95$ percent, and then rapidly reduces to zero. The definition of **speed ratio** is

$$\text{Speed ratio} = \frac{\omega_t}{\omega_p} \qquad (12\text{-}77)$$

The fluid torque converter is similar to the fluid coupling, except for a series of fixed guide vanes which are constructed between the pump impeller and the turbine runner. In this case, the liquid exerts a torque on these stationary guide vanes as it flows from pump to turbine. The transmission of fluid torque to the guide vanes causes a different torque and speed output from the driven unit than occurs with a fluid coupling. The torque relationship is

$$T_t = T_p + T_v \qquad (12\text{-}78)$$

in which

T_t is the torque of the turbine runner;

T_p is the torque of the pump runner;

T_v is the torque of the fixed guide vanes.

Actually, T_v can be positive or negative in Eq. 12-78. That is, if the stationary guide vanes are designed to receive a torque from the fluid, which is in the opposite direction to that exerted on the driven shaft, an increased torque output results. Conversely, if the vanes are designed to receive a torque, which is in the same sense as that of the driven shaft, a torque reduction results.

To obtain a large reduction in speed, $\omega_p \gg \omega_t$, and a large torque magnification, fluid torque converters are designed which utilize two or more sets of turbine runners and fixed vanes, the fixed vanes being located between the turbine runners.

The variation of **torque convertor efficiency** with speed ratio ω_p/ω_t is illustrated qualitatively in Fig. 12-15.

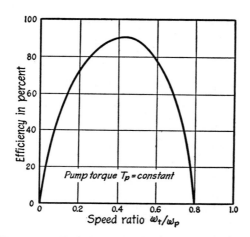

Fig. 12-15. Variation of torque convertor efficiency.

From the efficiency curve, it is apparent that the efficiency of the torque converter is better at smaller speed ratios than that of the fluid coupling.

The advantages of both the fluid coupling and the torque converter can be obtained when necessary, such as in the automotive industry, by designing the guide vanes between the pump impeller and turbine runner so that, when $\omega_p > \omega_t$, the torque on the vanes holds them in a fixed position, and when $\omega_p \rightarrow \omega_t$, the guide vanes free wheel—taking no part in the transmission of power—, and hence the unit then performs as a fluid coupling.

- ## SUMMARY

1. Hydraulic machinery can be **classified** as:
    a. Turbomachinery, or
    b. Positive displacement machinery

2. **Turbomachinery** consists of:
    a. The *rotating element*, which is usually called the impeller if it is a part of a pump, and a runner if it is a part of a turbine, and
    b. The *casing or housing* which encloses the rotating element.

3. By means of rotation and the resulting *centrifugal force*, a centrifugal-type or turbine-type *pump* overcomes the attraction of gravity to add kinetic and/or potential energy to the fluid system.

4. A *turbine* acts as a pump in reverse to subtract energy from a fluid system.

5. The theoretical head added to or subtracted from a fluid by a pump or a turbine can be calculated from the Bernoulli equation.

$$\frac{V_1^2}{2g} + \frac{p_1}{\gamma} + z_1 = \pm H + \frac{V_2^2}{2g} + \frac{p_2}{\gamma} + z_2 + h_L$$

6. The *brake horsepower* required to drive a *pump* is

$$\text{BHP} = \frac{\gamma QH}{550e_p} \tag{12-12}$$

7. The *horsepower delivered* by a *turbine* is

$$\text{HP} = \frac{e_t \gamma QH}{550} \tag{12-18}$$

8. The *draft tube* is an expanding section of conduit downstream from a turbine, which helps to increase the efficiency of the turbine by converting more available energy to mechanical energy.

9. The *characteristics* of turbomachines are described by the *affinity laws* given in Eq. 12-28.

10. The specific speed N_s is used to classify turbomachines with respect to their type and characteristics. When N_s is small the turbine is a high head, small capacity machine, and when N_s is large the turbine is a low head, high capacity machine.

The *specific speed* for *pumps* is

$$N_s = \frac{NQ^{1/2}}{H^{3/4}} \tag{12-29}$$

The *specific speed* for *turbines* is

$$N_s = \frac{NP^{1/2}}{H^{5/4}} \tag{12-30}$$

11. The *impulse turbine* which has a small N_s consists of a series of *buckets* mounted on a disk, and a *nozzle* which discharges the fluid at high speed against the buckets.

12. Analysis of *propellers and windmills* is based on the momentum principle.

13. A *fluid coupling* consists of a *pump impeller* which is attached to a driving shaft and a *turbine runner* attached to a driven shaft.

14. The *torque converter* is similar to the fluid coupling except that it has a series of fixed guide vanes placed between the impeller and runner.

15. *Gear pumps* are a positive-displacement, constant discharge type of pump which utilize various arrangements of two geared rotors and a close-fitting housing.

References

1. Addison, Herbert, *Applied Hydraulics*. New York: John Wiley & Sons, Inc., 1945.

2. Binder, R. C., *Advanced Fluid Mechanics and Fluid Machinery*. Englewood Cliffs, N. J.: Prentice-Hall, Inc., 1951.

3. Binder, R. C., *Fluid Mechanics*, 3rd ed. Englewood Cliffs, N. J.: Prentice-Hall, Inc., 1955.

4. Church, A. H., *Centrifugal Pumps and Blowers*. New York: John Wiley & Sons, Inc., 1944.

5. Daily, James W., "Hydraulic Machinery," Chapter XII of *Engineering Hydraulics*, ed. Hunter Rouse. New York: John Wiley & Sons, Inc., 1950.

6. Daugherty, R. L. and A. C. Ingersoll, *Fluid Mechanics*. New York: McGraw-Hill Book Co., Inc., 1954.

7. Francis, J. R. D., *Fluid Mechanics*. London: Edward Arnold, Ltd., 1958.

8. Hunsaker, J. C. and B. G. Rightmire, *Engineering Applications of Fluid Mechanics*. New York: McGraw-Hill Book Co., Inc., 1947.

9. Rouse, Hunter, *Elementary Mechanics of Fluids*. New York: John Wiley & Sons, Inc., 1946.

10. Russell, G. E., *Hydraulics*, 5th ed. New York: Henry Holt & Co., 1941.

11. Shepherd, D. G., *Principles of Turbomachinery*. New York: The MacMillan Company, 1956.

12. Stepanoff, A. J., *Centrifugal and Axial Flow Pumps*. New York: John Wiley & Sons, Inc., 1948.

13. Wislicenus, G. F., *Fluid Mechanics and Turbomachinery*. New York: McGraw-Hill Book Co., Inc., 1947.

Problems

12-1. Referring to Fig. 12-2, prove that $V \cos \alpha = u + v \cos \beta$.

12-2. Referring to Fig. 12-2, prove that $v^2 = V^2 + u^2 - 2uV \cos \alpha$.

12-3. A centrifugal pump which has an impeller 18 in. in diameter and which rotates at 1000 rpm has a relative velocity $v = 100$ fps and a bucket angle $\beta = 160°$ at exit from the impeller. At the point of exit, compute: (**a**) the magnitude of the absolute velocity V, (**b**) the magnitude of α_2, (**c**) the tangential and radial components of V.

12-4. If a centrifugal pump with an impeller diameter of 15 in. discharges 5 cfs of water when $N = 1700$ rpm, $\beta_2 = 150°$, $v_2 = 40$ fps, and $\alpha_1 = 90°$, compute: (**a**) the angular velocity of the impeller in radians per second and the linear velocity of the impeller u, (**b**) the theoretical head added to the water by the pump.

12-5. Calculate the available N.P.S.H. for a 3000 gpm centrifugal pump if the pump is set 5 ft above the water surface, $T = 60°F$, its altitude is 5000 ft above sea level, and the suction piping consists of 20 ft of 12-in. cast-iron pipe, one 90° elbow, and one open 12-in. gate valve.

12-6. Calculate the efficiency of a 6-in. centrifugal pump which has an 8-in. suction pipe when $Q = 1500$ gpm, p_1 in the suction pipe at the pump is 6-in. of mercury vacuum, the pressure p_2 in the discharge line at an elevation 5 ft above the axis of the pump is 18 psi, and the input to the pump is 30 horsepower.

12-7. A centrifugal pump delivers 600 gpm of water at 1200 rpm against a head $H = 50$ ft when the input to the pump is 10 BHP. If this pump is operated at 1800 rpm, calculate its discharge Q in gpm, the theoretical head H imparted to the water, and the input to the pump in horsepower, assuming its efficiency is constant.

12-8. List the information which is essential to the proper selection of a centrifugal pump from the standpoint of its suction characteristics.

12-9. The total head of a pump, known as total dynamic head, is the sum of the static suction and static discharge heads, the loss of head due to friction and minor losses in the suction and discharge lines, and the velocity head in the discharge line if significant. Compute the TDH of a pump which discharges 500 gpm into the bottom of a tank. Suction piping consists of one 90° elbow, and 20 ft of 4-in. steel pipe; the discharge piping consists of 200 ft of 3-in. steel pipe and one 3-in. gate valve (open.) The sum of the static suction and discharge heads is 50 ft.

12-10. The impeller of a centrifugal pump is 22 in. in diameter and rotates at 2000 rpm. If the absolute velocity $V_2 = 185$ fps, the tangential velocity $u_2 = 200$ fps and the relative velocity $v_2 = 35$ fps at exit from the impeller, calculate the magnitude of the bucket angle β_2 and the angle α_2.

12-11. A centrifugal pump which utilizes 30 horsepower for its operation when rotating at 1750 rpm delivers 1000 gpm of water against a head of 40 ft. Calculate the pump's probable discharge, developed head, and required power if its speed N is reduced to 1500 rpm.

12-12. A centrifugal pump which uses 10 horsepower at a speed of 1750 rpm when delivering 500 gpm against a head of 75 ft is modified by replacing its original 10-in. impeller with an 8-in. impeller. With the new impeller what is the pump's estimated discharge capacity, head and power requirement?

12-13. Compute the specific speed of a pump which delivers 1650 gpm against a head of 30 ft when its speed N is 1200 rpm. What type of centrifugal pump is this most apt to be?

12-14. What speed N should a pump operate at if its specific speed $N_s = 8000$, $Q = 62,500$ gpm when working against a head of 100 ft?

12-15. Develop equations in terms of discharge in gpm, total head H in ft, and the specific gravity of the liquid being pumped, for: (a) the horsepower delivered by a pump, and (b) the brake horsepower required to drive the pump.

12-16. Compute the power applied to the shaft and the turbine efficiency e_t of a reaction turbine if $\alpha_1 = 25°$, $\beta_2 = 160°$, $V_1 = 49$ ft/sec, $\phi = 0.70$, $H = 60$ ft, and $Q = 160$ cfs. The area A_1 measured at right angles to V_1 is equal to 0.8 of a_2, which is the area measured at right angles to v_2 and the radius $r_1 = 1.15\ r_2$. See Fig. 12-5.

12-17. A reaction turbine discharges 2000 cfs when $\alpha_1 = 22°$, $\alpha_2 = 90°$, $\beta_1 = 105°$, $A_1 = 30$ ft^2, and $r_1 = 4$ ft. With these conditions, compute (a) the horsepower applied to the shaft, and (b) the speed of the wheel.

12-18. Considering flow through the runner of a reaction turbine, prove that

$$\frac{v_1^2}{2g} - \frac{u_1^2}{2g} + \frac{p_1}{\gamma} + z_1 = \frac{v_2^2}{2g} - \frac{u_2^2}{2g} + \frac{p_2}{\gamma} + z_2 + C_L\frac{v_2^2}{2g}$$

by starting with the Bernoulli equation

$$\frac{V_1^2}{2g} + \frac{p_1}{\gamma} + z_1 = \frac{V_2^2}{2g} + \frac{p_2}{\gamma} + z_2 + H + h_f$$ utilizing the trigonometric reactions be-

tween V_1, V_2, u_1, u_2, v_1, and v_2 and assuming that $h_f = C_L\dfrac{v_2^2}{2g}$.

12-19. Using the equation derived in Problem 12-18, calculate the pressure head at exit from a reaction turbine runner if the loss coefficient $C_L = 0.2$, $v_1 = 30$ fps, $r_1 = 3.0$ ft, $r_2 = 2.20$ ft, $p_1 = 30$ psi, $v_2 = 60$ fps, $N = 250$ rpm, and $z_1 = z_2$.

12-20. If a turbine develops 20,000 horsepower when the effective head is 52 ft and the relative speed N = 150 rpm, (a) what is the turbine's specific speed N_s? (b) Referring to Fig. 12-4, what type of turbine is it?

12-21. A tangential turbine, as illustrated in Fig. 12-7 (a), has a diameter of 100 in. and develops 2000 horsepower when rotating at 200 rpm. Compute the magnitude of the average force exerted by the jet on the buckets.

12-22. Water is discharged against the buckets of a tangential turbine through a 10-in. nozzle which is attached to a 24-in. pipe. If the pressure at the base of the nozzle is 350 psi, $C_v = 0.98$ and $C_c = 0.65$, compute: (a) the effective head, (b) the energy in the jet, (c) the head loss in the nozzle, and (d) the power lost in the nozzle.

12-23. Considering a 100-in. tangential turbine, if the effective head is 1100 ft, the diameter of the jet is 6 in., the coefficient of velocity is 0.97, the relative speed $\phi = 0.45$, the bucket angle $\beta = 165°$, the overall efficiency $e = 0.85$ and the coefficient $k = 0.9$, compute: (a) power input to shaft, (b) the turbine speed in rpm, (c) the horsepower output, and (d) the pressure at the base of the nozzle.

12-24. Compute the horsepower input to the shaft of a tangential turbine if the water is discharged to the turbine from an 8-in. nozzle with a coefficient of contraction $C_c = 0.70$. The effective head H is 1500 ft, $\phi = 0.47$, $C_v = 0.98, \beta = 160°$ and $k = 0.88$.

12-25. Calculate the relative speed ϕ, the efficiency of the turbine e_t, and the power input to the shaft of an impulse turbine if the diameter of wheel $D = 70$ in., the diameter of the jet $d = 6$ in., the speed is 400 rpm, the effective head is 1000 ft, the bucket angle $\beta = 165°$, $C_v = 0.97$, $C_c = 0.70$, and $k = 0.90$.

12-26. A 100 horsepower tangential turbine with a diameter of 30 in. has a speed of 30 radians per sec, an over-all efficiency $e = 0.80$, an effective head of 200 ft, a relative speed $\phi = 0.47$, and a coefficient of velocity $C_v = 0.97$. Compute the specific speed N_s, the discharge Q and the diameter of the jet from the nozzle.

12-27. Compute the specific speed N_s of a turbine which developes 15,000 horsepower under a 1600 ft head at 350 rpm. What is its over-all efficiency if $C_v = 0.98$, $\phi = 0.47$, and $D/d = 12$.

12-28. An impulse turbine develops 5500 horsepower when $N = 250$ rpm, $Q = 80$ cfs, $C_v = 0.97$, $\phi = 0.46$, $k = 1$, $\beta = 170°$, and the effective head $H = 800$ ft. Compute N_s, V_1, u_1, e, e_t and e_m.

12-29. Select a pump and state your reasons for selecting this type of pump to deliver (a) 500 gpm is of corrosive liquid against a varying head with an average value of 60 psi, and (b) a very viscous liquid at a uniform rate of 20 gpm at a pressure of 150 psi.

12-30. A motor with an efficiency of 90 per cent which draws 5000 watts drives a gear pump which has an efficiency of 80 per cent. Calculate the pressure drop across the pump.

12-31. An airplane engine delivers 6,000 horsepower to its propeller, which has an 8 ft diameter, when it is flying at an absolute velocity of 200 miles per hour into a head wind of 30 mph. If the velocity of the slipstream in the plane of the propeller is 440 fps, calculate (a) the specific weight of the air in which it flies, and (b) the efficiency of the propeller.

12-32. Determine the horsepower input and the thrust required of a propeller to drive an airplane at a speed of 200 mph if the diameter of the propeller is 8 ft, atmospheric pressure is 12 psia, air temperature is 50°F and the propeller efficiency is 80 per cent.

12-33. An airplane flies at a speed of 250 mph through air with a specific weight of 0.075 lb/ft³ when the engine is delivering 1000 horsepower to an 8 ft diameter propeller. Calculate (**a**) the velocity of the air in the plane of the propeller, (**b**) the rate of flow of air in the slipstream, (**c**) the thrust of the propeller, and (**d**) the pressure drop across the propeller.

12-34. Assuming incompressible flow, determine the power input and the power output considering a windmill which has a 10 ft diameter and an efficiency of 50 per cent if the wind velocity is 45 mph, atmospheric pressure is 32 ft of water and the air temperature is 70°F.

Index

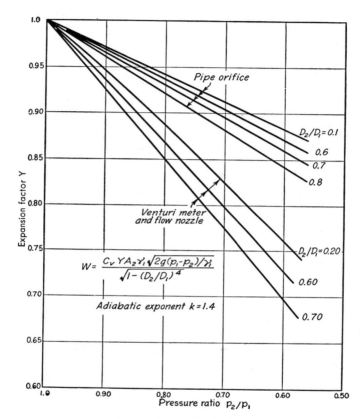

Fig. I. Expansion factors.

Table I. PARTIAL LIST OF SYMBOLS

Symbol	Quantity	Units	Dimensions M-L-T	Dimensions F-L-T
a	Acceleration	ft/sec²	L/T^2	L/T^2
A	Area	sq ft	L^2	L^2
c	Celerity of wave	ft/sec	L/T	L/T
C	Chezy coefficient	ft$^{1/2}$/sec	$L^{1/2}/T$	$L^{1/2}/T$
D,d	Diameter	ft	L	L
E	Modulus of elasticity	lb/sq ft	M/LT^2	F/L^2
e	Height of roughness	ft	L	L
f	Resistance coefficient	None	—	—
F	Force	lb	ML/T^2	F
Fr	Froude number	None	—	—
g	Gravitational acceleration	ft/sec²	L/T^2	L/T^2
G	Mass discharge	lb-sec/ft	M/T	FT/L
h	Piezometric head	ft	L	L
H	Specific head	ft	L	L
k	Adiabatic exponent	None	—	—
L	Length	ft	L	L
M	Mass	slug	M	FT^2/L
Ma	Mach number	None	—	—
n	Manning coefficient	ft$^{1/6}$	$L^{1/6}$	$L^{1/6}$
p	Pressure	lb/sq ft	M/LT^2	F/L^2
pv	Vapor pressure	lb/sq ft	M/LT^2	F/L^2
P	Wetted perimeter	ft	L	L
Q	Discharge	cu ft/sec	L^3/T	L^3/T
q	Discharge per unit width	cfs/ft	L^2/T	L^2/T
r	Radius	ft	L	L
R	Hydraulic radius	ft	L	L
Re	Reynolds number	None	—	—
S	Slope	None	L/L	L/L
t	Time	sec	T	T
v	Velocity at a point	ft/sec	L/T	L/T
v	Specific volume	ft³/lb	L^2T^2/M	L^3/F
V	Average velocity	ft/sec	L/T	L/T
$\tilde{V}$	Volume	cu ft	L^3	L^3
W	Weight discharge	lb/sec	ML/T^3	F/T
We	Weber number	None	—	—
y	Depth and distance	ft	L	L
Y	Expansion factor	None	—	—
z	Vertical distance	ft	L	L

Symbol	Name	Quantity	Units	Dimensions M-L-T	Dimensions F-L-T
Γ	Gamma	Circulation	sq ft/sec	L^2/T	L^2/T
γ	Gamma	Specific weight	lbs/cu ft	M/L^2T^2	F/L^3
δ	Delta	Thickness of boundary layer	ft	L	L
δ'	Delta	Thickness of laminar sublayer	ft	L	L
ϵ	Epsilon	Kinematic eddy viscosity	sq ft/sec	L^2/T	L^2/T
κ	Kappa	Karman constant	None	—	—
η	Eta	Dynamic eddy viscosity	lb-sec/sq ft	M/LT	F-T/L^2
μ	Mu	Dynamic viscosity	lb-sec/sq ft	M/L-T	F-T/L^2
ν	Nu	Kinematic viscosity	sq ft/sec	L^2/T	L^2/T
ρ	Rho	Mass density	lb-sec^2/ft^4		FT^2/L^4
			slugs/ft^3	M/L^3	
σ	Sigma	Surface energy	lb/ft	M/T^2	F/L
τ	Tau	Shearing stress	lb/sq ft	M/LT^2	F/L^2
ω	Omega	Angular velocity	radians/sec	$1/T$	$1/T$

Table II. CONVERSION TABLE

Length	30.48 cm = 1 ft
Volume	7.48 gal = 1 cu ft
	35.31 cu ft = 1 cu m
Velocity	1.467 fps = 1 mph
	1.689 fps = 1 knot
Discharge	449 gpm = 1 cfs
	1.98 af/day = 1 cfs
Mass	453.6 gms = 1 lb*
	32.2 lbs = 1 slug*
Force	4.45 $\times$ 10^5 dynes = 1 lb
	2.205 lb = 1 kg*
Energy	778.3 ft-lb = 1 Btu
Power	550 ft-lb/sec = 1 HP
	746 watts = 1 HP
Viscosity	478.8 poise = 1 lb-sec/sq ft
	929 stokes = 1 sq ft/sec

* Non-homogeneous equation which relates mass to force (weight) by means of Newton's second law for which the acceleration is $g = 32.2$ ft/sec^2 = 980.7 cm/sec^2.

Table III. PROPERTIES OF WATER*

Temperature °Fahrenheit	Specific Weight γ lb/ft³	Mass Density ρ lb-sec²/ft⁴	Dynamic Viscosity $\mu \times 10^5$ lb-sec/ft²	Kinematic Viscosity $\nu \times 10^5$ sq ft/sec	Surface Energy† $\sigma \times 10^3$ lb/ft	Vapor Pressure p_v lb/sq in	Bulk Modulus $E \times 10^{-2}$ lb/sq in
32	62.42	1.940	3.746	1.931	5.18	0.09	290
40	62.43	1.940	3.229	1.664	5.14	0.12	295
50	62.41	1.940	2.735	1.410	5.09	0.18	300
60	62.37	1.938	2.359	1.217	5.04	0.26	312
70	62.30	1.936	2.050	1.059	5.00	0.36	320
80	62.22	1.934	1.799	0.930	4.92	0.51	323
90	62.11	1.931	1.595	0.826	4.86	0.70	326
100	62.00	1.927	1.424	0.739	4.80	0.95	329
110	61.86	1.923	1.284	0.667	4.73	1.24	331
120	61.71	1.918	1.168	0.609	4.65	1.69	333
130	61.55	1.913	1.069	0.558	4.60	2.22	332
140	61.38	1.908	0.981	0.514	4.54	2.89	330
150	61.20	1.902	0.905	0.476	4.47	3.72	328
160	61.00	1.896	0.838	0.442	4.41	4.74	326
170	60.80	1.890	0.780	0.413	4.33	5.99	322
180	60.58	1.883	0.726	0.385	4.26	7.51	318
190	60.36	1.876	0.678	0.362	4.19	9.34	313
200	60.12	1.868	0.637	0.341	4.12	11.52	308
212	59.83	1.860	0.593	0.319	4.04	14.7	300

* Most of this table was taken from "Hydraulic Models," A.S.C.E. Manual of Engineering Practice, No. 25, A.S.C.E., 1942.

† In contact with air.

Table IV. PROPERTIES OF AIR*

Temperature T °Fahrenheit	Specific Weight γ lb/ft³	Mass Density ρ lb-sec²/ft⁴	Dynamic Viscosity μ lb-sec/ft²	Kinematic Viscosity ν ft²/sec.
−40	0.0946	0.00294	3.12×10^{-7}	1.06×10^{-4}
−20	0.0903	0.00280	3.25	1.16
0	0.0863	0.00268	3.38	1.26
20	0.0827	0.00257	3.50	1.36
40	0.0795	0.00247	3.62×10^{-7}	1.46×10^{-4}
60	0.0764	0.00237	3.74	1.58
80	0.0735	0.00228	3.85	1.69
100	0.0709	0.00220	3.96	1.80
150	0.0651	0.00202	4.23	2.07
200	0.0602	0.00187	4.49×10^{-7}	2.40×10^{-4}

* Approximate values at standard atmospheric pressure.

Table V. PROPERTIES OF COMMON LIQUIDS*

Liquid	Specific Gravity sp. gr.	Viscosity μ lb-sec/ft²	Surface Energy† σ lb/ft	Vapor Pressure p_v psi
Benzene	0.90	1.4×10^{-5}	2.0×10^{-3}	1.5
Gasoline	0.68	6.2×10^{-6}		
Glycerine	1.26	3.1×10^{-2}	4.3×10^{-3}	2×10^{-6}
Kerosene	0.81	4.0×10^{-5}	1.7×10^{-3}	
Mercury	13.55	3.3×10^{-5}	3.2×10^{-2}	2.33×10^{-5}
Olive Oil	0.92	1.8×10^{-3}	$2., \times 10^{-3}$	
SAE 10 Oil	0.92	1.7×10^{-3}	2.5×10^{-3}	
SAE 30 Oil	0.92	9.2×10^{-3}	2.4×10^{-3}	
Turpentine	0.87	3.0×10^{-5}	1.8×10^{-3}	7.7×10^{-3}
Water	1.00	2.1×10^{-5}	5.0×10^{-3}	0.34

Carbon Tetra 1.594

* Approximate values at standard atmospheric pressure and 68° F.
† In contact with air.

Table VI. PROPERTIES OF COMMON GASES*

Gas	Specific Weight γ lb/cu ft	Dynamic Viscosity μ lb-sec/sq ft	Gas Constant R	Molecular Weight m	Adiabatic Exponent k
Air	0.0753	3.72×10^{-7}	53.3	29	1.40
Carbon dioxide	0.114	3.34×10^{-7}	34.1	44	1.29
Helium	0.0104	4.11×10^{-7}	386	4	1.66
Hydrogen	0.00522	1.89×10^{-7}	767	2	1.40
Methane	0.0416	2.51×10^{-7}	96.4	16	1.26
Oxygen	0.0830	4.42×10^{-7}	48.3	32	1.40

* Approximate values at standard atmospheric pressure and 68°F.
* R, m, and k are unaffected by pressure and temperature.

Table VII. Properties of Areas and Volumes

	Sketch	Area or Volume	Centroid	Moment of Inertia
Ellipse		$\dfrac{\pi bh}{4}$	$\bar{y} = \dfrac{h}{2}$	$\bar{I} = \dfrac{\pi bh^3}{64}$
Parabola		$\dfrac{2}{3}bh$	$\bar{y} = \dfrac{3h}{5}$ $\bar{x} = \dfrac{3b}{8}$	$\bar{I} = \dfrac{2bh^3}{7}$
Semicircle		$\dfrac{\pi D^2}{8}$	$\bar{y} = \dfrac{4r}{3\pi}$	$\bar{I} = \dfrac{\pi D^4}{128}$
Cone		$\dfrac{\pi D^2 h}{12}$	$\bar{y} = \dfrac{h}{4}$	
Cylinder		$\dfrac{\pi D^3}{12}$	$\bar{y} = \dfrac{h}{2}$	
Hemisphere		$\dfrac{\pi D^2 h}{4}$	$\bar{y} = \dfrac{3D}{16}$	
Paraboloid of Revolution		$\dfrac{\pi D^2 h}{8}$	$\bar{y} = \dfrac{h}{3}$	
Sphere		$\dfrac{\pi D^3}{6}$	$\bar{y} = \dfrac{D}{2}$	

Table VIII. A STANDARD ATMOSPHERE

Altitude ft	Temperature °F	Pressure psia	Specific Weight lb/ft³
0*	59.0	14.7	0.0765
5,000	41.2	12.2	0.0657
10,000	23.4	10.1	0.0565
15,000	5.5	8.29	0.0480
20,000	−12.3	6.76	0.0407
25,000	−30.1	5.45	0.0343
30,000	−47.9	4.36	0.0286
35,000	−67.8	3.46	0.0237
35,330	−67.0	3.40	0.0233
40,000	−67.0	2.72	0.0187
50,000	−67.0	1.69	0.0115
60,000	−67.0	1.05	0.0072
70,000	−67.0	0.65	0.0045
80,000*	−67.0	0.40	0.0028
90,000†	−67.0	0.25	0.0017
100,000	−67.0	0.16	0.0011
105,000	−67.0	0.12	0.00084
110,000	−47.4	0.098	0.00063
120,000	−7.2	0.063	0.00037
140,000	73.3	0.029	0.00015
160,000	153.7	0.015	0.000067
180,000	170.0	0.0085	0.000036
200,000†	159.4	0.0046	0.000020

* U.S. Standard Atmosphere (0 to 80,000).
† N.C.A.A. Standard Upper Atmosphere (90,000 to 200,000).

Table IX. Areas of Circles

Diameter (in.)	AREA in.²	AREA ft²	Diameter (in.)	AREA in.²	AREA ft²
0.25	0.049	0.00034	7.50	44.18	0.307
0.50	0.196	0.00136	7.75	47.17	0.328
0.75	0.442	0.00307	8.00	50.27	0.349
1.0	0.785	0.00545	8.25	53.45	0.371
1.25	1.227	0.00852	8.50	56.74	0.394
1.50	1.767	0.0128	8.75	60.13	0.418
1.75	2.405	0.0167	9.00	63.62	0.442
2.00	3.142	0.0218	9.25	67.20	0.467
2.25	3.976	0.0276	9.50	70.88	0.492
2.50	4.909	0.0341	9.75	74.66	0.518
2.75	5.940	0.0412	10.00	78.54	0.545
3.00	7.069	0.0491	10.25	82.52	0.573
3.25	8.296	0.0576	10.50	86.59	0.601
3.50	9.621	0.0668	10.75	90.76	0.630
3.75	11.045	0.0767	11.00	95.03	0.660
4.00	12.57	0.0873	11.25	99.40	0.690
4.25	14.19	0.0985	11.50	103.87	0.721
4.50	15.90	0.110	11.75	108.43	0.754
4.75	17.72	0.123	12.00	113.10	0.785
5.00	19.63	0.136	15.00		1.23
5.25	21.65	0.150	18.00		1.77
5.50	23.76	0.165	21.00		2.41
5.75	25.97	0.180	24.00		3.14
6.00	28.27	0.196	30.00		4.91
6.25	30.68	0.213	36.00		7.07
6.50	33.18	0.230	48.00		12.57
6.75	35.78	0.249	60.00		19.63
7.00	38.48	0.267	72.00		28.27
7.25	41.28	0.287